SPECIAL PUBLICATION
of CARNEGIE MUSEUM OF NATURAL HISTORY

WINTER ECOLOGY OF
SMALL MAMMALS

edited by
JOSEPH F. MERRITT
Resident Director
Powdermill Nature Reserve
Star Route South
Rector, PA 15677

NUMBER 10   **PITTSBURGH, 1984**

SPECIAL PUBLICATION OF CARNEGIE MUSEUM OF NATURAL HISTORY
Number 10, pages 1–380

Issued 28 December 1984

Price $45.00 a copy

Robert M. West, *Director*

ISBN 0-935868-10-0

Library of Congress number 84-72213

CARNEGIE MUSEUM OF NATURAL HISTORY, 4400 FORBES AVENUE
PITTSBURGH, PENNSYLVANIA 15213

CONTENTS

FOREWORD

During the second International Theriological Congress held in Brno, Czechoslovakia, in the summer of 1978, a small group of scientists from the U.S.A., Canada, and Europe assembled at a small cafe to discuss their common interest—the winter ecology of small mammals. We retired that evening with the realization that indeed there were many scientists from many nations with interests in winter ecology of small mammals. They represented disciplines of botany, mammalogy, ethology, anatomy, ecology, and physiology. Because of their diverse disciplines they had never assembled at a common meeting—as we know, most meetings select for mammalogists, ornithologists, botanists or physiologists, rarely bridging these disciplines. Perhaps we could organize a colloquium to bring together scientists of many specialties and backgrounds with the common denominator of winter ecology of small mammals. If such a colloquium could be organized, what guidelines would be adhered to and who would host and ultimately fund such a major undertaking? Because of the high level of interest in such an "international" gathering, I felt inclined to outline objectives for such a conference, although no meeting site had been selected. Those objectives were as follows:

1) The colloquium must remain small in size, so that communication between all participants would be enhanced. It was felt that 50 scientists actively pursuing winter ecological research was an optimal number for greatest productivity. These scientists would be selected by reviewing their contributions to winter research and recommendations by their peers.

2) The colloquium must represent a multi-disciplinary group of scientists. This group would necessarily include animal and plant biologists, functional morphologists, behaviorists, physiologists, and ecologists. The "common denominator" for this assemblage would be recognition of the importance of snow to the ecology of winter-active small mammals.

3) Scientists attending the colloquium would be expected to present information derived from the ecosystem in their "home" geographic region. This information, therefore, would provide representative samples of those Holarctic ecosystems where snow acts as a major selective force in the lives of small mammals.

4) The structure of the colloquium must provide for both formal presentations by invited participants and round-table discussions. The round-table discussions would employ an "informal" format and address salient issues in winter ecology. It was felt that the colloquium would not be dominated by formal presentations of a unilateral nature as is common to many conferences. The presence of discussion groups would permit free exchange of ideas between participants from different parts of the world. Ideally, results stemming from these discussion groups would represent an integration of international viewpoints.

5) In order to circulate widely the information derived from this international colloquium, a volume of proceedings must be published and in a form readily available to both foreign and North American scientists.

The above objectives were established in 1978. However, no site had been chosen and no "volunteers" emerged. In spring 1979, I became Resident Director of Powdermill Nature Reserve, the biological field station of the Carnegie Museum of Natural History. I soon discovered that this field station would prove to be an ideal site for our proposed international colloquium. A precedent had been established in 1977 with the *Colloquium on the Ecology and Taxonomy of African Small Mammals* held at Powdermill Nature Reserve and organized by Duane A. Schlitter of the Section of Mammals, Carnegie Museum of Natural History. Proceedings of this colloquium were published in the *Bulletin of the Carnegie Museum of Natural History* in 1978. With this strong historical precedent, I pursued the plan of hosting another international colloquium at Powdermill with Craig C. Black (Director, Carnegie Museum of Natural History, 1975 to 1982) and received an enthusiastic response which then precipitated initiation of our colloquium. The aims and objectives outlined above were closely adhered to in order to provide for a successful and productive conference.

In August 1980, letters of invitation were mailed to 67 scientists actively pursuing research in the field of winter ecology of small mammals. On 14 October, 1981 our colloquium opened with 45 participants representing six nations of the world (USA, Canada, Finland, Sweden, Czechoslovakia, and USSR) and integrating many different research disciplines. Their ideas are presented in this volume of proceedings.

Joseph F. Merritt

v

ACKNOWLEDGMENTS

The International Colloquium on Winter Ecology of Small Mammals held at Powdermill Nature Reserve from 14 to 18 October 1981 was made possible by funds from the Carnegie Museum of Natural History, Pittsburgh, Pennsylvania. I am indebted to Craig C. Black (Director) for his encouragement and devotion to my idea of holding this Colloquium, for without his support it could not have become a reality. I thank Robert M. West (Director, Carnegie Museum of Natural History, 1983 to present) for making possible the publication of this volume of proceedings.

This Colloquium would not have been successful without the active support of many people. Chief among these are the local people who helped house and feed participants. I am indebted to William and Ingrid Rea for providing lodging and hosting a cocktail party for participants. I thank Graham and Jane Netting for providing sleeping accommodations for participants, and Sue and Wayne Hale for hosting dinner for some of the Finnish scientists that arrived early. I am grateful to Virginia Johnson, Joanie Merritt, and Tammy Zufall for cooking and serving many delicious meals at Raven's Roost Conference Center. Also, I thank Bobbie Thorne of Laughlintown for taking time to demonstrate capabilities of her Apple computer to members of our Colloquium, and to Robert Leberman and Robert Mulvihill of the Powdermill staff for presenting bird-banding demonstrations to participants.

Transportation to the Colloquium site from the Pittsburgh International Airport and return was achieved through help from Robert Rose, Tim Schumann, Roger Everton, and staff of the Section of Mammals. I thank the Powdermill staff, Gilbert and Albert Lenhart and Lloyd Moore for their enthusiasm and devotion to preparing the Powdermill facilities—this work began a year prior to the meeting date. In addition, I am indebted to them for providing logistical support during the Colloquium—with their able assistance everything ran quite smoothly.

I am grateful to the creative talents of Jim Senior of the Section of Exhibits, Carnegie Museum of Natural History, for designing the Colloquium logo. I thank Cliff Morrow (Chairman), Carol March, Nancy Perkins, and Gail Richards of the Section of Exhibits for preparing the Colloquium program and abstracts. Professional photographic coverage of the Colloquium was made possible by the efforts of Vince Abromitis (Section of Exhibits) and Howard P. Nuernberger (Pittsburgh free-lance photographer). I extend my appreciation for a fine job.

My warmest thanks must go to the individual participants for submitting their manuscripts promptly in nearly all instances. The burdensome job of proofreading manuscripts was spearheaded by Diane Schnupp with able assistance from the following "volunteers": Martin Friday, Carol Irwin, Terry Irwin, Robert Rose, and Jane Rowe. I extend my thanks to these dedicated people for a job well done.

Lastly, special thanks go to my colleague, Hugh H. Genoways, Publications Editor, Carnegie Museum of Natural History for editing this volume for publication.

Joseph F. Merritt

PARTICIPANTS

Cassie W. Aitchison
Department of Entomology
University of Manitoba
Winnipeg, Manitoba R3T 2N2, Canada

Craig C. Black
Los Angeles County Museum of Natural History
900 Exposition Blvd.
Los Angeles, California 90007

Ronald J. Brooks
Department of Zoology
University of Guelph
Guelph, Ontario N1G 2W1, Canada

Lee Christianson
Department of Biological Sciences
University of the Pacific
Stockton, California 95204

Jack A. Cranford
Department of Biology
Virginia Polytechnic Institute
and State University
Blacksburg, Virginia 24061

Roger Everton
Department of Biological Sciences
Old Dominion University
Norfolk, Virginia 23508

Dale D. Feist
Institute of Arctic Biology
University of Alaska
Fairbanks, Alaska 99701

William A. Fuller
Department of Zoology
University of Alberta
Edmonton, Alberta T6G 2E9, Canada

Lennart Hansson
Department of Wildlife Ecology
Swedish University of Agricultural Sciences
S-750 07 Uppsala, Sweden

Kalevi Heikura
Zoological Museum
University of Oulu
Kasarmintie 8
Oulu, Finland

Heikki Henttonen
Kilpisjärvi Biological Station

Institute of Zoology
P. Rautatiekatu 13
SF-00100 Helsinki 10, Finland

Tom B. Herman
Department of Biology
Acadia University
Wolfville, Nova Scotia B0P 1X0, Canada

Robert S. Hoffmann
Museum of Natural History and
    Department of Systematics and Ecology
University of Kansas
Lawrence, Kansas 66045

Heikki Hyvärinen
Department of Biology
University of Joensuu
Box 111
80100 Joensuu 10, Finland

Frederick J. Jannett, Jr.
Department of Biology
The Science Museum of Minnesota
30 E 10th Street
St. Paul, Minnesota 55101

Asko Kaikusalo
Ojajoki Field Station
Forest Research Institute
SF-12700 Loppi, Finland

Emil Kucera
Manitoba Department of Mines, Resources and En-
    vironmental Management
Winnipeg, Manitoba R3N 0H6, Canada

Kari Laine
Department of Botany
University of Oulu
P.O. Box 191
SF-90101, Oulu 10, Finland

Daniel R. Ludwig
Forest Preserve District of Dupage County
881 W. St. Charles Road
Lombard, Illinois 60148

Stephen F. MacLean, Jr.
Institute of Arctic Biology
University of Alaska
Fairbanks, Alaska 99701

Participants in the International Colloquium on Winter Ecology of Small Mammals, 14–18 October 1981, Powdermill Nature Reserve of Carnegie Museum of Natural History. Photograph by Howard P. Nuernberger.

Front row (left to right): M. Graham Netting, Robert S. Hoffmann, William A. Fuller, Dale M. Madison, Stephen F. MacLean, Jr., Frank B. Salisbury, Ronald J. Brooks.

Second row: Heikki Henttonen, Penny S. Reynolds, Polley A. McClure, Joseph F. Merritt, Cassie W. Aitchison, Dale Feist, Wilber B. Quay.

Third row: Asko Kaikusalo, Keikki Hyvärinen, Ingrid Rea, Daniel R. Ludwig, Stephen D. West, Steve Mihok, Vladimir A. Yaskin.

Fourth row: Jan Zejda, Peter J. Marchand, Johan Tast, Tom B. Herman, Warren P. Porter, Jerry O. Wolff, Bruce A. Wunder, Emil Kucera, Jack A. Cranford, John S. Millar, Ronald W. Pauls.

Fifth row: Jussi Viitala, Paul H. Whitney, Kari Laine, Lee Christianson, William D. Schmid, Kalevi Heikura, Albert W. Spencer, Frederick J. Jannett, Jr., Lennart Hansson, Roger Everton, Douglas C. Ure, Robert K. Rose.

Lucius L. Stebbins and Tim Schumann participated in the Colliquium but missed the group photograph.

Dale M. Madison
Department of Biologial Sciences
State University of New York
Binghamton, New York 13901

Peter J. Marchand
Department of Environmental Studies
Johnson State College
Johnson, Vermont 05656

Polley A. McClure
Department of Biology
Indiana University
Bloomington, Indiana 47401

Joseph F. Merritt
Powdermill Nature Reserve of Carnegie
Museum of Natural History
Star Route South
Rector, Pennsylvania 15677

Steve Mihok
Whiteshell Nuclear Research Establishment
Atomic Energy of Canada Limited
Pinawa, Manitoba R0E 1L0, Canada

John S. Millar
Department of Zoology
University of Western Ontario
London, Ontario N6A 5B6, Canada

M. Graham Netting
Star Route South
Rector, Pennsylvania 15677

Warren P. Porter
Department of Biological Sciences
University of Wisconsin
Madison, Wisconsin 53711

Ronald W. Pauls
Syncrude Canada Ltd.
10030-107 Street
Edmonton, Alberta T5J 3E5, Canada

Wilbur B. Quay
2003 Ida Street
Napa, California 94558

Penny S. Reynolds
Department of Zoology
University of Guelph
Guelph, Ontario N1G 2W1, Canada

Robert K. Rose
Department of Biological Sciences
Old Dominion University
Norfolk, Virginia 23508

Frank B. Salisbury
Department of Plant Science
UMC 48
Utah State University
Logan, Utah 84322

William D. Schmid
University of Minnesota
108 Zoology Building
Minneapolis, Minnesota 55455

Tim Schumann
110 Roberts Drive
Clairton, Pennsylvania 15025

Albert W. Spencer
Department of Biology
Fort Lewis College
Durango, Colorado 81301

Lucius L. Stebbins
Department of Biology
University of Lethbridge
Lethbridge, Alberta T1K 3M4, Canada

Johan Tast
Korvenk 44
SF-33300 Tampere 30
Finland

Douglas C. Ure
Department of Zoology and Entomology
Colorado State University
Fort Collins, Colorado 80523

Jussi Viitala
Institute of Biology
University of Jyvaskyla
Hameekatu 3
SF-40100, Jyvaskyla 10
Finland

Stephen D. West
Wildlife Sciences Group
College of Forest Resources
University of Washington
Seattle, Wsshington 98195

Paul H. Whitney
BEAK Consultants Inc.
317 SW Alder
Portland, Oregon 97204

Jerry O. Wolff
Department of Biology
University of Virginia
Charlottsville, Virginia 22901

Bruce A. Wunder
Department of Zoology and Entomology
Colorado State University
Fort Collins, Colorado 80523

Vladimir A. Yaskin
Institute of Plant and Animal Ecology
8 Marta Street 202
620008 Sverdlovsk, USSR

David Zegers
Department of Biology
Millersville University of Pennsylvania
Millersville, Pennsylvania 17551

Jan Zejda
Czechoslovak Academy of Sciences
Institute of Vertebrate Zoology
603 65 Brno
Kvetna 8, Czechoslovaka

SNOW AND SMALL MAMMALS

W. O. PRUITT, JR.

ABSTRACT

Snow is the all-pervasive factor that integrates various aspects of the winter ecology of small mammals. This paper discusses some ecological aspects of snow that are important to small mammals. Other physical factors of the winter environment are also considered.

SNOW AND SMALL MAMMALS

The ways by which small mammals survive winter are difficult to outline because "winter" varies so much in different ecological regions.

In a desert it is usually a wet period. In temperate deciduous forest or in steppes, it is a period of fluctuating temperatures and snow conditions when the soil freezes and thaws continually; very intemperate. In taiga and tundra it is the time between hiemal threshold and hiemal termination. It is also a period of short days, although most of the reduction of photoperiod usually occurs before the hiemal threshold. True winter occurs usually with a short or lengthening photoperiod. Low temperature, the criterion of winter in usual human terms, is important only for a relatively few species of supranivean mammals. For small mammals the one critical and important characteristic of winter is the snow cover.

We should be careful to distinguish between the questions "How do small mammals meet winter?" and "How have small mammals evolved adaptations to the North?" Some southern species (for example, deciduous forest forms, steppe forms) have evolved adaptations to short-term winter that are entirely unsuitable for northern winter (for example, torpor or increased metabolism).

In Canada, winter begins officially on the median date of the first snow cover of 2.5 cm or more and terminates on median date of the last snow cover of the same thickness. "Spring thaw" is the two week period following the termination of winter. "Spring shoulder" begins with the termination of spring thaw and terminates with the start of high summer. "High summer" begins on the date that the mean daily maximum temperature rises above 18°C and terminates on the date that the mean daily maximum temperature drops below this figure. "Autumn shoulder" begins with the termination of high summer and terminates with the onset of winter. The official definitions of seasons are quite removed from biological reality, as we shall see later.

One definition we should agree on is "small mammal." For my purposes I define "small mammal' as one that must retreat beneath the snow cover at some time during the winter. With this definition I, therefore, include *Tamiasciurus* as a small mammal but not *Lepus* or *Glaucomys*.

For small mammals the characteristics of the snow cover of duration, thickness, hardness and density are the important features. Hardness and density are governed primarily by wind and winter thaws or freeze-thaw cycles. We can display these characteristics in a 2 by 2 table with resulting combinations that agree remarkably well with major natural landscape divisions (Table 1). Thus we have (1) steppes and coastal regions with freeze-thaw and wind, (2) tundra with wind but no freeze-thaw, (3) inland southern and some maritime regions with freeze-thaw but no wind and (4) taiga with no freeze-thaw and no wind. For mammals, high mountains constitute a separate ecological realm (Shvarts, 1963), but because of variations in height and latitude, and because wind is such an important influence on the snow cover at high elevations, it seems best for the present discussion to consider the snow cover of high mountains as special cases of (1) and (2).

We may also recognize two general zones of snow influence in North America, one where snow cover

Table 1.—*Relationships between thaws and wind as affecting snow cover.*

	Thaws	No thaws
Wind	Wind and thaws (steppes and coastal regions)	Wind and no thaws (tundra)
No wind	Thaws and no wind (inland southern and some coastal regions)	No thaws and no wind (taiga)

is infrequent or of such short duration as to be unimportant in the evolution of mammals and one where snow cover has influenced mammalian evolution. These two zones were shown by Hall (1951) in the case of *Mustela frenata*. In the northern region on his map, weasels always moult into a white winter pelage; in the southern region their winter pelage is brown. Because obliterative coloration is the most logical explanation for white winter pelage (Hammel, 1956), we can conclude that snow cover has influenced the evolution of this particular species.

API AND SMALL MAMMALS

The simplest relations of small mammals to winter occur where the general picture is not complicated by freeze-thaw or wind. As snow flakes fall onto the moss cover or needle litter of the taiga floor, they are subject to metamorphosis which modifies their shape and characteristics and thus affects the internal physical properties of the snow cover.

The metamorphosis of taiga api is caused by heat and moisture that rise from the earth below, pass through the api and escape to the cold, dry supranivean air.

The heat gradient results in the lowest layer of snow crystals being warmer than the one just above it which, in turn, is warmer than the one above it. Water molecules flow from the attenuated tips of the arms of the warmer crystals and attach themselves to the cooler crystals. The lower part of the api is thereby transformed into a series of fragile, interconnected columns composed of many hollow, pyramidal crystals. This columnar basal layer of api is called *pukak* and may be 10 or 20 cm thick with individual pukak pyramids as large as 10 mm across (Fig. 1).

The metamorphosis of api is a continuing but irregular process. The pukak expands upward from the soil surface, governed by (1) the amount of heat and moisture flowing from the earth and (2) the lack of heat and moisture in the supranivean air. The colder and drier the supranivean air, the more intense and quicker is pukak expansion. One result of the process is a decrease in hardness of the pukak layer. Hard layers or even layers of vesicular ice in the lower half of the api are progressively eroded and softened. The progressive softening is important to *Rangifer* and probably to small mammals as well.

In southern or maritime regions which do not have continuous cold, dry supranivean conditions, the heat and moisture gradients are broken. Sometimes heat may even flow downward through the api. Under these conditions pukak cannot form and the api becomes an amorphous, dense mass resting directly on the substratum. Such conditions are inimical to small mammal survival.

For example, in the snowy, but maritime climate of Newfoundland I have found true pukak in only a few spots in the Long Range Mountains. I suspect the lack of pukak is just as important a factor governing the depauperate small mammal fauna of the island as is the barrier of the Strait of Belle Isle. The hard, dense maritime api presses closely against the ground with weights of several tens of grams per square centimetre.

When small mammals tunnel through the api, they invariably use a slightly harder layer for a floor, actually penetrating a layer of lesser hardness (Fig. 2). The hard, dense api over most of Newfoundland would be very difficult to tunnel.

The temperature regime of api, in all its regional variations, is fairly well known by now. The temporal categories of Thermal Overturn, Critical Period, Hiemal Threshold, True Winter, Hiemal Termination, Spring Critical Period, and Spring Thermal Overturn have served as useful check points for descriptions and discussion. The fact that various workers, based on field data, have disagreed as to relative importance of the time periods, shows they are actual natural phenomena. It is also clear that a great deal of work needs to be done to elucidate geographic variation, secular climatic variation and variation associated with population cycles.

The moisture regime under and in api is relatively simple. Virtually all measurements show atmospheric saturation (Bader et al., 1954). This is relative, of course. There may be departures from saturation associated with temperature changes, but we know nothing about their effect on small mammals.

The light regime under api is less well known. Evernden and Fuller (1972) confirmed Geiger (1961) that the pukak space under 30 to 50 cm of api is essentially dark. They also showed that when light does penetrate to the pukak level it is mostly in the red end of the spectrum. Because red light, even with increasing photoperiod, does not stimulate female *Clethrionomys gapperi* to ovulate, the api over these voles ensures that they do not breed until the api is broken in spring. These are exceedingly im-

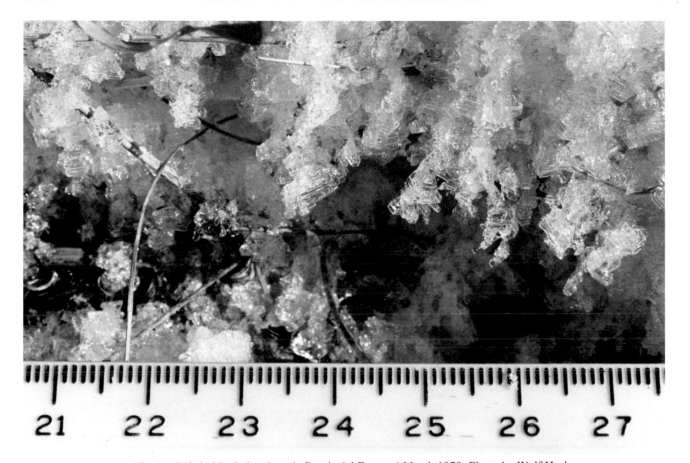

Fig. 1.—Pukak. Manitoba, Agassiz Provincial Forest, 4 March 1978. Photo by Wolf Heck.

portant observations. We need a series of studies extending Evernden and Fuller's work to encompass other geographic zones and other species throughout population cycles. The studies of Richardson and Salisbury (1977) and Salisbury et al. (1973) in southern mountains show the types of variation to be found.

Subnivean activity by small mammals, ectothermal invertebrates, and bacteria, as well as some physical processes, all release carbon dioxide into the pukak space. Penny (1978) found that CO_2 accumulated each winter in certain habitats but not in others, and in concentrations up to five times ambient. She found that the accumulation was affected by density and hardness of the api. The CO_2 accumulation sometimes was accompanied by changes in small mammal distribution. The animals shifted their home ranges away from the affected site and returned later when subnivean CO_2 fell to ambient concentrations. Subnivean CO_2 thus affects small mammal distribution. In the case of *Clethrionomys gapperi,* concentration of subnivean CO_2

must be considered as an additional dimension to the ecological niche of this species. The discovery of a new niche-dimension means that all the ecological relations of this species must now be reconsidered. In other words, conclusions about population declines, overwintering survival, subnivean breeding and movements, etc. previously attributed to the effects of temperature, moisture or light must all be re-evaluated. (For example, CO_2 accumulation might be an explanation for the midwinter habitat shift and aggregation in *Clethrionomys rutilus* observed by West (1977). Probably one of the most pressing needs in the study of winter ecology of small mammals is clarification of the subnivean CO_2 problem—variation by geographic region, vegetation type, topography, and api characteristics, as well as source and movement of the subnivean CO_2, and its effects on physiology and behavior of the species concerned.

Api has other effects on small mammals. As mentioned earlier, hardness of different layers governs tunnelling and burrowing through the api mass. Sub-

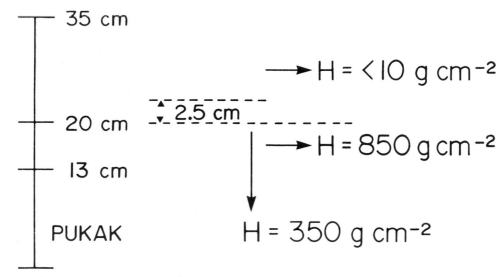

Fig. 2.—Api tunnels of *Microtus pennsylvanicus*. Manitoba, Whiteshell Provincial Park, 31 December 1970.

nivean mammals almost all weigh less than 250 g. Not only does the api protect these mammals from the vagaries of the supranivean environment, but it also governs the pattern of predation on them (Formozov, 1946). In general, large "pouncing" predators (fox, lynx, and raptorial birds) can catch subnivean animals only with difficulty. However,

because subnivean mammals require bulky insulated nests, they construct these in the friable pukak on top of the frozen soil. Nests in such a location are more vulnerable to predation by "tunnelling" predators (weasels, mink, and others) than are nests surrounded by frozen soil.

UPSIK AND SMALL MAMMALS

Table 1 shows that wind is the most important environmental feature uniting the snow cover of the tundra and steppe. The wind causes two great groupings of snow characteristics; understanding these is necessary for understanding many aspects of tundra ecology (Pruitt, 1966, 1970).

The first dichotomy is that of upsik-siqoq (Pruitt, 1978). There are two phases, in the physical sense, to wind-blown tundra or steppe snow. The wind-reworked snow cover has become consolidated into a hard mass called upsik. Above the snow cover in the air is another phase, the moving snow or siqoq (Fig. 3). This phase either moves along, governed by wind force and direction, or is consolidated into a succession of drift forms (Pruitt, 1978). The drift forms and their succession seem not to have any known effects on subnivean mammals, although the different thicknesses of upsik result in spatial variations in pukak thickness and hardness. Siqoq, however, is the mechanism whereby topographic concavities becomes filled with upsik.

The difference between concave and convex land surfaces governs the second dichotomy of snow cover (and the distribution of small mammals). Concavities tend to be filled by siqoq and are called zaboi, while convex surfaces are blown clear and are called vyduv (Fig. 4). Valleys of small streams may be occupied completely by zaboi that may not melt until late the following summer. From the periphery of the zaboi to the center, the growing season is progressively shorter and, as a consequence, the number of species and degree of ground cover is reduced.

Zaboi affect small mammals just as they do plants. For example, I found (Pruitt, 1966) that *Microtus gregalis* occurred only in zaboi sites and was the only small mammal found in these sites. Numerous other workers have reported tundra and steppe small mammals being associated closely with topographic concavities. The reasons for this association are now known. In the case of steppes, moisture during the snow-free season is usually postulated. In the case of tundra, accumulation of subnivean CO_2 may be a factor.

Fig. 3.—Upsik and Siqoq. Alaska, Ogotoruk Valley, 14 February 1960.

In tundra, quite large mammals such as *Vulpes* and *Gulo* become subnivean at times. Avoidance of wind chill is probably the cause of such behavior even though tundra mammals are larger than those in other landscape zones and are thus better able to withstand the supranivean environment (Hagmeier and Stults, 1964; Baker et al., 1978).

As far as small mammals are concerned, the tundra environment is more predictable than that of forested regions. Snow cover arrives and accumulates in a regular and predictable sequence. Year after year zaboi and vyduvi form and disappear in the same sequence. Thus topographic concavities may have sufficient snow cover to pass the hiemal threshold early each autumn, whereas a few meters distant convex ground surfaces never acquire this protection. Such regularity in space and time is extremely important for small mammals (Parker, 1974). Thus, tundra small mammals are markably restricted spatially. Densification of the upsik in-

creases its transmissivity to heat and light. Tundra small mammals are also subjected to lower and more fluctuating bioclimate temperatures than are taiga small mammals. Because light transmission through upsik is relatively high, tundra small mammals may be exposed to increasing photoperiod in spring at the same time that supranivean mammals are.

The hardness of tundra upsik affects small mammals by restricting their movements. Sutton and Hamilton (1932) observed that, while *Lemmus* excavated tunnels in the pukak at the base of upsik, only *Dicrostonyx* with its temporarily enlarged winter claws, dug tunnels up through the hard mass. Even *Dicrostonyx* use softer layers of upsik. For instance, on 28 May 1972 on Devon Island I measured *Dicrostonyx* tunnels through a 200 g/cm² layer floored by 6,000 g/cm². In another spot the tunnel was through 90 g/cm² roofed by 400 g/cm² and floored by 3,000 g/cm² layers (Fig. 5).

We see that although taiga subnivean small mam-

Fig. 4—Zaboi and vyduv. Alaska, Ogotoruk Valley, 17 April 1960.

mals live in conditions that are quite different from those obtaining in the supranivean realm, tundra small mammals live in conditions that still are very reminiscent of the supranivean realm. The tundra subnivean environment differs from the supranivean primarily in having a saturated atmosphere and in the lack of wind. The snow cover of the tundra is not only more variable in thickness and may even be discontinuous but is also denser and harder than in the taiga. The pukak layer is less developed than in the taiga. Thus, tundra subnivean mammals must expend more energy moving from place to place than do those in the taiga.

Miller and Kiliaan (1980) recently described a High Arctic phenomenon whereby meltwater trickles through upsik and refreezes against the ground. This phenomenon is known to Lappish reindeer herders as "čuokki" (Eriksson, 1976; Pruitt, 1979). Miller outlined the effects of čuokki on large mammals, but the effects on small mammals are unknown. Quite likely the effects are similar to those

of naledi in subarctic taiga—complete denial of the area to subnivean small mammals (Pruitt, 1968).

Pruitt (1957, 1970, 1978) and Payne (1979), among others, have shown that the popular and traditional concept of "winter" is insufficient for biological purposes. Pruitt (1957) defined "True Winter" for taiga small mammals as occurring from the hiemal threshold in the autumn until an analogous time in the spring when the snow cover thins sufficiently for diel temperature variations to be felt once more in the underlying soil. (We call this time in the spring the "Hiemal Termination.") During "true winter" subnivean small mammals live in an environment quite different from that above the snow cover. The degree of difference varies with the thickness, hardness and density of the snowcover and the number and thickness of layers of vesicular ice in the api or upsik. In general, the thicker the cover and the less its hardness and density, the more differences there will be between the subnivean and supranivean environments.

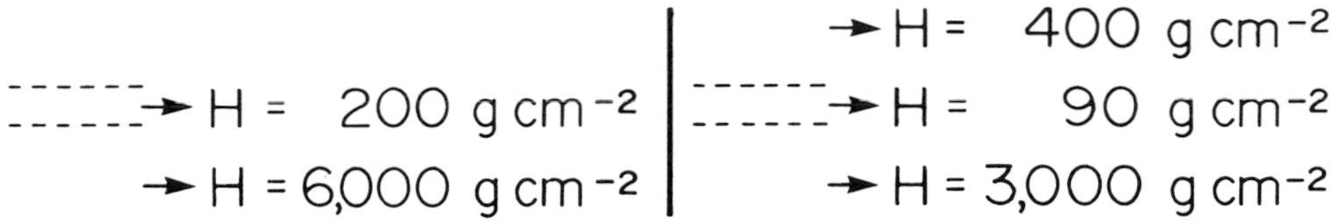

$$- - - - - \to H = 200 \text{ g cm}^{-2} \qquad \to H = 400 \text{ g cm}^{-2}$$
$$\to H = 6{,}000 \text{ g cm}^{-2} \qquad - - - - - \to H = 90 \text{ g cm}^{-2}$$
$$\to H = 3{,}000 \text{ g cm}^{-2}$$

Fig. 5—Upsik tunnels of *Dicrostonyx groenlandicus* N.W.T., Devon Island, 28 May 1972.

In the taiga, true winter comes to most of a region at about the same time. The major exception is in qamaniq. In contrast, in steppes, alpine tundra or arctic tundra (or any site affected by wind) true winter arrives, and persists, at different times depending on degree of convexity or concavity of the substrate surface, or presence and orientation of any obstruction to the wind. In other words, in a tundra region the zaboi achieve the hiemal threshold rather quickly, in contrast to vyduv (even close by) that never achieve it.

Thus, we arrive at a semantic impasse. In this progression we see that subnivean true winter occurs as a period of temperatures that are relatively warm and stable for an extended period of time, a period of saturated air, very little air movement and very little sound transmission. True winter occurs uniformly over large regions (almost the only exceptions being qamaniq) of the transcontinental taiga. To the south in the steppes, to the north in the tundra and at high elevations in alpine tundra, the areas experiencing true winter become more and more fragmented because of the influence of the wind.

We see, then, that the region we traditionally define as having the most severe winter (the tundra) actually has this season only in isolated patches (if we maintain our original definition). What occurs between the patches of true winter conditions is environmentally rigorous, indeed, but is actually a greatly extended or continuous "Critical Period." Payne (1979) pointed out that in the Newfoundland maritime taiga, with its fluctuating snowcover, the entire "calendar winter" was a continuous Critical Period for small mammals. Numerous authors have discussed winter restrictions of tundra and steppe small mammals to topographic concavities or to zaboi sites where true winter comes early and stays late. Pruitt (1953, 1957, 1959a, 1959b) compared microclimates and local distribution of small mammals in two regions having these different regimes of winter.

Much work remains to be done, especially integrated field and laboratory work. Far too much laboratory work has been done without a foundation of field natural history observations. Especially needed are long-term field studies encompassing extremes of population cycles and local temporal variations in winter.

LITERATURE CITED

BADER, H. ET AL. 1954. Snow and its metamorphosis. SIPRE Transl. no. 14 of Bader, H. et al. 1939 Der Schnee und seine Metamorphose. Beitrage zur Geologie der Schweiz. Geotechnische Serie. Hydrologie, Lieferung 3. Bern.

BAKER, A. J. ET AL. 1978. Statistical analysis of geographic variation in the skull of the Arctic hare (*Lepus arcticus*). Canadian J. Zool., 56:2067–2082.

ERIKSSON, O. 1976. Snöförhållendenas inverkan på renbetningen. Meddelanden från Vaxtbiologiska institutionen, Uppsala, 2:1–22.

EVERNDEN, L. N., and W. A. FULLER. 1972. Light alteration caused by snow and its importance to subnivean rodents. Canadian J. Zool., 50:1023–1032.

FORMOZOV, A. N. 1946. Snow cover as an environmental factor and its importance in the life of Mammals and birds. Moscow Society of Naturalists. Materials for fauna and flora USSR, Zoology Section, new series, 5:1–152. Reprinted in English translation in Occas. Papers no. 1, Boreal Institute, Univ. Alberta.

GEIGER, R. 1961. The climate near the ground. Harvard Univ. Press Cambridge, Massachusetts, 611 pp. (translation from 4th German editor).

HAGMEIER, E. M., and C. D. STULTS. 1964. A numerical analysis of the distributional patterns of North American mammals. Syst. Zool., 13:125–155.

HALL, E. R. 1951. American weasels. Univ. Kansas Publ., Mus. Nat. Hist., 4:1–466.

HAMMEL, H. T. 1956. Infrared emissivity of some Arctic fauna. J. Mamm., 37:375–378.

MILLER, F. L., and H. P. L. KILIAAN. 1980. Some observations on springtime snow/ice conditions on 10 Canadian high arctic islands—and a preliminary comparison of snow/ice conditions between eastern Prince of Wales Island and Western Somerset Island, N.W.T. 5 May–2 July 1979. Canadian Wildlife Service, Progress Notes, 116:1–11.

PARKER, G. R. 1974. A population peak and crash of Lemmings and Snowy Owls on Southampton Island, Northwest Territories. Canadian Field-Nat., 88:151–156.

PAYNE, L. E. 1979. Comparative ecology of introduced *Clethrionomys gapperi proteus* (Bangs) and native *Microtus pennsylvanicus terraenovae* (Bangs) on Camel Island, Notre Dame Bay, Newfoundland. Unpublished M.Sc. thesis, Univ. Manitoba, Winnipeg, 136 pp.

PENNY, C. E. 1978. Subnivean accumulation of CO_2 and its effect on winter distribution of small mammals. Unpublished M.Sc. thesis, Univ. Manitoba, Winnipeg. 106 pp.

PRUITT, W. O. JR. 1953. An analysis of some physical factors affecting the local distribution of the shortail shrew (*Blarina brevicauda*) in the northern part of the Lower Peninsula of Michigan. Miscel. Publ. Mus. Zool., Univ. Michigan, 79:1–39.

———. 1957. Observations on the bioclimate of some taiga mammals. Arctic, 10:130–138.

———. 1959a. A method of live-trapping small taiga mammals in winter. J. Mamm., 40:139–143.

———. 1959b. Microclimates and local distribution of small mammals on the George Reserve, Michigan. Miscel. Publ. Mus. Zool., Univ. Michigan, 109:1–27.

———. 1966. Ecology of terrestrial mammals. Chapt. 20, *in* Environment of the Cape Thompson region, northwestern Alaska, U.S. Govt. Printing Office, Washington, D.C., pp. 519–564.

———. 1968. Synchronous biomass fluctuations of some northern mammals. Mammalia, 32:172–191.

———. 1970. Some ecological aspects of snow. Pp. 83–99, *in* Proceedings of 1960 Helsinki Symposium on "Ecology of Subarctic Regions" UNESCO series "Ecology and Conservation," Paris, 1:1–364.

———. 1978. Boreal ecology. Edward Arnold Publ. Ltd. London, Studies in Biology, 91:iv + 1–73.

———. 1979. A numerical "Snow Index" for reindeer (*Rangifer tarandus*) winter ecology (Mammalia, Cervidae) Ann. Zool. Fennici, 16:271–280.

RICHARDSON, S. G., and F. B. SALISBURY. 1977. Plant responses to the light penetrating snow. Ecology, 58:1152–1158.

SALISBURY, F. B., S. L. KIMBALL, B. BENNETT, P. ROSEN, and M. WEIDNER. 1973. Active plant growth at freezing temperatures. Space Life Sci., 4:124–138.

SHVARTS, S. S. 1963. Adaptations of terrestrial vertebrates to the environmental conditions of the Subarctic. Acad. Science USSR, Ural Affiliate, Inst. Biol., Trudy, 33:1–19.

SUTTON, G. M., and W. J. HAMILTON, JR. 1932. The mammals of Southampton Island. Mem. Carnegie Mus. 7 (Part II, Section 1):1–109.

WEST, S. D. 1977. Midwinter aggregation in the northern red-backed vole, *Clethrionomys rutilus*. Canadian J. Zool., 55:1404–1409.

Address: Department of Zoology, University of Manitoba, Winnipeg R3T 2N2, Canada.

SMALL MAMMALS IN WINTER: THE EFFECTS OF ALTITUDE, LATITUDE, AND GEOGRAPHIC HISTORY

ROBERT S. HOFFMANN

ABSTRACT

Boreal and montane parts of the Northern Hemisphere are, because of the present distribution of continents, the principal regions of the globe where the winter environment has exerted a strong selective influence on the nature of adaptations in small mammal species. This is also true of the evolution of small mammal "communities" during the Late Cenozoic (Plio-Pleistocene). The common evolutionary experience of these lineages that have radiated in the same geographic area, especially during the Pleistocene, has produced associations of mammalian species (as well as other organisms) whose similarities we recognize by grouping them in an Holarctic faunal realm. These small mammal communities share many closely related species by virtue of common ancestry, but other similarities result from convergent adaptations. Various strategies have been utilized by Eurasian and North American boreal and montane small mammals in adapting to different sorts of winter environments; these adaptive strategies are compared, with particular reference to arctic tundra and temperate alpine small mammals, and related to the evolutionary histories of the taxa.

HISTORY

For a topic such as "winter ecology of small mammals" to have meaning, there must be winter, and there must be mammals. Therefore, I wish to review some general aspects of the evolutionary history of Holarctic mammals. In the evolutionary history of the ancestral mammals, it is at the end of the Paleozoic Era, about 250 million years ago, that the therapsids appeared, in a world in which strong seasonality and probably winter snow were important environmental features. While therapsids are, in traditional classifications, placed in the Class Reptilia (Hopson, 1970; McKenna, 1975; Vaughan, 1978), this allocation is based on evolutionary grade, and they must clearly be considered stem Mammalia in a phylogenetic classification (Reed, 1960; Crusafont Pairo, 1962).

Recent reconstructions of several therapsids and their environment (Anon., 1977) are notable for three probable, though speculative, points: therapsids were active, fur-covered homeotherms; they were perhaps social; and they lived in a seasonally cold, snowy environment. Many were fairly large carnivores (Romer, 1966), although later Mesozoic mammals were all small (Lillegraven, 1979).

The world when therapsids first appeared in the Permian was Pangaea, and it was south-polar in orientation with widespread glaciation (Cox et al., 1976). Subsequently during the Mesozoic Pangaea broke up, and the resulting continental plates drifted northward. By the Late Mesozoic (Cretaceous), 100 million years ago, major northern and southern land masses (Laurasia and Gondwana) had been created by continental drift, and were separated by the Tethys Sea and the developing Atlantic Ocean (Cox et al., 1976; McElhinny, 1973). The Mesozoic saw the evolution of the "ruling reptiles"—including dinosaurs, but it was also the time when prototherian and therian mammals evolved from therapsids (Lillegraven et al., 1979). These Mesozoic mammals ranged from shrew- to rat-size, and were probably nocturnal, furry, homeothermic omnivores. They also probably lactated and, in seasonally cold habitats, remained capable of controlled torpor.

All of this is, of course, speculative, as was the reconstruction of the Permian therapsids (see above). Nevertheless, the broad outline of a major evolutionary and ecological dichotomy can be sketched. The basic reptilian adaptations (no insulative body covering, poikilothermy, and so forth) are better adapted to at least moderately warm, equable climates, such as occur equatorward. The basic mammalian adaptations (fur, homeothermy, and so forth) are better adapted to cool, seasonal climates, such as occur poleward. The divergence of the synapsid lineage (theriodont therapsids) leading to "true" mammals is thus more likely to have occurred in the temperate south of Pangaea, whereas the lineages leading to the various reptilian groups is more likely to have occupied the more northerly, subtropical to tropical parts of Pangaea. Indeed, virtually all of the early theriodont therapsids (gorgonopsids) have been found in South Africa, as have most of the cynodont therapsids that succeeded them, until the Triassic, when they appear in the fossil record

9

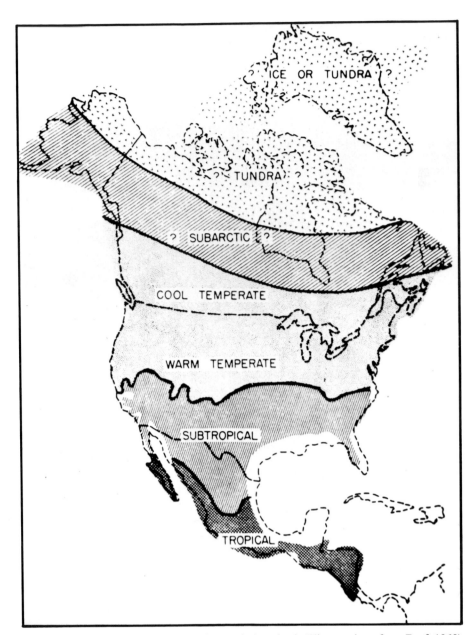

Fig. 1.—Zonal climate (vegetation belts in North America in Pliocene time, from Dorf, 1960).

of other continents (Romer, 1966). In Middle and Late Triassic, fossils "just on the line of transition between the [traditional] classes [Reptilia, Mammalia] in grade of evolution" have been found in South America (Clemens et al., 1979), and these were followed by the earliest prot?therians (morganucodontids) in South Africa, Europe, and Asia (at least), dated as latest Triassic to earliest Jurassic (Rhaeto-Liassic). The divergence between the sister groups Prototheria (including the surviving monotremes) and Theria (including marsupials and placental mammals) must have occurred about this time, for the earliest and most primitive therian fossil, *Keuhneotherium,* is known from the same Rahaeto-Liassic beds as *Morganucodon,* which is among the earliest and most primitive prototherian yet discovered (Clemens et al., 1979).

In short, I suggest that the mammalian clade be-

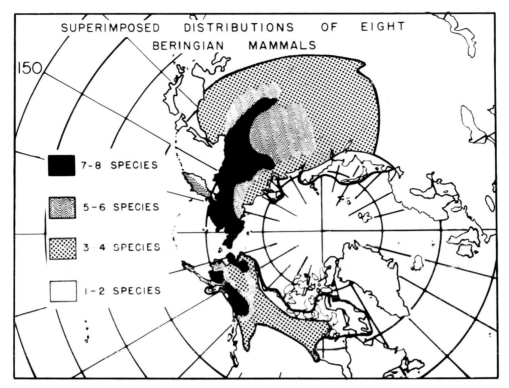

Fig. 2.—Superimposed ranges of eight species of boreal mammals with Beringian distributions (modified from Hoffmann and Peterson, 1967).

gan to evolve about 250 million years ago in a seasonally cold, snowy environment which placed a selective premium on higher metabolic heat production and conservation, and active control of body heat. This led directly to changes in jaw suspension and specialized heterodonty (for quick food processing and rapid digestion), development of a secondary palate (in order to breath while chewing), and other, non-fossil-forming characters such as insulative pelage, specialized mammary glands, and a muscular diaphram. During the Mesozoic, from about 230 to 70 million years ago, mammals continued to evolve. However, for most of the Mesozoic, the global climate was warmer than in the Permian-Early Triassic, though often more arid. Throughout this period of warm, equable climate, reptiles dominated the world fauna, and Mesozoic mammals were small and apparently rare. They were probably nocturnal because their thermogenic and thermoregulatory abilities would have conferred on them an advantage over reptiles during the cooler night hours. This nocturnal habit has important implications for the evolution of the typical sensory modes of modern mammals; namely, hearing and olfaction. However, in the Late Cretaceous, there was a marked cooling, an increase in the poleward climate gradient, and decrease in equability in most parts of the globe. In this period when seasonally cold, perhaps snowy climates became widespread, most Mesozoic reptile groups declined to extinction.

The late Mesozoic gave way to the early Cenozoic, and, once more, to climatic amelioration. Dinosaurs had become extinct, and mammals now diversified in size, food habits, and other ways, while the continental plates were approaching their present positions. Only on the extreme northern or southern margins of the continents were mammals likely to be subject to seasonally cold, snowy winters. However, beginning in the mid-Cenozoic (Miocene) a marked global cooling trend set in, and by the Pliocene (5 million years ago) the northern continents in particular had strongly zoned climatic belts, with circumboreal taiga and tundra (Fig. 1).

This cooling trend culminated, in the last 2 million years, in the periodic advances of glacial ice in the Pleistocene (Hoffmann and Taber, 1968; Hoff-

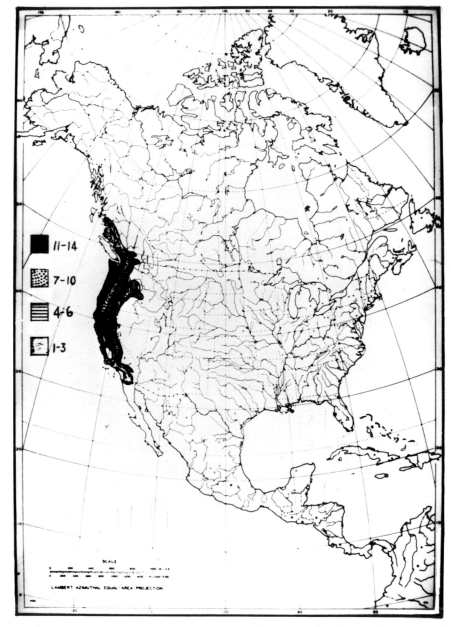

Fig. 3a.—Superimposed ranges of taiga- and tundra-adapted mammals in North America: Pacific coastal taiga mammals (from Crewe, 1965).

mann, 1981). This was accompanied by the rapid evolution of certain mammalian lineages and the frequent shifting and restructuring of the recently evolved boreal communities. A geographic key to understanding the recent evolution of such communities is the "Beringian connection." This was not a tenuous "land bridge," but a 1,500 mile wide union of Eurasia and North America, uniting the faunas of the Holarctic (Palearctic plus Nearctic).

Beringia possessed several significant attributes (Hoffmann, 1981):

1. It was intermittant, alternating with a water barrier, the Bering Sea.
2. Its western and eastern portions differed; West Beringia was confluent with tundra, taiga and steppe communities to the west in the rest of Eurasia, and was larger, whereas East Beringia

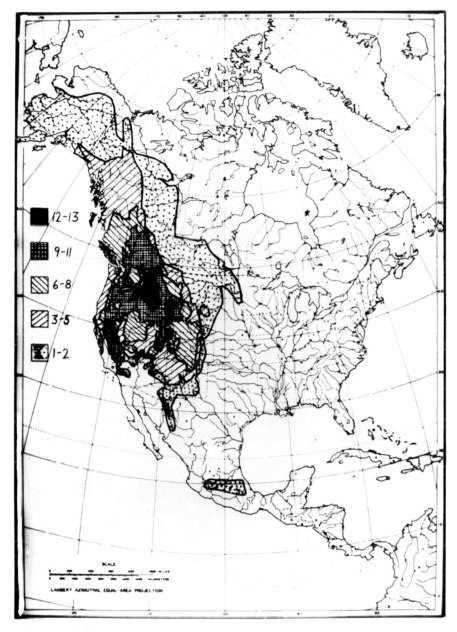

Fig. 3b.—Superimposed ranges of taiga- and tundra-adapted mammals in North America: montane taiga mammals (from Crewe, 1965).

was often cut off from contact with communities elsewhere in North America by a barrier of glacial ice, or at best was in contact via "ice-free corridor" and was smaller.

Beringia was an important area in the late Cenozoic for the evolution of boreal-adapted mammal communities, often with unique characteristics. Many of these species became extinct at the end of the Pleistocene, about 10 thousand years ago, but if

the distributions of surviving mammal species are superimposed, they illustrate nicely the "amphiberingian" distributions of many taxa of Holarctic mammals (Fig. 2).

The end of the Pleistocene was characterized by rapid global climatic warming, melting of much of the glacial ice in the Northern Hemisphere, and major reorganization of many biotic communities; for example, the "steppe tundra" of the Pleistocene (Guthrie, 1968; Kowalski, 1967) almost disap-

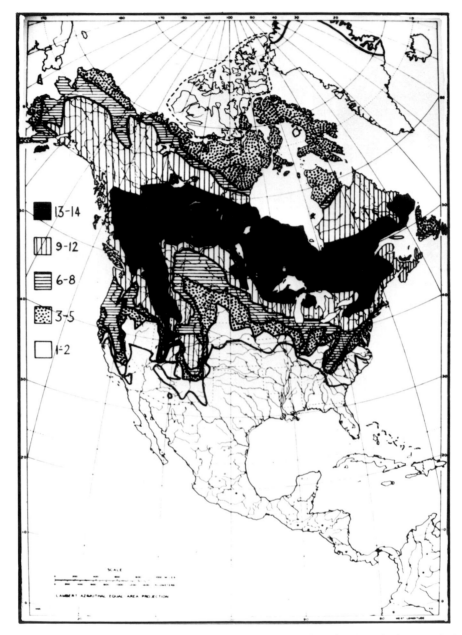

Fig. 3c.—Superimposed ranges of taiga- and tundra-adapted mammals in North America: boreal taiga mammals (from Crewe, 1965).

peared. The superimposed distributions of taiga- and tundra-adapted mammal species provide insight into the post-Pleistocene boreal communities and their varying environments, from equable, mild coastal taiga (Fig. 3a), through progressively more continental, colder montane and boreal taiga (Figs. 3b, c), to strongly seasonal alpine fellfield and polar desert (Fig. 3d).

CHARACTERISTIC MAMMALS

Among the mammals characteristic of these boreal communities are the red-toothed shrews (subfamily Soricinae; Repenning, 1967), especially the Holarctic genus *Sorex* (Fig. 4), and the microtine, or arvicolid rodents, the voles and lemmings. Shrews are relatively unmodified descendants of early Cenozoic eutherian mammals (see above), but the lineage leading to modern *Sorex* diverged in the

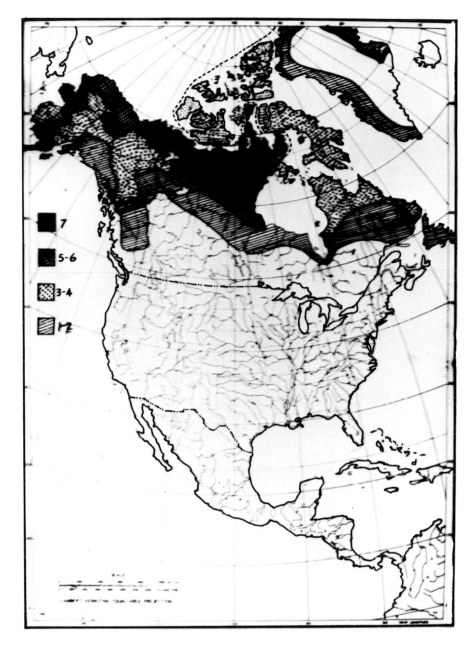

Fig. 3d.—Superimposed ranges of taiga- and tundra-adapted mammals in North America: tundra mammals (from Crewe, 1965).

Miocene, with the emergence of the boreal communities of which it is a characteristic member (Repenning, 1967). Arvicolids are of even more recent origin, and their remarkable adaptive radiation has occurred in the Plio-Pleistocene (Chaline and Mein, 1979; Martin, 1979) (Fig. 5). Different lineages have adapted to a wide variety of temperate (Fig. 6a) and boreal (Fig. 6b) habitats. One of the most specialized lineages is the collared lemmings (*Dicrostonyx*), the most northerly of small mammals. Among other adaptations, they develop "winter claws," presumably useful in digging packed snow. Collared lem-

mings are only distantly related to other lemmings (*Lemmus, Synaptomys, Mictomys*) (Chaline and Mein, 1979). Oddly, the only other arvicolid reported to develop winter claws is *Synaptomys* (=*Mictomys*) *borealis,* the northern bog lemming (Peterson, 1966). Winter claws in *Mictomys* are poorly known. They have been reported from specimens taken 26 January and 14 April, but winter-caught examples of the species are so scarce in collections that one cannot determine how widespread the phenomenon is. Even if winter claws are developed only occasionally by *Mictomys,* they de-

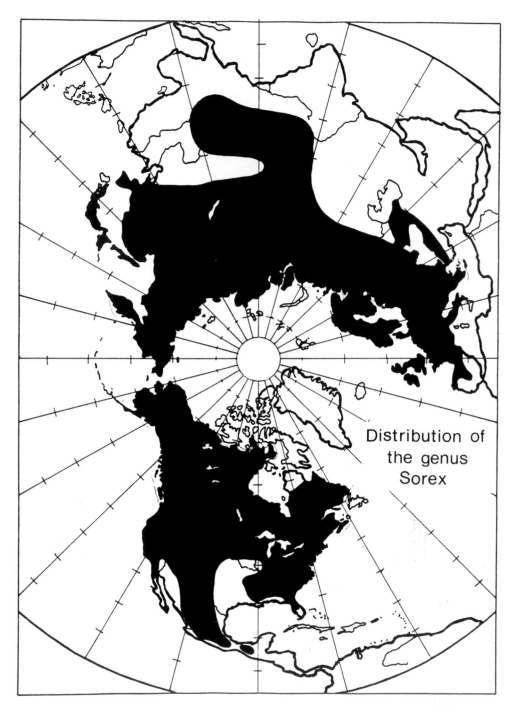

Fig. 4.—Distribution of the Holarctic, boreal, red-toothed shrews (genus *Sorex*).

serve careful comparison with those of *Dicrostonyx*. Evidence from patterns of dental enamel evolution suggest that *M. borealis* is the most recently derived of its lineage (Koenigswald and Martin, in press). The presence of winter claws in these two distantly related lemmings is, then, either a case of persistence of a primitive character in two long-divergent taxa, or, more likely, one of convergent evolution of specialized derived characters.

The story is even more confusing, however, because collared lemmings also have a highly specialized system of sex determination, in which a

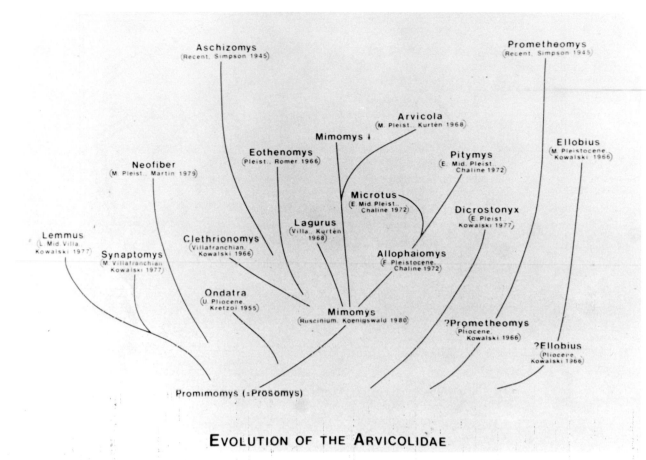

EVOLUTION OF THE ARVICOLIDAE

Fig. 5.—Adaptive radiation of arvicolid rodents (from Terry, 1981).

normal Y chromosome is not present (Gileva, 1980). Males have only an X chromosome, while females have either two X chromosomes or only one; females outnumber males by about 2:1 at times. One hypothesis to account for these observations is that the X chromosome bears either a female (f) or male (m) regulatory function. Theoretically, this sex ratio is 4 ♂♂ : 7 ♀♀. The Eurasian wood lemming, *Lemmus* (=*Myopus*) *schisticolor,* exhibits what appears to be a very similar mechanism of sex determination (Fredga et al., 1977), even though, like *Mictomys,* it is only distantly related. Since this is clearly a derived character state, as are winter claws, there appears to be strong selective pressure for this sort of convergence.

Collared lemmings are Holarctic in distribution and fill a unique ecological niche in the arctic tundra (Fig. 7), occupying well-drained areas (terrace, pingo, high-centered polygon, ridge) with grasses, sedges and lichens, such as in cottongrass meadows (Bee and Hall, 1956; Batzli et al., 1980). In the taiga and the alpine, the lemming role is filled in the Palearctic by *L. schisticolor,* and in the Nearctic by *M. borealis,* both of which share what may be derived characters with *Dicrostonyx,* as noted above, but have a very different ecological niche. These species occur in bogs and wet meadow areas, both in openings in the taiga, and in the alpine and subalpine zones (Fig. 8), although they may move to drier habitats in winter (Cowan and Guiguet, 1956; Corbet, 1966). For all three lemmings, winter snow cover is probably very important, but we know little as yet concerning the details of their nival ecology.

Other species of Holarctic small mammals have niches which may be occupied by the same species (for example, *Sorex tundrensis*) (Junge et al., 1981), sister species (for example, *Spermophilus undulatus, S. columbianus*) (Nadler et al., 1975; Vorontsov and Lyapunova, 1976), or unrelated, convergent lineages (for example, *Myospalax,* Geomyids).

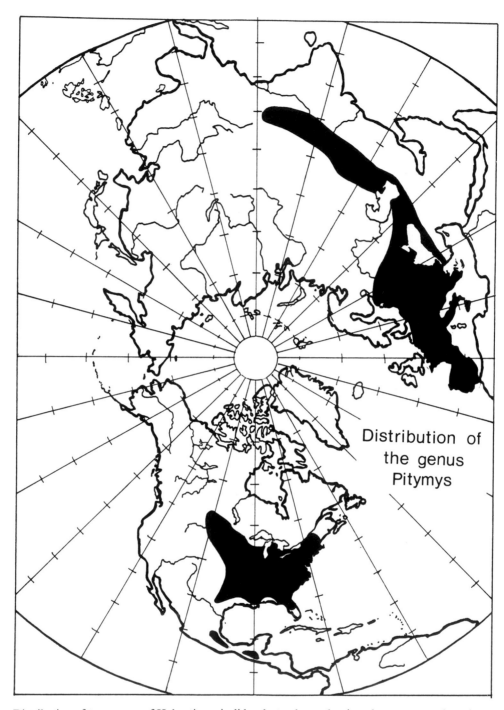

Fig. 6a.—Distribution of two genera of Holarctic arvicolid rodents: the predominantly temperate-adapted genus *Pitymys*.

ADAPTATIVE STRATEGIES

A detailed comparison of niche relationships in Holarctic mammals has not yet been made, but a brief example will perhaps be useful. The fossorial rhizovore niche, an important one in subalpine and alpine communities in the Holarctic (Hoffmann, 1974), is absent in boreal taiga and arctic tundra,

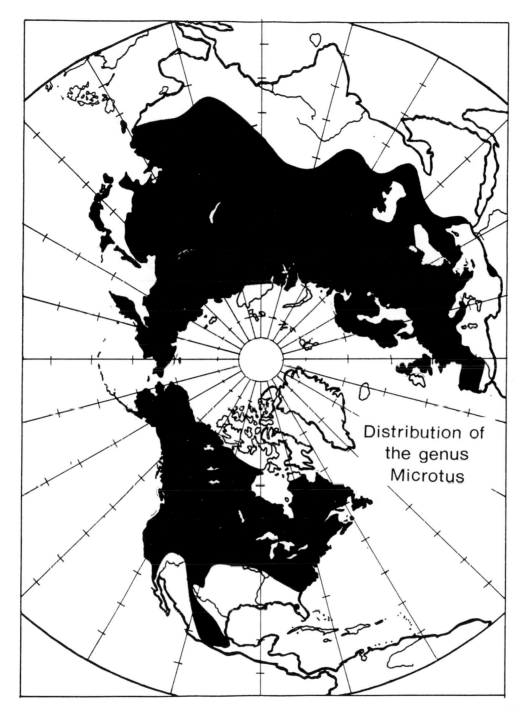

Fig. 6b.—Distribution of two genera of Holarctic arvicolid rodents: the predominantly boreal-adapted genus *Microtus*.

apparently mostly because of structural niche constraints. In the alpine and subalpine, fossorial rhizovores are restricted to stand types with sufficient winter snow-cover to inhibit deep freezing of the soil surface, yet well-drained enough to permit burrowing such as meadows and terraces (Fig. 8). These habitats are the analogs of the arctic tundra habitats occupied by collared lemmings, and a comparison of the winter ecology of arctic *Dicrostonyx* with temperate alpine forsorial rodents might be enlightening. In the Nearctic, several species of the geomyid *Thomomys* occupy this niche (Hoffmann, 1974),

ARCTIC TUNDRA STAND TYPES

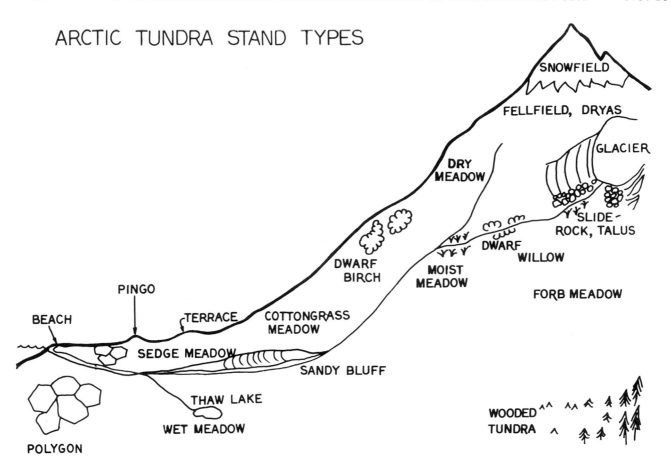

Fig. 7.—Schematic representation of arctic tundra stand types. Collared lemmings are typically found in cottongrass meadows and other well drained habitats.

while in the eastern Palearctic are found several species of the cricetid *Myospalax,* and other muroid genera are found elsewhere in Eurasia (Ognev, 1947; Yudin et al., 1979).

Lemmings and voles in boreal habitats such as we have noted often undergo high amplitude fluctuations in population density commonly called "cycles." A classical lemming cycle occupies three to four years, and includes a "crash," followed by a "low" phase, an "increase," and a "peak," after which another crash occurs (Batzli, et al., 1980). Such fluctuations have been documented in many

arvicolid populations in the Holarctic. "Cycles", are not, however, ubiquitous. They appear to be absent, or at most, strongly damped in some alpine populations (Hoffmann, 1974). Moreover, Terry (1981) has noted that high-amplitude fluctuations have not been recorded in many species, especially specialists from "complex" habitats (Fig. 9) with relatively low reproductive rates and high-nutrient food habits. The role of snow cover in "cycles" has been much debated, but there are many other ecological and evolutionary factors that must be considered as well.

SUMMARY AND CONCLUSIONS

Seasonally cold and snowy environments have been a part of the evolution of mammalian lineages since their beginning. Thus, many adaptations to winter snow may constitute shared primitive characters. Eutherian mammals radiated during a period

when seasonal cold was restricted to the regions of the poles, however, and intensification of globally zoned climates did not become significant until the Miocene. Recent mammals that most clearly exhibit adaptations to winter snow belong for the most part

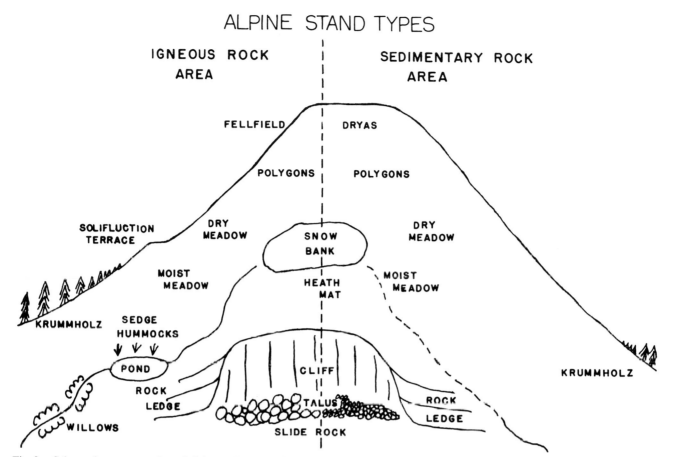

Fig. 8.—Schematic representation of alpine and upper taiga stand types. Wood and northern bog lemmings are typically found in taiga (krummholz and below), and in subalpine and alpine bogs and wet meadows.

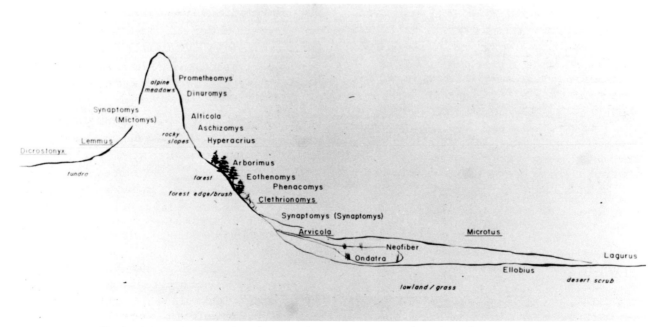

Fig. 9.—Schematic distribution of genera of arvicolid rodents by stand-type (from Terry, 1981).

to lineages that have evolved during or after the Miocene in the Northern Hemisphere. North America and Eurasia were often united during this time through Beringia, and a Holarctic fauna developed. The stand-types inhabited by Holarctic mammals are often relatively simple (compared to warm temperate or tropical), and similarities in ecological niches between Nearctic and Palearctic, between arctic tundra and alpine zone, or between different sorts of taiga, offer many opportunities for comparative studies of nival ecology.

LITERATURE CITED

ANONYMOUS. 1977. The linked voyages of land and life. Pp. 64a–64n, in Science Year (W. H. Nault, ed.), Field Enter. Educ. Corp., Chicago, Illinois.

BATZLI, G. O., R. G. WHITE, S. F. MacLEAN, JR., F. A. PITELKA, and B. D. COLLIER. 1980. The herbivore-based trophic system. Pp. 355–410, in An arctic ecosystem (J. Brown, P. C. Miller, L. L. Tieszen and F. L. Bunnell, eds.), US/IBP Synthesis Series, Downden, Hutchinson and Ross, Stroudsburg, Pennsylvania, 12:xxv + 1–571.

BEE, J. W., and E. R. HALL. 1956. Mammals of northern Alaska. Misc. Publ. Mus. Nat. Hist., Univ. Kansas, 8:1–309.

CHALINE, J., and P. MEIN. 1979. Les rongeurs et l'evolution. Doin, Paris, xi + 235 pp.

CLEMENS, N. A., J. A. LILLEGRAVEN, E. H. LINDSAY, and G. G. SIMPSON. 1979. Where, when, and what—a survey of known Mesozoic mammal distribution. Pp. 7–58, in Mesozoic mammals . . . (J. A. Lillegraven, Z. Kielan-Jaworowska and W. A. Clemens, eds.), Univ. California Press, Berkeley, California, x + 311 pp.

CORBET, G. B. 1966. The terrestrial mammals of western Europe. Dufours, Philadelphia, xi + 264 pp.

COWAN, I. McT., and C. J. GUIGUET. 1956. The mammals of British Columbia. British Columbia Prov. Mus., Victoria, Handbook, 11:1–413.

COX, C. B., I. N. HEALEY, and P. D. MOORE. 1976. Biogeography: an ecological and evolutionary approach. John Wiley, New York, 2nd. ed., ix + 194 pp.

CREWE, G. V. E. 1965. An attempt to delineate mammal faunal groups in North America. Univ. British Columbia, Vancouver, 2 pp. + 27 maps, processed.

CRUSAFONT PAIRO, M. 1962. Constitucion de una nueva chase (Ambulatilia) para los llamados "Reptiles mamiferoides." Not. Comm. Inst. Geol. Min. España, 66:259–266.

DORF, E. 1960. Climatic changes of the past and present. Amer. Sci., 48:341–364.

FREDGA, K., A. GROPP, H. WINKING, and F. FRANK. 1977. A hypothesis explaining the exceptional sex ratio in the wood lemming (Myopus schisticolor). Hereditas, 85:101–104.

GILEVA, E. A. 1980. Chromosomal diversity and an aberrant genetic system of sex determination in the arctic lemming, Dicrostomyx torquatus Pallas (1779). Pp. 99–103 in Animal genetics and evolution (N. N. Vorontsov and J. M. Van Brink, eds.), W. Junk, B. V., The Hague, ix + 383 pp.

GUTHRIE, R. D. 1968. Paleoecology of the large mammal community in interior Alaska during the late Pleistocene. Amer. Midland Nat., 79:346–363.

HOFFMANN, R. S. 1974. Terrestrial vertebrates. Pp. 475–568, in Arctic and alpine environments (J. Ives and R. Barry, eds.), Methuen, London, xxiii + 999.

———. 1981. Different voles for different holes: environmental restrictions on refugial survival of mammals. Pp. 25–45, in Evolution today (G. G. E. Scudder and J. L. Reveal, eds.), Proc. 2nd Int. Congr. Evol. Biol., Vancouver, British Columbia. iv + 486 pp.

HOFFMANN, R. S., and R. S. PETERSON. 1967. Systematics and zoogeography of Sorex in the Bering Strait area. Syst. Zool., 16:127–136.

HOFFMANN, R. S., and R. D. TABER. 1968. Origin and history of Holarctic tundra ecosystems, with special reference to their vertebrate faunas. Pp. 143–170 in Arctic and alpine environments (H. E. Wright, Jr. and W. H. Osburn, eds.), Indiana Univ. Press, Bloomington, xii + 308 pp.

HOPSON, J. A. 1970. The classification of nontherian mammals. J. Mamm. 51:1–9.

JUNGE, J. A., R. W. DeBRY, and R. S. HOFFMANN. 1981. Morphometric analysis of Sorex arcticus and Sorex tundrensis. Abst. Tech. Pap., 61st ann. meet. Amer. Soc. Mamm. No. 45.

KEONIGSWALD, W. VON, and L. D. MARTIN. 1984. Revision of the fossil and Recent Lemminae (Rodentia, Mammalia). Spec. Publ., Carnegie Mus. Nat. Hist., 9:122–137.

KOWALSKI, K. 1967. The Pleistocene extinction of mammals in Europe. Pp. 349–364 in Pleistocene extinctions. The search for a cause. (P. S. Martin and H. E. Wright, Jr., eds.), Yale Univ. Press, New Haven, Connecticut, x + 453 pp.

LILLEGRAVEN, J. A. 1979. Introduction. Pp. 1–6 in Mesozoic mammals . . . (J. A. Lillegraven, Z. Kielan-Jaworowska, and W. A. Clemens, eds.), Univ. California Press, Berkeley, Calif.

LILLEGRAVEN, J. A., Z. KIELAN-JAWOROWSKA and W. A. CLEMENS, EDS. 1979. Mesozoic mammals . . . Univ. California Press, Berkeley, California, x + 311 pp.

MARTIN, L. W. 1979. The biostratigraphy of arvicoline rodents in North America. Trans. Nebraska Acad. Sci ., 7:91–100.

McELHINNY, M. W. 1973. Paleomagnetism and plate tectonics. Cambridge Univ. Press, Cambridge, x + 358 pp.

McKENNA, M. C. 1975. Toward a phylogenetic classification of the Mammalia. Pp. 21–46, in Phylogeny of the primates (W. P. Luckett and F. S. Szalay, eds.), Plenum Publ. Corp., New York, xiv + 483 pp.

NADLER, C. F., E. A. LYAPUNOVA, R. S. HOFFMANN, N. N. VORONTSOV, and N. A. MALAGINA. 1975. Chromosomal evolution in holarctic ground squirrels (spermophilus). I. Giemsa-band homologies in Spermophilus columbianus and S. undulatus. Z. Saugetierk, 40:1–7.

OGNEV, S. I. 1947. Zveri SSR i prilezhashchik stran. Vol. 5. Glires. AN SSSR, Moscow-Leningrad, 809 pp. English translation, Off. Tech. Serv., Washington, D.C., 1963.

PETERSON, R. L. 1966. The mammals of eastern Canada. Oxford Univ. Press, Toronto, xxxii + 465 pp.

REED, C. A. 1960. Polyphyletic or monophyletic ancestry of mammals, or: What is a Class? Evolution, 14:314–322.

REPENNING, C. A. 1967. Subfamilies and genera of the Soricidae. Geol. Surv. Prof. Papers, 565:1–74.

ROMER, A. S. 1966. Vertebrate paleontology. Univ. Chicago Press, Chicago, Illinois, 3rd ed. ix + 468 pp.

TERRY, C. J. 1981. Population fluctuations in the family Arvicolidae (Rodentia) and their relationship to "microtine cycles." Unpublished Ph.D. diss., Univ. Kansas, Lawrence, [v] + 128 pp.

VAUGHAN, T. A. 1978. Mammalogy. W. B. Saunders, Philadelphia, 2nd ed., x + 518 pp.

VORONTSOV, N. N., and E. A. LYAPUNOVA. 1976. Genetika i problemy transberingiiskikh svyazei Golarkticheskikh mlekopitayushchikh. Pp. 337–353 in Beringiya v Kainozoe (V. L. Knotrimavichus, ed.), Acad. Sci., Valdivostok, 594 pp.

YUDIN, B. S., L. I. GALKINA, and A. F. POTAPKINA. 1979. Mlekopitayushchie Altae-Sayanskoi gornoi strany. [Mammals of the Altai-Sayan montane region.] Nauka, Novosibirsk, 296 pp.

Address: Museum of Natural History, and Department of Systematics and Ecology, University of Kansas, Lawrence, Kansas 66045.

MATERIALS AND METHODS OF SUBNIVEAN SAMPLING

WILLIAM D. SCHMID

ABSTRACT

Snow cover is an important ecological factor that affects microclimate, activity and survival of organisms living below it and creates many unique but interesting sampling problems. Permanent trap shelters that allow access to subnivean habitat from above (for trapper) and below (for trappee) have been constructed cheaply, easily and successfully from plastic wastebaskets. Successful capture-recapture studies require live traps with dry nest chambers. The Longworth, Sheffer, Sherman and tube with attached plastic box are easily adapted to this requirement by packing with dry wood shavings or alfalfa hay. Pitfall traps work well for winter-active invertebrates.

Long term measurements of temporal variation in microclimate parameters; for example, temperature, soil frost depth, are best done with permanent recording stations or sensor-transmitters that broadcast to a central receiver and recorder. Extensive spatial sampling requires portable instruments with sensors on wands or spears for temporary placement into or below the snow. A variety of such sampling devices will be described; for example, temperature, light and carbon dioxide. A preliminary model for prediction of subnivean temperatures based on snowfield characteristics will be presented.

Experimental manipulation of natural snowfields; for example, compaction with snowmobiles, removal with snowblowers, icing with water spray, and even artificial snow makers, may provide us with further insights to the roles of snow as a natural factor in the lives of organisms that live below it.

TRAPPING SHELTERS

Winter sampling techniques, because of lower temperatures and snow, must be modified from methods used during the summer. Capture of subnivean mammals is accomplished by using a type of cover that protects from snow and provides easy access to the trap from above for the trapper and from below for the trappee. Chimneys of plywood (Pruitt, 1959; Brown, 1973) or of roofing paper (Merritt and Merritt, 1978) were built for this purpose, whereas Beer (1961) used no trap shelters during a snow-free winter. Iverson and Turner (1969) set shallow chimneys and dug down to them if snow fell or drifted over trap sites. Christiansen (personal communication) and Jannett (1980) simply placed traps in empty waxed cardboard milk cartons. I have used plastic waste baskets (Fig. 1) for trap shelters through many seasons. They are relatively inexpensive, lightweight, portable, and durable. It is important to establish and prebait the trap station shelters before snow accumulates in the fall for the most successful winter capture of small mammals. These shelters also have been used to protect instruments, like thermographs, as well as pit traps for winter-active invertebrates. When snow accumulates deeper than the heights of the shelters, it is necessary to flag their locations in some manner and to dig down to their tops.

TRAPS

Most types of traps used during the summer months work well for live captures of small mammals during the winter (Fig. 2). The addition of bedding materials, such as wood shavings, dry straw or hay, or cotton batting, greatly enhances survival when traps are checked only once or twice a day. Extra bait and a piece of potato or carrot can be used as food and water sources for the mammals during periods of entrapment. Pitfall traps (Nasmark, 1964; Aitchison, 1974, 1979; Marshall, 1976) have been used to capture winter-active invertebrates that may be important food items of small mammals. Successful capture techniques for small mammals throughout the year provide only a partial answer to questions of winter ecology. Microclimatic factors may play important roles in population regulation and seasonal timing of physiological changes and behavior.

CHARACTERISTICS OF THE SUBNIVEAN ENVIRONMENT

Problems of winter sampling of microclimate when there is snow cover are analogous to those faced by aquatic biologists; that is, one must sample into spaces that one cannot directly observe. Unlike aquatic sampling, however, the microclimate space can be severely disrupted by the investigator's ac-

TRAP SHELTER
(PLASTIC WASTEBIN)

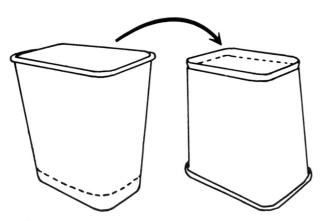

Fig. 1.—Plastic wastebaskets used as shelters to prevent snow from inactivating trap mechanisms and to provide easy access for both trapper and trappee. The inverted wastebaskets were pinned to the ground at all four corners by 20-penny nails. The shelters were closed by inverting the cut-off bottoms to friction-fit tightness into the top of each chimney. Drawn from Jarvinen and Schmid, 1971.

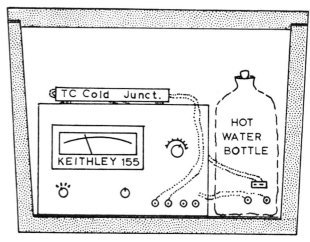

Fig. 3.—Reinforced styrofoam ice chest with hotwater bottle for protection of electronic instruments from extreme cold temperatures. Double pane window allows direct reading of meter and thermocouple plugs set into and extending through the wall allow direct coupling of sensors to the enclosed instruments.

tivity and measuring devices. An additional problem exists because the circuits of many electronic instruments may not function properly when exposed to severely cold temperatures. Some modification of summer methods of measurement and a few new techniques unique to sampling the subnivean microclimate are generally required and are described below.

THE THERMAL ENVIRONMENT

Snow cover is important to the survival of many subnivean animals in north temperate and subarctic

latitudes (Mail, 1930; Formozov, 1946; Beer, 1961) because of the protection it affords them from stresses of direct exposure to the severe winter climate and predation (Pruitt, 1957, 1970; Coulianos and Johnels, 1962; Beer, 1961; Geiger, 1965; Vose and Dunlap, 1968; Schmid, 1971; Jarvinen and Schmid, 1971). Evaluation of snowpack parameters can be a study in itself (Kingery, 1963; Richens and Madden, 1973); however, measurements of the microclimate per se are more directly related to physiology and behavior of subnivean animals.

The thermal environment of subnivean animals is one of the most important parameters to be measured. Recording thermometers can be set in shelters with sensing probes extended onto the ground

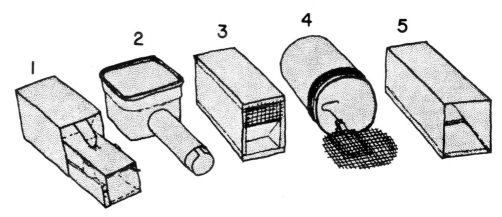

Fig. 2.—Live traps used in winter capture of small mammals all have a closed nest chamber to be filled with dry bedding material: (1) Longworth (Chitty and Kempson, 1949), (2) tube attached to plasticware box (Brown et al., 1969), (3) Burt trap (Burt, 1940), (4) Sheffer (Davis, 1956), and (5) Sherman.

FROST DEPTH GAGE

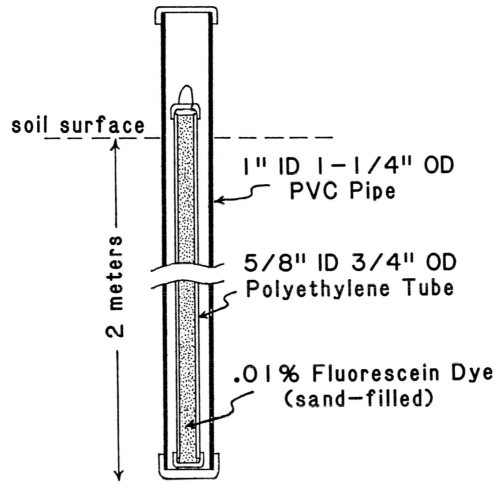

Fig. 4.—Frost depth is read directly from the inner tube: fluorescein dye solution produces a sharp color difference between frozen and liquid phase, while sand increases rigidity, decreases expansion upon freezing and prevents thermal mixing. The outer tube should extend higher above ground than expected snow depth. The ground surface level is marked on the inner tube as zero reference, and frost depth is directly read from below that mark. After Harris, 1970.

surface in the vicinity of permanent trapping sites (Marshall, 1976; Merritt and Merritt, 1978; Schmid, 1982; Marchand, 1982). Such recording stations should be established prior to the winter season because later placement of instruments would disrupt the natural snow cover. Radio telemetry is an alternative to field recorders for periodic temperature readings (Aitchison, 1974). However, longer range devices (Winter et al., 1978) could allow for reception and automatic recording of temperature data from a number of sites, each on a different carrier frequency. Glass-bulb alcohol thermometers, thermistors, and thermocouples have been used suc-

cessfully for direct thermometry. Low cost, reliability, and versatility favor thermocouples if a high precision microvoltmeter is available for readout of the thermoelectric potential (about 40 μV/°C for copper-constantan junctions).

Thermocouples constructed of fine wire have very low thermal inertia and therefore quickly reach equilibrium with their environment. A series of thermocouples mounted with junctions spaced at measured intervals along the surface of a fiberglass arrow shaft and with connecting wires within the tube to separate plugs are useful for quick and accurate measurements of thermal gradients from the

HEAVY DUTY GAS SAMPLING SPEAR

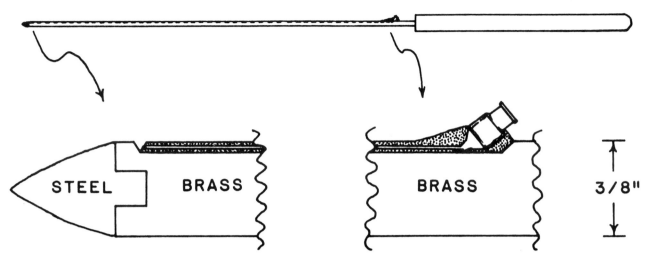

Fig. 5.—Spear for sampling subnivean gases. A three-foot brass rod is grooved to contain polyethylene tubing (I.D. = 0.047 inch, O.D. = 0.067 inch) that runs from the point to just below the wooden handle where an 18-gauge needle is fitted into it. Epoxy holds the tubing and needle on the rod. Gas samples are slowly withdrawn through the tubing by an attached syringe.

ground upward through the snowpack. This wand of sensors works well for instantaneous thermometry in different sites when used with an electronic cold junction, for example, Omega CJ-T or MCJ-T, Omega Engineering, Stamford, Connecticut. The electronic cold junction eliminates the usual necessity for a reference thermocouple at 0°C. A fiberglass-epoxy reinforced styrofoam ice chest can be used to house the electronic cold junction and the precision voltmeter, such as a Keithly 155 (Fig. 3).

A series of thermocouples can be set vertically into the ground to monitor soil temperature and depth of the frost line. However, I have also used a very inexpensive and reliable gauge (Fig. 4) that permits direct reading of the frost line or bonded frost depth (Harris, 1970). These gauges have been used to measure the frostline under natural snow cover and under snow that has been packed by a snowmobile: −19 cm versus −58 cm for a threefold increase in depth of frozen soil after mechanical compaction of the snow (Marshall, 1976). Amelioration and stability of both subnivean and soil temperatures are dependent upon an adequate snow cover (Marchand, 1982), but snow affects other microclimate parameters as well as temperature.

LIGHT

Light, long recognized as a zeitgeber for seasonal changes in many plants and animals, has been studied as a component of subnivean microclimates in precious few instances (Evernden and Fuller, 1972; Richardson and Salisbury, 1977). Photometry, like thermometry, can be executed through permanent installations of instruments or by temporary placement of sensors, wands or spears, into subnivean habitats. Cadmium sulfide (CdS) cells, and photodiodes or silicon blue cells behave as photoresistors and photovoltaic transducers respectively, and require different readout instruments. A simple wheatstone bridge circuit can be readily built for the CdS cell in which a 10-turn dial potentiometer is used to balance the circuit against the variable resistance, depending on light energy absorbed. The CdS cell, as a variable resistor, can take the place of the thermistor in radio-transmitters for remote sensing of light. The photodiodes and silicon blue cells produce current or voltages proportional to light energy absorbed and can be read with the same high precision voltmeters employed to read thermocouples.

The use of a quantum sensor like LI-190S (Lamb-

Table 1.—*Thermal diffusivities of water, snow, and ice.*

Substance	Conductivity k (cal/cm sec°C)	Density γ (g/cm³)	Specific heat C (cal/g °C)	Diffusivity α (cm²/sec)
Water	1.32×10^{-3}	1.00	1.01	1.31×10^{-3}
Ice	5.3×10^{-3}	.92	.5	11.52×10^{-3}
*Old snow	2.12×10^{-3}	.50	.5	8.5×10^{-3}
*Old snow	$.61 \times 10^{-3}$	.30	.5	4.08×10^{-3}
*New snow	$.68 \times 10^{-4}$	.10	.5	1.36×10^{-3}
*New snow	$.17 \times 10^{-4}$	.05	.5	$.68 \times 10^{-3}$

* Values of specific heat were approximated according to the equation $C = .506 + .00186\,T$ (°C). $\gamma = .5$ is the maximum (critical value) for densification of snow by packing.

Table 2.—*Values of ω for cycles of different periods.*

Period	ω (frequency)	I°
1 Day	7.272×10^{-5} radians/sec	.151
4 Days	1.818×10^{-5} radians/sec	.389
7 Days	1.039×10^{-5} radians/sec	.490
14 Days	5.194×10^{-6} radians/sec	.604
28 Days	2.597×10^{-6} radians/sec	.700

I° as a function of variable periods of oscillation beneath 20 cm of old snow ($\gamma = .30$ g/cm³).

da Instrument Corporation, Lincoln, Nebraska) can accommodate the calibration of many home-made instruments. New developments in solid state electronic components provide the potential construction of sensing, transmitting and recording devices. For example, the millivolt output from thermocouples or silicon photo cells could be converted by a voltage controlled oscillator into an audio tone that could then be recorded on a small portable tape recorder (La Barbera and Vogel, 1976). The frequency of the audio tone can be converted subsequently to units of measurement, for example, temperature or radiant energy flux. As an alternative to sensing light beneath the snow, it is now possible to bring subnivean light to the surface, and even into the laboratory, by means of fiber optics. Once the light has been piped to the surface via either portable probes or permanent installations of fiber optics, its spectral composition and energy can be evaluated. Photographic recording of subnivean light becomes feasible and offers an interesting possibility. The degree to which both qualitative and quantitative variations in light energy at the base of snowpacks correlate with seasonal and daily cycles of physiology and behavior can give a more complete understanding of the winter ecology of small mammals.

QUALITY OF AIR

The quality of subnivean air in relation to small mammal ecology has been little studied. Only one report of carbon dioxide accumulation, as high as four times ambient (although the tabulated values read as four percent) has been correlated with subnivean mammal nests and activity (Bashenina, 1956). It is unlikely that snow, as a porous, permeable ground cover, could effectively retard carbon dioxide diffusion from subnivean spaces (Bergen,

1968; Kelly et al., 1968) to a degree that would result in respiratory stress to small mammals (Fenn, 1948; Hayward, 1966; Dejours, 1966). However, a solid crust of ice in or on the snow could effectively seal the subnivean air from the ambient atmosphere and allow accumulation of abnormal concentrations of respiratory gases (Rakitina, 1965). Active muskrat houses, frozen solid during the winter, commonly have between five and ten percent carbon dioxide (Huenecke et al., 1958; and my own measurements) that would be stressful to nondiving mammals (Scott, 1917; Giaja and Markovic, 1953).

The role of snow as a moderator of gas concentrations in subnivean air is proably not very important under normal weather conditions, but is worthy of more study, especially in cases of heavy ice crustation by winter rain. The relatively thin and apparently porous ice crusts formed during periodic thaws do not seem to be effective barriers to gas exchange. I have not found more than 0.14% carbon dioxide in over 20 samples taken from beneath thaw crusts with a gas sampling spear (Fig. 5). The gas samples were withdrawn by a five milliliter syringe that was sealed after removal from the spear and returned to the lab for analysis by microtitration of barium hydroxide solution used as a quantitative absorbant (Conway, 1963). Gas samples taken from beneath uncrusted snow varied over the same range of carbon dioxide values. The effects of variation in subnivean gases might be more completely evaluated by artificially icing snow cover with water spray.

HEAT FLUX

As a final note on sampling techniques in winter ecology, I present the following nonequilibrium model for heat flux between subnivean habitat and its above-snow atmosphere. This example is not meant to be a final predictor of subnivean temperature fluctuations, but rather, an illustration of some major parameters that we must consider for the development of useful models.

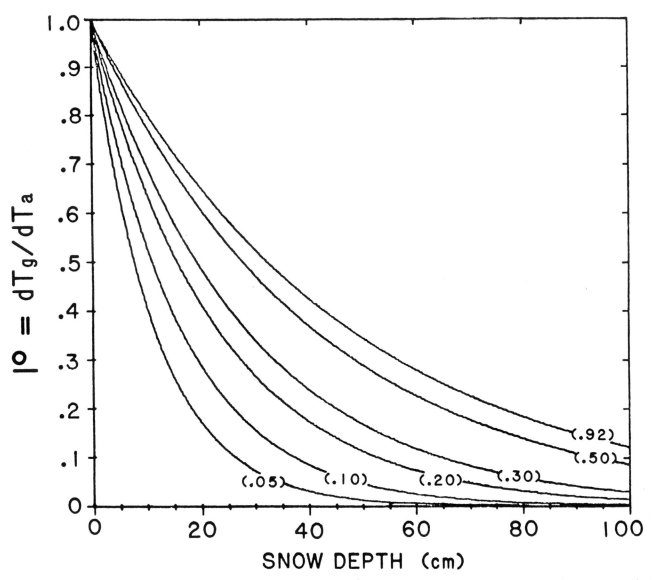

Fig. 6.—Effect of depth and density upon the ability of snow cover to damp temperature fluctuations. All curves are for the same period of oscillation (seven days), and each curve is for a different snow density given by numerals in brackets (0.05 to 0.50 g/cc) and ice (0.92 g/cc).

The basic Fourier equation for conductive heat flow is useful for estimation of instantaneous rates and is applicable to steady state systems. However, given that environmental temperatures normally oscillate throughout the winter, a more useful relationship is based on nonequilibrium heat transfer (Ingersoll et al., 1954). This equation describes the degree to which periodic variations in temperature are reduced by (1) a given thickness (x, in cm) of material of (2) known or estimated thermal diffusivity** when (3) the period of oscillation can be approximated.

Let: dT_g = amplitude of variation of subnivean temperature beneath × cm of snow

dT_a = amplitude of variation of air temperature at the snow surface

$$I^\circ = dT_g/dT_a$$

So: I° = 1.0 when there is no damping by snow, and

I° = 0.0 if there is complete damping of the periodic variations of temperature at the snow-air interface.

$$I^\circ = \exp(-x(\omega/2\alpha)^{1/2}) \equiv e^{-x(\omega/2\alpha)^{1/2}}$$

Where: e = base of natural logarithms
(2.71828+)

x = snow depth (cm)

α = thermal diffusivity (cm²/sec)

ω = frequency of temperature cycle (radians/sec)

**Note: Thermal diffusivity of a conducting material, snow for example, is equal to the ratio: thermal conductivity divided by the product of density and specific heat.

$$\alpha = (k/\gamma c)cm^2/sec$$

Where: k = cal/cm sec °C

c = cal/gm °C

γ = gm/cm³

Table 1 is a summary of some thermal diffusivity values for water, ice and snow of variable densities. By inspection of the basic equation for the degree of damping, we see that (1) increases of snow depth, (2) decreases of thermal diffusivity, and (3) increases of frequency (=decreases of period) of temperature oscillations will all promote a decrease in I°; that is, the subnivean temperaure cycle will be more completely damped.

The frequency (ω) of temperature oscillation is given as the ratio of $\theta = 2\pi$ radians for a complete cycle, to the time period, in seconds. For example, a 24 hour cycle would have its frequency defined as:

$$\omega = \theta/t$$
$$= 2\pi/24 \text{ hr}$$

= 2(3.14159)/86,400 sec

= 7.272 × 10⁻⁵ radians/sec

Values of ω for cycles of different periods are shown in Table 2.

Further extensions of the basic heat transfer relationships can be constructed to show the effects of changing depth of snow and changing density of snow as illustrated in Fig. 6. This model explicitly defines snow depth and density as measures important to determination of stability of subnivean temperature. Thermal diffusivity is a better statistic than thermal conductivity because the temperature of insulative snow cover changes as heat passes through it from the ground to the colder atmosphere. The frequency (or period) of ambient temperature fluctuation is another important factor, and relatively rapid temperature changes (circadian) are much more strongly damped than longer term changes that occur through exchange of major winter weather systems. Yet this model is incomplete because continued heat flux from the earth has not been factored into the basic equation, but it serves as a predictor of relative change under variable snow and weather conditions. Marchand (1982) has published the best index to the effect of snow cover on subnivean temperature, but with some of the considerations presented here it, like the model itself, might be expanded as we collect more data and develop a more thorough understanding of winter microclimatology.

SUMMARY

I have briefly summarized some of the techniques that have worked for me and my students of winter ecology during the past ten years. If this material stimulates any new research work, then I will have been successful in my writing. I close by dedicating this paper to two pioneers in winter ecology. William O. Pruitt, Jr. and the late A. N. Formozov.

LITERATURE CITED

AITCHISON, C. W. 1974. A sampling technique for active subnivean invertebrates in southern Manitoba. Manitoba Entom., 8:32–36.

———. 1979. Winter-active subnivean invertebrates in southern Canada. I. Collembola. Pedobiol., 19:113–120.

BASHENINA, N. V. 1956. Influence of the quality of subnivean air on the distribution of winter nests of voles. Zool. Zhur., 35:940–942.

BEER, J. R. 1961. Winter home ranges of the red-backed mouse and white-footed mouse. J. Mamm., 42:174–180.

BERGEN, J. D. 1968. Some measurement of air permeability in a mountain snow cover. Bull. I.A.S.H., 13:5–13.

BROWN, E. B., W. R. SAATELA, and W. D. SCHMID. 1969. A compact, light-weight live trap for small mammals. J. Mamm., 50:154–155.

BROWN, E. B. 1973. Changes in patterns of seasonal growth of *Microtus pennsylvanicus*. Ecology. 54:1103–1110.

BURT, W. H. 1940. Territorial behavior and populations of some small mammals in southern Michigan. Misc. Publ. Mus. Zool., Univ. Michigan, 45:1–58.

CHITTY, D. H., and D. A. KEMPSON. 1949. Prebaiting small mammals and a new design of live trap. Ecology, 30:536–542.

CONWAY, E. J. 1963. Microdiffusion analysis and volumetric error. Chemical Publ. Co., Inc., New York, 467 pp.

COULIANOS, C. C., and A. G. JOHNELS. 1962. Note on the subnivean environment of small mammals. Arkiv. Zool., 15:363–370.

DAVIS, D. E. 1956. Manual for analysis of rodent populations. Edward Bros., Ann Arbor, Michigan, 82 pp.

DEJOURS, R. 1966. Respiration. Oxford Univ. Press, New York, 244 pp.

EVERNDEN, L. N., and W. A. FULLER. 1972. Light alteration caused by snow and its importance to subnivean rodents. Canadian J. Zool., 50:1023–1032.

FENN, W. D. 1948. Physiology of exposures to abnormal concentrations of respiratory gases. Proc. Amer. Phil. Soc., 92:144–154.

FORMOZOV, A. N. 1946. The covering of snow as an integral factor of the environment and its importance in the ecology of mammals and birds. Material for fauna and flora of USSR, new series, Zool., 5:1–152. (English edition, occasional paper, no. 1, Boreal Institute, Univ. Alberta, Edmonton, Alberta, Canada, 197 pp.)

GEIGER, R. 1965. Climate near the ground. Harvard Univ. Press, Cambridge, Massachusetts, 611 pp.

GIAJA, J., and L. MARKOVIC. 1953. L'hypothermie et la toxicité du gaz carbonique. Compt. rond. Acad. SC, 236:2437–2440.

HARRIS, A. R. 1970. Direct reading frost gage is reliable, inexpensive. Res. Note NC-89, North Central Forest Expt. Sta., Forest Service, Folwell Ave., St. Paul, Minnesota 55101.

HAYWARD, J. S. 1966. Abnormal concentrations of respiratory gases in rabbit burrows. J. Mamm., 47:723–724.

HUENECKE, H. S., A. B. ERICKSON, and W. H. MARSHALL. 1958. Marsh gases in muskrat houses in winter. J. Wildlife Mgmt., 22:240–245.

INGERSOLL, L. R., O. J. ZOBEL, and A. C. INGERSOLL. 1954. Heat conduction. Univ. Wisc. Press, Madison, Wisconsin, 325 pp.

IVERSON, S. L., and B. N. TURNER. 1969. Under-snow shelter for small mammal trapping. J. Wildlife Mgt., 38:722–723.

JANNETT, F. J. 1980. Social dynamics of the montane vole, *Microtus montanus,* as a paradigm. The Biologist, 62:3–19.

JARVINEN, J. A., and W. D. SCHMID. 1971. Snowmobile use and winter mortality of small mammals. Pp. 131–141, *in* Proc. 1971 Snowmobile and off-the-road vehicle research (M. Chubb, ed.), Symp. Tech. Report 8, College of Agric. and Natural Res., Michigan State Univ., East Lansing, Michigan, 196 pp.

KELLY, J. J., D. F. WEAVER, and B. P. SMITH. 1968. The variation of carbon dioxide under snow in the Arctic. Ecology, 49:358–360.

KINGERY, W. D. (Ed.), 1963. Snow and Ice. The M.I.T. Press, Cambridge, Massachusetts, 684 pp.

LA BARBERA, M., and S. VOGEL. 1976. An inexpensive thermistor flow-meter for aquatic biology. Limnol. Oceanogr., 21:750–756.

MAIL, G. A. 1930. Winter soil temperature and its relation to subterranean insect survival. J. Agric. Res., 41:571–592.

MARCHAND, P. J. 1982. An index for evaluating the temperature stability of a subnivean environment. J. Wildlife Mgmt., 46:518–520.

MARSHALL, O. 1976. Seasonal changes in a grassland mesofaunal invertebrate community. Unpublished Ph.D. thesis, Univ. Minnesota, Minneapllis, 169 pp.

MERRITT, J. F., and J. M. MERRITT. 1978. Population ecology and energy relationships of *Clethrinomys gapperi* in a Colorado subalpine forest. J. Mamm., 59:576–598.

NÄSMARK, O. 1964. Vinteraktivitet under snon hos landlevande evertebrater. (Winter activity of terrestrial invertebrates under the snow.) Zool. revy., 26:5–15.

PRUITT, W. O. 1957. Observations on the bioclimate of some taiga mammals. Arctic, 10:130–138.

———. 1959. A method of live-trapping small taiga mammals in winter. J. Mamm., 40:139–143.

———. 1970. Some ecological aspects of snow. Pp. 83–100, *in* Ecology of the subarctic regions, UNESCO, Paris, France, 364 pp.

RAKITINA, Z. G. 1965. The permeability of ice for oxygen and carbon dioxide in connection with a study of reasons for winter cereal mortality under an ice crust. Fiziol. Rast., 12:909–919.

RICHARDSON, S. G., and F. B. SALISBURY. 1977. Plant responses to the light penetrating snow. Ecology, 58:1152–1158.

RICHENS, V. B., and C. G. MADDEN. 1973. An improved snow study kit. J. Wildlife Mgmt., 37:109–113.

SCHMID, W. D. 1971. Modification of the subnivean microclimate by snowmobiles. Pp. 251–257, *in* A. O. Haugen (Editor), Proc. of the snow and ice in relation to wildlife and recreation symposium (A. O. Haugen, ed.), Iowa Coop. Wildl. Res. Unit, Iowa State Univ., Ames, Iowa, 280 pp.

———. 1982. Survival of frogs in low temperature. Science, 215:697–698.

SCOTT, R. W. 1917. Carbon dioxide intake and volume of inspired air. Amer. J. Physiol., 44:18–21.

VOSE, R. N., and D. G. DUNLAP. 1968. Wind as a factor in the local distribution of small mammals. Ecology, 49:381–386.

WINTER, J. D., V. B. KUECHLE, D. B. SINIFF, and J. R. TESTER. 1978. Equipment and methods for radio tracking freshwater fish. Ag. Expt. Sta. Misc. Report 152, Univ. of Minnesota, St. Paul, 152:1–18.

Address: University of Minnesota, 108 Zoology Building, Minneapolis, Minnesota 55455.

LIGHT EXTINCTION UNDER A CHANGING SNOWCOVER

PETER J. MARCHAND

ABSTRACT

The transmission of white light (400–700 nm) through a snow-cover has been reported in the literature as both increasing and decreasing as snow density becomes greater, in some cases following, while in others not following the Bouguer-Lambert Law, and in either case described by extinction coefficients which vary by a factor of five. Such confusion is partly the result of differences in experimental procedure, but much can be reconciled by accounting for differences in snowpack conditions over the range of densities for which light measurements have been reported.

Studies conducted here show that for constant depth, light penetration initially decreases as snowpack density increases. However, as snow is compacted beyond a density of about 0.5 g cm^{-3} light transmission begins to increase again, possibly the result of increased growth of ice crystals and decreased light refraction in a snowpack of lower total porosity. Thus, a curve of light transmission versus snow density approximates a parabolic function rather than the simple exponential function often observed for narrow ranges of snow density. These results are discussed in relation to the subnivean light regime as a snowpack undergoes natural changes during the winter.

INTRODUCTION

The disposition of solar radiation over snow-covered ground has been measured, modeled, and described in great detail by hydrologists, glaciologists and engineers. It could be said, without overstating the case, that the influence of snow on surface energy exchange is reasonably well understood. Yet the light regime of the subnivean environment, in which many plants and animals of northern and mountainous regions function for as much as half the year, remains poorly understood. This would seem a major void in our understanding of subnivean ecology, since even very low light levels may influence life processes in significant ways at this time of year. There is some evidence, for example, that plants may be able to utilize light energy penetrating deep snowcover in the synthesis of compounds, possibly through non-photosynthetic biochemical pathways (Richardson and Salisbury, 1977). It has also been suggested that the light environment under snow may influence sexual maturation and reproductive behavior of small mammals in the subnivean environment (Evernden and Fuller, 1972). Thus, a clearer understanding of how light extinction is influenced by changing snow conditions may substantially improve our understanding of the ecology of subnivean mammals.

Snow has been treated by some investigators as a homogeneously diffusing medium attenuating light according to the Bouguer-Lambert Law (for example, Thomas, 1963), realistic perhaps for a new snowfall (even then only for narrow wavebands and not for the visible spectrum as a whole), but unlikely the case for an older snowcover. Alternatively, Curl et al. (1972) proposed a logarithmic equation of the general form $y = ax^n$ to describe light extinction, n, at depth x, where y is percentage of light transmitted and a is the amount of light penetrating the surface after reflection. By either treatment, light is reported to behave in essentially the same way, with absorption decreasing as snow density increases. However, extinction coefficients given in the literature by different authors (summarized by Mantis, 1951, and Thomas, 1963) vary by a factor of five and may change with snow depth (Giddings and LaChapelle, 1961).

The conclusion that light extinction decreases with increasing snow density is itself problematical if we choose not to abandon human experience altogether. A snowball in the hand or a slab of wind-packed snow certainly appears to transmit less (is more opaque) than a handful of loose snow held to the light. Yet only one investigator known to this author (Mellor, 1964) has provided experimental evidence that such is the case.

How can these discrepancies in the literature be reconciled? It may be that light extinction in natural snowcovers follows not an exponential function as previously described for narrow ranges of snow density, but rather approximates a parabolic function in which light extinction first increases, then decreases as snow density ranges from initially low values to values approaching the density of ice. This hypothesis is tested here, with the implications of such changing light extinction explored as they may relate to subnivean mammal ecology.

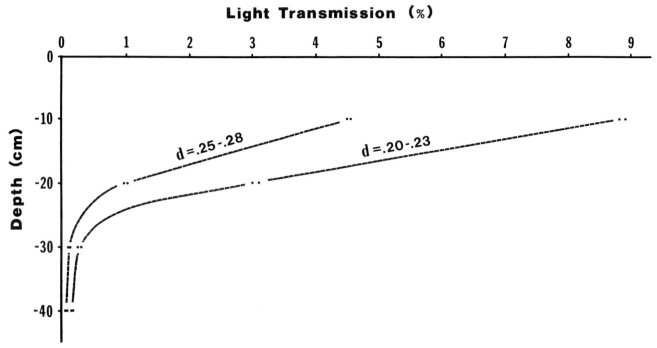

Fig. 1.—Light transmission (400–700 nm) as a function of depth in snow of different densities.

METHODS

Light penetration through snow was measured with a LICOR Model LI-190S quantum sensor. This instrument utilizes a silicon photodiode with a relatively flat spectral response in the region 400 to 700 nm (photosynthetically active radiation). The sensor is cosine-corrected to reduce errors in measuring flux densities through a plane surface under diffuse radiation conditions.

For measurements of light extinction as a function of depth, snow was collected, undisturbed, in corrugated cardboard boxes set out during a snowfall. The boxes were initially 40 cm in height, fitted with a hole in the bottom center to accept the quantum sensor. Light measurements were made under diffuse skylight, first measuring incident radiation, then reflected light by inverting the sensor over the snow surface, and finally transmitted light by inserting the sensor into the hole. Light transmission was calculated as a percentage of light penetrating the surface of the snow (net shortwave radiation) rather than as a percentage of

incident radiation in order to accomodate changes in albedo from one measurement series to another.

Following each series of light measurements the box with snow was weighed to obtain density (g cm^{-3}) and was then reduced in height by cutting with a razor and removing snow down to the new level. Light measurements were then repeated as described above. In this manner the depth of snow was systematically reduced while snow density remained constant.

The effects of changing density on light extinction were studied by packing snow to different densities in a reinforced plywood box, 15 cm in height. In order to attain high densities, natural sintering of ice (growth of bonds between particles) was enhanced by warming the packed snow to near-freezing temperatures for 2–3 hr. The highest snow density thus attained was 0.7 g cm^{-3}. Light measurements were repeated as previously described.

RESULTS

The extinction of photosynthetically active radiation (PAR) with depth in snow of different densities is shown in Fig. 1. At an average density of 0.215 g cm^{-3}, 3% of net PAR (approximately 1% of incident) was transmitted to a depth of 20 cm. Less than 0.3% of net PAR reached a depth of 30 cm and only 0.15% penetrated to a depth of 40 cm. An increase in average density to 0.265 g cm^{-3} resulted in a marked change in light extinction with light

levels reduced by 50% or more at depths between 10 and 40 cm. An increase in density of only 0.05 g cm^{-3} in the surface 10 cm would, in this case, have approximately the same effect as the addition of 8 to 10 cm of new snow (Fig. 1).

The relationship between light extinction and density over the range 0.08 to 0.70 g cm^{-3} was not a simple exponential one (Fig. 2). For a constant depth (15 cm) and initially low snow density light

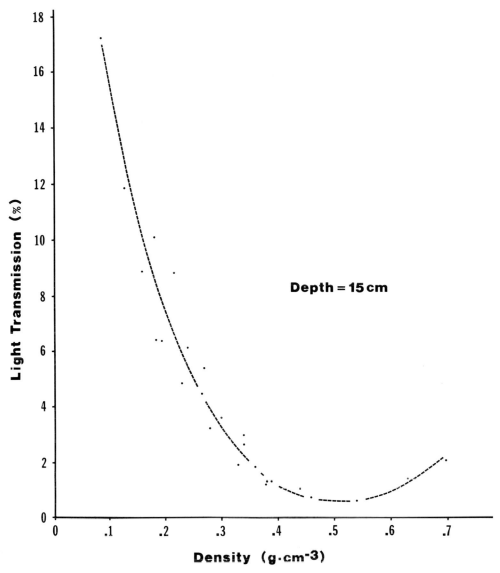

Fig. 2.—Light transmission (400–700 nm) as a function of snow density.

penetration decreased sharply as the snowpack was compacted and density increased. This decrease in transmission, per unit change in density, was greatest for densities between 0.1 and 0.3 g cm^{-3}. At densities greater than 0.3 g cm^{-3} the slope of the light transmission curve decreased rapidly, and maximum extinction was reached at a snow density of about 0.5 g cm^{-3}. Compaction and sintering of snow to still higher densities resulted in an increase in light transmission, possibly the result of increased growth of individual ice grains and decreased refraction in snow of lower total porosity.

DISCUSSION

Anderson (1976) has argued, on theoretical grounds, that in order for extinction coefficients to decrease as snow density increases, grain size (specifically the square root of grain diameter) must increase faster than the density. In essence, it is the relationship between grain diameter and density that determines the behavior of light penetrating the snowpack. An increase in snow density resulting from closer packing of small grains (maximum density 0.5 g cm^3) could be expected to reduce light

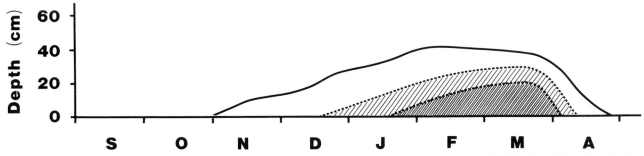

Fig. 3.—Changing light penetration with seasonal changes in snow depth, density, and solar radiation (44° N latitude) for a hypothetical snowpack. Light shading represents <1.0% of incident radiation and dark shading <0.1%. Average snowpack densities of 0.21, 0.26, and 0.30 g cm^{-3} for January, February, and March, respectively, were used for calculations and a constant albedo of 0.75 was assumed.

transmission, whereas an increase in snow density resulting from metamorphosis and growth of individual grains would increase light transmission. According to the theoretical arguments of Berger (1979), the absorption and scattering of infrared radiation is also linearly related to snowpack density and inversely related to grain size for snowpack densities up to 0.4 g cm^3.

This relationship between grain size and density could explain the change in light transmission reported here, as it may have been influenced by experimental procedures. Light measurements at low densities were taken immediately after hand packing the snow, allowing little opportunity for metamorphosis of snow grains to occur. Thus, density was increased through a reduction in pore space, as would be the case with wind packing of new snow, and a reduction in light transmission resulted. In contrast, the only way that snow densities above 0.5 g cm^{-3} could be attained experimentally was to promote increased bonding by warming the snow and allowing sufficient time for sintering to take place. An increase in effective grain diameter rather than a decrease in pore size could then explain the increase in light transmission at densities above 0.5 g cm^{-3}. It is interesting to note here that Mellor (1964) also found the density of maximum extinction to be about 0.5 g cm^{-3}.

How then is the light regime of a subnivean environment altered as the snowpack undergoes seasonal changes in depth and density? New fallen snow undergoes a vary rapid change (destructive or equitemperature metamorphism) as thermodynamically unstable crystals are transformed into aggregates of rounded ice grains. Within a few hours snow density may increase from less than 0.1 g cm^{-3} to nearly 0.2 g cm^{-3}. This is followed by settling, often aided by wind, which then reduces the size of snowpack interstices and further increases snow density. Light transmission would thus decrease sharply with time after a new snowfall.

As soon as a temperature gradient is established in the snowpack, upward moving water vapor causes a dissipation of ice crystals in the bottom layers and a growth of grains in the upper, colder layers of the snowpack (constructive metamorphism). This would favor a decrease in slope of the light extinction curve, but seasonal snow accumulation and packing result in a continued increase in density through destructive metamorphism to values approaching 0.4 g cm^{-3}. A plot of light penetration with time would then be asymmetrical with respect to seasonal changes in both snow cover and total incoming solar radiation (Fig. 3), because the increase in incident radiation through the winter is offset by steadily increasing snow density. Less than 0.1% of incident light may reach the ground from mid January to early April where snow depth is 40 cm or more and average density exceeds 0.25 g cm^{-3}.

At mid-latitudes the lowest absolute subnivean light levels would likely occur from late-February to mid-March (Fig. 3) except in extreme deep snow areas. During this period less than 0.3 J cm$^{-2d^{-1}}$ may reach the ground surface under 40 to 50 cm of snow. This may be sufficient light energy to support plant activity (Curl et al., 1972; Salisbury et al., 1973; Richardson and Salisbury, 1977; Priddle, 1980), but the response of animals to such low light levels is unknown and raises several interesting questions: To what extent might social interactions of subnivean mammals be influenced by changing light levels under snow? Does light quality play a role in sexual maturation and reproductive activity in the subnivean environment, as previously suggested, or do mammals perceive only total darkness under 30 or 40 cm of snow? Do circadian rhythms drift out of phase with "external" time setters through the winter or are the low light levels under snow sufficient to regulate diel activity patterns in subnivean mammals?

Questions of subnivean animal behavior become a little better focused with a clearer understanding of the light environment under snow, and it now appears possible to reconcile apparent discrepancies in the literature on the basis of differences in snowpack characteristics and the degree to which metamorphism has proceeded. Investigators working with highly metamorphosed, permanent snowcover in arctic and alpine snowfields may indeed expect de-creasing light extinction with increasing density. However, researchers working with seasonal snow-cover in northern forest regions can expect to see an increase in light extinction with time as both depth and density increase through the winter. Only with the approach of the melt season, when constructive metamorphism has proceeded to the extreme, will snow density, now a function of increasing grain size and lower total porosity, increase to the point where light extinction begins to decrease. Changing snowpack albedo as well as decreasing depth will also contribute to increasing subnivean light levels at this time.

ACKNOWLEDGMENTS

I am indebted to Ann Luginbuhl and Sheryl Crockett for their efforts in collecting the light transmission data reported here. The study was conducted during the 1981 Winter Ecology Program at The Center for Northern Studies, Wolcott, Vermont.

LITERATURE CITED

ANDERSON, E. A. 1976. A point energy and mass balance model of a snow cover. NOAA Tech. Rep. NWS 19.

BERGER, R. H. 1979. Snowpack optical properties in the infrared. USA CRREL Res. Rept. 79 11. Hanover, New Hampshire, 9 pp.

CURL, H., J. T. HARDY, and R. ELLERMEIER. 1972. Spectral absorption of solar radiation in alpine snowfields. Ecology, 53:1189–1194.

EVERNDEN, L. N., and W. A. FULLER. 1972. Light alteration caused by snow and its importance to subnivean rodents. Canadian J. Zool., 50:1023–1032.

GIDDINGS, J. C., and E. LaCHAPELLE. 1961. Diffusion theory applied to radiant energy distribution and albedo of snow. J. Geophys. Res., 66:181–189.

MANTIS, H. T. (ed.). 1951. Review of the properties of snow and ice. Eng. Exp. Sta. Inst. Techn., Univ. Minnesota, Minneapolis, 156 pp.

MELLOR, M. 1964. Properties of snow. USA CRREL Monogr. III-A1, Hanover, New Hampshire, 105 pp.

PRIDDLE, J. 1980. The production ecology of benthic plants in some Antarctic lakes I. *In Situ* production studies. Ecology, 68:141–53.

RICHARDSON, S. G., and F. B. SALISBURY. 1977. Plant responses to the light penetrating snow. Ecology, 58:1152–1158.

SALISBURY, F. B., S. L. KIMBALL, B. BENNETT, P. ROSEN, and M. WEIDNER. 1973. Active plant growth at freezing temperatures. Space Life Sci., 4:124–138.

THOMAS, C. W. 1963. On the transfer of visible radiation through sea ice and snow. J. Glaciology, 4:481–484.

Address: Department of Environmental Studies, Johnson State College, Johnson, Vermont 05656.

LIGHT CONDITIONS AND PLANT GROWTH UNDER SNOW

FRANK B. SALISBURY

ABSTRACT

Soil temperatures often drop well below freezing in autumn, but once snow is 20 to 50 cm deep, heat is conducted upward, thawing the soil and raising the temperaure at the soil/snow interface to 0°C. After this, soil is saturated with melt water and slightly above freezing. Penetration of light depends on snow quality—blue wavelengths predominate when snow is fresh and when the sun is low in a clear blue sky, but reflected light from fresh snow accounts for less penetration to a given depth. Penetration is highest through old snow, but accumulated dust reduces penetration, especially blue wavelengths.

Spring ephemerals grow and flower immediately after snow melt and become dormant when other plants begin to overgrow them. They include several wild plants (such as *Erythronium* spp., *Claytonia* spp.), ornamentals (such as *Croccus* spp., *Galanthus* spp.), winter cereals (including spring wheat or rye), and their associated weeds (such as *Veronica campylopoda, Ranunculus testiculatus*). Species have been observed to stay green all winter without growing, grow to the soil/snow interface and then stop growing, penetrate the snow pack, or even flower within the snow pack (some actually opening flowers in snow). Most are perennials, a few are annuals (such as the winter cereals), and some are not ephemerals but continue growth all summer (such as *Brodiaea douglassii*). Seeds have germinated under 2 m of snow, and a few apparently required the light penetrating snow. In one trial, seeds placed under deep snow were eaten by animals. Chlorophyll production is promoted by light penetrating snow, but one species did not respond phototropically; that is, did not bend toward light coming from one side. A metal "snow tunnel" with its top at the soil surface was used in these observations.

INTRODUCTION

There are various questions that might motivate a study of plant growth under snow. One of these is simple curiosity about a part of the natural environment that is relatively unknown because it is difficult to study. Another might be a desire to understand a natural (or agricultural) ecosystem during *all* its seasons. Are there seasonal niches occupied by plants?

The official basis for our studies of plant life under snow was an interest of the National Aeronautics and Space Administration in life under extreme conditions. What are the environmental limits to physiological function? Specifically, in our case, how cold can it be before the most cold-hardy plants can no longer grow? Our work was initiated in the mid 1960's and continued into the early 1970's. At that time, it still seemed likely that life might exist on Mars, although Mars was known to be an extremely cold planet by Earth's standards (ground surface exceeds the freezing point of water only for brief intervals in the afternoon during summer, close to the Martian equator).

Each of us who took part in the study (see authors' names given below) was also motivated by the ecological questions as well as questions of interest to NASA. As it turned out, our contributions were mostly in the ecological field, since situations are known in which organisms grow and develop at temperatures much colder than the soil-snow interface, which seldom drops below 0°C. Rarely fungi are found growing in deep-freeze food lockers, for example, and certain antarctic lichens have temperature optima for photosynthesis close to 0°C and can photosynthesize at −18°C (Lange and Kappen, 1972).

There were several aspects to our study (Salisbury et al., 1973; Salisbury, 1976), including the following:

1. Observations of plants in place (Kimball et al., 1973; Richardson and Salisbury, 1977; and Salisbury, 1976). In our preliminary studies, we simply dug down through the snow, sometimes two meters or more in depth. After the soil was reached, we dug into it several centimeters, looking for underground, actively growing roots or shoots. We continued this approach throughout the years of our study, examining conditions under snow at various locations in the Wasatch Mountains of Utah. In the early years we also examined winter wheat growing under snow on Utah State University's experimental farms (Kimball and Salisbury, 1971).

2. Measurement of the environment. This was also continued throughout the years of our study. We first buried thermocouples to different depths in the soil and located them on a pole at different heights above the soil surface, allowing us to measure temperature gradients from the air through the snow pack and down into the soil (see Fig. 2-B, discussed below). It became evident, especially in relation to our attempts to measure light under snow,

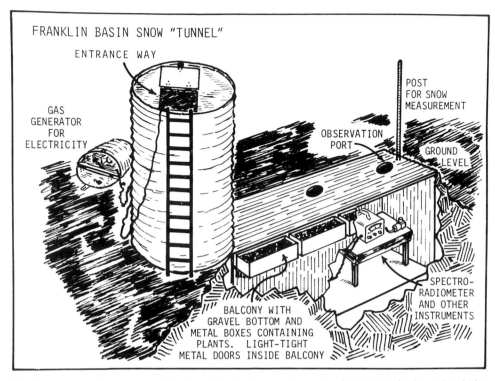

Fig. 1.—A schematic drawing of the Franklin Basin Snow Tunnel as it appeared in the early 1970's.

that digging the snow to place our instruments would never give us reliable data. Hence, we built two snow tunnels, as we called them. The first year, we supported the sides of our tunnel with plywood and wood cross-braces. It collapsed in the spring when the surrounding soil was saturated with water. We then constructed a tunnel (Fig. 1) of corrugated metal and with angle-iron cross braces (Salisbury et al., 1973; Richardson and Salisbury, 1977). The top of the tunnel at the level of the soil surface had observation ports so that, by removing a plexiglas shield, we could measure light penetration of the snow without disturbing the pack in any way. We were especially interested in the spectral distribution of light penetrating snow as well as the depth to which it might penetrate. Getting into the tunnel was somewhat of a problem the first year, so we added a tall entranceway. Along the side of the tunnel was a balcony where large metal containers of soil could be placed so that we could easily observe plant growth in them under snow in winter. The balcony was covered inside the tunnel with a light-tight cover.

3. Physiological and biochemical studies of frost hardening and of active growth at low temperatures. Some studies were done with native species, but most of our work compared two closely related cultivars of wheat (one more frost hardy then the other), or compared two sets of rye plants, one of which had been hardened at temperatures close to freezing for one to five days with another set that had not been hardened (Kimball and Salisbury, 1971; Salisbury et al., 1973; Weidner and Salisbury, 1974).

4. Light and electron microscope studies. In some observations, plants were collected at intervals from under the snow and examined microscopically (Kimball and Salisbury, 1974). In another series, ultrastructures of frost-sensitive plants were compared with ultrastructures of plants of intermediate and extreme hardiness (Kimball and Salisbury, 1973). Still other observations were made of hardened and non-hardened winter rye plants (Kimball, Bennett, and Salisbury, unpublished data).

5. Experimental studies of plants placed on the balconies of the snow tunnel. In one study (Rosen, 1972), mineral balances were examined, and experimental plants were treated with various concentrations of certain mineral ions with and without a cytokinin to see if growth and hardiness were influenced. In another study (Salvesen, 1977) plants were treated with various plant-growth regulators, and their growth under snow was studied. In a third

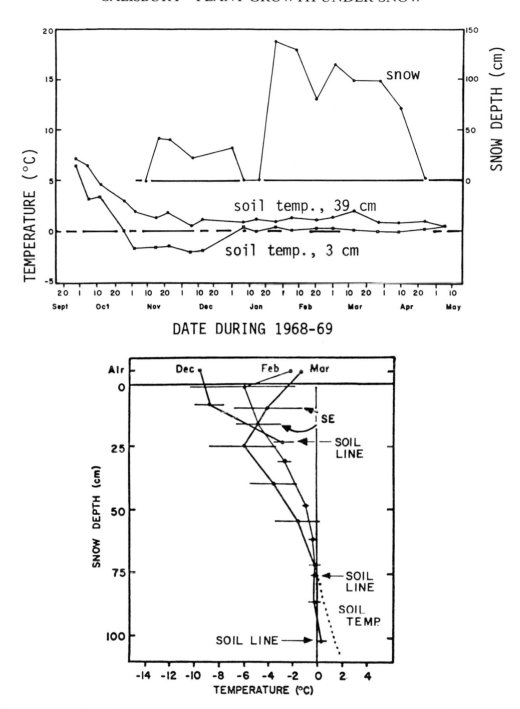

Fig. 2.—(A) Some representative soil temperatures along with snow depths at the Beaver Mountain study area during the winter of 1968–1969. (Measurements of S. L. Kimball; modified from Salisbury et al., 1973.) (B) Temperatures as a function of snow depths at Beaver Mountain (1968–69). The data are plotted with temperatures (abscissa) as a function of distance below the snow surface. Because snow depths were different, soil line occurs at different places for the three sets of measurements. The points represent averages for the months as shown, so soil line is also somewhat of an average value. Below about 50 to 75 cm, temperatures in either snow or soil were close to 0°C; the dotted line suggests that temperatures below the soil surface were slightly above freezing (no measurements actually made in this case). Horizontal lines indicate standard errors of the mean of the several measurements made during the month. (Measurements of S. L. Kimball; from Kimball et al., 1973.)

study (Kimball et al., 1973; Richardson and Salisbury, 1977; Salisbury, 1976), germination and other plant responses under snow were studied as they might be influenced by the light penetrating snow.

In this paper, I shall review only the measurements of the environment, some general observations of plants, and some plant responses to the light penetrating snow.

THE ENVIRONMENT UNDER SNOW

TEMPERATURE

Because average air temperatures in the mountains may be tens of degrees below freezing for days to weeks, we had thought that plant growth beneath the snow (which we had already observed) must go on at temperatures well below freezing. Thus, it came as a surprise to us (and to other laymen—but certainly not to anyone who has considered conditions under the snow—for example, Merritt and Merritt, 1980) that temperatures at the soil-snow interface are close to 0°C for most of the winter. The ground freezes in the autumn before snowfall and may stay

frozen for a few weeks while the layer of snow is rather shallow. When snow depth reaches 20 to 75 cm, however, the ground is insulated from all but extremely low air temperatures (Fig. 2). Once the insulating snow layer has formed, heat from deep in the soil is gradually conducted upward toward the surface, thawing the soil. Since the temperature of snow cannot exceed 0°C without the snow melting, the soil-snow interface then stays close to 0°C until the snow melts in spring. Soil temperatures warm above 0°C once the snow is gone, reaching warm levels in summer and then cooling with the coming of autumn until soils are frozen to a depth of a few centimeters before they thaw and warm

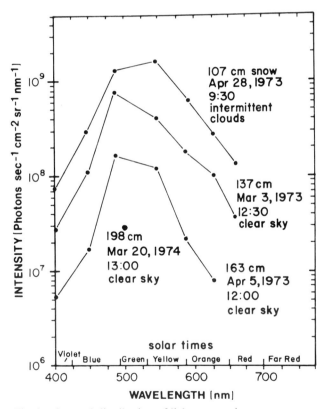

Fig. 3.—Spectral distribution of light penetrating snow as measured on three occasions in 1973 and one in 1974. Although measurements in 1974 were taken at 450, 500, and 550 nm, light levels through almost 2 m of snow could be successfully measured only at 500 nm. (Measurements of S. G. Richardson; from Salisbury, 1976.)

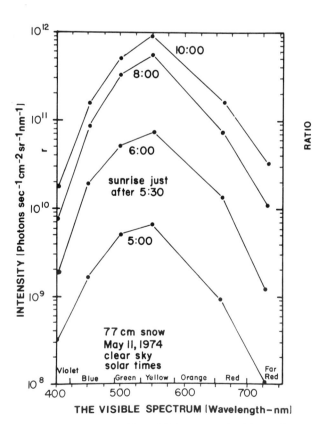

Fig. 4.—Spectral composition of light penetrating 77 cm of snow at various times during and after sunrise. (Measurements of S. G. Richardson; from Richardson and Salisbury, 1977.)

Fig. 5.—Left: *Ranunculus adoneus* (snow buttercups) growing in a snowbank at about 2,600 m elevation southwest of the cable-car unloading station at Snowbird, Utah, on 19 September, 1975. Note that flowers are opening. Right: Snow buttercups growing at the same location and on the same date but several days after being released from a snowbank.

under the insulating blanket of snow. Once the soil has thawed and snow begins to melt at the soil-snow interface, the soil becomes saturated with water and stays at the field capacity until days to weeks after the snow has melted. In western mountains in the United States, soils may become quite dry during mid to late summer and fall.

Although our studies were limited to northern Utah, the conditions just described must prevail in most mountain environments where a blanket of snow covers the ground during most of the winter. Alpine tundras are different. Snow blows off of much of the tundra during much of the winter, so only a few centimeters of snow are held around the dead and frozen vegetation. There are deep snowbanks where the blown snow collects and where conditions are much as in the montane situations, however, except that there is usually a much shorter warm

season in summer, meaning less stored heat in the soil and perhaps a longer time required for thawing to occur after the snowbank forms in autumn. Snow-free areas in the tundra have solidly frozen soil during most of the winter, although an insulating layer of usually wet snow may accumulate during spring. There is no active growth of plants on exposed and frozen tundra during the winter, but even the frozen plants must have some metabolic activity, since dormancy is increasingly broken in mountain avens (*Geum turbinatum* or *G. rossii*) with longer exposures to −10°C in a deepfreeze (Spomer and Salisbury, 1968).

Light

In our studies (summarized in Richardson and Salisbury, 1977), we were concerned with depths to which light might penetrate through snow of differ-

Table 1.—*Plants observed under snow.**

Species	Common name	Perennials, grow from storage organs	Annuals, germinate each year under snow	Turn green under snow	Grow all summer (not ephemeral)
A. Stay green under snow but do not grow					
Achillea millefolium	Yarrow	+			+
Fragaria vesca	Strawberry	+			+
Phacelia heterophylla	Virgate phacelia	+			+
Rudbeckia occidentalis	Western coneflower	+			+
Sedum debile	Stonecrop	+			+
B. Grow only to soil-snow interface; usually form flowers					
Broadiaea douglassii	Wild hyacinth	+			+
Camassia quamash	Camas	+			
Erythronium grandiflorum	Dogtooth violet	+			
Lithophragma glabra	Woodland star	+			
Orogenia linearifolia	Indian potato	+			
C. Grow into snow; usually form flowers					
Achillea millefolium	Yarrow	+			+
Allium acuminatum	Wild onion	+			
Claytonia lanceolata	Spring beauty	+		+	
Descurainia richardsonii	Tansy mustard		+		
Erythronium grandiflorum	Dogtooth violet	+		+	
Geum rossii	Alpine avens	+		+	+
Nemophila breviflora	Great Basin nemophila		+	+	
Ranunculus testiculatus	Bur buttercup		+	+	
Rudbeckia occidentalis	Western coneflower	+		+	+
Secale cereale	Winter rye		+	+	
Triticum sativa	Winter wheat		+	+	
Veronica campylopoda	Snow Speedwell		+	+	
D. Opens flowers in snow					
Erythronium grandiflorum	Dogtooth violet	+		+	
Ranunculus adoneus	Snow buttercup	+		+	
Ranunculus jovis	Sagebrush buttercup	+		+	

* Duplications imply observations on separate occasions.

ent physical conditions, although we did not measure snow densities (see Marchand, 1984). We were also concerned with the quality or spectra of light that had penetrated various depths of snow. Our early work (Kimball et al., 1973) used a spectroradiometer based on a wedge interference filter. In general, we confirmed reports of others (Curl et al., 1972; Gerdel, 1948). We found less penetration through fresh snow with its extremely high albedo and more penetration through older snow that had compacted and partially melted. We observed less penetration through dirty snow (blown dust late in spring), as might be expected.

In our snow tunnel we could see light penetration through two meters of snow, but at that time our spectroradiometer was not sensitive enough to measure it (Kimball et al., 1973). Measurements by Curl et al. (1972) indicated that most radiation penetrating snow was between 450 and 600 nm with peak transmission at about 475 nm. They reported their data on an energy basis (watts) instead of numbers of photons. Plotting on the basis of photons nm^{-1} instead of watts nm^{-1} shifts the peak toward longer wavelengths.

Richardson and Salisbury (1977) found peaks of maximum penetration (quantum flux density) around 550 nm when snow was rather shallow, with shifts toward the blue (500 nm) through deeper snow (Fig. 3). These spectral measurements used a sensitive photomultiplier tube with interference filters that transmitted at several wavelengths, as indicated in the figure.

Table 2.—*Some species that germinated under snow on the balcony of the Franklin Basin snow tunnel.*

		First season (light)	Second season (dark)
1.	*Agropyron cristatum*	Wheatgrass	+	
2.	*Avena fatua*	Oats		+
3.	*Balsamorhiza macrophylla*	Balsam root	+	−
4.	*Brassica caber*	Mustard		+
5.	*Brassica nigra*	Mustard	+	+
6.	*Camelina microcarpa*	False flax		+
7.	*Chrysothamnus nauseosus*	Rabbit brush	+	
8.	*Chrysothamnus viscidiflorus*	Rabbit brush	+	
9.	*Cynoglossum officinale*	Houndstongue	+	
10.	*Galium aparine*	Bedstraw	+	
11.	*Isatis tintoria*	Woad		+
12.	*Lepidium perforliatum*	Peppergrass		+
13.	*Ranunculatus testiculatus*	Bur buttercut		+
14.	*Roemeria refracta*	Field poppy		+
15.	*Secale cereale*	Rye		+
16.	*Sysymbrium altissimum*	Tumble mustard		+
17.	*Veronica campylopoda*	Speedwell		+
18.	*Wyethia amplexicaulis*	Mule ear	+	−

* Twenty-five species that did not germinate either year are not shown. Unfortunately, not all species planted the first year were available to plant the second year.

On a few occasions, spectral measurements were made at different times during sunrise (Fig. 4). Snow depths were intermediate. Early during dawn twilight on a clear day, blue wavelengths predominate, but as the sun rises, light penetrating the snow becomes enriched in the red part of the spectrum. This is because early light before sunrise is the light from a blue sky. After sunrise, direct rays of the sun increase the proportion of red wavelengths.

CARBON DIOXIDE

Kelley et al. (1968) measured increased levels of CO_2 under arctic snow. See also several reports given at this colloquium, including those of Schmid (1984) and Penny and Pruitt (1984). This implied

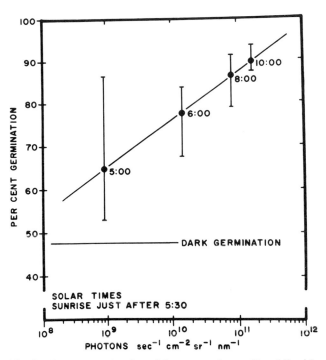

Fig. 6.—Percent germination of *Lactuca sativa* cv. Grand Rapids lettuce seeds as a function of light levels at 660 nm (means of four replications of 200 seeds, vertical lines show range). Seeds were exposed for 10 min beneath 77 cm of snow at various times as shown during the morning of 11 May 1974. (From Richardson and Salisbury, 1977.)

that decay processes are continuing under the snow, or perhaps that plants are respiring. Such measurements are highly significant for an understanding of plant and animal activities under the snow.

Oxygen levels under the snow should be measured. Other environmental factors that need to be investigated include pH and mineral ion concentrations in the soil solution. These factors would not be difficult to measure, although I am unaware of attempts to measure them.

PLANT GROWTH UNDER SNOW

Several authors have observed plants emerging from underground storage organs while the ground is still covered with snow. During mid-summer of 1938, Russell and Wellington (1940) studied the vegetation of Jan Mayen Island, situated at about 70° north latitude in the Greenland Sea north of Iceland. They found several angiosperm species with normal green leaves, apparently growing under about 50 cm of snow, the lower 5 to 10 cm being hard ice.

Light levels under the snow pack were reduced to 2% of daylight. Billings and Bliss (1959) also noted growing plants under 50 cm of snow in June and July in the Snowy Range of the Medicine Bow Mountains of Wyoming. Furthermore, "*Geum turbinatum, Carex elynoides,* and *Deschampsia caespitosa* frequently grow under 1 to 5 cm of snow near the edge of the melting bank."

Mooney and Billings (1960) mentioned that cer-

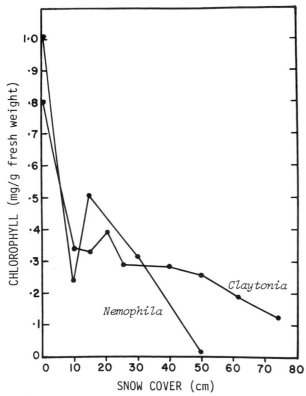

Fig. 7.—Chlorophyll contents of two plant species as a function of snow depth during May 1969. The "out-of-line" point for *Nemophila* (10 cm) occurred in an area of relatively heavy shade from nearby trees. (Measurements of S. L. Kimball; from Kimball et al., 1973.)

tain alpine species grow actively under snow. They measured sharp drops in carbohydrates during the period of growth under snow, although carbohydrates never became seriously depleted in these plants. Vezina and Grandtner (1965) studied spring beauties (*Claytonia caroliniana*) in Quebec, but found that they appeared only on bare ground or on ground nearly free of snow, whereas avalanche lilies (*Erythronium americanum*) penetrated the snow. Both of these species had reached their peak of leafing and flowering before leaves began to expand on near-by sugar maple (*Acer saccharum*) seedlings; *C. caroliniana* and *E. americanum* had almost disappeared by the time the maple leaves had fully expanded. Caldwell (1969) observed etiolated avalanche lilies (sometimes called dogtooth violets: *Erythronium grandiflorum*) under 1.5 m of snow at two sites in Colorado and Wyoming (elevations of about 3,350 and 3,390 m). Some plants were green and as tall as 10 cm at release from the snow.

Risser and Cottam (1967) studied effects of temperature on breaking of dormancy in five species

that grow in early spring in Wisconsin. They noted that some plants initiate growth before snow melt. Four of the species (*Erythronium albidum, E. americanum, Dicentra cucullaria,* and *D. canadensis*) required extended periods of cold to break their winter dormancy and begin active growth; minimum times required depended on species. *Claytonia virginica* grew in the greenhouse without a cold treatment. The longer the cold treatment for all species, the more rapid was spring growth once it was able to begin. Risser and Cottam (1968) also studied carbohydrate cycles in bulbs of the four species that required cold treatment to break dormancy. The primary storage material was found to be starch, which accumulated rapidly and to high levels by early June. The bulbs underwent some development in autumn, at which time starch contents dropped and soluble sugar increased. The high level of sugar was maintained throughout the winter and subsequent growing season, although molecular species of the sugars changed somewhat throughout the year. In any case, these bulbs should provide nutrient food sources for animals that consumed them.

The plants that are most likely to grow actively under a snowpack are the *spring ephemerals* (sometimes called geophytes), of which the *Claytonia* and *Erythronium* species are good examples. These are usually small plants that do most of their rapid growing, often flowering and setting seed, in the days to weeks immediately after the snow melts and before other plants form a dense canopy over the soil. Growth under the snow should give the spring ephemerals a head start. Such plants occupy a special niche in seasonal time, drawing on resources of light, water, and nutrients that will be used later by taller, larger, slower growing species. Usually by the time the larger species have grown, the spring ephemerals have dried up, continuing to exist only as underground bulbs, corms, or rhizomes. A few are annuals, existing during most of the summer and fall as seeds.

Some of our most important crop species are spring ephemerals. The winter cereals (wheat, barley, oats, and rye) are winter annuals sown in the fall, germinating either before snowfall or under the snow. The winter cultivars require the low temperatures of winter to promote rapid flowering when spring and summer come. This hastening of flowering is called vernalization. Spring cultivars do not require vernalization and are thus sown in the spring at northern latitudes too cold for the winter cultivars. The winter cereals usually yield more than spring

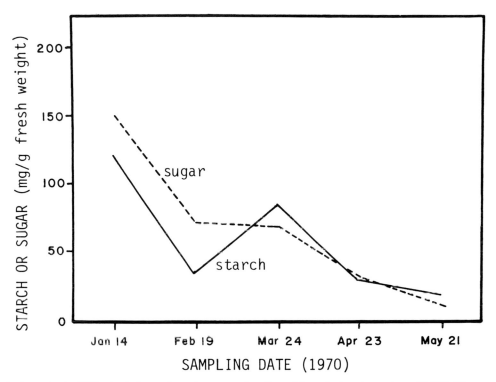

Fig. 8.—Carbohydrate contents of *Claytonia lanceolata* plants as a function of time during the spring of 1980. The surprising increase in starch in March was replicated in other data (Measurements of B. Bennett; from Kimball et al., 1973.)

cultivars. This is because they grow slowly during the entire winter and then grow rapidly immediately after snow melt, maturing a crop within 60 to 70 days. They have a head start on spring cultivars.

The winter cereals are infested by important weeds that are also spring ephemerals. Snow speedwell (*Veronica campylopoda*) introduced from Asia is a problem on many Western dry farms, and buttercup (*Ranunculus testiculatus*) from Europe has invaded ranges and alfalfa fields as well as winter wheat. Its burs damage the hides and tongues of animals as well as automobile tires. June grass (*Bromus tectorum*) is another introduced spring ephemeral that covers much of several western states and poses a serious fire hazard because it dries out so early in the season. Crocuses, snowdrops, and to a lesser extent tulips, hyacinths, and daffodils are perennial spring ephemerals, much like the native *Erythronia* and *Claytonia* species that grow in the forests and on the plains of North America.

Our observations of wild plants are summarized in Kimball et al. (1973), Richardson and Salisbury (1977), and Salisbury (1976). Species that we observed growing under the snow are listed in Table 1. They are separated into the four groups according to their level of activity under the snow. Some species

stay green all winter under the snow, but put on no noticeable growth. None of the species in this group is a true spring ephemeral, since these plants continue to grow actively during the summer months. Nevertheless, they initiate growth almost as early in spring as the true spring ephemerals, and their green conditions during winter might make them suitable foods for animals that are active under the snow.

The second group includes perennial plants that overwinter as a storage organ (bulb, corm, root, rhizome), from which shoots slowly elongate in the soil under the snow. In this group, elongation stops when the soil-snow interface is reached. Studying such plants requires digging in the soil under the snow and finding them almost fortuitously, although the species listed in Table 1 are so common that it was possible to find samples with virtually every digging.

It would be interesting to understand the physiological response that stops shoot growth when the snow is reached. Either these plants some way detect the slightly different texture or temperature of the snow pack compared to soil, or they respond to the weak light penetrating the snow pack. We made some initial attempts to test the latter possibility by covering the soil with black plastic, sometimes with a layer of insulation between the soil and the plastic, but results were inconclusive. The impermeable plastic was too unlike the soil-snow interface.

The third group of plants grows into the snow pack after elongating up through the soil and from a deep storage organ below. Nearly all the plants that we have examined form flowers during the winter months under the snow, and often these flowers are carried into the snow pack (Kimball and Salisbury, 1974).

The fourth group includes three species that have been observed to actually open their flowers in the snow pack so that they appear in full flower when the snow melts (Fig. 5). This occurs uniformly with the two buttercups, but erratically with the dogtooth violet. We usually observed unopened flower buds of the dogtooth violet when we dug carefully into the snow, but occasionally we encountered open flowers instead of buds. Note the late date for the snow buttercups of Fig. 5. These plants must grow rapidly in the few remaining days of autumn before they will again be covered with snow.

As indicated in Table 1, most of the native plants that we have observed growing or at least staying green under the snow are perennials. Two native spring ephemerals are annuals, as are the two cereals (wheat and rye) that we have studied along with their associated weeds. Table 1 also indicates plants that have been observed to turn green under the snow. Most of the ones in the table do, but a few others either were green when the snow fell and remained green and alive all winter, or their new parts were more or less colorless, usually a pale yellow, until snowmelt in the spring.

EXPERIMENTAL STUDIES ON GERMINATION UNDER SNOW

A number of higher plant species have been observed to germinate under snow (Hull, 1960; Williams and Cronin, 1968; Bleak, 1970). With these studies as background, we performed a number of experiments to test germination under snow. This was easily accomplished by placing seeds on the soil surface in the containers set on the balcony of our snow tunnel. Since many seeds are known that require light for germination (general discussion in Salisbury and Ross, 1978, pages 294–297), we were particularly interested in whether the light penetrating snow might lead to germination. Richardson (1975) placed a number of seeds on the balcony of the snow tunnel during the season of 1972–1973. All were exposed to the light penetrating the snow, and several species germinated (Table 2). In the 1973–1974 season, Richardson placed seeds of several species that germinated plus a number of others on the balcony of the snow tunnel, this time covering half the seeds with black plastic and leaving the other half exposed to the light coming through the snow. Unfortunately, all the seeds left exposed to the light disappeared, presumably eaten by rodents. Two species that germinated the first year in the light failed to germinate the second year in the dark: balsam root and mule ear. One species, the mustard *Brassica nigra,* germinated the first season in the light and also the second season in the dark.

Richardson and Salisbury (1977) did a number of experiments with two seed species known to require light for germination: lettuce and mustard. If seeds were properly treated to make them more light sensitive (24 hours imbibition in water at 20 or 30°C) before exposure to the light coming through the snow, clear promotion effects of the light could be observed. Fig. 6 shows results of an experiment in which plants were exposed for 10 min to the light penetrating 77 cm of snow at various times during a morning. When germination is plotted as a function of light level to which seeds were exposed, a promotion with increasing light levels is evident. Light is expressed as levels at 660 nm, the wavelength most effective in activating the phytochrome pigment system, which is highly important in controlling many plant responses to light (Salisbury and Ross, 1978:290–303). In view of these studies, it seems quite likely that the light penetrating snow is effective in promoting germination of some native seeds.

RESPONSES OTHER THAN GERMINATION

Richardson and Salisbury (1977) examined phototropism under snow by covering plants of *Camassia quamash* with cans in which a slit had been cut on one side. There was no indication of stem bending toward the light entering the can through this slit. It is likely that the gravitropic response under those conditions is stronger than the phototropic response, or that lateral light levels were too low.

A number of our studies considered the possibility

that chlorophyll forms in response to light penetrating the snow. In angiosperms (flowering plants) in general, light is required for the conversion of protochlorophyll to chlorophyll (Salisbury and Ross, 1978:127–128). We were able to obtain considerable circumstantial evidence that the light penetrating snow is capable of carrying out this photoconversion, at least in those species that were observed to become green under the snow (Table 1). On several occasions the ground was covered with black plastic before snowfall, and plants growing and germinating beneath this plastic were colorless after snowmelt in the spring. Richardson and Salisbury (1977) covered some camas plants with cans from which either one end had been removed or both ends had been removed and one end covered with clear plastic. Plants in the dark cans elongated (became etiolated) about twice as much as plants in the cans with one end covered with clear plastic. Plants in the dark remained colorless, while plants in the dim light were beginning to turn green, and

their leaves were unrolling when they were examined, still under 40 cm of snow (11 May 1974). Measurements of Kimball et al. (1973) also showed that chlorophyll content of two spring ephemerals, collected by digging away the snow, was indirectly proportional to snow depth and therefore roughly proportional to the light reaching the plants (Fig. 7). It seems highly likely that chlorophyll synthesis under snow is a response to the light penetrating the snow.

Kimball et al. (1973) studied starch and soluble carbohydrate contents of spring beauties under the snow from mid January until snowmelt in late May. As Fig. 8 shows, stored food dropped radically as midwinter progressed into early spring, with an apparent increase in starch from February to March. This could be of significance to this colloquium, since it suggests that plants under snow would be less nutritious for animals that ate them, as winter progressed into spring.

LITERATURE CITED

BILLINGS, W. D., and L. C. BLISS. 1959. An alpine snowbank environment and its effects on vegetation, plant development, and productivity. Ecology, 40:388–397.

BLEAK, A. T. 1970. Seed response under snow on a subalpine range in central Utah. Unpublished M.S. thesis, Utah State Univ., Logan, 49 pp.

CALDWELL, M. L. H. 1969. Erythronium: comparative phenology of alpine and deciduous forest species in relation to environment. Amer. Midland Nat., 82:543–558.

CURL, H., JR., J. T. HARDY, and R. ELLERMEIER. 1972. Spectral absorption of solar radiation in alpine snowfields. Ecology, 53:1189–1194.

GERDEL, R. W. 1948. Penetration of radiation into the snow pack. Trans. Amer. Geophysical Union, 29:366–374.

HULL, H. A., JR. 1960. Winter germination of intermediate wheatgrass on mountain lands. J. Range Mgtm., 13:257–260.

KELLEY, J. J., JR., D. F. WEAVER, and B. P. SMITH. 1968. The variation of carbon dioxide under the snow in the arctic. Ecology, 49:358–361.

KIMBALL, S. L., B. D. BENNETT, and F. B. SALISBURY. 1973. The growth and development of montane species at near-freezing temperatures. Ecology, 54:168–173.

KIMBALL, S. L., and F. B. SALISBURY. 1971. Growth and development of hardy and nonhardy wheat varieties (Triticum aestivum L.) at near-freezing temperatures. Agronomy J., 63:871–874.

———. 1973. Ultrastructural changes of plants exposed to low temperatures. Amer. J. Botany, 60:1028–1033.

———. 1974. Plant development under snow. Botanical Gazette, 135:147–149.

LANGE, O. L., and L. KAPPAN. 1972. Photosynthesis of lichens from Antarctica. Pp. 83–95, in Antarctic terrestrial biology (George A. Llano, ed.), Antarctic Res. Ser., 20.

MARCHAND, P. J. 1984. Light extinction under a changing snowcover. This volume.

MERRITT, J. F., and J. M. MERRITT. 1980. Population ecology of the deer mouse (Peromyscus maniculatus) in the front range of Colorado. Ann. Carnegie Mus., 49:113–130.

MOONEY, H. A., and W. D. BILLINGS. 1960. The annual carbohydrate cycle of alpine plants as related to growth. Amer. J. Botany, 49:594–598.

PENNY, C. E. and W. O. PRUITT, JR. 1984. Subnivean accumulation of CO_2 and its effects on winter distribution of small mammals. This volume.

RICHARDSON, S. G. 1975. Plant light responses under snow. Unpublished M.S. thesis, Utah State Univ., Logan, 60 pp.

RICHARDSON, S. G., and F. B. SALISBURY. 1977. Plant responses to the light penetrating snow. Ecology, 58:1152–1158.

ROSEN, P. 1972. Plant adaptation to cold I. Chlorophyll II. Minerals. Unpublished M.S. thesis, Utah State Univ., Logan, 72 pp.

RISSER, P. G., and G. COTTAM. 1967. Influence of temperature on the dormancy of some spring ephemerals. Ecology, 48:500–503.

———. 1968. Carbohydrate cycles in the bulbs of some spring ephemerals. Bull. Torrey Bot. Club, 95:359–369.

RUSSELL, R. S., and P. S. WELLINGTON. 1940. Physiological and ecological studies on an arctic vegetation. I. The vegetation of Jan Mayen Island. J. Ecology, 28:153–179.

SALISBURY, F. B. 1976. Snow flowers. Utah Sci., 27:35–41.

SALISBURY, F. B., S. L. KIMBALL, B. BENNETT, P. ROSEN, and M. WEIDNER. 1973. Active plant growth at freezing temperatures. Space Life Sci., 4:124–138.

SALISBURY, F. B., and C. W. ROSS. 1978. Plant physiology. Wadsworth Publ. Co., Belmont, California, Second ed., 422 pp.

SALVESEN, M. D. 1977. The influence of certain growth hormones on plant growth at cold temperatures. Unpublished M.S. thesis, Utah State Univ., Logan, 40 pp.

SCHMID, W. D. 1984. Materials and methods of subnivean sampling. This volume.

SPOMER, G. G., and F. B. SALISBURY. 1968. Eco-physiology of *Geum turbinatum* and implications concerning alpine environments. Botanical Gazette, 129:33–49.

VEZINA, P. E., and M. M. GRANDTNER. 1965. Penological observations of spring geophytes in Quebec. Ecology, 46:869–872.

WEIDNER, M., and F. B. SALISBURY. 1974. The temperature characteristics of ribulose-1,5-diphosphate carboxylase, nitrate reductase, and pyruvate kinase from seedlings of two spring wheat varieties. Z. Pflanzenphysiologie, 71:398–412.

WILLIAMS, M. C., and E. H. CRONIN. 1968. Dormancy, longevity, and germination of seed of three larkspurs and western false hellebore. Weed Sci., 16:381–384.

Address: Plant Science Department, UMC 48, Utah State University, Logan, Utah 84322.

DEMOGRAPHY OF A SUBARCTIC POPULATION OF *CLETHRIONOMYS GAPPERI*: CAN WINTER MORTABILITY BE PREDICTED?

WILLIAM A. FULLER

ABSTRACT

A population of *Clethrionomys gapperi* was sampled in nine successive autumns and again in the following May. Regression analysis was used to evaluate fall numbers, mean fall weight, mean fall body length, fall age ratio, and mean minimum October temperature as possible predictors of either May numbers or an estimate of the 28-day survival rate from fall to spring. None of the variables tested yielded statistically significant results when all years were included in the regression. Most failed to predict high survival in the pre-peak winter and often also low survival in the post-peak winter. Those that predicted the peak failed to account for low survival in 1977–1978. Thus, there is still reason to believe that winter ecology exerts a strong influence on the demography of small mammals in regions with long, cold, and snowy winters.

INTRODUCTION

Formozov (1946) categorized northern endotherms according to their ability to cope with snow. At one extreme chionophobes, such as small, ground-feeding birds, must avoid areas with long snowy winters. At the other extreme, chionophiles either have morphological adaptations, such as "snowshoes" or white pelage (plumage) or are dependent on the snow cover for winter survival (small subnivean rodents and shrews). Since virtually all the classically cyclic birds and mammals are chionophiles, and since summer studies had not solved the riddle of multi-annual cycles, it seemed reasonable, in the mid 1960's, to postulate that the solution may lie hidden beneath the winter snow (Fuller, 1967).

In an attempt to test these ideas I established a year-round base of operations just south of the western end of Great Slave Lake, N.W.T., Canada in 1965. From that base my students and I have attempted to study the demography of an assemblage of small mammals exhibiting a variety of techniques for winter survival. Greatest emphasis was placed on the North American red-backed vole (*Clethrionomys gapperi*) and its parapatric Eurasian relative (*C. rutilus*), both of which remain active under the snow. *Peromyscus maniculatus* is sympatric with the voles, but undergoes bouts of torpor during cold weather (Stebbins, 1971). Shrews (primarily *Sorex cinereus*) are active all winter, but are carnivores. Chipmunks (*Eutamias minimus*) are hibernators, and red squirrels (*Tamiasciurus hudsonicus*) remain active above the snow at moderately low temperatures, but retreat to the warmer subnivean environment during periods of deep cold (Pruitt and Lucier, 1958; Zirul and Fuller, 1971). In addition to biological studies, air and subnivean temperatures were recorded at the laboratory, and snow surveys were done twice monthly at the lab and six other stations.

This report deals only with *C. gapperi*. In it I examine a number of features of the September population and temperature in the fall critical period (Pruitt, 1957) that might serve as predictors of spring numbers or overwinter survival rates. If spring numbers are a simple function of some characteristic of the autumn population or environment, there is no role for winter in the causation of multi-annual cycles. If, on the other hand, no predictors can be found, a role for winter is at least strongly implied.

METHODS

Attempts to sample the population at regular intervals during winter proved to be unsuccessful. Midwinter catches were obviously underestimates of population size because they were lower by a factor of 8–10 than catches in April and May (Fuller, unpublished data). I thus had to rely on comparisons of autumn (September, except October in 1971) and spring (May) catches. All calculations are based on mean catch per 100 trap nights ($\bar{C}$/100 TN), but I have shown that such estimates are strongly correlated with estimates based on capture-mark-release techniques (Fuller, 1977). The possible disadvantage of using relative numbers as a census technique is at least partly offset by the wealth of information on age, size and breeding condition obtained from autopsy of carcasses. Trapping and autopsy methods have been described (Fuller 1977).

Four age categories, based on morphology of the second upper molar (M^2), are used in this study. Animals that have overwin-

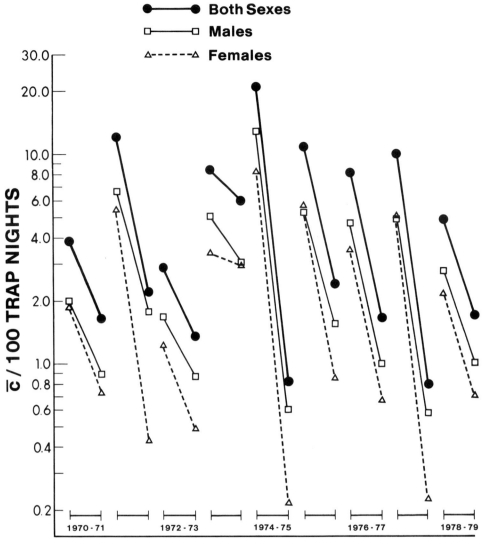

Fig. 1.—Fall and spring population estimates for the study population, September 1970 to May 1979.

tered have well-developed roots on M^2, usually more than 1 mm long in May and more than 1.5 mm long in September. Only a few overwintered animals survive until September, and essentially none live through two winters (three possible cases in 15 winters). Individuals born in early summer have roots on M^2 ranging in length from 0.1 to 0.5 mm (occasionally more) in September. They will be referred to as age class 3 or individuals with rooted molars. Those lacking roots (rootless) in September fall into two categories. In the younger category (age class 1) the anterior labial groove is open for the entire length of the tooth, whereas in age 2 the groove is closed at the alveolar end. Mihok (1980) has shown that the groove closes at about 32 to 35 days of age and that measurable roots are formed at about 62 to 72 days. Animals with roots on M^2 in mid-September were thus born by mid-July.

Estimates of 28-day survival rates (S) are based on the assumption that winter declines are exponential. For each winter, the mean date of capture in both September and May and the

number of days between those dates were determined. Survival was calculated from the equation

$$S = \exp \frac{\ln \dfrac{N_t}{N_0}}{T} \qquad (1)$$

where N_0 is relative number of animals in September (October in 1971), N_t is the relative number in May, and T is the number of 28-day periods between the mean dates of capture in fall and spring.

STUDY AREA AND STUDY POPULATION

The study area lies in the southern part of the Mackenzie District of the Northwest Territories, Canada, near the northern limit of closed-canopy boreal forest. A mosaic of forest types occurs in the region as a result of differences in such factors as parent materials, soils, moisture, and recent fire history. Sampling

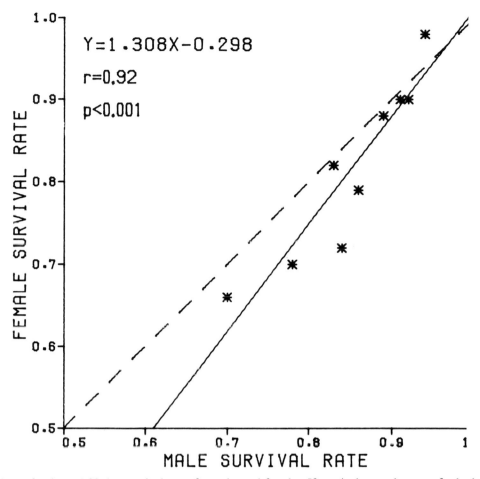

Fig. 2.—Comparison of estimated 28-day survival rates for males and females. If survival were the same for both sexes, the points would fall along the broken line.

was essentially confined to upland sites in mesic or drier habitats, and to forest stands at least several decades old. Trap lines were not permanently marked, but, in general, the same sites were trapped each May. However, sites of September samples varied from year to year.

The study population underwent marked annual fluctuations and reached at least one multi-annual peak (in 1974, Fuller, 1977). Winter mortality varied from year to year (Fig. 1). The smallest decline occurred in the pre-peak winter (1973–1974), and the largest in the post-peak winter (1974–1975). In addition

to the post-peak winter, survival was very poor in 1977–1978. In the analyses that follow, the pre- and post-peak winters and 1977–1978 appear as recurring problems.

In general males survived somewhat better than females. The lone exception occurred in the pre-peak winter of 1973–1974. When the 28-day survival rate was higher than 0.80, there was little difference in survival between the sexes (Fig. 2); when the rate was less than 0.80, the difference between the sexes was accentuated. The greatest disparity was in 1971–1972, and it may merely reflect the small size of the 1972 sample.

RESULTS

Test of Hypotheses

H_1: Spring numbers are a function of fall numbers.—This hypothesis is obviously not true for all nine winters, although it may be true for six of them (Table 1, cases 1 and 2). If the pre- and post-peak winters and the winter of 1977–1978 are removed

from the regression, fall numbers explain 80% of the variance in May numbers (sexes combined) or 76% of the variance for males only. However, for females fall numbers have no value as predictors of spring numbers.

H_2: 28-day survival rates are a function of fall

Table 1.—*Regression statistics for various predictors of spring numbers and winter survival rates of* Clethrionomys gapperi *in northwestern Canada 1970–1971 to 1978–1979.*

Case	Dependent variable	Independent variable	Sex	a[1]	b[2]	r[3]	P[4]	Not in regression[5]
1	Log May no.	Log fall no.	♂♀	−0.003	0.325	0.89	<0.02	1974, 1975, 1978
2			♂	−0.193	0.471	0.87	<0.03	1974, 1975, 1978
3		Mean Sep. weight	♂	−0.868	0.049	0.76	<0.05	1974, 1978
4		Mean Sep. length	♂♀	−4.450	0.053	0.75	<0.04	1978
5			♂	−4.044	0.046	0.76	<0.04	1978
6		Mean min. Oct. temp. (°F)	♂♀	−1.079	0.054	0.75	~0.05	1974, 1978
7			♂	−1.736	0.075	0.73	<0.05	1978
8	28-day survival	Log fall no.	♂♀	1.052	−0.257	−0.92	<0.002	1974
9			♂	0.983	−0.228	−0.90	<0.003	1974
10			♀	0.964	−0.308	−0.93	<0.001	1974
11		Mean sep. weight	♂♀	1.217	−0.019	−0.86	<0.03	1974, 1975, 1978
12			♀	1.455	−0.032	−0.91	<0.02	1974, 1975, 1978
13		Log age ratio[6]	♂♀	1.069	0.327	0.78	<0.03	1972
14		Mean min. Oct. temp. (°F)	♂♀	1.348	−0.020	−0.80	<0.03	1974, 1975

Notes: [1]Intercept. [2]Slope. [3]Pearson correlation coefficient. [4]Probability that slope is equal to zero. [5]Winters identified by year of the spring, that is, 1974 means winter of 1973–74. [6]Ratio of number in age class 3 to number in age classes 1 and 2 combined.

numbers.—At first sight this hypothesis seems to be substantiated for all winters except the pre-peak winter, for both sexes (Table 1, case 8), for males only (case 9), and for females only (case 10). The negative slope suggests a density-dependent effect and appears to be at odds with the positive slope linking September numbers to May numbers (cases 1 and 2). Complete resolution of this paradox will be postponed (see Discussion), but it can be pointed out here that the regressions in Figs. 6–8 are suspect because fall numbers enter into calculation of 28-day survival rates so that the assumption of independence is violated.

H₃: May numbers are a function of the size of individuals in September.—Two measures of size are available—body weight and body length. There is a weak, but statistically insignificant, positive correlation between weight and numbers for the six "normal" years when sexes are combined. This relationship reaches statistical significance for males alone *if* the otherwise intractable winter of 1974–1975 is included in the regression (Table 1, case 3). For females alone there is no correlation.

A somewhat different picture is presented when body length (total length—tail length) is used as the criterion of size. Marginally significant ($P \leq 0.05$) correlations exist for both sexes (Table 1, case 4) and for males only (case 5), but only if *both* the pre- and post-peak winters are included. There is no correlation between fall length of females and number of females in the following May.

H₄: 28-day survival is a function of size of individuals in September.—For sexes combined (Table 1, case 11) and especially for females alone (case 12), September weight accounts for a sizable proportion of variance in survival for the six "normal" years, but fails to predict either sharp increases or sharp declines. Mean weight does not predict survival for males, even for the "normal" years.

Body length, as an index of size, has no value as a predictor of 28-day survival rates.

H₅: Spring numbers are a function of the age of the fall population.—Estimates of absolute age based on tooth morphology are only reliable for age-class 1 and immature members of age-class 2 (Mihok 1980); therefore, I used various estimates of the relative age of each sample. Obviously, the mean age of a sample dominated by age-class 1 must be younger than that of a sample dominated by age-class 2, for example. No measure of relative age of fall samples that I was able to devise was significantly correlated with numbers the following May.

H₆: 28-day survival is a function of age of fall populations.—The log-transformed ratio of the number of individuals with rooted molars to the number with rootless molars, that is, the number in age-class 3 divided by the sum of the numbers in classes 1 and 2, is the only measure of relative age that predicts 28-day survival rates (Table 1, case 13). Since there is no way to "back-date" the October sample of 1971 to make it comparable with September samples of the other years, 1971–1972 cannot be included in the regression, which accounts for about 61% of the variance in the other eight years. Fall age ratio, however, fails to predict survival for either sex considered separately.

H_7: *May numbers are a function of October mean minimum temperature.*—October temperature was used as a crude approach to temperature during the fall critical period (Pruitt, 1957). A more refined definition of the fall critical period, which will undoubtedly vary from year to year, awaits detailed analysis of snow survey data. A marginally significant relationship is evident if 1973–1974 and 1977–1978 are not included in the regression (Table 1, case 6) and if sexes are combined. A significant relationship occurs if males are considered separately (case 7), which appears to account for all winters except 1977–1978.

H_8: *28-day survival is a function of October mean minimum temperatures.*—Once again, by selectively ignoring the pre- and post-peak winters, it is possible to obtain a significant correlation (sexes combined) for the remaining years, including 1977–1978 (Table 1, case 14).

DISCUSSION

In the foregoing analysis I used four features of the autumn population and mean minimum October air temperature as possible predictors of either May numbers or estimated 28-day survival rates. With two exceptions predictive power was no better than chance if all the available data were used. One could suggest at this point that these results are consistent with my hypothesis that a factor (or factors) governing multi-annual cycles operates during winter. However, the analyses show two reasonably consistent patterns. First, six or more of the nine winters rather frequently form a pattern, and one or more of the pre-peak winter, the post-peak winter, or 1977–1978 cannot be accounted for. Second, associations, if any, between the various predictors and May numbers are always positive, whereas associations, if any, between predictors and 28-day survival rates are negative with one exception.

With respect to point one, I suggest that over a range of subnivean conditions winter ecology may play only a small role, so that May numbers may in fact be predictable given sufficient knowledge about the fall population. Since this suggestion does not favor my hypothesis, I am clearly not using *ad hoc* manipulation of the data to support my own position. On the contrary, it is only by taking liberties with statistics that I find any support for alternative hypotheses. Even then, the support is partial, and the alternative hypotheses mainly fail to predict the high survival in the pre-peak winter or the crash decline following the peak. Since a peak and a decline are essential features of a cycle, the alternative hypotheses, even where placed in the most favorable light, are not fully adequate.

The paradox raised by the second point can be resolved in the following way. It is obvious from Fig. 1, and it has been documented statistically (Mihok and Fuller, 1981) that May relative numbers showed few significant differences from year to year. Furthermore, the number of days between September and May sampling periods was nearly constant (except 1971–1972). Substitution of constants for N_t and T in equation (1) immediately reveals that survival is inversely proportional to N_0 (fall numbers). What is biologically significant in this relationship is the observed constancy of May numbers, not the apparent density-dependence of the 28-day survival rate. A nearly constant number of survivors suggests a more or less fixed amount of winter survival habitat, which is consistent with the ideas of Fuller (1967) and Anderson (1970, 1980), and with the observation of West (1977) for *C. rutilus* in Alaska. The problem, then, is to explain occasional departures from the mean spring density. Two obvious possibilities are change in the quality of habitat or change in the quality of the animals. A third possibility, which could only explain higher than normal "survival," is that some individuals captured in May were born in winter and thus inflated the apparent survival rate.

Two lines of evidence suggest that winter breeding occurs sporadically in the study population. The first line is occasional capture of a female in May with placental scars on the uterus, and the second is capture of individuals with roots on M^2 so short that they must have been born no earlier than 1 January. The mean length of roots on M^2 was shorter in 1974 than in any other year because of the presence of a considerable proportion of individuals with very short roots, and the May sample included four females with visible placental scars (Fuller, unpublished data). Thus, the 1974 May numbers and survival rates are both biased upwards. However, I have shown that the proportion of marked animals that survived the winter of 1973–1974 was exceptionally high (Fuller, 1977), so that even removal

Table 2.—*Variation in proportions of individuals in age-classes 1 and 2 in selected fall samples. Years are compared that have similar survival rates and/or similar age ratios.*

Year	S	Age* ratio	No. in Age 1	No. in Age 2	G	P
1970–1971	.91	.17	27	9	19.02	≪0.005
1978–1979	.88	.18	31	63		
1975–1976	.83	.15	36	19	15.54	≪0.005
1976–1977	.83	.27	9	28		
1974–1975	.69	.12	35	60	20.94	≪0.005
1977–1978	.74	.12	10	91		

* No. in age-class 3/sum of numbers in age-classes 1 and 2.

of the bias would not bring that winter into line with the other years.

Size may be an index to quality of the individual (Chitty, 1958, 1967). Mean length and mean weight of individuals varied from fall to fall, but there was no consistency as to which parameter (weight or length) has predictive value for one sex or the other, or for the sexes combined, even when one or more years were arbitrarily omitted from the regressions (Table 1, cases 3, 4, 5, 11, 12).

Changes in frequency of alleles at the transferrin locus, another possible index to quality, were monitored in this population from 1966–1968 and 1974–1978 (Mihok et al., 1983). Overwintered animals in 1974 were distinguished by a low frequency of Tf^J (0.29) and a high frequency of $TF^{M/M}$ (0.50). The frequency of TF^J was 0.49 in 1975, 0.52 in 1978, two years following winters of low survival, and 0.50 and 0.62 in 1976 and 1977, which were years with identical survival rates. Overall, there is little to suggest that winter survival is related to the kind of transferrin that an individual possesses.

Mihok and Fuller (1981) investigated changes in skull morphology of males in May as still another possible index to quality. A canonical analysis of discriminance based on 19 skull measurements, plus body length and weight, revealed year-to-year changes in skull shape and several significant correlations between canonical axes and environmental parameters, but not demographic parameters. However, bivariate skull-body regressions showed differences in slope or elevation across years, and deviations from the common regression were correlated with one demographic index. Overall, my data provide little support for the hypothesis that spring numbers, or winter survival rates, are influenced by the quality of individuals in fall populations.

The only predictor that is positively correlated with 28-day survival rate (September age ratio, Table 1 case 13) is also the only one that accounts for the pre- and post-peak years and 1977–1978. In spite of these positive features, the relationship is suspect for four reasons. First, it does not hold for either sex considered separately. Second, age ratio in September does *not* predict age (as reflected in length of roots on M^2) in May (r = 0.42, df = 6, P ≈ 0.25). Third, there is no consistency in the proportions of age-class 1 and age-class 2 in samples with similar ratios of age 3 to age (1 + 2), or in samples with similar 28-day survival rates (Table 2). Finally, September age ratio does not predict actual May numbers.

Although the major conclusion from this analysis is negative, I cannot claim to have eliminated all possible predictors. However, elimination of logical candidates such as number, size and age of the fall population suggests that it is still worth pursuing some factor or factors associated with snow that may be controlling the demography of *C. gapperi* and, by implication, other microtines.

ACKNOWLEDGMENTS

Financial support for this work came from grants to the author from the National Research Council of Canada and the National Science and Engineering Research Council, and from support for the Heart Lake Biological Station provided by the Department of Zoology of the University of Alberta. Assistants and graduate students too numerous to mention individually assisted in collection of the material.

LITERATURE CITED

ANDERSON, P. K. 1970. Ecological structure and gene flow in small mammals. Symp. Zool. Soc. London, 26:299–325.
———. 1980. Evolutionary implications of microtine behavioral systems on the ecological stage. The Biologist, 62:70–88.
CHITTY, D. H. 1958. Self-regulation of numbers through changes in viability. Cold Spring Harbor Symp. Quant. Biol., 22:277–280.
———. 1967. The natural selection of self-regulatory behaviour in animal populations. Proc. Ecol. Soc. Australia., 2:51–78.
FORMOZOV, A. N. 1946. Snow cover as an integral factor of the environment and its importance in the ecology of mammals

and birds. Materials for Fauna and Flora of the USSR, new series, Zool. 5:1–152. (In Russian.) English translation, in Boreal Institute, University of Alberta, Occasional Publication 1:1–176, 1964.

FULLER, W. A. 1967. Ecologie hivernale des lemmings et fluctuations de leurs populations. La Terre et la Vie., 21:97–115.

———. 1977. Demography of a subarctic population of *Clethrionomys gapperi*: numbers and survival. Canadian J. Zool., 55:42–51.

MIHOK, S. 1980. Ageing young *Clethrionomys gapperi*: with M² tooth characteristics. Canadian J. Zool., 58:2280–2281.

MIHOK, S., and W. A. FULLER. 1981. Morphometric variation in *Clethrionomys gapperi*: are all voles created equal? Can. J. Zool., 59:2275–2283.

MIHOK, S., W. A. FULLER, R. P. CANHAM, and E. C. McPHEE. 1983. Genetic changes at the transferrin locus in the red-backed vole (*Clethrionomyus gapperi*). Evolution, 37:332–340.

PRUITT, W. O., JR. 1957. Observations on the bioclimate of some taiga mammals. Arctic, 10:131–138.

PRUITT, W. O. JR., and C. V. LUCIER. 1958. Observations on winter activity of red squirrels in interior Alaska. J. Mamm., 39:443–444.

STEBBINS, L. L. 1971. Seasonal variations in circadian rhythms of deer mice, in northwestern Canada. Arctic, 24:124–131.

WEST, S. D. 1977. Midwinter aggregation in the northern red-backed vole, *Clethrionomys rutilus*. Canadian J. Zool., 55: 1404–1409.

ZIRUL, D. L., and W. A. FULLER. 1971. Winter fluctuations in size of home range of the red squirrel (*Tamiasciurus hudsonicus*). Trans. N.A. Wildlife and Nat. Res. Conf., 35:115–127.

Address: Department of Zoology, University of Alberta, Edmonton, Alberta T6G 2E9 Canada.

WINTER SUCCESS OF ROOT VOLES, *MICROTUS OECONOMUS,* IN RELATION TO POPULATION DENSITY AND FOOD CONDITIONS AT KILPISJÄRVI, FINNISH LAPLAND

Johan Tast

ABSTRACT

Winter success of cycling subarctic populations of the root vole, *Microtus oeconomus,* was not density-dependent at Kilpisjärvi (69°3'N, 20°50'E). The weights of wintering immature root voles varied significantly from year to year and were clearly correlated with the amount of winter food available. Their survival seemed to depend on their weights. Obviously, food was the most important factor regulating their winter success. *Eriophorum angustifolium* forms the principal food of root voles in most habitats throughout the year. In winter their diet consists of nutrient-rich, first-year rhizomes, the numbers of which fluctuate greatly. The sizes of root vole populations in spring depended mostly on their winter food conditions, the numbers of first-year rhizomes of *Eriophorum.* However, the size of the population in the preceding fall also had an effect. Spring populations could not be high, if voles were scarce already in the fall.

The culmination of population development often occurred during the winter, but sometimes it took place in summer. Thus, winter food was not the only factor limiting population increase or leading to "crash."

INTRODUCTION

Populations of the root vole *Microtus oeconomus* fluctuate in fairly good synchrony with those of four other microtine rodents, *M. agrestis, Clethrionomys rufocanus, C. rutilus* and *Lemmus lemmus,* according to investigations carried on at Kilpisjärvi (69°3'N, 20°50'E) since 1946 (Kalela, 1962; Tast and Kalela, 1971; Lahti et al., 1976). However, lemmings were not numerous during every peak, but all lemming peaks coincided with cyclic highs of other species (Tast and Kalela, 1971; Lahti et al., 1976; Henttonen and Järvinen, 1981).

The niches of the two *Microtus* species largely overlap in the birch forest region. During years characterized by very high root vole numbers, *M. agrestis* populations may decline a little earlier than the others (Tast, 1966, 1968a, 1968b; Henttonen et al., 1977).

Root voles together with *C. rufocanus* show most regular patterns of cycles. However, population increase in root voles leads to dispersal and occupation of suboptimal habitats from where individuals of *M. agrestis* are expelled (Tast, 1966, 1968a, 1968b; Henttonen et al., 1977; Viitala, 1977). Thus, comparisons of population development in several areas are needed, in order to get a reliable view of root vole cycles, because their numbers do not increase beyond a maximum which in optimal habitats may be reached already before the ordinary population peak.

METHODS

Snap trapping was carried out in three study plots with different altitudes (Table 1): 1) an open bog area with surrounding thickets in the subalpine birch forest region close to the shores of Lake Kilpisjärvi, 2) a peatland area at the timber line consisting of both open bog areas and willow and dwarf shrub thickets, and 3) an area in the low alpine willow thicket subzone consisting of small meadows and dwarf shrub thickets. The plot 1 is an optimal habitat of root voles where they have been caught during 22 of 23 years of study, the only year without successful trapping being 1979. In 1973 maximum numbers of root voles were attained on this plot a year before the peak (Lahti et al., 1976; Henttonen et al., 1977). In plot 2 root voles were captured during 15 of 19 years, and in the alpine plot 3 during 10 of 23 years.

Kalela (1957) developed a line trapping method where traps are set in straight lines up the mountain slopes beginning from Lake Kilpisjärvi and continuing to the tree limit using spacing of 7 to 8 m between successive traps.

RESULTS AND DISCUSSION

As the proportion of root voles in catches is very small especially during cyclic lows, here in winter success calculations combined figures of all five species are used (Table 2, 3 and 5). Winter success indices are obtained by comparing successive fall and spring catches. When winter success is com-

Table 1.—*Mid-summer fluctuations in relative numbers of root voles and all small rodents in three study plots with different altitudes. Figures given denote the numbers per 100 trap-nights. 1 = study plot in the subalpine birch forest region, 2 = study plot at timber line, 3 = study plot in the low alpine willow thicket subzone. A = root voles, B = all small rodents 4 = indices for rodent cycles = 1 + 2 + 3. Results of trappings in the same plots in 1959–1962 are given for example in Tast (1968a).*

Year	1		2		3		4	
	A	B	A	B	A	B	A	B
1963	12.9	12.9	8.0	11.3	0.0	3.4	20.9	37.6
1964	33.3	34.1	—	—	1.7	15.5	35.0	49.6
1965	—	—	0.8	1.2	0.2	1.4	1.0	2.6
1966	4.5	4.5	—	—	0.0	0.0	4.5	4.5
1967	8.1	10.8	1.0	4.0	0.0	3.2	9.1	18.0
1968	7.5	7.5	2.1	4.9	0.0	1.4	9.6	13.8
1969	18.9	18.9	13.7	19.2	0.5	22.1	33.1	60.2
1970	49.3	53.7	6.3	24.0	1.3	11.4	56.9	89.1
1971	1.6	1.6	0.0	0.0	0.0	0.0	1.6	1.6
1972	8.0	8.0	0.6	0.6	0.0	0.0	8.6	8.6
1973	50.0	50.0	5.4	12.4	0.6	17.0	56.0	79.4
1974	25.3	26.6	27.2	45.3	34.2	41.8	86.7	113.7
1975	18.2	20.8	7.3	13.8	6.0	13.6	31.5	48.2
1976	1.5	1.5	0.0	0.0	0.0	0.5	1.5	2.0
1977	4.7	6.2	11.3	13.3	0.0	4.4	16.0	23.9
1978	14.4	25.2	13.8	27.5	8.0	38.7	36.2	91.4
1979	0.0	1.2	0.0	0.0	0.0	0.5	0.0	1.7
1980	10.0	10.0	0.6	0.6	0.0	0.0	10.6	10.6
1981	34.3	34.3	6.0	10.8	6.7	10.0	47.0	55.1

pared with the initial size of fall populations, the probable effects of population density on the winter success are seen. Root voles winter in immature stage, as do also the other species. During 11 winters root voles have been trapped at Kilpisjärvi. Rather soon it was established that the weights of immature root voles vary annually from winter to winter (Tast, 1972). The weights are compared with the winter success.

The weight variations are most easily explained on the basis of food conditions. The principal winter food of root voles at Kilpisjärvi in most habitats consists of first-year rhizomes of the cotton grass *Eriophorum angustifolium* (Tast, 1966, 1974; Tast and Kalela, 1971). The numbers of different-aged shoots of the species have been counted annually since 1963 on some study plots. The numbers of sterile shoots including the first-year rhizomes vary considerably from year to year (Tast and Kalela, 1971). The variation of the young *Eriophorum* rhizomes will be used as a measure of winter food resources of the root vole. Thus, in this paper winter success of root voles is compared with: (1) the population size of the preceding autumn, (2) the weights

of immature wintering root voles, and (3) the numbers of sterile cotton grass shoots.

WINTER SUCCESS AND POPULATION DENSITY

In Table 2 the results of line trappings performed in 1963 through 1981 are presented for spring and fall trapping periods. Highest spring populations have been in the following years (corresponding figures of the previous fall are given in parentheses). Figures given denote the number of small rodents per 100 trap nights: 1974, 28.3 (20.0); 1978, 15.6 (14.6); 1964, 14.2 (14.1); 1975, 10.6 (16.3); 1970, 8.4 (29.3); 1969, 6.5 (4.8); 1973, 5.5 (9.7); 1977, 4.6 (0.9). Highest fall populations on the other hand were (spring figures are given in parentheses): 1978; 31.5 (0.0); 1969, 29.3 (8.4); 1973, 20.0 (28.3); 1964, 19.7 (1.1); 1974, 16.3 (10.6); 1977, 14.6 (15.6); 1963, 14.1 (14.2). When the above figures are compared, it is easily seen that the sizes of spring populations after high fall populations may be high (as in 1964, 1970, 1974, 1975 and 1978) or low (as in 1965, 1971 and 1979). There seems to be no correlation between the size of fall population and winter success.

In seven cases fall populations were at the time of snap trapping (August–early September), smaller than during the following spring. Interpretation is not the same for all years. In 1968/69, 1972/73, 1977/78 and 1980/81, winter reproduction or at least attempts of it were observed (Tast and Kaikusalo, 1976; Kaikusalo and Tast, 1984). In 1963/64 no winter trapping was yet performed, but the phase of population cycles was the same, as when winter breeding has occurred. Thus, that year the possibility of winter breeding is not excluded. In 1971/72 and 1976/77, fall populations were very small. In years of small populations, breeding may continue longer than during high populations, as has been established in the case of *C. rufocanus* (Kalela, 1957; Viitala, 1977) and *L. lemmus* (Kalela, 1970; Tast and Kalela, 1971; Henttonen and Järvinen, 1981). Probably trapping ceased before the youngest voles reached trappable age, and hence the density indexes were higher in spring of 1972 and 1977 than in the previous falls.

In 1973 root vole populations were already high in optimal habitats (Table 1) in subalpine open bog areas. Seasonal change of habitat in fall (Tast, 1966) led to occupation of suboptimal habitats both in alpine and subalpine birch forest region. This immigration from peatlands to forests interprets the differences between density indexes in fall 1973 and

Table 2.—*The combined numbers of small rodents captured in line trapping performed in subalpine birch forest region at Kilpisjärvi. Figures given denote the number of rodents per 100 trap nights. Spring trapping was performed in June and autumn trapping in mid-August to mid-September.*

Year	Spring	Autumn
1963	6.7	14.1
1964	14.2	19.7
1965	1.1	0.7
1966	0.0	0.8
1967	0.6	2.1
1968	0.3	4.8
1969	6.5	29.3
1970	8.4	9.6
1971	0.0	0.1
1972	0.7	9.7
1973	5.5	20.0
1974	28.3	16.3
1975	10.6	3.2
1976	0.0	0.9
1977	4.6	14.6
1978	15.6	31.5
1979	0.0	0.3
1980	0.2	2.1
1981	3.0	18.0

Table 3.—*The mean weights of immature wintering root voles at Kilpisjärvi in different years.*

Winter	Weight (g)	N	Density indexes for previous autumn and following spring
1964/1965	21.1	30	19.7—1.1
1965/1966		0	0.7—0.0
1966/1967	25.6	9	0.8—0.6
1967/1968	17.3	4	2.1—0.3
1968/1969	24.2	23	4.8—6.5
1969/1970	26.2	20	29.3—8.4
1970/1971		0	9.6—0.0
1971/1972	no trapping		0.1—0.7
1972/1973	winter reproduction		9.7—5.5
1973/1974	30.2	13	20.0—28.3
1974/1975	24.6	12	16.3—10.6
1975/1976		0	3.2—0.0
1976/1977	no trapping		0.9—4.6
1977/1978	29.0	2	14.6—15.6
1978/1979	21.4	9	31.5—0.0
1979/1980		0	0.3—0.2
1980/1981	winter reproduction		2.1—3.0

spring 1974, as the change of habitat took mostly place after the fall trapping period.

Spring populations were smallest in: 1966, 0.0 (previous fall 0.7); 1971, 0.0 (9.6); 1976, 0.0 (3.2), 1979, 0.0 (31.5); 1980, 0.2 (0.3); 1968, 0.3 (2.1); 1967, 0.6 (0.8); and 1972, 0.7 (0.1). From the above figures it is clearly seen that the only correlation between population densities in spring and fall is that if there are few voles in the fall, they are not numerous in spring either.

The population culmination did not in every case take place during winter. In three years, 1965, 1974 and 1975, spring populations were higher than autumn populations. In 1965 there was a distinct decrease in rodent numbers in the previous winter already (19.7–1.1), and the decrease then continued. But in 1974 after slight summer decrease, voles survived very well during the following winter, spring populations in 1975 being the fourth largest during the study period. Then a decrease again continued in the summer of 1975 and the winter of 1975/76. Good winter survival of 1974/75 is clearly seen also in all samples of the actual root vole study plots (Table 2).

WEIGHTS OF WINTERING ROOT VOLES

Root voles have been trapped in winter since 1964. Four winters (1965/66, 1970/71, 1975/76 and 1979/

80) were characterized by such low root vole populations that no specimens were caught in spite of trapping of same intensity as in successful winters. As seen from Table 3, all of these winters were followed by very small populations; in 1970, fall population was rather high. In two of the winters (1971/72 and 1976/77), no trapping was performed.

The bulk of overwintering root voles consists of animals born late in the preceding season and remaining at the immature stage. However, the catches of early winter, especially, still may contain some mature specimens from the previous summer, which may weigh up to 20 g more than the immature ones (Tast, 1966, 1972). In order to eliminate the effect of these voles, individuals belonging to the immature cohort only are treated in this connection.

The weights of wintering root voles of earlier years, 1964–1973, have been treated already (Tast, 1972; Tast and Kaikusalo, 1976). It was established that there did not exist differences in weights between sexes at immature stage. Correspondingly, in *M. agrestis* there are generally no detectable size differences between sexes from the onset of winter before early spring (Myllymäki, 1977). Hence, males and females are treated jointly in this paper. In two of the winters, 1972/73 and 1980/81, root voles bred in the area (Tast and Kaikusalo, 1976; Kaikusalo and Tast, 1984). In these years the samples consist of heterogenous groups, there being immatures born the previous summer and comparable with the voles of other winters, and immatures born during the

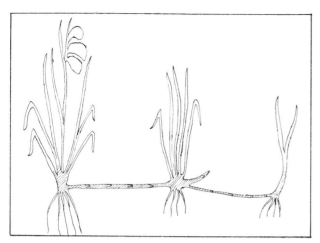

Fig. 1.—Part of a plant system of the narrow-leaved cotton grass, *Eriophorum angustifolium*. The fertile shoot on the left dies after flowering. In the middle a sterile shoot that will flower the following summer. To the right a first-year shoot that has developed from a rhizome in the current spring. The bud from which a new rhizome will grow is seen between the two sterile shoots. The growing of new rhizomes mostly takes place in July and August.

winter and being of various ages. Hence, in Table 3 the two winters with breeding have been excluded.

The weights of root voles vary significantly from winter to winter. During three of the winters with successful trapping, 1964/65, 1967/68 and 1978/79, voles have weighed less than 22 g. Among these winters are the two with the heaviest drop in the rodent numbers from fall to spring. After the third winter, 1967/68, rodent populations also were small, but as the initial population already was small, the drop could not be great. In winter 1973/74 root voles were at their heaviest, and next spring the populations were highest during the study period. In 1977/78 the root voles weighed almost as much as in 1973/74, and in 1978 spring populations were the second highest. The only spring with small population after a winter with rather heavy root voles was 1967. This exception is quite easy to interpret, as fall population was the smallest (index 0.8) among years with successful winter trapping. Thus, voles were scarce already in the fall. Hence, they could not be numerous in the spring either.

As a conclusion, the winter success of root voles seems to be positively correlated with their winter weight. On the other hand, the winter weights do not seem to be correlated with population density. Root voles have been large-sized after high fall populations for example in 1973/74, 1977/78 and 1969/70 and small-sized in 1964/65 and 1978/79. And

after small fall populations, root voles were small in 1967/68 and rather heavy in 1966/67.

ANNUAL VARIATION IN WINTER FOOD

The root vole is almost exclusively herbivorous. Its feeding habits were investigated by observations made in the field (Tast, 1966), by cafeteria tests (results not yet published) and by analyzing contents of 307 root vole stomachs (Tast, 1974). The winter food consists mostly of succulent storage organs lying underground. Although the species was recorded to be able to feed on dozens of plant species, in subarctic ecosystems only few plant species are found, and hence the potential food items are rather few. In most of their wintering habitats at Kilpisjärvi, root voles mainly feed on underground organs of *Eriophorum angustifolium* (Tast, 1966, 1972).

The development of a cotton grass shoot takes about four years (Fig. 1) at Kilpisjärvi. Rhizomes develop in the late summer and may grow some 40 cm (Fig. 2). Next spring they reach the surface and usually two leaves grow during the first summer. The numbers of leaves increase during the following summers. Different-aged leaves are recognizable as the tops of overwintered leaves become reddish due to temperatures below 0°. If circumstances are suitable, the flowering takes place during the fourth summer, but when conditions are bad, the flowering may be postponed to later years or the shoot may remain sterile.

Among sterile shoots there are different-aged shoots belonging to at least four age categories: (1) first-year shoots with two–three leaves, (2) two-year old ones with about 4 to 6 leaves, (3) third-year shoots, and (4) older ones which do not flower owing to adverse circumstances. In counting the numbers of sterile shoots presented in Table 4, all age groups have been treated jointly. The absolute numbers of sterile shoots do not give a right view of the food conditions of root voles because first-year rhizomes, which appear as first-year shoots during the summer, make up the principal food item of the species in winter. As seen from the following, they are nutrient-richer and have higher caloric values than older rhizomes do. The following figures denote 10 m lengths of rhizomes: first-year rhizomes—dry weight, 8.75 g, nitrogen content, 73.5 mg, caloric value, 39.4 kcal; older rhizomes—dry weight, 6.00 g, nitrogen content, 23.4 mg, caloric value, 30.6 kcal.

The changes in the numbers of sterile shoots from one year to the next reflect the numbers of first-year shoots. The increase in the numbers no doubt in-

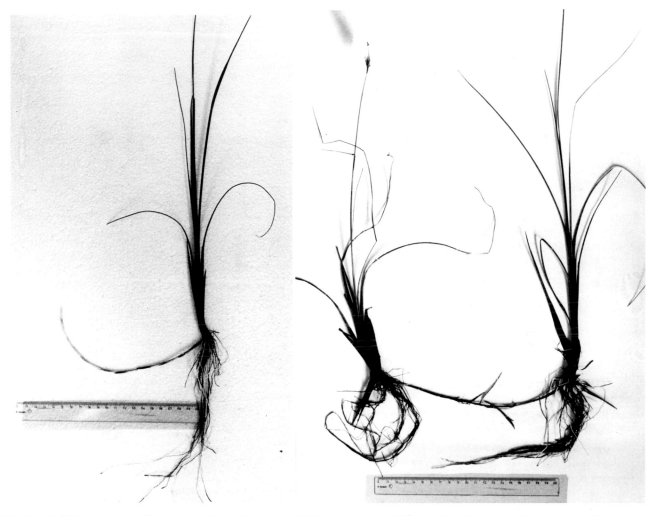

Fig. 2.—(Left:) Lower parts of a two-year-old sterile shoot of *Eriophorum angustifolium* with a first-year rhizome some 40 cm long. Photo taken on 15 August by Ritva Hiltunen. At the beginning of July no signs of new rhizomes were yet visible. (Middle and right) Lower parts of a three year old sterile shoot of *Eriophorum angustifolium* and those of a fertile one with an old rhizome between. The colour of the rhizomes and roots changes with time. Roots of different ages can be seen. In the fertile shoot there are no light-coloured young roots.

dicate increasing of first-year rhizomes during the previous winter. But decrease shows that more older shoots have died off than new have been developed.

This fact ought to have been pointed out more clearly already in our earlier papers (Tast and Kalela, 1971; Tast, 1972). Krebs and Myers (1974:329–330) plotted the weights of root voles with the total number of *Eriophorum* shoots in my study plots, while only first-year rhizomes ought to have been taken into comparison. A second point of view, which is partially misunderstood by Krebs and Myers, results from the fact that bogs are ecosystems where ecological succession changes the situation continu-

ously. Hence, the plant species composition is changing. In the case of my study plots, human activities have increased the speed of ecological succession, as the drainage of the area has been changed. That is why the total number of *Eriophorum* shoots has decreased to about half of their number at the beginning of my study. The number of fertile shoots has diminished even more. Nevertheless, at present the sharp annual fluctuations in the amount of sterile shoots can be seen (Table 4). Since 1971 another study plot has been investigated in order to get reliable and comparable figures after *Eriophorum* has been replaced by other graminids

Table 4.—*Annual variations in the numbers of sterile shoots of the cotton grass* Eriophorum angustifolium *at Kilpisjärvi in 1963 through 1981.*

Year	No. sterile shoots
1963	4,111
1964	6,171
1965	4,686
1966	4,224
1967	4,575
1968	3,544
1969	4,183
1970	4,647
1971	3,087
1972	2,710
1973	2,737
1974	2,787
1975	3,216
1976	2,051
1977	3,039
1978	3,618
1979	2,003
1980	2,490
1981	1,846

Table 5.—*Annual variations in the numbers of sterile shoots of* Eriophorum angustifolium *at Kilpisjärvi compared with the weights of wintering root voles and their winter success.* — *equals winters, when no root voles were obtained.*

Year	No. of shoots	Change in shoot number	Weight	Density indices Autumn	Spring
1963	4,111		no trapping		6.7
1964	6,171	+2,060	no trapping	14.1	14.2
1965	4,686	−1,485	21.1	19.7	1.1
1966	4,224	−462	—	0.7	0.0
1967	4,575	+351	25.6	0.8	0.6
1968	3,544	−1,031	17.3	2.1	0.3
1969	4,183	+639	24.2	4.8	6.5
1970	4,647	+464	26.2	29.3	8.4
1971	3,087	−1,560	—	9.6	0.0
1972	2,710	−377	no trapping	0.1	0.7
1973	2,737	+27	winter breeding	9.7	5.5
1974	2,787	+50	30.2	20.0	28.3
1975	3,216	+429	24.6	16.3	10.6
1976	2,051	−1,165	—	3.2	0.0
1977	3,039	+988	no trapping	0.9	4.6
1978	3,618	+579	29.0	14.6	15.6
1979	2,003	−1,615	21.4	31.5	0.0
1980	2,490	+487	—	0.3	0.2
+1981	(1,846)	(−644)	winter breeding	2.1	3.0

in my ordinary study plot. The fluctuations have been quite synchronous in the two areas with the exception of the last two years. In 1981 the amount of sterile shoots increased in the study plot, which at present is optimal to *Eriophorum.* Hence in Table 5, the figures for 1981 are in parentheses.

CONCLUSIONS AND DISCUSSION

In Table 5 annual variations in the numbers of sterile *Eriophorum* shoots are compared with the winter weights of root voles and their winter success. Winter success is clearly correlated with mid-winter weight. When root voles have weighed less than 22 g, great decrease in their numbers has taken place by spring. All of these winters, 1964/65, 1967/68 and 1978/79 were years with a very sharp dip in the numbers of sterile *Eriophorum* shoots. The other winters with marked diminishing in sterile shoots were such that no root voles were trapped in mid-winter. Obviously, populations had collapsed already in early winter before trapping in 1970/71 and 1975/76. On the other hand, good winter survival and large-sized root voles coincided with increasing of sterile shoot numbers in 1966/67, 1968/69, 1969/70, 1973/74, 1974/75 and 1977/78. So, winter success seems to be in fairly good correlation with winter food available. The annual variation in winter food, the first-year rhizomes of *Eriophorum,* is obviously the basic factor responsible for the winter survival of root voles. It is also one of the major

factors leading to rodent cycles. However, the size of spring population also depends on the size of the initial fall population, which mostly is a sequence of reproduction rate in the preceding summer. No doubt the rodent cycles are affected by several factors. This is seen also from the fact that population development has culminated sometimes in mid-summer.

The results do not support different kinds of theories interpreting rodent cycles by factors deriving from population density, including for example stress or shock disease (Christian, 1950; Christian and Davis, 1964), and selection towards certain genotypes in the course of different phases of the cycle (for example Voipio, 1950; Chitty, 1960; Freeland, 1974; Haukioja and Hakala, 1975). Observations of 1973–1975 especially do not fit with these hypotheses. In 1973 rodents were already relatively numerous, and in 1974 they reached the highest peak densities during the whole study period. After a slight decrease in numbers by fall, the rodents survived well the winter of 1974/75. Spring populations in 1975 were among the highest during the study period. These high spring populations released breeding in their main avian summer predators. For example the rough-legged buzzard, *Buteo lagopus,* bred in 1975 more successfully than in 1973 (Lagerström et al., in preparation). The chemical defense of plants against herbivores (Freeland and Janzen, 1974; Haukioja and Hakala, 1975) may be a factor influencing preference of root voles for first-year rhizomes to the older ones. Negus and Berger (1977) state that phenolic compounds which inhibit reproduction, for example in *Microtus montanus,* increase in concentration in grasses as these become older.

The mean body weight of wintering root voles was not influenced by population density. Similarly, Fuller et al. (1969) found that population density at the beginning of winter did not correlate with winter weights of *Clethrionomys gapperi, C. rutilus* and *Peromyscus maniculatus.* The importance of winter food to successful wintering has been pointed to for example by Wolff and Lidicker (1980), who found that wintering of *Microtus xanthognathus* succeeded only where a good supply of storable rhizomes was available.

The synchrony of population cycles of different rodent species suggests co-variation in several food plants. Since 1971 at Kilpisjärvi, annual variations in the plant biomass, net production, flowering, and berry and seed crops have been investigated (Kyllönen and Laine, 1980; Eurola et al., 1984). Significant variations from year to year have been found in fairly good synchrony between different species and at different altitudes.

ACKNOWLEDGMENTS

I would like to express my gratitude to the organizers of the Powdermill Colloquium on Winter Ecology of Small Mammals, for making it possible for me to participate and have opportunities to meet so many winter ecologists and discuss current research efforts with them. For financial support I am thankful to the Finnish Academy of Science and the Nordic Council for Ecology.

LITERATURE CITED

CHITTY, D. 1960. Population processes in the vole and their relevance to general theory. Canadian J. Zool., 38:99–113.

CHRISTIAN, J. J. 1950. The adreno-pituitary system and population cycles. J. Mamm., 31:247–259.

CHRISTIAN, J. J., and D. E. DAVIS. 1964. Endocrines, behavior and populations. Science, 146:1550–1560.

EUROLA, S., H. KYLLÖNEN, and K. LAINE. 1984. Plant production and its relation to climatic conditions and small rodent density in Kilpisjärvi region (69°05'N, 20°50'E), Finnish Lapland. Spec. Publ., Carnegie Mus. Nat. Hist., 10:127–136.

FREELAND, W. J. 1974. Vole cycles—another hypothesis. Amer. Nat., 108:238–245.

FREELAND, W. J., and D. H. JANZEN. 1974. Strategies in herbivory by mammals: the role of plant secondary compounds. Amer. Nat., 108:269–289.

FULLER, W. A., L. L. STEBBINS, and G. R. DYKE. 1969. Overwintering of small mammals near Great Slave Lake Northern Canada. Arctic, 22:34–55.

HAUKIOJA, E., and T. HAKALA. 1975. Herbivore cycles and periodic outbreaks. Formulation of a general hypothesis. Rep. Kevo Subarctic Res. Stat., 12:1–9.

HENTTONEN, H., A. KAIKUSALO, J. TAST, and J. VIITALA. 1977. Interspecific competition between small rodents in subarctic and boreal ecosystems. Oikos, 29:581–590.

HENTTONEN, H., and A. JÄRVINEN. 1981. Lemmings in 1978 at Kilpisjärvi: population characteristics of a small peak. Memoranda Soc. Fauna Flora Fennica, 57:25–30.

KAIKUSALO, A., and J. TAST. 1984. Winter breeding of microtine rodents at Kilpisjärvi, Finnish Lapland. This volume.

KALELA, O. 1957. Regulation of reproduction rate in subarctic populations of the vole *Clethrionomys rufocanus* (Sund.). Ann. Acad. Scient. Fennicae A, 4(34):1–60.

———. 1962. On the fluctuations in the numbers of arctic and boreal small rodents as a problem of production biology. Ann. Acad. Scient. Fennicae A, 4(66):1–38.

———. 1970. Movements of the Norwegian lemming (*Lemmus lemmus*) in a year with extremely large populations. Paper read at Kilpisjärvi Biol. Station during the excursion of IBP Meeting on Secondary Productivity in Small Mammal Populations, Helsinki, 24–26 August 1970.

KREBS, C. J., and J. H. MYERS. 1974. Population cycles in small mammals. Advanc. Ecol. Res., 8:267–399.

KYLLÖNEN, H., and K. LAINE. 1980. Tunturikasvillisuuden vuotuisista vaihteluista Kilpisjärvellä. (Annual variation in the plant biomass, net production, flowering, and berry and seed crops in the Kilpisjärvi fjeld area, Finnish Lapland). Luonnon Tutkija, 84:19–23.

LAGERSTRÖM, M., A. KAIKUSALO, and J. TAST. In prep. Breeding biology of the rough-legged buzzard, *Buteo lagopus,* and rodent cycles at Kilpisjärvi, Finnish Lapland.

LAHTI, S., TAST, J., and H. UOTILA. 1976. Pikkujyrsijöiden kannanvaihteluista Kilpisjärvellä vuosina 1950–1975. (Fluctuations in small rodent populations in the Kilpisjärvi area in 1950–1975.) Luonnon Tutkija, 80:97–107.

MYLLYMÄKI, A. 1977. Demographic mechanisms in the fluctuating populations of the field vole, *Microtus agrestis.* Oikos, 29:468–493.

NEGUS, N. C., and P. J. BERGER. 1977. Experimental triggering of reproduction in a natural population of *Microtus montanus.* Science, 196:1230–1231.

TAST, J. 1966. The root vole, *Microtus oeconomus* (Pallas), as an inhabitant of seasonally flooded land. Ann. Zool. Fennici, 3:127–171.

———. 1968a. Influence of the root vole, *Microtus oeconomus*

(Pallas), upon the habitat selection of the field vole, *Microtus agrestis* (L.), in northern Finland. Ann. Acad. Scient. Fennicae A, 4(136):1–23.

———. 1968b. The root vole, *Microtus oeconomus* (Pallas), in man-made habitats in Finland. Ann. Zool. Fennici, 5:230–240.

———. 1972. Annual variations in the weights of wintering root voles, *Microtus oeconomus,* in relation to their food conditions. Ann. Zool. Fennici, 9:116–119.

———. 1974. The food and feeding habits of the root vole, *Microtus oeconomus,* in Finnish Lapland. Aquilo Ser. Zool., 15:25–32.

TAST, J., and A. KAIKUSALO. 1976. Winter breeding of the root vole, *Microtus oeconomus,* in 1972/1973 at Kilpisjärvi, Finnish Lapland. Ann. Zool. Fennici, 13:174–178.

TAST, J., and O. KALELA. 1971. Comparisons between rodent cycles and plant production in Finnish Lapland. Ann. Acad. Scient. Fennicae A, 4(186):1–14.

VIITALA, J. 1977. Social organization in cyclic subarctic populations of the voles *Clethrionomys rufocanus* (Sund.) and *Microtus agrestis* (L.). Ann. Zool. Fennici, 14:53–93.

VOIPIO, P. 1950. Evolution at the population level with special reference to game animals and practical game management. Papers on Game Res., 5:1–176.

WOLFF, J. O., and W. Z. LIDICKER, JR. 1980. Population ecology of the taiga vole, *Microtus xanthognathus,* in interior Alaska. Canadian J. Zool., 58:1800–1812.

Address: Kilpisjärvi Biological Station, University of Helsinki, P. Rautatiek. 13, SF-00100 Helsinki 10, Finland.

Present address: Korvenkatu 44, SF-33300 Tampere 30, Finland.

FOOD HABITS, GRAZING ACTIVITIES, AND REPRODUCTIVE DEVELOPMENT OF LONG-TAILED VOLES, *MICROTUS LONGICAUDUS* (MERRIAM) IN RELATION TO SNOW COVER IN THE MOUNTAINS OF COLORADO

ALBERT W. SPENCER

ABSTRACT

The winter food habits, size and weight distribution, reproductive cycle, and signs of winter activity of long-tailed voles, *Microtus longicaudus* (Merriam), in various subalpine and montane habitats were studied over several years at locations in southwestern and north-central Colorado. Mean body length and weight declined over winter, primarily because of disappearance of older age classes. Fully grown post-reproductive animals lost proportionately more weight than animals whose maturation had been delayed. The voles' choice of food plants changed as the winter progressed. Peak consumption of bark coincided with the period of thaw. Distribution of bark-gnawing exhibited altitudinal zonation. Initiation of reproductive development appeared to be related to temperature and precipitation in late February and March in southwestern Colorado. Winter ranges of voles were determined by mapping the distribution of cropped herbs and gnawed stems. The dimensions of the ranges seemed to be related to the characteristics of the vegetation, particularly the amount of mechanical support provided, which in turn influenced the amount of snow-free space available at the ground surface. The observations support a hypothesis that long-tailed voles are deterred from burrowing through snow of density greater than about 0.15 g/cm³ and avoid doing so by adaptive behavior including choice of habitats, versatility in feeding, and possibly changes in social interactions.

INTRODUCTION

Few aspects of the biology of long-tailed voles, *Microtus longicaudus* (Merriam), have been extensively studied or reported. Cranford (1981) has monitored activity and movements under snow of long-tailed voles tagged with radioisotope-labelled wire. Sleeper et al. (1976) recorded reproductive condition and food habits throughout several years. The present report reviews information on those aspects and presents new information on winter range and feeding activities.

STUDY SITES

The two principal areas in which the studies reported below have been carried out were located in the San Juan Mountains of southwestern Colorado and near Rocky Mountain National Park in north-central Colorado. Several sites along Missionary Ridge (37°30′N, 107°45′W) 15–30 km northeast of Durango, La Plata County, Colorado, in townships 36–38 North, Range 8 West, New Mexico Prime Meridian, at elevations between 2,500 m and 3,500 m were sampled. Detailed descriptions of the biotic components of this area may be obtained from Steinhoff (1976). An important characteristic of these sites is the general east-to-west trend of the valleys and the steepness of the terrain. Long-tailed voles were taken from three principal habitats. On south slopes, aspen groves with an understory dominated by snowberry (*Symphoricarpos* spp., *S. oreophilus* commonly) or stands of scrub oak (*Quercus gambelii*) were the important habitats of voles. On north slopes, voles were found in numbers only in clear-cut areas within spruce-fir forest. Montane voles (*M. montanus*) replaced long-tailed voles in fescue parks on south slopes. The characteristics of the vegetation at the Wallace Lake grid (S32T37NR8W) where the extent of bark-gnawing was assessed are outlined in Table 2.

The second area was situated at 2,900–3,000 m along Trap Creek (40°34′N, 105°49′W) approximately 58 km west of Fort Collins, Larimer County, Colorado, in sections #17 and #18, T7NR75W, Sixth Principal Meridian. Spencer and Pettus (1966) include a detailed description and map of the site. Long-tailed voles inhabit bogs, fens, and willow shrub-carr as well as upland clearing, especially a checkerboard array of clear-cuts in lodgepole pine, spruce, and fir forest. The vegetation of the clear-cuts in 1972 is characterized by the results of a 120 m transect summarized in Table 6. Most early stages of secondary succession back to lodgepole pine forest are represented in the clear-cuts, which were established over the period between 1950 and 1967. Clear-cuts on Missionary Ridge, although of similar age, support a very different vegetation dominated by shrubs and dicotyledonous herbs. There are no sedges similar to *Carex rossii* and few conifer saplings.

CLIMATE

Microclimates at four sites on Missionary Ridge were monitored through five years (Steinhoff, 1976). Climatic data were gathered only during summer at the Trap Creek locality. The climate of the Trap Creek sites is very much like that described by Merritt and Merritt (1978), a continental climate with snow persisting generally from mid-October to late May or early June. The climate of the Missionary Ridge area is generally similar, but differs in several important aspects. Temperatures are about 1° to 3°C warmer than at comparable elevations in north-central Colorado, primarily because winter minimums are less extreme. Wind is very much less important in the San Juans (personal evaluation). A larger proportion of the precipitation falls during the period October through April (Barry and Bradley, 1976). The intensity of solar radiation is very much higher than in typical boreal climates. Sunny days in early January may exceed 400 langleys/day. The average for January was about 250 langleys/day. LaChapelle (Armstrong et al., 1974) calls it a "radiation snow climate" and ascribes several important consequences to the coincidence of high snowfall and intense solar radiation. Mean weekly temperatures during the three years of the study did not dip below −13°C at 3,200 m. Yearly precipitation (1 October–30 September) at 3,200 m ranged from 56 cm in 1971–1972, a dry year, to 114 cm in the 1972–1973 season. The snow pack in three seasons ranged from 69 percent of normal (1970–1971) to 136 percent of normal (1972–1973). It exceeded 250 cm depth in April of 1973 in clear-cuts at 3,200 m (Sleeper et al., 1976). Perhaps the most significant characteristic of the climate of the subalpine is the high variability between the seasons and the high diversity in conditions on north-facing versus south-facing slopes. This is reflected in the variability of duration of the snow pack. It persisted from mid-November in 1971 to late February of 1972 on south-facing aspen slopes at 3,200 m (3.5 months), compared to 6.5 months from late October 1972 to early May 1973 (Sleeper et al., 1976). On the north slope in those same years, snow cover persisted from late October in both years to early May 1972 (5.75 months) and late June 1973 (7.5 months). The winter of 1972–1973 was a year of heavy snow early and late in the season. Nevertheless, under the impact of the sunny weather of late January and February, the snow pack on south slopes dwindled markedly. Some steep south slopes up to 3,100 m were bared by mid-March (Sweeney and Steinhoff, 1976).

METHODS

STANDARD MEASUREMENTS AND REPRODUCTIVE CYCLE

I attempted to obtain a monthly sample of voles by means of museum special traps. A typical trapping foray involved 300 to 450 traps set for two or three nights. To avoid unduly depressing vole populations in any one area, trapping was conducted in a different location each month. The same locations were trapped in the same months each year. During July, animals captured in the course of Standard Calhoun Censuses conducted by Steinhoff and his aides were made available to us. Standard measurements of voles, including body weights, were taken at autopsy. We also recorded testis length, mean length of the seminal vesicle, weight of testis stripped of epididymis, condition of the uterus and vulva, number and length (crown-rump) of embryos, uterine scars, and number of corpora lutea or corpora albicantia.

MAPPING THE DISTRIBUTION OF VOLE ACTIVITY

Grids were established by laying out quadrats 30 m square and staking off north and south lines at intervals of 3 m. Measures were stretched between the stakes. Two sticks, each 3 m long and marked off in 1 m increments, laid across the measures defined 1 m square quadrats. The sticks and measures were moved by leap-frogging one past the other as each rank of the grid was recorded. The diameters in inches of sedge tussocks were estimated to the nearest even integer. I felt more precise measurements were unproductive because the tussocks were not perfectly circular, and the boundary of a clump can be only arbitrarily defined without excavating. If a tussock were only partly consumed I recorded the estimated percentage removed. In calculating the area cropped, the tussocks were regarded as perfect circles. Initially I recorded all tussocks, whether cropped or not, plus other vegetation on the quadrat. However, this required too much time to sustain throughout.

Mapping of gnawed stems was conducted in essentially the same manner. Lines marked off in meter increments proved easier to manage in thickets. The length and average breadth of the area of bark removed and the mean diameter of the stem in that strip were found by comparison to a meter stick held alongside. If the stem forked, each branch was recorded separately. The mean diameter of the stem was taken to be its thickness midway along the area debarked. Small stems were often completely girdled. The area in such a case was approximated by the product, mean diameter times the length times three.

Two methods were employed in measuring or mapping the extent of runways in bluegrass. One system (#17 and #18, Table

5), described below, was very open and linear. It was convenient to measure and sketch this system in its entirety. The area grazed was obtained by multiplying the total length of runways by the modal runway width. In two others, the runways were very tightly knit, and it would have been tedious to map them precisely. I laid two metric measures parallel to one another at a fixed distance; a tape was stretched vertically between them at 30 m intervals, and the intercepts of the edges of the casts of discarded material were recorded. The area within the outer limits was approximated by Simpson's Rule. The area grazed was calculated by dividing the sum of the intercepts by the sum of the vertical lines within the system. This can be only a crude approximation because the edge of the casts rarely coincided with the edge of the areas actually grazed, if indeed, that can be precisely defined.

ESTIMATION OF FOOD CONSUMED

Two methods were used to find the mass of green tissue available in the sedge cropped. Shortly after the snow melted from the tussocks, we clipped a measured area of sedge approximately to the level the voles did, air dried them for several weeks and weighed the total mass of leaves obtained. Then we separated them into dead or living fractions and weighed these. If less than one-half the length of the leaf was green, it was placed in the dead pile. Most leaves containing green sections had dry tips 15 to 30 mm long, equal to one-tenth to one-third their lengths. In a second trial, we lifted out a measured area of sod along the diameter of a tussock. The individual stems were then separated. The number of stems and number of living and dead leaves per stem were recorded. This was repeated for nine tussocks. An arbitrary sample of living leaves was extracted.

For each leaf, the total length and the length and the width of the green portion were recorded. Each leaf was air-dried, weighed, and divided into green and brown fractions, which were weighed to obtain estimates of the weight of green material per cm² (product of stems per cm times mean number of leaves with green sections per stem times average dry weight of green tissue per leaf).

Samples of stems of several species of shrub gathered in March 1973 were sectioned with a razor, and the thickness of phloem and cambium combined were measured with an ocular micrometer. I also tried scraping the green underbark from ten square centimeters of stems of oak and snowberry. The weight of the dried scrapings was close to 0.1 g/cm² in both cases. I felt, however, that the scrapings included some of the outer layers of xylem in each case, and that the mass of readily digested material was overestimated by this method.

ANALYSIS OF STOMACH CONTENTS FOR FOOD HABITS

Stomachs from voles were removed, placed in paper folders and air dried. During the summer of 1974, two undergraduate assistants collected the stomachs into samples by date, site of capture and sex of the vole. They weighed each stomach, sectioned it longitudinally, and reweighed the half utilized. The half stomachs from all voles of the same sample were blended together, and the resultant suspension was sieved through a 128 mesh screen. The particles were then dispersed in Modified Hoyer's solution on microscope slides, labelled, and sent to the Composition Analysis Laboratory at Colorado State University for analysis. The procedures, including analysis, are described more completely in Hansen and Flinders (1969).

RESULTS AND DISCUSSION

BODY WEIGHT IN WINTER

The relationship of body length to weight in winter is compared to that for summer in Table 1 and Fig. 1. Females were not compared because pregnancy in females confounds the relationship. As the figure indicates, winter weights of females followed the same trend as for males. I calculated a regression of body weight against body lengths for males caught in two periods (Table 1). The summer samples included all age classes. Sexually immature subadults and post-reproductive adults made up the winter samples. The model implicit in linear regression is inappropriate for the general relationship of body weight to length. However, over the short intervals considered (102–114 mm in winter, 104–112 mm in summer), the errors introduced are small. The differences in the magnitude of the regression coefficients for summer and winter reflect the greater severity of the stress for the larger voles. As Fig. 1 illustrates, weight loss increases by about 0.3 g per mm of length but amounts to only about 2–4 g (10–20%) for voles 100 mm in length, whereas it is close to 30% for males of 115–120 mm length. The difference was not so great for immature versus post-reproductive females.

The basis of the differential loss could not be determined, but its age dependency may explain the disappearance of older age classes (Fuller et al., 1969; Sealander, 1972; Kaikusalo, 1972). I infer that large animals, mostly post-reproductive, encountered more nutritional stress than the small animals born after the solstice whose maturation had been delayed (Brown, 1973). Larger animals may have been somewhat more severely handicapped in moving under the snow, and therefore may have suffered more acute limitations in foraging. It would be of interest to compare relative metabolic rates and efficiencies of post- and pre-reproductives in winter. The actual extent of the loss is biased on the low side. All winter-caught voles had greatly distended stomachs, whereas most summer-caught voles had relatively little in their stomachs. This appears to be an adaptation to increase the absolute rate of assimilation albeit at reduced efficiency. Parra (1978)

Table 1.—*Comparison of body weights and regression of body weight on body length in long-tailed voles caught on Missionary Ridge in winter and summer. Means and regression coefficients plus or minus 95% confidence intervals (from Sleeper et al., 1976).*

Season	n	Regression coefficient	Mean weight	Confidence interval	Mean length	Confidence interval
1972						
Winter	30	.353 ± .163	26.77	0.54	110.0	0.98
Summer	38	.528 ± .126	33.08	1.31	111.87	2.04
1973						
Winter	11	.068 ± .490	24.45	1.74	109.64	2.61
Summer	28	.672 ± .235	33.88	2.56	116.64	2.28

points out that small herbivores, because of their limited fermentation capacity, generally choose the most readily digestible foods that can be processed quickly. Goldberg et al. (1980) showed that the beach vole adheres to this pattern in its selection of food. Harlow (1981) considered the strategy of digestion. Rapid passage at reduced efficiency may yield more energy/unit time than slow passage with increased efficiency. The residence time of most food in the intestine of voles is on the order of 3 to 6 hrs (Kostelecka-Myrcha and Myrcha, 1964). Passage of time may be more important in *Microtus* than in many small mammals because of the limited surface area of their intestine relative to body weight (Barry, 1976).

BARK-GNAWING ACTIVITY

Nutritional stress is further indicated by changes in food habits in winter. Snow melt in 1972 revealed extensive damage to shrubs by voles. This was particularly evident in a transect along a west-north-west-facing ridge at the 2,500 m contour in a stand of mixed oak brush, aspen, and conifers (Wallace Lake site, S32T37NR8WNMPM). At higher elevations, significant amounts of bark had been removed from the lower boles of aspens on south-facing slopes. This was established as the work of long-tailed voles by the characteristic toothmarks, abundant feces mingled with waste bark, and by frequent sightings of long-tailed voles in this area as the damage was being surveyed. Table 2 indicates the composition of the woody vegetation at this site and the extent to which different species had been utilized by voles. Apparent preferences are reflected in the relative consumption of aspens compared to their frequency in the vegetation. Oaks were also preferred over snowberry or other shrubs. Shrubs

such as chokecherry and serviceberry, whose bark releases a strong odor of bitter almonds, were not used in proportion to their abundance. The aspens gnawed were, for the most part, stems that had been felled by wood gatherers, but some boles had been attacked at the base. Only oak stems less than 6 cm in diameter that had been weighed down under heavy accumulations of snow (Fig. 2C) were attacked. The stems were typically attacked on their lower surface, from which the snow tends to settle, leaving a space. Terminal twigs were rarely attacked. They settle with the snow and therefore are accessible only by tunneling. The twig pointed out in Fig. 2A is an example. Likewise, boles were not often attacked close to the soil surface, but usually on their downhill or southern surface 20 to 50 cm above the ground where slope creep and shrinkage of the snow pack after solar heating had opened a gap between the snow pack and the bole. Snowberry stems, however, were typically stripped close to the ground in the base of the bush. Their distal ends, bowed down by the weight of the pack, support an arch of snow above the bush, leaving the heart free of snow or only loosely filled. The space formed in this way provides a sizable area often overlapping those in other bushes in which a small mammal might move freely. Note that the biggest range was in snowberry. Snowberry was used lightly relative to its abundance, but herbs may have been more accessible in the area sheltered by the shrub.

The voles discarded the corky layers of the bark and ingested only the living cambium and phloem. They consumed both the green layer under the periderm of aspens and the vascular cambium within the phellum. Use of aspen boles was more prevalent among trees in which the outer green layer was still vital, a characteristic which varied with site and age of tree. Partially girdled stems of oaks suffered a high frequency of mortality (about 70% overall) in the dry spring and summer months that followed (Sleeper et al., 1976).

Fig. 3 illustrates the dispersion of the bark-gnawing activity on the plot mapped. The intensity of consumption is obviously clumped at a few foci. Applying the premise that the outlying foci represented the ranges of single voles and assuming that ranges of two or more voles may have converged on exceptional food sources such as fallen aspens, I conjectured that 18 vole ranges were included in the 1.14 ha area mapped. The average dimensions of the range calculated on this basis would be 424.06 m² (range = 87–1,134) or approximately 25 m by

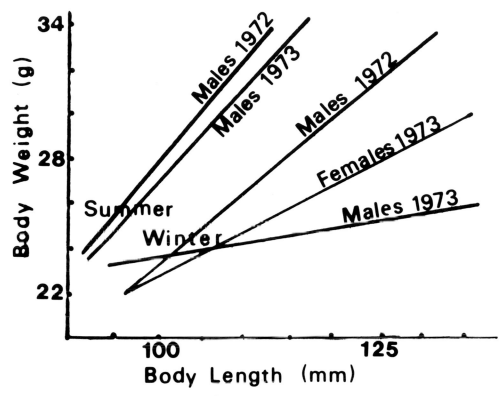

Fig. 1.—Regression of body weight on body length in long-tailed voles caught in winter compared to summer (after Sleeper et al., 1976).

23 m (range = 12 by 12 to 63 by 39), within the range of areas catalogued by Hayne (1950). The average volume of cambium and phloem consumed would be 617.4 cm³/vole. If this volume were assumed to be 50% water, it would approximate 309 g of dry matter per vole; if it were 75% water, then only 154 g dry matter would be available per vole. The smallest range was associated with a felled aspen. The largest was in a stand of large diameter aspen with an understory primarily of snowberry.

Girdling of shrubs in 1972 was associated with a relatively high density of voles during an increasing phase of their cycle. The activity seemed to be most intense along a belt around 2,500 m on west-facing slopes and in aspens along the 3,000 to 3,200 m contour on south slopes. These observations suggest to me a weather-related factor, perhaps the zone in which the first persistent snowfall was wet, sleety, partially melted after falling, or otherwise denser and harder to penetrate than typical new fallen snow.

Table 2.—Composition of woody vegetation at Wallace Lake site and relative utilization of cambium by long-tailed voles.

Measurements	Gambels oak	Snowberry	Aspen	Rose	Service-berry	Choke-cherry	Conifer	All species
Density (stems/m²)	4.8	8.2	0.2	4.2	0.2	2.3	0.01	23.8
Frequency of utilization by voles (percent of total stems of that species)	7.2	2.1	1.3	4.1	trace	0.03	—	—
Mean thickness of phloem and cambium (mm)	0.36	0.18	0.63	0.205	—	0.29	—	—
Mean area of bark consumed per unit area (cm²/m²)	20.80	5.75	9.36	0.36	0.06	0.18	0.02	36.6
Mean volume of consumption (cm³/m²)	0.749	0.104	0.590	0.008	—	0.005	—	1.456

Fig. 2.—Examples of bark gnawing activity. (A) Newly fallen aspens were disproportionately utilized by voles in comparison to other woody vegetation. Note that the small twig below the lower stem is untouched. (B) Voles attacked the upper branches of oak and other large shrubs only when they were bowed down by heavy accumulations of snow, as shown here. This photo, however, was taken the following year at a location 8 km south and 600 m lower than the Wallace Lake site in the winter following the one in which the damage shown in Fig. 2C occurred. (C) Several stems of oak brush gnawed by long-tailed voles are shown. The largest is about 1.6 m tall and 5 cm in diameter at the base. The intensity of bark gnawing activity shown was measured at approximately 2,000–2,100 cm² of phloem per 81 m². This photo was taken at the far left tip of the grid depicted in Fig. 3.

Although I have not encountered such extensive girdling in the years since 1972, the situations where I have most frequently observed bark-gnawing included many sites with extra heat input, for example, under steep south-facing escarpments, around cave openings, near flowing springs, and along streams. Such areas tend to have denser snow pack or flooded ground. Snow tends to partially thaw and refreeze in such places and so becomes dense quickly. They are also areas where herbaceous vegetation remains green longest into the fall, perhaps attracting voles in numbers greater than can be supported later on. The stomach samples (Table 4) indicated use of shrubs on south slopes increased after the snow pack began to thaw at the surface and to collapse. I am assuming that these may have been the circumstances under which bark-gnawing occurred at sites with no particular differences from the hab-

itat in general. The basis of the erratic occurrence of bark-gnawing could be the extra effort involved in obtaining the edible portion of the bark. These observations are in accord with the appraisal of Stenseth et al. (1977) of optimal utilization of foodstuffs by field voles. Voles would resort to shrubs only in response to reduced availability of their preferred foods. Tast (1972) recorded such a shift when root voles exhausted their stock of cotton-grass bulbs. Danell et al. (1981) have suggested that the occurrence of bark-gnawing is a useful indicator of peak vole numbers. Frischknecht and Baker (1972) thought heavy snow and high population density were both necessary correlates of extensive girdling damage. My observations suggest, however, that still a third condition, density of the snow at the ground surface or some other site-dependent factor, may be involved also.

Fig. 3.—Distribution of bark gnawing activity by long-tailed voles over the winter of 1971–1972 in a 1.14 ha area of mixed oak-aspen-conifer stand (Wallace Lake site). Each cell represents a quadrat of nine square meters. The number in each cell represents the total area of bark removed by the voles in that quadrat (in units of 400 cm²). Dots signify trace amounts (up to 400 cm²) of bark consumed on the quadrat. A hairpin curve in the Missionary Ridge access road bounds the northern two-thirds of the plot.

Winter Food Habits

Table 3 summarizes the analysis of stomach contents of long-tailed voles. Through confusion, the samples for January and May 1973 were never analyzed. Thirty-one kinds of foods were recognized among the stomach contents. The relative consumption of some of the genera of plants identified varies among the samples in agreement with their relative abundance, as subjectively determined, on the slopes where the voles were captured. *Festuca* and *Berberis,* for example, were more important on the south slopes where *F. thurberi,* Thurber's fescue, and *B. repens,* Oregon grape, are abundant, but comprised a smaller portion of the contents from north slopes where they are rare.

Herbs, not unexpectedly, were the most important species in the food of long-tailed voles during the winter and early spring period. Fescue comprised nearly 34% of the total contents and 52% of the stomach contents of voles from south slopes. Sedges, *Carex* spp., (11%) and yarrow, *Achillea lanulosa,* (16%) were other important constituents. Herbaceous species combined accounted for almost 72% of the contents. Oregon grape and moss were included with the herbs because of their similar size and comparable growth forms. They made up 4 and 2% of the components, respectively. Shrubs and trees comprised most of the remaining 28% of the material. Aspen, *Populus tremuloides,* accounted for 11%, red berried elder, *Sambucus pubens,* 6%, and spruce, *Picea* spp., almost 10%. Eight different kinds of plants made up 88% of the foodstuffs identified. Eight of the samples included only one stomach each. Of these, three contained essentially only one species (99–100%) of plant and in the other five, two plant species constituted 95–99% of the stomach contents. Among all samples on an average, one kind of plant represented 68 ± 2.6% of the foodstuffs identified, indicating, I believe, a tendency of long-tailed voles to feed to repletion on one species per bout of foraging during the winter. The selection broadened following snow melt. Neither sedge nor yarrow occurred by themselves, nor did they make up more than 70% of an individual stomach's content, perhaps to dilute their toxic components (Hansen and Johnson, 1976; Bergeron, 1980).

Oak and snowberry appeared only as minor components. Other shrubs attained importance primarily at or following the thaw. As Table 4 shows, shrubs constituted a more important fraction of the diet on north slopes than south in 1972 and on both slopes in 1973, the year of heavier snow pack. Aspen was a major component of the food on south slopes in February and March 1972, on north slopes in late

Table 3.—*Comparison of the relative importance of different plants identified in the stomach contents of long-tailed voles from Missionary Ridge at two different periods of the winter. The table indicates percent volume for each species based on 80 field samples. To obtain the percentages shown several samples were combined. The contribution of each sample to the average value was weighted in accord with the number of stomachs included in the sample.*

Plant species	Males—South				Males—North			
	1972 16 January– 26 March	1972 1 April– 18 May	1973 17 March– 3 April	1973 23–27 April	1972 16 January– 26 March	1972 1 April– 18 May	1973 8 April	1973 23–27 April
Number of stomachs in sample	26	21	8	2	13	29	1	4
Festuca	82.0	33.7	12.4	5.9	.36	.77	0	0
Carex	2.9	15.7	10.4	72.9	12.07	14.47	0	51.4
Achillea	2.2	33.4	0	4.2	12.34	33.11	0	2.5
Lupine	0.8	0.86	.3	0	4.44	0.29	0	0
Oenothera	0	tr	0	2.59	10.26	3.89	0	0
Moss	0	0.1	0	0	20.50	.57	0	0
Berberis	7.4	tr	25.9	0	1.12	0	0	0
Sambucus	0	0	0	14.32	0.73	0.23	100.0	44.9
Rosa	0	0.5	0	0	0	0	0	0
Populus	0	tr	45.5	0	10.96	7.57	0	0
Picea	0	6.6	0	0	28.32	32.54	0	0
Pinus	2.3	0	0	0	0	0	0	0
Arthropods	0.5	0.7	tr	0	.99	1.85	0	0.5
Other herbs	0	0	0	0	0	0	0	0
Symphoricarpos	0	0	0	0	0	0	0	0

Plant species	Females—South				Females—North			
	1972 16 January– 26 March	1972 1 April– 18 May	1973 17 March– 8 April	1973 24–27 April	1972 15 February– 4 April	1972 2 May– 18 May	1973 17–18 March	1973 23–27 April
Number of stomachs in sample	24	12	8	5	23	8	3	4
Festuca	66.8	31.2	79.3	0	35.3	8.9	0	0
Carex	1.8	22.9	4.4	21.1	4.0	25.6	8.4	0
Achillea	0.8	14.3	0.2	0	35.7	9.1	0	0
Lupine	8.3	0.2	0.2	0	0.2	0	0	0
Oenothera	0	0.1	0	2.3	0	0	0	14.4
Moss	0	0	11.3	tr	0.8	tr	0	0
Berberis	8.4	4.6	0.8	0	1.8	0	0	0
Sambucus	0.1	0	0	73.6	0.6	0	0	85.6
Rosa	0	2.3	0	0	0	0	0	0
Populus	5.2	0.2	0.2	0.6	5.0	2.5	91.2	0
Picea	0	11.7	0	0.4	.87	7.4	0	0
Pinus	2.4	0	0	0	tr	0	0	0
Arthropods	0.5	0.8	0	0	2.6	2.9	.4	0
Other herbs	0	0	0	0	12.5	42.9	0	0
Symphoricarpos	0	0	0	0	tr	0	0	0

March and early April 1972, and on south slopes in mid-March and early April 1973. The time in each case coincided with a period when the snow pack was melting and settling rapidly. Red-berried elder was a prominent component of the diet of only the late April sample of 1973, a period of both maximum snow depth and accelerating snow melt. There was a progressive shift from fescue and other herbs in mid-winter to shrubs during snow melt and then to forbs as the surface of the ground was uncovered and new growth began (Table 3). The trends are similar to those portrayed by Larsson and Hansson (1977) for the field vole. At no time during the winter did the samples include the variety of foods utilized or any of the species most commonly consumed in summer by the montane vole (Vaughan, 1974), a species with food habits similar to the long-tailed vole (Clark, 1973). The shifts from one species

Table 4.—*Percentage of identified contents of stomachs of long-tailed voles (pooled samples) arranged by vegetative form, habitat, sex of vole, and date of capture. Based on 80-field samples (from Sleeper et al., 1976).*

| | Males | | | | | | Females | | | | | |
| | South Aspect | | | North Aspect | | | South Aspect | | | North Aspect | | |
Date	n	Herbs	Shrubs and trees	n	Herbs	Shrubs and trees	n	Herbs	Shrubs and trees	n	Herbs	Shrubs and trees
9 I 72	10	91.56	7.17	—	—	—		90.79	8.82	—	—	—
15 II 72	5	98.56	1.16	4	91.51	7.72	2	36.78	62.90	9	91.30	3.42
12 III 72	3	100.00 **	—	5	32.15	65.90	6	100.00	—	8	97.59	0.96
26 III 72	8	97.79	2.21	4	56.97	41.63	2	98.93	1.07	—	—	—
1 IV 72	—	—	—	1	35.22	64.32	—	—	—	1	—	100.00
4 IV 72	2	88.10	11.90	5	16.59	83.41	1	72.49	27.51	5	94.68	5.32
8 IV 72	6	87.32	11.97	—	—	—	2	99.68	0.32	—	—	—
17 IV 72	4	98.01	0.39	4	74.04	20.73	3	98.43	1.26	—	—	—
2–5 V 72	8	90.76	8.87	12	47.83**	49.96	5	71.93	28.07	4	85.06	14.41
16–18 V 72	1	98.82	1.18	7	98.24	0.88	1	100.00	—	4	89.32	5.34
17–19 III 73	6	57.70	42.30	—	—	—	2	100.00	—	3	8.41	91.15
1–3 IV 73	2	44.62	55.37	—	—	—	5	99.75	0.25	—	—	—
8 IV 73	—	—	—	1	—	100.00	1	100.00	—	—	—	—
23–27 IV 73	2	85.68	14.32	4	54.41	44.90	3	17.43	82.57	4	14.40	85.60
23–27 IV 73	—	—	—	—	—	—	1	49.05	50.95	—	—	—
		**			**							

** Period of thaw.

to another doubtless are related, at least in part, to changes in quality of the plants (Goldberg et al., 1980). Fescue, however, as well as the other plants in the winter diet, change little in appearance from late fall to snow melt. The direction and timing of the changes support the interpretation that their inaccessibility is reduced progressively as the snow pack deepens, ages, and increases in density.

WINTER CROPPING OF HERBS

On arriving at the Trap Creek site in 1972, I found very extensive systems of vegetation had been cropped by voles over winter. Sedge tussocks (Fig. 4) in clear-cut areas had been particularly hard hit. The system seemed clearly demarcated and easier to define than workings in sod or marsh that I had earlier attempted to quantify. Having enjoyed some success with mapping the distribution of bark-gnawing activity earlier in the year, I decided to apply the same methods to mapping the distribution of cropped sedge tussocks. The results of this impromptu effort are depicted in Fig. 5.

Sedge and grasses were the only plants in the systems obviously clipped by voles. The sedge, presumed to be of a single species, has characteristics close to those of *Carex rossii* (Herman, 1970). It forms round tussocks from 5 to 50 cm in diameter and is the only abundant herb on the clear-cuts which retains green leaves throughout the winter. The leaves and culms of grasses, chiefly *Agrostis scabra* and *Trisetum spicatum,* were withered and dry at snow melt. Mosses and *Vaccinium scoparium* had not been noticeably utilized. Lodgepole pine seedlings were the important woody species in the stand, with a few spruce and fir saplings scattered among them. Hence, the diet of the voles over the winter must have been predominantly the one species of sedge since it was so closely cropped. However, the growing points must not have been damaged, for the tillers which had been buried in the dried stems of the tussocks sent up new growth, and cropped tussocks were difficult to distinguish from untouched tussocks in the following season.

The voles characteristically mined out the center

$\longrightarrow$

Fig. 4.—Typical tussocks of sedge at the Upper Lily site. Note how the clump at left in A and that shown in B have been completely clipped except for a fringe of peripheral leaves which had been pressed flat to the ground by over-lying snow pack before the voles moved in to feed. The discarded portions have been shaped into casts of the runways. The lens cover in each photo has a diameter of 58 mm.

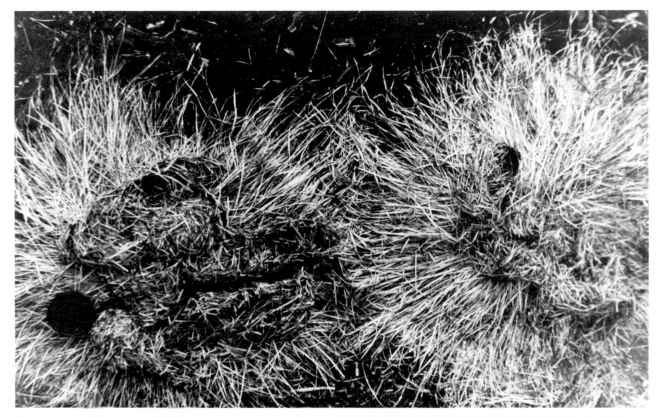

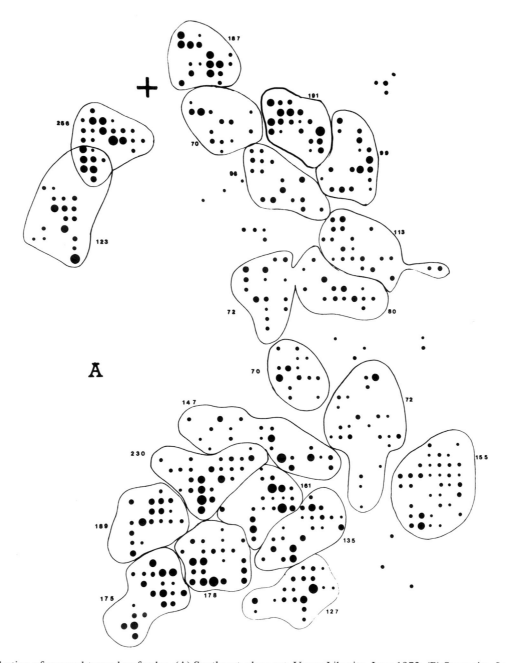

Fig. 5.—Distribution of cropped tussocks of sedge. (A) Southeast, clear-cut, Upper Lily site, June 1972. (B) Same site, June 1975. The cross indicates the same coordinate in both maps. (C) Clear-cut, Mosquito site, June 1973. The size of the circles is proportional to the area of sedge cropped in each m² quadrat of the grid. An X indicates the location of a winter nest. Numbers indicate the area cropped in units of 100 cm² within the area enclosed by the irregular closed lines. North in each map is at the top. The arms of the crosses (+) are one meter long.

of the tussock. They apparently selected the green parts and arranged the discarded dead and dry parts of the plant into a lining for their burrow in the snow. This appears to have been extended as the vole moved systematically through the tussock (Fig.

4). My interpretation is that the vole was nearly continuously sheltered in this lining while feeding and perhaps in this way escaped some of the thermal stress it would have otherwise encountered. It may have been able to avoid the dilemma posed by

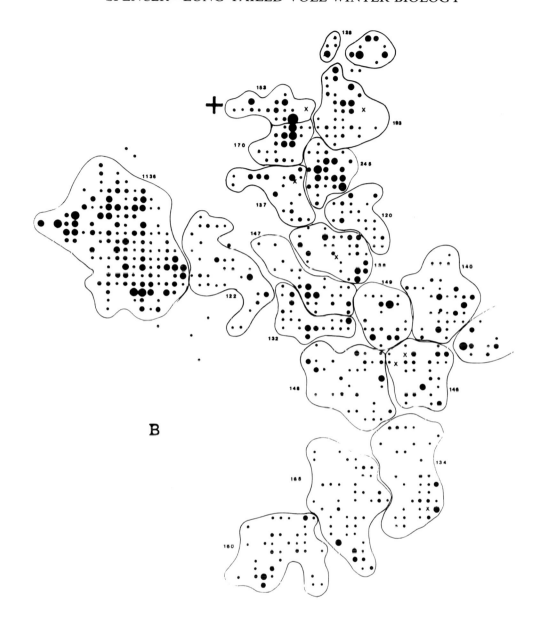

B

McLean and Thomsen (1981) by not leaving its "nest" to forage, but more-or-less carrying the nest with it continuously, thereby conserving the warmth built up in the mass. Perhaps the scarcity of nests located on the systems is related to this activity.

Voles did not attempt to utilize the leaves of sedge around the periphery of the tussock (Fig. 4). The space they did utilize was probably free of snow close to the ground or only loosely filled. I found this to be the case in late November 1972, my only observation of sedge tussocks under unthawed, unsaturated snow.

I examined "isolated" systems believed to be the work of single voles (Table 5). The combined area of cropped tussocks incorporated into one of these isolated systems was only a fraction of the area contained in the complexes illustrated in Fig. 5. None of the systems were in fact truly isolated. Each was within 10 to 30 m of another system and, in some cases, connected to it by a discernible runway. On one occasion, I was able to trace a runway over 90 m across a boggy meadow between two systems. Some "isolated" systems, upon further investigation, proved to be extensions of complexes quite

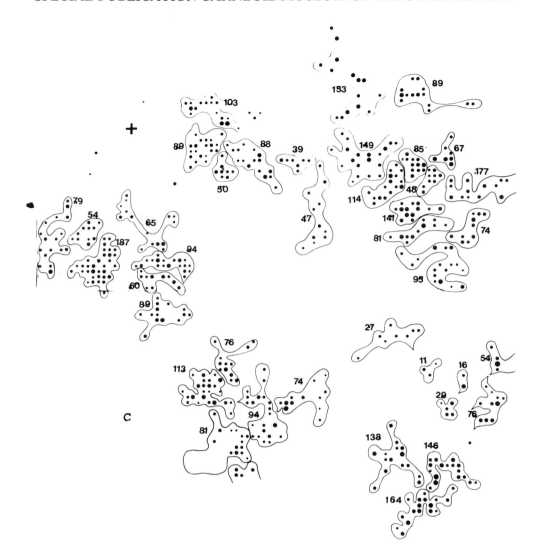

comparable to the three mapped in Fig. 5. The several clusters comprising System #1 in Table 5, one of the earliest to be examined, were found, when snow melt was complete, to be a continuation of a complex extending downslope for another 60 m. The next 13 systems in Table 5 were more distinctly separate and probably were each the work of a single vole. Systems #5 and #6 may have been the range of a single vole surviving into a second winter. They were situated in the same place, had similar outline, and nearly identical dimensions.

It must be understood that I did not observe the voles directly during winter at the Trap Creek sites, but only the sign they left behind. I could not be certain even of the vole species responsible for the sign. Long-tailed voles were captured during the fol-

lowing summers. The inferences drawn below depend on the validity of my inference that Systems #2–#14 (Table 5) each represent the winter range of a single vole. Their dimensions are similar in extent to the winter ranges determined by Cranford (1981 and this volume) and to ranges of other species of *Microtus* (Ambrose, 1969; Hayne, 1950; Getz, 1961). Evidence against accepting them as ranges includes a disparity between the number of systems and number of winter nests, the high density, close proximity, and minimum area of equivalent systems in the complexes, and the deficiency in energy available compared to estimates for similar animals by other authors. I attempt to explain these discrepancies below. If the correspondence of these systems to ranges of individual voles can be confirmed, they might provide more accurate approximations of the winter range of these small mammals because the statistical bias inherent in repeated telemetric fixes or captures could be avoided (Mares et al., 1980). The boundaries of the distribution of the cropped tussocks are definable with little error.

The distribution of cropped tussocks was not random. Their dispersion approximated a negative binomial rather than Poisson distribution. That is, they were contagiously distributed. Each system appeared to be made up of two or more lobes or clusters. The larger complexes, mapped in Fig. 5, incorporated thirty to forty such clusters. I have arbitrarily attempted to group them, as shown by the closed lines. Part of the pattern can be related to characteristics of the terrain or habitat. The large void in the center of Fig. 5C is the location of a slight rise swept free of most snow by the prevailing westerly winds. The voids in Figs. 5A and B correspond to a precociously reforested area of the clearcut. The saplings here were up to 2.5 m tall in 1972, compared to the seedling status of the majority of conifers on the remainder of the grid. The boundaries of the complexes in general did not seem to relate to any discontinuities in character of vegetation or terrain. I suspect intrinsic factors, perhaps social structure, are more significant than extrinsic factors.

Some of the isolated systems were aligned in accord with features where early drifts might be expected to form. Several were located in the lee of banks transverse to the prevailing wind or in depressions below road shoulders. Others showed a strong correlation between feeding activities and the location of felled trees and slash (Fig. 6). Some of the strung-out series of tussocks shown in Fig. 5C were associated with felled tree tops or logs. The particularly dense aggregations in the northwest quadrants of Figs. 5A and B lay on the site where a trailer, belonging to a CSU graduate student, had been parked at times in previous years. Household waste waters may have enriched that area with nutrients or salts.

The voles were selecting a richer set of quadrats than a random choice of quadrats would have afforded. A vegetational analysis, based on a 120 m strip transect of the southeast clear-cut, indicated a basal area of 492 cm²/m² (Table 6); for sedge in quadrats where cropping occurred, the average was 1,074.2 cm²/m². The voles consumed 74.2% of the sedge present in quadrats where some sedge had been cropped. 84% of the remaining 25.8% of the

Table 5.—*Characteristics of isolated systems of cropped herbs. Systems 1–14 were comprised of sedge tussocks in clear-cut areas within spruce-fir forest at the Trap Creek sites. Systems 15–18 were located in bluegrass lawns on the campus of Fort Lewis College.*

#	Year observed	Habitat	Greatest Length	Width (m)	Estimated area cropped
1	1972	Clear cut	28	7	15,161 cm²
2	1972	Clear cut	30	16	65,290
3	1972	Clear cut	33	21	19,355
4	1974	Clear cut	24	6	20,368
5	1974	Clear cut	34	6	16,084
6	1975	Clear cut	34	6	7,845
7	1974	Clear cut	24	12	16,406
8	1974	Clear cut	28	3	7,310
9	1974	Clear cut	21	13	15,323
10	1975	Clear cut	20	20	20,658
11	1975	Clear cut	15	15	21,832
12	1975	Clear cut	10	5	16,897
13	1977	Clear cut	23	19	14,135
14	1977	Clear cut	25	18	14,535
15	1980	Bluegrass lawn	7	5	158,940
16	1980	Bluegrass lawn	8	3	42,540
17	1980	Bluegrass lawn	15	11	46,845
18	1980	Bluegrass lawn	25	10	14,000

total available fell into the 15 to 30 cm diameter size classes extensively used by the rodents. Few tussocks in the 35 to 50+ cm classes escaped use by the voles. The 10 cm or smaller classes, however, were used less than would be expected on a random basis, although some tussocks of this class were superior in density, and perhaps quality, to the larger tussocks when sampled in 1981. Perhaps they did not permit the construction of sufficient protective cover. Tussocks tended to be either consumed or untouched (Fig. 4A, right). I interpret this as an indication that the voles tended to exhaust one tussock before moving to another.

The sample of clippings from three tussocks in 1972 indicated that the tussocks included, on a dry weight basis, 41% leaves which were green for more than half their length. These averaged 0.028 g dry weight green matter/cm² of tussock cropped. A 25 cm tussock had a greater density (0.039 g/cm²) than a 20 cm one (0.026 g/cm²), and a 15 cm tussock had less (0.023 g/cm²). In 1981 an estimate obtained by counting stems per cm², number of leaves per stem, and millimeters of green blade per leaf yielded an estimate of .022 ± .072 g/cm² (range = .018–0.0638 g/cm²). Combining this estimate with that for an average mean area cropped per isolated system of 17,342 cm² (mean of Systems #2–#14, excluding

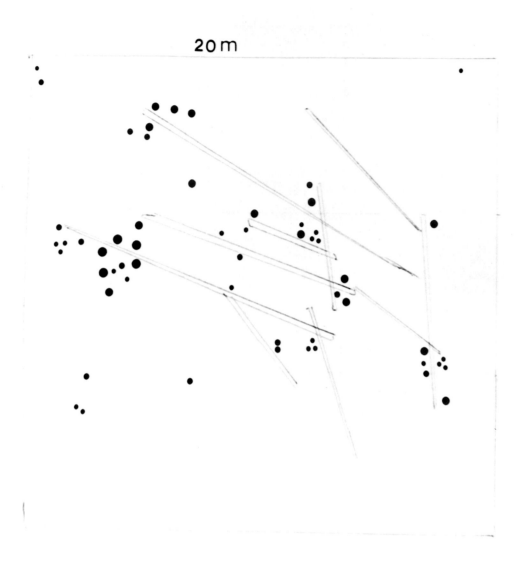

Fig. 6.—A map of system 10 in Table 5 showing the association of cropped tussocks to fallen logs.

#6 and #8), we obtain a mean consumption of 381.5 g/vole/winter. This mass, if the caloric value is taken to be 4.7 kcal/g (Hadley and Bliss, 1964; Batzli, 1975) at 80% digestibility (Drozdz, 1968), would support a daily energy budget of 10.8 kcal/22 g vole (Grodzinsky et al., 1970) for 166 days (1973.0 kcal/vole/winter). The winter snow pack at the Trap Creek sites persists from 180 to 240 days or more. The prior estimate obtained in 1972 indicated 485.6 g/vole/winter or 2,282.2 kcal/vole/winter, which would support a vole at 10.8 kcal/vole/day for 211 days. These estimates must be regarded with skepticism, but they hold out a possibility of measuring with fair precision the actual consumption of an individual vole over winter by further studies of such systems. The tussocks sampled in 1981 had been grazed the previous year and were surrounded by much taller lodgepole saplings than had been the case in 1972. Either of these factors might have reduced the vigor of the plants.

Several alternatives are proposed to account for the caloric deficiency. The estimate is probably minimal because voles may not have remained on the sites throughout the winter, instead migrating or dying. Use of other plants may have supplemented the sedge; conifer bark and seeds and vaccinium are possible sources. Long-tailed voles may cache foods as does the meadow vole on occasion (Gates and Gates, 1980). They may also be able to increase digestive efficiency or be able to use some of the brown portions of the leaves through reingestion of fecal pellets as Ouellette and Heisinger (1980) have found for the meadow vole. They may have been able to reduce their caloric requirements to less than 10.8 kcal per 22 g vole per day by reducing the time spent in active movement or conserving heat more effectively through the feeding behavior suggested above (based on Wunder's 1975 analysis of energy allocations). Cranford (1981) noted that activity of long-tailed voles was reduced in the winter compared to autumn or spring.

The small dimensions of systems in the complexes, their close proximity, and the exceptional intensity of grazing in a few systems lead me to conjecture that the voles must have shared space to some degree. Vole ranges do overlap, as captures of second and third voles at the same site on consecutive nights demonstrate. Although long-tailed voles are more solitary than most species in summer, their aggressiveness may be reduced in winter as in some other microtines (West, 1977; Turner et al., 1975). I can see several advantages or significant implications of aggregation. If snow does restrict movements of voles to existing burrows, access to neighboring systems would afford more possibilities for escape from short-tailed weasels, a predator able to follow them under the snow. It would integrate the foraging efforts of the group over a patchy, non-homogeneous environment. In effect, it might be comparable to the flocking behavior of spermophilous birds (Cody, 1971). Cooperation in conservation of heat by communal nesting (Trojan, 1969) would be facilitated. Aggregation might perform epideictic functions. It would at least increase the opportunities for social interactions, whether agonistic or mutualistic, with consequences for the dynamics of the population in the following season. Complexes were evident only in the peak years, and their epicenters were not used in years of lower density. Perhaps only cooperative aggregations could exploit these areas lacking cover, providing extensive preexisting runways. As Hansson (1977) has pointed

Table 6.—Composition of winter vegetation at Lily Pond site.

	Conifer saplings stems/m²*	Component: undershrubs** percent ground surface occupied	"Carex rossii"*** basal area (cm²)/m²	Other sedges and grasses basal area (cm²)/m²
Mean and confidence interval	0.84 ± 0.36	10.9 ± 20.8	492.06 ± 109.51	70.06 ± 23.39

* Primarily seedlings of lodgepole pine 0.1 m to 1 m tall.
** Almost entirely Vaccinium spp. rather sparse and stunted by their exposed situation.
*** See text for discussion of identity of this sedge.

Table 7.—*Comparison of the intensity of cropping activity in different years or at different sites in the Trap Creek locality.*

Site	Mean[a] (cm²)	SE	n[b]	Area mapped (ha)
Sedge				
Upper Lily Southeast 1972	706.6	129.29	430	0.24
Upper Lily Southeast 1975	690.2	47.35	597	0.35
Mosquito 1973	660.5	52.00	524	0.60
Shrubs				
Wallace Lake 1972	48.9[c]	3.38[d]	8,244	1.14

[a] Mean area (cm²) cropped per meter square quadrat in which cropping occurred.
[b] Number of meter-square quadrats in which some evidence of feeding activity was recorded.
[c] Calculated by dividing mean consumption per (3 m)² quadrat by 9.
[d] Calculated by dividing SE of mean consumption per (3 m)² quadrat by 9.

out, they may be invadable only during winter because of the lack of sufficient cover in summer.

The Trap Creek clear-cut sites are exceptionally barren in comparison to most vole habitats, particularly those seen in photos shown at the colloquium. This may be a consequence of the impoverished nutrient supply characteristic of lodgepole pine ecosystems in this area (Fahey, 1983) and/or of their early seral status. Whatever the cause, less than 20% (approximately 1955 cm²/m², including 320 cm² for pine saplings) of the ground was covered by vegetation. Since the evidence indicates no depth hoar existed over bare ground surface, vole movements on these clear-cuts may be exceptionally restricted. Nevertheless, the clear-cut habitat is richer than the arctic tundra. Batzli (1975) estimated 4.2×10^5 kcal/ha was available in coastal tundra. The sedge in the clear-cuts would have provided $5.1–6.5 \times 10^5$ kcal/ha at a basal area of 492 cm²/m². The voles, by

selecting richer quadrats, could reduce the amount of movement under or through snow even further.

The isolated systems showed the broadest dispersion of cropped tussocks. The Upper Lily complexes exhibited the greatest degree of aggregation; the Mosquito complex was intermediate. The density of cropping per meter quadrat in which some feeding activity occurred, however, was similar for all systems (Table 7). When the quadrats are grouped by area cropped, the frequencies falling into each interval are similar for all three complexes (Table 8). The voles seem to have been sampling the vegetation in the same manner in all three years on both sites.

As the main difference between sites was the relative absence of logs and slash on the Upper Lily grid, I attribute the greater dispersion on the Mosquito site to the freedom of movement provided by the rigid cover. Other differences between the two sites seemed unimportant. They included the aspect and the importance of *Vaccinium* in the vegetation. The Mosquito site has a gentle slope to the south and lies on the windward side of the ridge; only vestiges of *Vaccinium* remain. The Upper Lily site lies on the slope behind a wooded ridge, slopes gently to the northeast, and has retained a little more *Vaccinium*. The isolated systems included a wider variety of situations. Among them I could see no consistent patterns related to either exposure or the presence of the plant.

The winter of 1979–1980 was heavier than usual for Durango (140% of the long-term average). When the snow melted away, I found several systems inscribed in the bluegrass lawns of Fort Lewis College (elevation 2,090 m). These could have been the work

Table 8.—*Comparison of the intensity of cropping at different Trap Creek sites or in different years. Percentage of meter-square quadrats falling into each area-class.*

Statistics	Area class[1]										n
	1	2	3	4	5	6	7	8	9	10	
Upper Lily Southeast (1972)											
n	28	59	156	77	55	30	16	5	4	—	430
%	6.5	13.7	36.3	17.9	12.8	7.0	3.7	1.2	0.9	—	
Upper Lily Southeast (1975)											
n	27	79	198	94	67	47	6	6	—	—	524
%	5.2	15.1	37.8	17.9	12.8	9.0	1.1	1.1	—	—	
Mosquito (1973)											
n	53	69	228	110	66	46	14	8	2	1	597
%	8.9	11.6	38.2	18.4	11.1	7.7	2.3	1.3	0.3	0.2	

[1] Midpoint of area (cm²) in density class: 1—64.5, 2—193.6, 3—419.4, 4—709.7, 5—1,129.0, 6—1,612.9, 7—2,064.5, 8—2,580.6, 9—3,225.8, 10—>3,548.5.

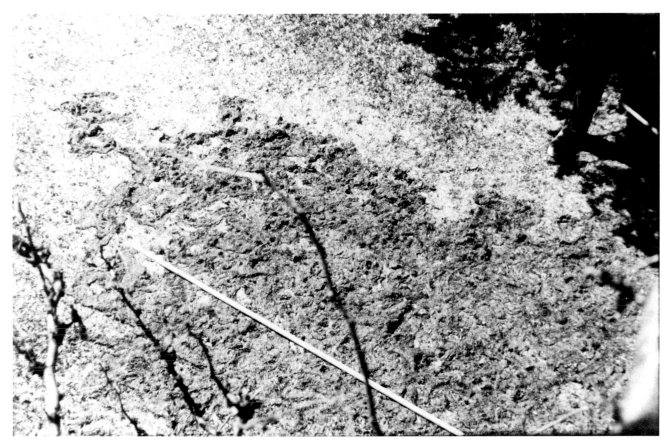

Fig. 7.—A view looking down from an adjacent locust tree onto approximately 35% of System #16, Table 5. Notice how the bluc-grass turf all around the system is pressed close to the ground.

of either long-tailed or montane voles since both mountain shrub and pinyon-juniper-sage communities abut the campus. Systems #15 and #16 (Table 5) were located in places where early snows had persisted as isolated drifts—#15 in the shadow of pinyon and juniper trees; #16 in a swale adjacent to a culvert below the surrounding parking lots and lawns. A portion of #15 is shown in Fig. 7. Persistent snow cover in these areas predated establishment of general snow cover by 30 days.

The third system, #17 and #18 in Table 5 combined, spread across 75 m of a playing field used in winter as a cross country ski course. It consisted of a single smoothly-arcing main trunk for most of that distance, with widely spaced laterals and short parallel runs clumped toward the two ends. I initially interpreted #17 and #18 as separate entities, but the geometric continuity of their path and the presence of a well defined path across a broad, denuded area between the two sections convinced me they were related. The arcs of the main trunk and parallel runs

resembled paths of the snowmobile used to lay out ski trails or of individual ski tracks. There was no feature of the terrain which corresponded to the arc. This system had an inordinate number of well developed nests, two in the northern section and five (three of them less than three meters apart) in the southern subunit. I supposed that so many nests were constructed because the comparatively thin thatch in the area was insufficient to construct adequately sheltered runways.

The three systems span the gamut of dimensions. I explained the contrasts among them in terms of a reconstructed history of the development of the snow pack in each case. I believe the two cramped systems were established in remnant patches of early snow. Perhaps wetness of the subsequent snow deterred their enlargement when a continuous cover was established. Temperatures during the early part of the storm that initiated the permanent pack hovered around 0°C, melting occurred, and the density of the snow deposited was about 0.25 g/cm^3. I think

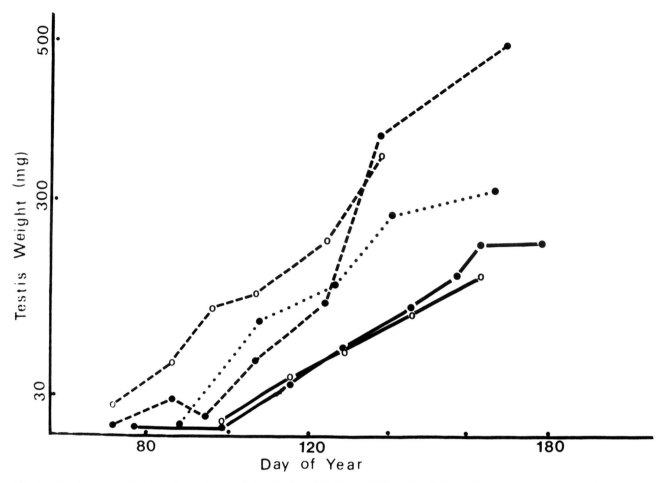

Fig. 8.—Development of testis weight in long-tailed voles from Missionary Ridge. The difference between slopes and years is correlated with the time of disappearance of the snowpack (after Sleeper et al., 1976).

the linear system, on the other hand, was formed after this storm, when a vole, possibly displaced from a nearby roadside ditch by meltwater from the road, moved into a temperature-gradient layer under the compacted snow of ski or snowmobile tracks. The vole would have adjusted its range in accordance with the paths of least resistance.

Bluegrass lawns provide comparatively poor pasture for voles. The density of living stems available amounted to only 0.0026 g/cm². Thus, #15, the system with the greatest total area clipped, would have furnished fodder for only about 133 days (again assuming 80% digestibility and DEB of 10.8 kcal/22 g vole/day). The duration of the pack was actually about 145 days.

The dimensions of the systems in oak brush, sedge, and bluegrass decline in association with the decreasing availability of rigid supporting structures.

The largest systems occur in habitats with the greatest amount of sheltered space, for example, snowberry.

REPRODUCTIVE CYCLE

On Missionary Ridge, testes weight began to increase in late winter of each year (Fig. 8 and Table 9). Testes growth, once initiated, proceeded at about the same rate in all years on both north and south slopes. The date of initiation, however, appeared to be related to the disappearance of snow cover. Early snow melt in 1972 left much of the south slopes clear by late February to mid-March. The voles responded by precociously attaining sexual maturity and pregnancies before mid-April. Voles on north slopes still covered by snow in March lagged by three weeks or more. In 1973, the year of deepest and longest enduring snow cover, deep snow covered both slopes through March and April; timing and culmination of reproductive development on both slopes were essentially identical. Growth of the testes began in late March while the ground was still deeply covered. Maturation was complete, and some females were newly pregnant by late May. The timing of development in 1973 closely approximated the latest expected return of favorable feeding (and breeding) conditions at that elevation (Sleeper et al., 1976). A flexible strategy such as this could be adaptive only if some factor of the environment provided a dependable forecast of snow conditions seven weeks in the future, when reproductive maturation is completed. A preliminary analysis of various climatic factors indicates that such a reliable correlation exists between, for instance, March precipitation and the occurrence of favorable conditions in April.

BASIS OF VOLE RESPONSE TO SNOW

New fallen snow often has a density around 0.05–0.10 g/cm^3. It can be easily penetrated; the flakes "cog" together readily to support and maintain the shape of an opening. Within a few days the density doubles and quadruples, and resistance to penetration increases greatly (Yosida, 1963). My samples of snow conditions at the ground surface are too limited to generalize broadly, but I found that snow lying directly on the ground surface or on flattened herbs resisted penetration with a finger. A crude penetrometer indicated a force in excess of 50 g/cm^2 was required to enter three-week-old snow 30 to 50 cm deep and with a density of 0.3 g/cm^3. In the same pack, cavities or weak cobweb-like snow occupied the space among the basal stems of shrubs or the leaves of tufted herbs, so long as they were still upright. It appeared to me that in some cases an arch of hardened snow, capable of conveying the weight of the pack to the ground, had formed above the tussock. Culms, upon collapsing, continued to support snow; they bridged obliquely down to the ground surface or across to other tufts leaving a space beneath. These observations are in accord with those of Coulianos and Johnels (1963). Snow immediately above the bare soil had a higher density, on the order of 0.25 to 0.35 g/cm^3, than snow immediately higher in the pack in midwinter. Schmid (1971) found voles came to the surface to cross snowmobile tracks rather than burrow through snow compacted to 0.4 g/cm density.

Reference to snow profiles included in Armstrong et al. (1974), and Sweeney and Steinhoff (1976) reveals that the layer at the ground surface in the San

Table 9.—*Comparison of testis weights (mg) of long-tailed voles caught on north and south aspects of Missionary Ridge (from Sleeper et al., 1976).*

Date	North aspect			South aspect		
	n	Mean	SE	n	Mean	SE
15 II 72	4	6.0	0.82	5	9.2	2.85
12 III 72	4	8.8	2.43	3	31.7	2.73
26 III 72	4	44.8	9.96	8	89.6	8.75
1–8 IV 72	6	18.2	2.73	8	160.8	16.30
16 IV 72	7	94.9	13.10	6	176.1	25.49
3 V 72	11	165.3	6.55	8	243.6	17.90
17 V 72	7	365.9	9.02	4	340.7	35.24
18 VI 72	2	348.5	111.5	1	515.0	—
18 III 73	—	—	—	7	10.3	1.53
1–8 IV 73	—	—	—	4	9.75+	2.25
25 IV 73	2	46.0	14.00	4	73.5	14.41
5–13 V 73	10	104.2	10.85	15	115.9	9.48
22–31 V 73	3	140.0	10.50	3	183.3	3.48
6–16 VI 73	5	234.8	10.44	2	220.5	68.50
22–30 VI 73	8	242.8	10.78	—	—	—
10 V 74	17	201.7	10.93	17	220.1	9.35
21 V 74	4	226.0	24.91	—	—	—
19 VI 74	8	307.1	16.72	—	—	—

Juans is often harder than snow lying above it in the pack. "Pukak" (recrystallized snow) is more often found on the surface or in the middle (representing a former surface) of the pack rather than at the bottom. The comparatively mild air temperatures and deep snow accumulations prevent the establishment of steep temperature gradients required for formation of pukak (more than 0.3°C/cm) deep within the pack. Powerful solar radiation, however, heats the snow below the surface, while the surface itself remains cold because it is in balance with the air and sky. This makes possible a very steep gradient (up to several degrees C/cm). Accompanying high humidity and a porous matrix facilitate vapor transport to cooler parts of the pack (Armstrong et al., 1974). When only a shallow layer lies above the soil surface, the ground may be heated similarly with the result that a layer of this "temperature-gradient snow" (Armstrong et al., 1974), or possibly a space, forms between the soil and a crust of snow above.

Vole behavior is consistent with the foregoing account of snow characteristics. I noticed networks of burrows through shallow (one inch) new snow at several places on Missionary Ridge in late October 1971. These systems recorded the movements of mice overnight. They were very extensive; in some places, they spread continuously over areas of several acres. I was unable to detect any boundaries between systems. They were probably the work of many individuals, and perhaps of several species. The systems were networks of frequently interconnected runs spaced from 1 to 10 m. Frequently a run ended in a dendritic pattern. At these places, laterals about one meter in length branched off the axial run at about 30 to 60 cm. These seemed to be located in the most favorable habitat. Perhaps each early storm stimulates the mice to lay out a system of passages as a locus for their range nightly, as they usually do during the snow-free period. It may be that new, light snow at night evokes the same response throughout the winter. Although I almost never saw tracks of voles on the surface of the snow while trapping, I encountered several microtine tracks crossing the road within two miles during a nighttime snowmobile trip at snail's pace in a heavy snowstorm. Likewise, burrows within the snow pack observed at snow melt always lay just above the interace of two layers. The animals make holes to the surface of the snow, which they apparently visit often, for I found several newly formed during very heavy snow (Spencer et al., 1973). Their function could be to maintain access to upper levels and to associated possibilities of lateral movement.

CONCLUSIONS

My interpretation of the observations detailed above is that long-tailed voles expend very little energy burrowing through snow in search of food. They escape doing so by preferentially choosing habitats which provide some free passage, by creating at least a skeleton network of passages very early in the history of the snow pack while its density is low and penetration is easy, by resorting to less preferred, but more accessible foods rather than burrowing through the snow in search of the considerable reserves of previously preferred foods still present, by changing their social behavior, by sacrificing weight or growth, and by timing reproduction to coincide with favorable conditions. They exhibit considerable flexibility in their responses, but there are indications of innate adaptations to recurrent, and therefore predictable, patterns in their habitat.

ACKNOWLEDGMENTS

This study was supported in part by a grant from the Bureau of Reclamation, Dept. of Interior, to Colorado State University. Many people helped me with the field or laboratory work, analysis, or preparation of the manuscript, among them Kevin Hekrdle, Caroline Engel, Lonnie Renner, Eric Rechel, William Overcast, Bill Rottenhause, Mary Austin, Harold Lewis, and Tony Ferdinando. Ted Jacques, in particular, deserves praise for his exceptional efforts in the first two years of the study. Completion of the winter work would not have been possible without the voluntary help, often under the most arduous conditions, of my son Frederick, daughters Margaret and Ann, colleague Preston Somers, and friend Scott Bell. Theresa Foppe of the Compositions Analysis Laboratory carried out the analysis of stomach contents. Dr. Richard Hansen provided helpful advice. Dr. David Armstrong verified my identification of the voles encountered. Billie McNeely typed the final draft of the manuscript. Dr. Roger Sleeper generously shared his unpublished data on reproduction in red-backed voles. Discussions with Drs. Harold Steinhoff, Herbert Owen, and Randolph Constantine importantly influenced my interpretations of the data. However, I must assume responsibility for the final form the interpretations took. I reserve my deepest gratitude for my wife, Georgiana. She not only tolerated long months of isolation and deprivation at the Trap Creek site, but also took an active role in some of the most strenuous field work, shouldered the burden of typing endless revisions of the manuscript, and managed to do so with patience and tolerance one step above the earthly level.

LITERATURE CITED

AMBROSE, H. W., III. 1969. A comparison of *Microtus pennsylvanicus* home ranges as determined by isotope and live trap methods. Amer. Midland Nat., 81:535–555.

ARMSTRONG, R. L., E. R. LaCHAPELLE, M. J. BOVIS, and J. D. IVES. 1974. Development of methodology for evaluation and prediction of avalanche hazard in the San Juan Mountains of southwestern Colorado. INSTARR Occas. Papers, Univ. Colorado, 13:1–141.

BARRY, R. E. 1976. Mucosal surface areas and villous morphology of the small intestines of small mammals: functional interpretations. J. Mamm., 57:273–290.

BARRY, R. G., and R. S. BRADLEY. 1976. Historical climatology. Pp. 43–67, *in* Ecological impacts of snow pack augmentation in the San Juan Mountains of Colorado: final report of the San Juan Ecology Project (H. W. Steinhoff and J. D. Ives, eds.), Colorado State Univ., Fort Collins, Colorado, xii + 489 pp.

BATZLI, G. O. 1975. The role of small mammals in arctic ecosystems. Pp. 243–268, in Small mammals: their productivity and population dynamics (F. B. Golley, K. Petrusewicz, and L. Ryszkowski, eds.), IBPS Cambridge Univ. Press, London. xxv + 451 pp.

BERGERON, J. M. 1980. Importance des plants toxiques dans le régime alimentaire de Microtus pennsylvanicus á deux étapes opposées de leur cycle. Canadian J. Zool., 58:2230–2238.

BROWN, E. B. 1973. Changes in patterns of seasonal growth of Microtus pennsylvanicus. Ecology, 54:1103–1110.

CLARK, T. W. 1973. Local distributions and interspecies interactions in microtines, Grand Teton National Park, Wyoming. Great Basin Nat., 33:205–217.

CODY, M. 1971. Finch flocks in the Mojave Desert. Theor. Popul. Biol., 2:142–158.

COULIANOS, C. C., and A. G. JOHNELS. 1963. Notes on the subnivean environment of small mammals. Arkiv. Zool., 15:363–370.

CRANFORD, J. A. 1981. Activity patterns in subnivean and fossorial rodents. International Colloquium: Winter ecology of small mammals (Abstr.).

———. 1984. Population ecology and home range utilizations of two subalpine meadow rodents (Microtus longicaudus and Peromyscus maniculatus). This volume.

DANELL, K., L. ERICSON, and K. JAKOBSSON. 1981. A method for describing former fluctuations of voles. J. Wildlife Mgmt., 45:1018–1021.

DROZDZ, A. 1968. Digestibility and assimilation of natural foods in small rodents. Acta Theriol., 13:367–389.

FAHEY, T. J. 1983. Nutrient dynamics of above ground detritus in lodgepole pine (Pinus contorta ssp. latifolia) ecosystems, southeastern Wyoming. Ecol. Monogr., 53:51–72.

FRISCHKNECHT, N., and M. BAKER. 1972. Voles can improve sage brush rangelands. J. Range Mgmt., 25:466–468.

FULLER, W. A., L. L. STEBBINS, and G. R. DYKE. 1969. Overwintering of small mammals near Great Slave Lake, northern Canada. Arctic, 22:34–55.

GATES, J. E., and D. M. GATES. 1980. A winter food cache of Microtus pennsylvanicus. Amer. Midland Nat., 103:407–408.

GETZ, L. L. 1961. Home ranges, territoriality, and movement of the meadow vole. J. Mamm., 42:24–36.

GOLDBERG, M., N. R. TABROFF, and R. H. TAMARIN. 1980. Nutrient variation in beach grass in relation to beach vole feeding. Ecology, 61:1029–1033.

GRODZINSKI, W., B. BOBEK, A. DROZDZ, and A. GORECKI. 1970. Energy flow through small rodent populations in a beech forest. Pp. 291–299, in Energy flow through small mammal populations (K. Petrusewicz and L. Ryszkowski, eds.), Proc. IBD meeting on secondary productivity in small mammal populations, Oxford, England, 298 pp.

HADLEY, E. B., and L. C. BLISS. 1964. Energy relationships of alpine plants on Mt. Washington, New Hampshire. Ecol. Monogr., 34:331–357.

HANSEN, R. M., and J. T. FLINDERS. 1969. Food habits of North American hares. Colorado State Univ., Range Sci. Dept., Sci. Ser., 1:1–18.

HANSEN, R. M., and M. K. JOHNSON. 1976. Stomach content weight and food selection by Richardson's ground squirrels. J. Mamm., 57:749–751.

HANSSON, L. 1977. Spatial dynamics of field voles, Microtus agrestis, in heterogeneous landscapes. Oikos, 29:534–544.

HARLOW, H. J. 1981. Effect of fasting on rate of food passage and assimilation efficiency in badgers. J. Mamm., 62:173–177.

HAYNE, D. W. 1950. Apparent home range of Microtus in relation to distance between traps. J. Mamm., 31:26–39.

HERMAN, F. J. 1970. Carex of the Colorado Basin. Agriculture Handbook No. 374, Forest Service USDA, Government Printing Office, Washington, D.C., 397 pp.

LARSON, T. B., and L. HANSSON. 1977. Vole diet on experimentally managed afforestation areas in northern Sweden. Oikos, 28:242–249.

KAIKUSALO, A. 1972. Population turnover and wintering in the bank vole, Clethrionomys glareolus (Schrever) in southern and central Finland. Ann. Zool. Fennici, 9:219–221.

KOSTELECKA-MYRCHA, A., and A. MYRCHA. 1964. The rate of passage of foodstuffs through the alimentary tracts of certain Microtus under laboratory conditions. Acta Theriol., 9:37–53.

McLEAN, S. F., JR., and P. THOMSEN. 1981. The energetic value of the winter nest to lemmings and voles. International Colloquium: Winter ecology of small mammals (Abstr.).

MARES, M. A., M. R. WILLIG, and N. A. BITAR. 1980. Home range size in eastern chipmunks, T. striatus as a function of number of captures: statistical biases of inadequate sampling. J. Mamm., 61:661–669.

MERRITT, J. F., and J. M. MERRITT. 1978. Population ecology and energy relationships of Clethrionomys gapperi in a Colorado subalpine forest. J. Mamm., 59:576–598.

OUELETTE, D. E., and J. F. HEISINGER. 1980. Reingestion of feces by Microtus pennsylvanicus. J. Mamm., 61:366–368.

PARRA, R. 1978. Comparison of foregut and hindgut fermentation in herbivores. Pp. 205–229, in The ecology of arboreal folivores (G. G. Montgomery, ed.), Smithsonian Instit. Press, Washington, D.C., 574 pp.

SCHMID, W. 1971. Modification of the subnivean microclimate by snowmobiles. Pp. 251–257, in Proceedings of the Snow and Ice in Relation to Wildlife and Recreation Symposium (A. O. Haugen, ed.), Iowa Cooperative Research Unit, Iowa State Univ., Ames, Iowa, iv + 280 pp.

SEALANDER, J. A. 1972. Circumannual changes in age, pelage characteristics and adipose tissues in the northern red-backed vole in interior Alaska. Acta Theriol., 17:1–24.

SLEEPER, R., A. W. SPENCER, and H. W. STEINHOFF. 1976. Effects of varying snowpack on small mammals. Pp. 437–485, in Ecological impacts of snowpack augmentation in the San Juan Mountains of Colorado: final report of the San Juan Ecology Project (H. W. Steinhoff and J. D. Ives, eds.), Colorado State Univ., Fort Collins, Colorado, 489 pp.

SPENCER, A. W., and D. PETTUS. 1966. Habitat preferences of five sympatric species of long-tailed shrews. Ecology, 47:677–683.

SPENCER, A. W., H. W. STEINHOFF, and R. SLEEPER. 1973. Small mammals. Pp. 45–55, in Interim Progress Report, March 1973, The San Juan Ecology Project, Colorado State Univ., Fort Collins, Colorado, 173 pp.

STEINHOFF, H. W. 1976. Introduction to forest ecosystems projects. Pp. 297–310, in Ecological impacts of snowpack augmentation in the San Juan Mountains of Colorado (H. W. Steinhoff and J. D. Ives, eds.), Colorado State Univ., Fort Collins, Colorado, 489 pp.

STENSETH, N. C., L. HANSSON, and A. MYLLYMÄK. 1977. Food selection of the field vole, Microtus agrestis. Oikos, 28:511–524.

SWEENEY, J. M., and H. W. STEINHOFF. 1976. Elk movements and calving as related to snow cover. Pp. 415–436, *in* Ecological impacts of snowpack augmentation in the San Juan Mountains of Colorado: final report of the San Juan Ecology Project (H. W. Steinhoff and J. D. Ives, eds.), Colorado State Univ., Fort Collins, Colorado, 489 pp.

TAST, J. 1972. Annual variations in the weights of wintering root voles, *Microtus oeconomus,* in relation to their food conditions. Ann. Zool. Fennici, 9:116–119.

TROJAN, P. 1969. An ecological model of the costs of maintenance of *Microtus arvalis* (Pall.). Pp. 113–122, *in* Energy flow through small mammal populations. Small mammals: their productivity and population dynamics (K. Petrusewicz and L. Ryszkowski, eds.), IBPS Cambridge Univ. Press, London, 451 pp.

TURNER, B. N., M. R. PERRIN, and S. L. IVERSON. 1975. Winter coexistence of voles in spruce forests: relevance of seasonal changes in aggression. Canadian J. Zool., 53:1004–1011.

VAUGHAN, T. A. 1974. Resource allocation in some sympatric subalpine rodents. J. Mamm., 55:764–795.

WEST, S. D. 1977. Midwinter aggregation in the northern red-backed vole., *C. rutilus.* Canadian J. Zool., 55:1404–1409.

WUNDER, B. A. 1975. A model for estimating metabolic rate of active or resting mammals. J. Theor. Biol., 49:345–354.

YOSIDA, Z. 1963. Physical properties of snow. Pp. 485–527, *in* Ice and snow (W. D. Kingery, ed.), M.I.T. Press, Cambridge, Massachusetts, 684 pp.

Address: Department of Biology, Fort Lewis College, Durango, Colorado 81301.

LIFE HISTORY PROFILES OF BOREAL MEADOW VOLES (*MICROTUS PENNSYLVANICUS*)

Steve Mihok

ABSTRACT

Meadow voles were live-trapped every two weeks from 1968 to 1978 on a 3.24-ha oldfield grid in southeastern Manitoba, Canada. Both annual and multiannual density fluctuations were observed, with peak densities of about 100 voles/ha. Considerable winter dispersal, a high degree of reproductive success in immigrants, poor juvenile survival in midsummer and inhibition of sexual maturity in peak years were some of the unusual demographic features observed. Life expectancy (including both viability and dispersal) of voles after first capture was the major single factor directly related to multiannual density changes. Variation in the life expectancy of mature females born early in the breeding season may have determined the demographic history of this microtine. Further research on the proximate and ultimate causes of death is required to determine the underlying processes responsible for these changes.

INTRODUCTION

In 1974, Krebs and Myers challenged the existence of non-cycling populations of microtine rodents. Since then, various researchers have discovered that the classically cyclic genus *Microtus* does not always behave predictably (Gaines and Rose, 1976; Henttonen et al., 1977; Tamarin, 1977; Abramsky and Tracy, 1979; Baird, 1980). This complex demographic pattern in *Microtus* has led to an increasing proliferation of new and modified hypotheses (Krebs, 1979; Rosenzweig and Abramsky, 1980; Garsd and Howard, 1981) that attempt to explain each successive exception to the general rules developed from previous studies (Krebs and Myers, 1974).

To assess these new ideas properly, the study of microtine population dynamics requires a sound theoretical base (Stenseth, 1980) that can be tested with a similarly sound data base. Integrated, long-term live-trapping studies are required for the detailed testing of hypotheses, but such studies have yet to be reported for any species other than *Microtus townsendii* (Krebs, 1979). In this paper, I report preliminary results from another continuous live-trapping program, involving the meadow vole (*Microtus pennsylvanicus*), which was initiated in the spring of 1968 and continued for 10 years. The data were collected as control data for current radio-ecological studies of the meadow vole (*Microtus pennsylvanicus*), and are summarized here to provide a general picture of the life history of a common microtine rodent.

METHODS

The study area was a 3.24-ha, unmanipulated control grid located in a 32-ha oldfield at the Whiteshell Nuclear Research Establishment near Pinawa, Manitoba, Canada (50°N, 96°W). The grid consisted of a 10 by 10 arrangement of trap shelters (Iverson and Turner, 1969), spaced at 20-m intervals, that was trapped by standard mark-recapture techniques (Iverson and Turner, 1974) from the spring of 1968 to the summer of 1978. Sampling was conducted at 2-week intervals, with a 2-day sampling period, using two Longworth traps per shelter. Traps were locked open and left in place between sampling periods up to the summer of 1970; after that, they were removed from the field between sampling periods. Animals were not returned to the field until the day after they were captured (Turner and Iverson, 1973); hence, they were exposed to capture only once every two weeks. Weasels, skunks, and other predators were always removed from the grid when they interfered with trapping. Nearly all captured shrews were dead, and hence were continuously removed from the small-mammal community.

Various previously described data sets, pertinent to the Whiteshell area, were used for autopsy information, or for alternate density indices (see Iverson and Turner, 1973). Autopsy data were taken from captures in live and snap-traps on grids and on lines operated for various purposes. Spring density indices were taken from three to five permanently marked, 20-station lines in fields that were trapped with three Museum Special snap-traps per station for three nights at snowmelt. Summer density indices were taken from the FPG removal grid (Iverson and Turner, 1973), an 8 by 8 grid with 15-m trap spacing, located in the same oldfield as the control grid, that was trapped with two snap-traps per station for 30 consecutive days in August of each year.

Since breeding usually began in spring and ended in autumn, each year's data were summarized for a *summer recruitment period* and a *winter non-recruitment period*. Recruitment periods began four weeks after the capture of the first lactating female and ended four weeks after the capture of the last lactating female. Voles were classified as *summer recruits* (<30 g at first capture,

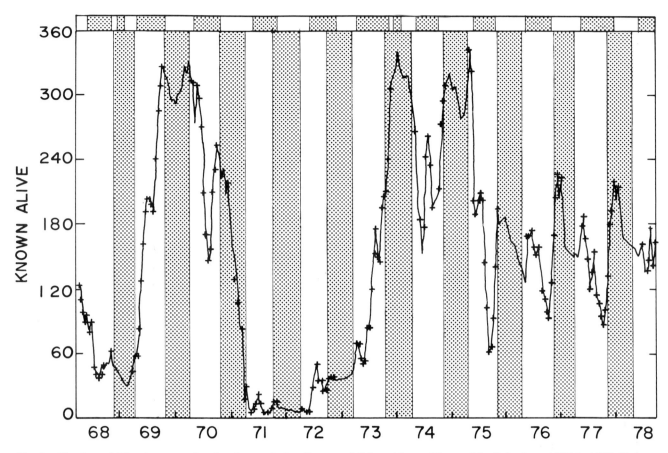

Fig. 1.—Number of *Microtus pennsylvanicus* known to be alive on a 3.2-ha grid near Pinawa, Manitoba from 1968 to 1978. Crosses represent points based on 60% or more of the animals known to be alive actually captured. Vertical shading represents the period of snow cover. Shaded bars at the top of the figure represent periods of recruitment.

first captured during the *summer recruitment period,* probably born on the grid), *summer immigrants* ($\geq$30 g at first capture, first captured during the *summer recruitment period,* probably not born on the grid), and *winter immigrants* (any weight at first capture, first captured during the *winter non-recruitment period*).

Males were classified as *reproductively active* if their testes were scrotal, females if their vagina was perforate, their nipples were medium to long (lactating), or they were obviously pregnant. Opening of the pubic symphysis was not recorded. Voles were permanently classified as *mature* if they were *reproductively active* in the year of their birth, and *immature* if they showed no signs

of *reproductive activity* in that year. Males were classified as *reproductively successful* if their testes were scrotal at any time from six weeks prior to the first capture of a lactating female to six weeks prior to the last capture of a lactating female (the approximate period when females would be in estrus). Females were considered to be *reproductively successful* only if they were recorded as lactating and were still alive two weeks later, demonstrating that a litter could potentially have been weaned. Note that for both males and females, reproductive success only signified a potential that may or may not have been realized.

RESULTS

POTENTIAL TRAPPING BIASES

Numerical fluctuations on the control grid (Fig. 1) and the other areas sampled by removal techniques once a year were similar (Fig. 2), with the possible exception of spring 1970. Numbers were high on the control grid in 1970, but few voles were caught on the five snap-trap lines that were set in the spring of 1970. Hence, spring line indices were

not significantly correlated with spring numbers on the control grid ($r = 0.52$, $N = 10$, NS), unless the 1970 data were excluded ($r = 0.83$, $N = 9$, $P < 0.01$). Numbers on the control grid in August were highly correlated with numbers on the FPG removal grid ($r = 0.87$, $N = 11$, $P < 0.01$).

There was no obvious multiannual cycle on the control (Fig. 1) and FPG (Fig. 2) grids from 1976

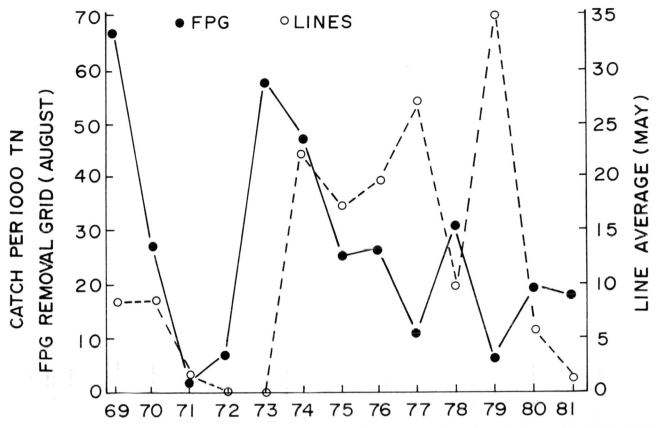

Fig. 2.—Density indices (catch per 1,000 trap nights) for *Microtus pennsylvanicus* captured near Pinawa, Manitoba from 1969 to 1981 on snap-trap lines set in spring and on a 1.1-ha removal grid set in summer.

on. This change in density patterns on the oldfield occurred well after the termination of occasional removal trapping on an adjacent live-trapping grid (summer 1973, the Depleted grid mentioned in Iverson and Turner, 1974). Small variations in plant productivity, species composition, and phenology on this particular oldfield (Turner and Iverson, 1980) also seemed insufficient to account for this shift in demography. Finally, a continuation of the live-trapping program on six artificial "ZEUS" meadows (Turner and Iverson, 1976) indicated a possible return to the increase phase in 1978, with densities of about 120 to 170 voles/ha observed in autumn 1978 on three grids (unpublished data). Spring snap-trap-line indices also showed an apparent return to a multiannual cycle with a strong peak and subsequent decline following 1979 (Fig. 2). Given these various trends, it seems unlikely that the unusual demographic pattern on the oldfield was a result of the trapping program itself.

The unusual practices of overnight removal of voles and all-season trapping did not appear to have

a significant effect on demography. In 24,615 captures, only 1.9% of the voles were captured dead or died overnight in captivity. Recruitment indices (grand average of 1.05 new voles captured at t + 4 weeks per lactating female at t weeks) were similar to what one would expect from a live-trapping study of *Microtus* (Krebs et al., 1969).

BREEDING

Initiation and termination of breeding was a rapid process, with a *separate* transition from 0 to nearly 100% reproductive activity in each sex in one or two trapping periods (Table 1). Males often came into reproductive condition before snowmelt, sometimes well in advance of the initiation of breeding in females. Termination of breeding was more variable than initiation. Males came out of reproductive condition at about the same time as females, sometimes well before permanent snow cover was established. Variations in these patterns produced breeding seasons of 25.9 ± 2.5 (range: 22–30) weeks for males, and 23.4 ± 2.6 (18–30) weeks for females.

Table 1.—*Dates for the median initiation and termination of breeding and the disappearance and establishment of permanent snowcover.*

| Year | Snow disappears | >50% of voles reproductively active | | Snow established | >50% of mature voles reproductively quiescent | |
		Males	Females		Males	Females
1968	April 1	April 15	April 15	December 3	November 25	October 28
1969	April 13	April 14	April 28	November 14	September 15	September 29
1970	April 17	March 30	May 13	November 17	October 13	October 13
1971	April 12	April 13	April 27	October 31	—†	November 9
1972	April 17	—	April 24	October 26	November 6	October 23
1973	March 26	March 26	March 26	November 4	October 22	October 22
1974	April 23	April 8	April 22	November 18	October 7	September 23
1975	April 25	April 7	May 20	November 19	September 22	October 20
1976	April 12	March 23	April 19	November 26	September 21	September 7
1977	April 1	March 28	May 2	November 16	October 18	October 3
1978	April 24	March 20	May 1	November 13	—	—

† The few males alive at this time retained scrotal testes all winter.

Notes: 1) Snow dates accurate to ±1 day, data from main meteorological tower in oldfield at Whiteshell Nuclear Research Establishment. 2) Breeding dates assumed to be accurate to ±1 week (given 2-week trapping rotation). 3) Reproductive activity in females is based on vagina perforation, pregnancy or lactation; hence, termination of breeding typically corresponds to the time when the last litters have been weaned, whereas initiation typically corresponds to the time when females come into estrus.

In females, the three longest breeding seasons occurred during the buildup in population density from 1971 to 1973 (Fig. 1). On a population basis, these year to year variations in breeding season lengths may have been far less variable than what one would assume from Table 1. Peaks in the recruitment of new voles occurred at roughly the same time each year, indicating that the majority of females were weaning litters at the same time each year.

Exceptions to this seasonal breeding pattern occurred in three years. In 1971–1972, the few males still alive in autumn retained scrotal testes through the entire winter. No reproductive activity was detected in the two or three females alive at this time. In January and February of 1969 and 1974, a short bout of breeding activity occurred in some of the overwintering voles. Nine and seven lactating females, respectively, were recorded during these winters.

DENSITY FLUCTUATIONS

Changes in density corresponded to two multiannual cycles followed by at least two annual cycles (Table 2, Fig. 1: 1968 (decline or low), 1969 (increase), 1970 (peak), 1971 (decline), 1972 (low), 1973 (increase), 1974 (peak), 1975 (decline or transition to annual phase), 1976 (annual), 1977 (annual), 1978 (annual or possible increase, data incomplete)). With the exception of the severe population decline in the winter of 1970–1971, winter was not a major factor in the decline of population density (Table 2). An increase in density over winter occurred in some years, largely due to immigration exceeding emigration in spring (Fig. 1). Major yearly declines occurred from late winter, just prior to the start of breeding, up to late summer. These breeding season declines resembled an extended form of the typical spring decline seen in many microtine populations. The declines included a small, temporary recovery in numbers in late June or early July when first-litter young of the year entered the trappable population. The declines extended well past the expected initial recruitment of young born to mature young of the year. These third generation voles did not survive well to the age of recruitment, unless they were born relatively late in the breeding season. Thus, except in increase years, numbers rose substantially only very late in the breeding season (Fig. 1).

Although net density changes were different in each breeding season (Table 2), patterns of recruitment were regular. All years had a distinct bimodal influx of new voles, with peaks near the end of June and the end of September. A third peak in the capture of new voles was also evident around snowmelt and the initiation of breeding. This peak was entirely due to the movements of overwintered voles. Meadow voles in Manitoba are known to move from grassland to forest in winter, and to return to grassland in spring (Iverson and Turner, 1972; Turner et al., 1975). Considerable dispersal within grassland habitats also occurs in spring. In autumn, neither phenomenon could be detected with certainty, as normal recruitment was occurring at the same time.

ATTAINMENT OF SEXUAL MATURITY

Based on autopsies of 1,076 voles captured during peak breeding activity (June and July), males attained sexual maturity at a median weight of 23.6

Table 2.—*Demographic statistics for* Microtus pennsylvanicus *tabulated for the summer recruitment period and the following winter non-recruitment period.*

Year	Start	Finish	Summer recruitment period Duration (weeks)	Rate of change††	Relative change§	Winter non-recruitment period† Duration (weeks)	Rate of change††	Relative change§
1968	June 24	November 25	22	−0.038	0.66	26	0.021	1.32
1969	May 26	November 10	24	0.112	3.82	28	−0.011	0.86
1970	May 26	November 9	24	−0.018	0.81	32	−0.168	0.07
1971	June 22	November 23	22	−0.057	0.53	32	0.099	4.88
1972	July 4	December 18	24	−0.007	0.92	20	0.036	1.44
1973	May 7	December 3	30	0.121	6.17	26	−0.046	0.55
1974	June 3	October 21	20	0.056	1.76	34	−0.002	0.65
1975	June 16	November 4	20	−0.009	0.92	30	−0.014	0.82
1976	May 31	November 1	22	0.026	1.34	30	−0.027	0.67
1977	May 30	December 12	28	0.015	1.23	26	0.005	1.07

† Some winter recruitment occurred from 1969 January 8 to February 17, and from 1974 January 14 to February 25.
†† Instantaneous rate of change in population density per 2 weeks.
§ Ratio of density at end of period over density at start of period.

g, and females at 18.8 g (based on probit analysis of log-transformed weights). The transition to sexual maturity occurred over a moderately wide range, with 13% of the males and 22% of the females mature at 18 g, and 79% of the males and 85% of the females mature at 24 g (actual data). Hence, even with this large sample size, 95% fiducial limits for the median weight at sexual maturity were wide— 20.2 to 27.4 g for males, and 17.7 to 20.1 g for females.

Assessment of the proportion of voles maturing in the live-trapped population was prone to numerous potential errors (possibility of differential survival, dispersal, trappability, and detection of maturity), and was, therefore, dependent on both biological and statistical features of the data. However, general patterns reflecting presumed biological processes were clearly evident. Most young of the year, first captured early in the breeding season (May–June), matured in the year of their birth (Fig. 3A) and did not survive to breed in the next year (Fig. 3B). The fraction of early-born voles that matured never quite reached 1.0, probably because of the inclusion of a few voles too young to show signs of sexual maturity, and the difficulties of scoring reproductive information in live voles. For example, by excluding voles caught once only and those weighing < 20 g or ≥ 30 g at first capture, there was only a 68% (males) and 54% (females) chance of detecting maturity at first capture (G = 8.2, P < 0.005) in voles that showed signs of maturity in the year of their birth. With an average of only 3.5 captures per vole overall, and many single captures, the chances of detecting maturity were not good. As would be expected, exclusion of voles weighing <20

g at first capture noticeably elevated the mature proportion (Fig. 3A).

Problems in assessing sexual maturity were greatest at the end of the breeding season, when immigrant voles coming out of breeding condition were similar in weight to legitimate recruits (Iverson and Turner, 1974). Exclusion of light-weight voles at this time strongly biased the data towards presumed immigrants (up to 80% of the data would be excluded as opposed to 20% early in the season). Consequently, the data suggest that some voles matured late in the breeding season (Fig. 3A). As in the case of less than 100% attainment of maturity early in the breeding season, I believe that these trends reflect methodological rather than biological features of the data, and conclude that all voles first captured late in the breeding season (September on) did not mature in the year of their birth (Fig. 3A).

Between these hypothesized periods of absolute attainment and non-attainment of sexual maturity, there was a transitional period that probably included most voles first captured in July and August (Fig. 3A). Voles captured at this time appeared to have an option as to whether they did or did not attain sexual maturity. Irrespective of attainment of maturity, all voles were equally likely to be reproductively successful in the following year (based on G tests by month of first capture for each sex). The power of these statistical comparisons, however, was poor, since the chances of being reproductively successful in the following year were low (Fig. 3B: 0.00 to 0.08 for July; 0.07 to 0.14 for August). Unlike patterns early and late in the breeding season, the attainment of sexual maturity in this transitional period was not independent of the population cycle.

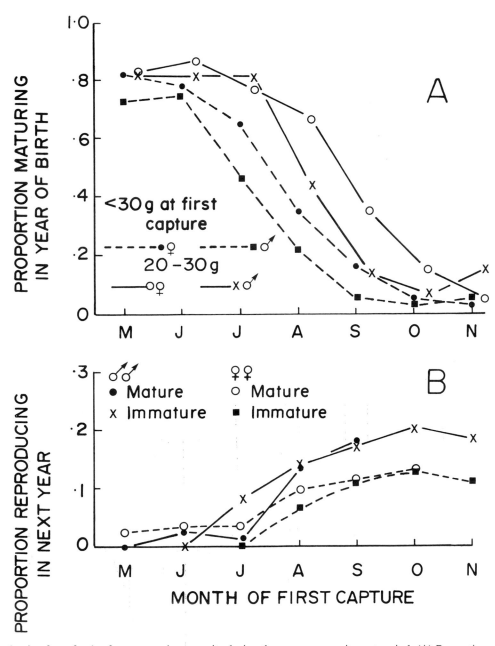

Fig. 3.—Reproductive fate of voles first captured as recruits during the summer recruitment period. (A) Proportions showing signs of sexual maturity in the year of birth. (B) Proportions reproductively successful in the following year of those weighing <30 g at first capture.

Voles born in years of low or increasing density were more likely to mature than voles born in years of high density (Table 3). Inhibition of sexual maturity was most pronounced in males during the peaks of 1970 and 1974, but also occurred inexplicably in 1976.

LIFE EXPECTANCY

Survival from birth to first capture was estimated by the usual juvenile recruitment index (voles < 30 g at first capture at time t + 4 weeks divided by lactating females at time t) divided by an estimate of the mean litter size of 5.6 (Iverson and Turner,

1976). Juvenile survival varied little from year to year, with an overall estimate of 18.8%. As noted previously, juvenile survival had a strong seasonal trend, with good early survival, poor mid-season survival, and good late-season survival. In contrast, survival after first capture showed strong multiannual trends. Life expectancies of recruits after first capture varied from 4.4 (1971) to 18.3 (1969) weeks with a grand average of 14.4 weeks (N = 3,009). Among population categories, mature females most clearly showed variations in life expectancy that paralleled the variations in population density (Fig. 4). Life expectancies for immature males and females tracked those of mature females until 1973; after that, their life expectancies remained relatively constant, between 14 and 18 weeks. In sharp contrast, life expectancies for mature males were uniformly low (8.9 weeks) in all years, with little, if any, variation related to population density.

LIFE HISTORIES

Population events were reduced to two simple indices: *survival* to the end of recruitment in the year of birth and to the beginning of recruitment in the following year, and *reproductive success* in the year of birth and in the following year (Table 4). From Table 4, one can see that immigrants accounted for a major portion (52%) of all voles captured. The large number of immigrants does not appear to be related to trapping efficiency, as mean weights at first capture averaged 20–25 g, and trappability was similar to that found in other studies (78% of known alive captured in summer, 54% in winter—see Krebs et al., 1969, for *M. pennsylvanicus* in Indiana).

YEAR OF BIRTH

Since weight and sexual maturity were closely related, by definition, all summer immigrants were necessarily mature. Hence, all immigrant males were naturally defined as reproductively successful. Female immigrants were as reproductively successful as female recruits (50 versus 51%). Even during the high densities of July and August of 1970 and 1974, 42% of immigrant females (N = 105) were reproductively successful. Immigrant females were, nevertheless, slightly less likely to survive to the end of recruitment than mature recruits (25% versus 30%, G = 4.2, P < 0.05). In males, there were more pronounced differences in the life histories of immigrants versus mature recruits. Immigrants were much

Table 3.—*Attainment of sexual maturity in the year of birth for recruits first captured in July or August.*

Year	Cycle phase	Males		Females	
		N	% mature	N	% mature
1970	peak	46	6.5	66	24.2
1974	peak	64	12.5	61	37.7
1976	annual	21	14.2	22	45.5
1969	increase	39	33.3	49	75.5
1975	decline/ transition	44	40.9	40	47.5
1977	annual	26	46.2	26	76.9
1973	increase	57	54.4	59	59.3
1972	low	15	73.3	10	90.0
1971	decline	9	77.7	4	75.0
1968	low/decline	16	87.5	30	76.7

less likely to survive to the end of recruitment (12% versus 25%, G = 19.7, P < 0.001), and hence accounted for only 40% of the mature males present at that time. Immature recruits of both sexes necessarily survived well to the end of recruitment (Fig. 3A), since most were born towards the end of the breeding season. Autumn populations at the end of recruitment were, therefore, composed largely of immature recruits (85% males, 63% females).

YEAR AFTER BIRTH

On average, 28% of the voles present at the end of recruitment were still present at the start of recruitment the following year. This corresponds to a 24-week minimum survival of 0.33 ± 0.08 (SE), assuming a constant rate of change, and adjusting each year's rate to a 24-week standard. Given the small declines in density seen in most winters (Fig. 1), the general lack of significant winter breeding, and the extremely large number of apparent winter immigrants (2,251), "survival" of marked voles was grossly underestimated. If survival is estimated by including all new immigrant voles present in spring in the numerator, and setting the maximum possible survival to 1.0, a maximum 24-week survival of 0.79 ± 0.10 is obtained.

The existence of considerable winter turnover in a vole population is unusual and requires further comment. Similar apparent winter dispersal has recently been confirmed by more efficient routine trapping of six isolated "ZEUS" meadows (unpublished data); hence, it appears to be a real process, and not an artifact of trappability or methodology. Data from three "ZEUS" grids trapped when the

LIFE EXPECTANCY FROM FIRST CAPTURE (WEEKS)

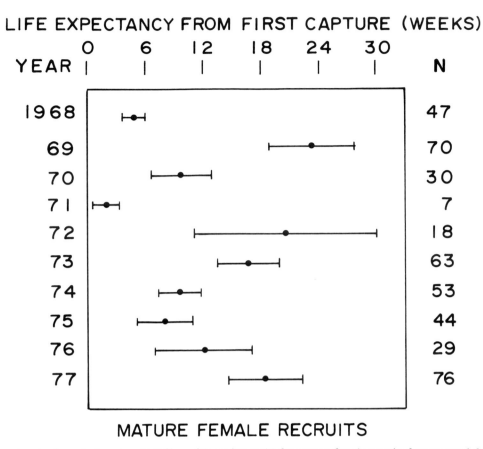

MATURE FEMALE RECRUITS

Fig. 4.—Life expectancies from first capture (±95% confidence intervals) for mature female recruits first captured during the summer recruitment period.

vole density was high (1978–1979), and the average of results from six grids trapped when the vole density was low (1980–1981), when combined with the oldfield control data, indicate a general negative relationship between survival rates of marked voles and the proportion of immigrants present in spring. In simpler terms, a general positive correlation exists between the number of voles, marked from the previous year, that disappear in winter, and the number of new voles that immigrate in winter and are still present in spring (Fig. 5; $r^2 = 0.54$). From this relationship, I conclude that roughly half of the voles that disappeared during winter on the control grid successfully dispersed to another area and did not die until the next breeding season. This generalization obviously did not apply to the winter of 1970–1971 (Fig. 5).

If a vole survived to the end of recruitment, its chances of "surviving" the non-recruitment period were good, regardless of its reproductive or residency status in the year of its birth (males: 18 to 21.5%,

$G = 1.4$, ns; females: 24 to 32%, $G = 6.3$, $P < 0.05$; Table 4). Hence, reproductive success in the year after birth depended mainly on when a vole was born (Fig. 3). Summer immigrants and mature recruits, born early in the season, had little chance of being reproductively successful the following year (2.5 to 5.7%). Immature recruits, born late in the season, had a better chance of being reproductively successful the following year (males: 16.6%, $G = 93.8$, $P < 0.001$; females: 9.6%, $G = 28.5$, $P < 0.001$). Similarly, reproductive success in winter immigrants was largely a function of surviving long enough to be present when breeding activity resumed in spring. Immigrant males first captured in the months from December through April had a 18, 17, 29, 46, and 69% chance, respectively, of reproductive success in the following year. Reproductive success was not clearly a function of survival to spring in female winter immigrants, as they had a 10, 6, 14, 18, and 31% chance of reproductive success after first capture from December through April,

Table 4.—*Voles present at the end of recruitment in the year of their birth and at the beginning of recruitment in the following year and reproductive success for all voles captured.*

Voles present	Year of birth		Following year		% reproductively successful of all captured	
	No. captured	No. present at end of recruitment	No. present at start of recruitment	% present at start of recruitment of those present at end of recruitment	Year of birth	Following year
Males						
Mature recruits	311	79	17	21.5	100.0	4.8
Immature recruits	1,329	760	186	24.5	0.0	16.6
Summer immigrants	408	50	9	18.0	100.0	2.5
Winter immigrants	1,301	—	505	—	—	51.0
Females						
Mature recruits	442	134	43	32.1	51.1	5.7
Immature recruits	927	484	170	35.1	0.0	9.6
Summer immigrants	603	148	36	24.3	49.9	3.0
Winter immigrants	950		340	—	—	23.6

Note: Trap deaths, animals caught during irregular trapping prior to spring 1968, and animals for which life histories were not followed to completion (voles caught in 1978), were excluded from tabulation.

respectively. Most winter immigration occurred around snowmelt at the start of breeding activity in males, but not in females. Forty-seven percent of all winter immigrant males and 33% of all winter immigrant females were first captured in March and April.

DISCUSSION

The data in this paper indicate that a *Microtus* population can exhibit both multiannual and annual density fluctuations. This is a new pattern that has not been reported to date. Given this apparent flexibility, it is of paramount importance that we separate population events that are annually repeatable from those that appear to be unique to certain phases of the multiannual cycle.

In this boreal vole population, many demographic events were naturally tied into a seasonal reproductive cycle. Sporadic and limited winter reproduction was noted in low and increase phases, but it did not contribute significantly to numerical changes, in contrast to what is observed in populations in more temperate climates (Krebs et al., 1969; Tamarin, 1977; Rose, 1979). Winter, with one important exception, was not a significant period of mortality, and hence, spring densities were often moderately high. Severe overwinter mortality in *M. pennsylvanicus* has been noted only in a Minnesota population that inhabits a seasonally flooded, wetland habitat, and hence, cycles annually (Baird, 1980). In Manitoba, the enigmatic winter decline of 1970–1971 was the only completely devastating population event in 11 years of trapping. The analogous decline in 1974–1975 did not start until early spring

and appeared to be only an accentuated form of the spring decline seen every year.

Outside of winter, population demography was strongly affected by moderately synchronous and intensive reproduction, with only minimal evidence for limitation of reproductive output. Breeding seasons were slightly flexible in length, as observed in other microtines (Krebs and Myers, 1974), but did not vary in length to the extent that they could account for the major density changes observed. Similarly, I found no evidence for reduced reproductive activity in overwintered voles or early season young of the year. The population was also highly permeable to mature immigrants in all years. Resident and immigrant females were apparently capable of raising litters at peak densities of up to 100 voles/ha.

In common with other boreal rodents (Viitala, 1977; Gyug and Millar, 1981; and others), inhibition of sexual maturity in midsummer recruits was one of the few factors potentially limiting the rate of increase at high density. Changes in median weights at sexual maturity across the population cycle in temperate climates likely represent an analogous process in continuously breeding populations. A potentially more effective limiting factor was the yearly occurrence of poor juvenile survival at mid-season,

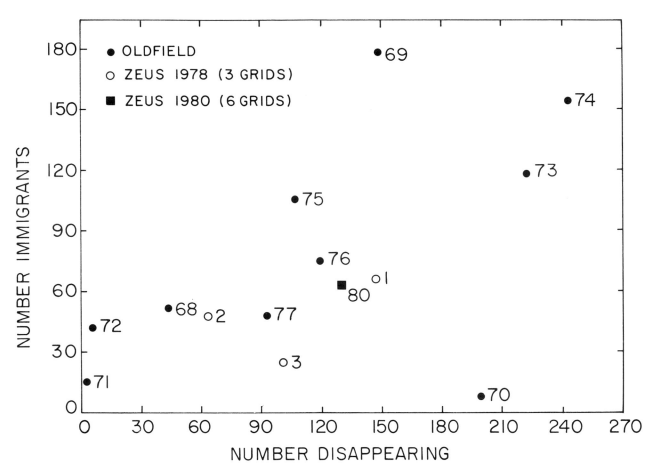

Fig. 5.—Relationship between the number of winter immigrant voles present at the start of recruitment in spring and the number of marked voles present at the end of recruitment in the previous autumn that disappear during winter (years refer to autumn, data for 3 grids at high density are presented separately for 1978, data for 6 grids at low density are combined in one point for 1980).

during hot and dry summer months. This phenomenon has also been observed in subarctic *M. agrestis*, and has been attributed to a nutritional midsummer crisis (Myllymäki, 1977). Alternative explanations, based on the effects of botfly parasites on reproduction (Boonstra et al., 1980), or the effects of mature voles on nestlings through infanticide (Mallory and Brooks, 1978), have also been suggested. Neither of these hypotheses appears to be appropriate for meadow voles (Iverson and Turner, 1968; Boonstra, 1980).

In general, the yearly midsummer crisis and the inhibition of sexual maturity in peak years had a limiting, rather than a regulating, effect on population density. Manitoba *M. pennsylvanicus* peaked at about 100 voles/ha, a density slightly lower than the peaks seen in southern populations (Krebs

et al., 1969; Tamarin, 1977; Rose, 1979). Most of the gross multiannual variation in density was accounted for by changes in the life expectancy of voles *after first capture*. Voles born in increase years survived well; voles born in peak and decline years survived poorly (Fig. 4). Without a major drop in life expectancy from 1975 on, numerical trends necessarily reflected only the annual patterns resulting from seasonal reproduction, recruitment and the attainment of sexual maturity. This observation is not particularly new, but has often been obscured by the complex statistical appraisal of demography in continuously breeding populations of *Microtus* (Krebs et al., 1969; Gaines and Rose, 1976; Tamarin, 1977).

Given this apparent relationship between life expectancy and multiannual density fluctuations, future studies should specifically examine the proximate and ultimate causes of changes in viability. Even with studies such as those of Hilborn and Krebs (1976) and Beacham (1980), we still know very little about the causes of death across the population cycle. In most studies, population events are inferred from conventional live-trapping, rather than observed directly. Although this problem cannot be completely eliminated, we can endeavour to both manipulate and observe populations in the field in such a way that we do not always have to rely on secondary statistics. The use of pitfall traps in optimal/suboptimal enclosures (Beacham, 1980) and intensive radiotelemetry of natural populations (Madison, 1980) are two of the currently available techniques that should be pursued. With a better understanding of both the proximate and ultimate causes of death in microtines, we can develop the potential to understand the processes that we can now only infer from conventional live-trapping techniques.

ACKNOWLEDGMENTS

I thank Bill Schwartz for much of the actual work involved in trapping the voles, and Stu Iverson and Brian Turner for their foresight in continuing the trapping program for as many years as they did.

LITERATURE CITED

ABRAMSKY, Z., and C. R. TRACY. 1979. Population biology of a "noncycling" population of prairie voles and a hypothesis on the role of migration in regulating microtine cycles. Ecology, 60:349–361.

BAIRD, D. D. 1980. Dispersal in meadow voles, *Microtus pennsylvanicus*. Unpublished Ph.D. dissert., Univ. Minnesota, Minneapolis, 111 pp.

BEACHAM, T. D. 1980. Survival of cohorts in a fluctuating population of the vole *Microtus townsendii*. J. Zool. London, 191:49–60.

BOONSTRA, R. 1980. Infanticide in microtines; Importance in natural populations. Oecologia (Berl.), 46:262–265.

BOONSTRA, R., C. J. KREBS, and T. D. BEACHAM. 1980. Impact of botfly parasitism on *Microtus townsendii* populations. Canadian J. Zool., 58:1683–1692.

GAINES, M. S., and R. K. ROSE. 1976. Population dynamics of *Microtus ochrogaster* in eastern Kansas. Ecology, 57:1145–1161.

GARSD, A., and W. E. HOWARD. 1981. A 19-year study of microtine population fluctuations using time-series analysis. Ecology, 62:930–937.

GYUG, L. W., and J. S. MILLAR. 1981. Growth of seasonal generations in three natural populations of *Peromyscus*. Canadian J. Zool., 59:510–514.

HENTTONEN, H., A. KAIKUSALO, J. TAST, and J. VIITALA. 1977. Interspecific competition between small rodents in subarctic and boreal ecosystems. Oikos, 29:581–590.

HILBORN, R., and C. J. KREBS. 1976. Fates of disappearing individuals in fluctuating populations of *Microtus townsendii*. Canadian J. Zool., 54:1507–1518.

IVERSON, S. L., and B. N. TURNER. 1968. The effect of *Cuterebra* spp. on weight, survival and reproduction in *Microtus pennsylvanicus*. Manitoba Entomologist, 2:70–75.

———. 1969. Under-snow shelter for small mammal trapping. J. Wildlife Mgmt., 33:722–723.

———. 1972. Winter coexistence of *Clethrionomys gapperi* and *Microtus pennsylvanicus* in a grassland habitat. Amer. Midland Nat., 88:440–445.

———. 1973. Habitats of small mammals at the Whiteshell Nuclear Research Establishment. Atomic Energy of Canada Limited Report AECL-3956, 51 pp.

———. 1974. Winter weight dynamics in *Microtus pennsylvanicus*. Ecology, 55:1030–1041.

———. 1976. Small mammal radioecology: Natural reproductive patterns of seven species. Atomic Energy of Canada Limited Report AECL-5393, 53 pp.

KREBS, C. J. 1979. Dispersal, spacing behaviour, and genetics in relation to population fluctuations in the vole *Microtus townsendii*. Fortschr. Zool., 25:61–77.

KREBS, C. J., B. L. KELLER, and R. H. TAMARIN. 1969. *Microtus* population biology: demographic changes in fluctuating populations of *M. ochrogaster* and *M. pennsylvanicus* in southern Indiana. Ecology, 50:587–607.

KREBS, C. J., and J. H. MYERS. 1974. Population cycles in small mammals. Adv. Ecol. Res., 8:267–399.

MADISON, D. M. 1980. Space use and social structure in meadow voles, *Microtus pennsylvanicus*. Behav. Ecol. Sociobiol., 7:65–71.

MALLORY, F. F., and R. J. BROOKS. 1978. Infanticide and other reproductive strategies in the collared lemming, *Dicrostonyx groenlandicus*. Nature, 273:144–146.

MYLLYMÄKI, A. 1977. Demographic mechanisms in the fluctuating populations of the field vole *Microtus agrestis*. Oikos, 29:468–493.

ROSE, R. K. 1979. Levels of wounding in the meadow vole, *Microtus pennsylvanicus*. J. Mamm., 60:37–45.

ROSENZWEIG, M. L., and Z. ABRAMSKY. 1980. Microtine cycles: The role of habitat heterogeneity. Oikos, 34:141–146.

STENSETH, N. C. 1980. Spatial heterogeneity and population stability: Some evolutionary consequences. Oikos, 35:165–184.

TAMARIN, R. H. 1977. Demography of the beach vole (*Microtus breweri*) and the meadow vole (*Microtus pennsylvanicus*) in southeastern Massachusetts. Ecology, 58:1310–1321.

TURNER, B. N., and S. L. IVERSON. 1973. The annual cycle of aggression in male *Microtus pennsylvanicus*, and its relation to population parameters. Ecology, 54:967–981.

———. 1976. Project ZEUS: A field irradiator for small-mammal population studies. Atomic Energy of Canada Limited Report AECL-5524, 22 pp.

———. 1980. Seasonal and annual plant production of a southeastern Manitoba oldfield. Atomic Energy of Canada Limited Report AECL-6792, 29 pp.

TURNER, B. N., M. R. PERRIN, and S. L. IVERSON. 1975. Winter coexistence of voles in spruce forest: Relevance of seasonal changes in aggression. Canadian J. Zool., 53:1004–1011.

VIITALA, J. 1977. Social organization in cyclic subarctic populations of the voles *Clethrionomys rufocanus* (Sund.) and *Microtus agrestis* (L.). Ann. Zool. Fennici, 14:53–93.

Address: Environmental Research Branch, Whiteshell Nuclear Research Establishment, Atomic Energy of Canada Limited Research Company, Pinawa, Manitoba R0E 1L0, Canada. *Issued as AECL-7469.*

WINTER ECOLOGY OF SMALL MAMMALS IN THE URAL MOUNTAINS

Vladimir N. Bolshakov

ABSTRACT

Observations were made of winter ecology of small rodents inhabiting the upper belts of the Ural Mountains. Animal transitions, fur insulation properties, changes in the relative weight of some inner organs and population age structure dynamics were examined.

INTRODUCTION

A very interesting region for a zoologist is the Ural Mountain chain, an area that stretches for over 2,000 km in a meridianal direction through several zones, from tundra in the North to steppes in the South. Floral belts are distinguished in the Urals as follows: a) Montane-steppe; b) montane-forest-steppe; c) montane-forest; d) subaltitudinal; e) altitudinal. A stable snow cover in the upper belts begins on approximately September 16. Its maximal thickness is reached between February and March. According to observations, snow cover thickness in the northern part of the middle Urals increases 17 to 18 cm with every 100 m increase in elevation. In the altitudinal belt of the Denezhkin Kamen Mountain, snow covered the ground for about 239 to 268 days in some years.

There are three groups of small mammals in the Urals: 1) Those characteristic of a certain altitudinal and landscape zone (in the Urals, these are mainly steppe and tundra species); 2) Dwellers of all landscape zones and corresponding altitudinal belts (for example, *Clethrionomys rutilus, Sorex* spp.); 3) Animals that are characteristic of upper belts only (for example, *Ochotona alpina hyperborea, C. rufocanus*).

In the upper belts, small mammals are active year-round. Our observations have shown that in the alpine belt (as distinguished from the plain) voles were almost never found on the snow cover, but actively moved below it among stones.

METHODS

We investigated small mammal winter ecology by mass marking with Rodamine-C and Rodamine-6G. The dyes were added to 1 cm³ pieces of bread or fried dough. Such bait was eaten no less readily than the undyed bait. No deaths were recorded during the experimental period. Animals succeeded in reproduction and many of them survived more than one year. Rodamine-C could be found in urine, intestines, and on snouts, chests and forefeet of the voles for seven days after the bait was eaten. Rodamine-6G was noticed only in the course of two days; integuments were less dyed. Experience has shown that dyed bait is better if spread in deep areas of subnivean paths. Phosphorus-32, Calcium-45 and Strontium-89 osteotropic isotopes were also widely used in marking. Addition of Sodium-22, Zinc-65, Ferrum-59, Cobalt-60, and Phosphorus-32 to the bait was also practiced.

The methods helped to reveal territorial distribution of animals in the mountains. Fig. 1 shows the line transects for registration of migration of rodents in the Ural Mountains. From the top of the Kukshik Mountain, a stone field extended far into the forest belt. We placed dyed bait on subnivean paths throughout the stone field. Traps were placed at a distance of 5 to 100 m and sheltered in ditches within the snow. Captures of 22 northern red-backed voles (*C. rutilus*) and grey-sided voles (*C. rufocanus*) from 2 to 9 February 1978 evidenced a marked concentration of animals in the stone field in winter. In spring, summer and autumn, voles moved up to 1.5 km away from the stone field.

Thawed patches were recorded in the top region of the Bakhty chain in late April 1979. The greater part of the slope was covered with deep snow, névé and ice crumbs. We tried to learn the role of the first thawed patches in the life of overwintered animals involved in reproduction. Dyed bait was spread on a 0.5 ha thawed patch. Four parallel trap lines were arranged below it (100 m distance between lines, 25 traps per line at a distance of 8 to 12 m). Traps were placed among stones under the snow. The shortest distance from the thawed patch to the nearest line was 50 m. The most remote line was near the fooded zone edge down to the stone field. A stream crossed the central part of the lines.

RESULTS

Twelve hours after the dye was spread, eight overwintered voles were caught (five *C. rufocanus*, three *Clethrionomys glareolus*). Only two females showed no dye at dissection, one being at a late stage of pregnancy, the other lactating. The dye was carried by animals no farther than 300 m. Fifteen voles were

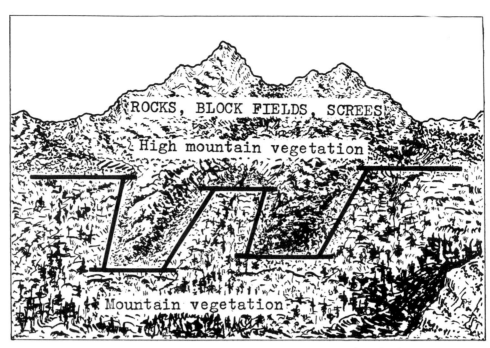

Fig. 1.—Scheme of line transects for registration of migration of rodents.

caught over a four day period—only four females lacked the dye (nursing or pregnant). All overwintered males caught in the lines had consumed the dyed food. Catches evidenced distribution on a territory of over 6 ha and nearly simultaneous visits to the 0.5 ha thawed patch.

DISCUSSION

Of special interest in studying small mammal winter ecology are their adaptive patterns. The thermoregulatory role of the snow cover in the life of small mammals in winter is well known. Measurements made in the Polar Urals have shown that in December, when a 100 m mantel of snow covered the mountains and surrounding air temperature fell to −45° to −48°C at night, subnivean air temperature was stable at −7° to −8°C. Daily fluctuations in air temperature below the snow cover were at most 3–4°C, while those above it were 10°C fluctuations. Although some authors believe that small mammal body sizes make their physical thermoregulation insignificant, some data indicate the opposite. Thus, it was shown on *Peromyscus maniculatus* that better insulation of their winter fur in comparison with summer pelage was the main adaptation of animals to prevent cooling in winter (Hart and Heroux, 1953). These authors also noted that as fur insulation increased, a lower metabolism was seen in mice. The best heat conserving properties of skins were recorded by Polish zoologists for the bank vole in higher altitudes (Kostelecka-Myrcha et al., 1970). The importance of physical as well as chemical thermoregulatory adaptations of small mammals to unfavorable surroundings was outlined by many authors.

We investigated insulation properties of skins of small rodents inhabiting higher altitudes with the help of an isolated temperature recorder. Fig. 2 reviews heat insulation of summer and winter fur for six species of rodents in terms of their heat transfer coefficient. Basically, similar insulation properties were recorded for all species (*C. glareolus, Microtus oeconomus, Microtus arvalis, Apodemus agrarius, Apodemus sylvaticus*) during summer. It is known that fur insulation qualities undergo significant seasonal changes. Thus, in summer *P. maniculatus* is 21.4% less insulated than in winter (Hart, 1956). Our data for mice and voles show an 11 to 12% difference. Besides good fur insulation, many highland rodent species have peculiar chemical thermoregulation (for example, constant body temper-

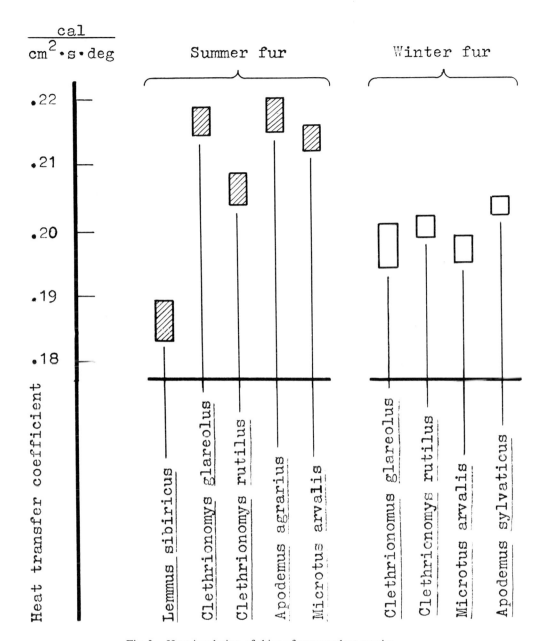

Fig. 2.—Heat insulation of skins of some rodent species.

ature), evidently connected with an improved physical thermoregulation. Experiments on *C. glareolus* from the upper belts of the Urals have shown that the critical level of their gas exchange is almost 45% lower in summer than in winter.

Studies of morphological structure of summer and winter skins have revealed regularities common for rodents inhabiting the Ural Mountains: increase of hair thickness towards winter and changes in the

hair structure and in hair root ratio. Thus, in summer, some 300 hair per 4 mm² can be counted on the back of a bank vole; in winter the number is 422 (sides = 280 and 366, respectively; belly = 220 and 280). In winter, guard hairs are more than two times thinner, but 1.5 to 1.8 times longer (from 8.4 mm to 11.5 mm on the back).

Method of morphophysiological indicators developed by Schwarz (1980) is widely used in the

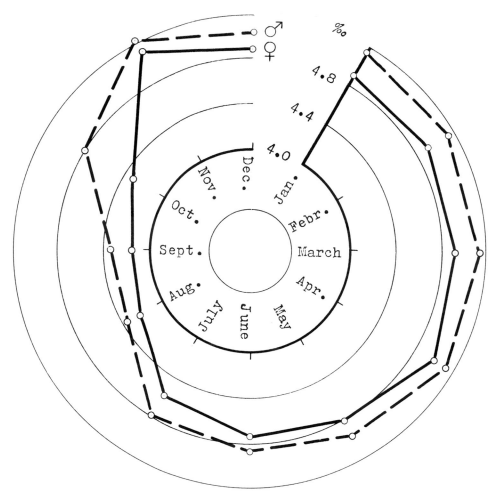

Fig. 3.—Seasonal changes of the relative weight of heart in *Clethrionomys rufocanus.*

Soviet Union for research on animal adaptive patterns. Briefly, the method is viewed as follows: relative weight of inner organs, blood composition and other indices are indicators of animal physiological adaptation. Our material on seasonal changes in relative heart weights indicates that variability of the bank and northern red-backed voles taken from the plain and upper belts roughly coincide. One can see reduction in relative heart weight from winter towards summer (minimum in summer months, rise in autumn and stabilization in the winter period). However, this regularity is displayed differently in subaltitudinal and altitudinal zones on one hand and in the plain forest zone on the other. Firstly, all changes are shifted in time, especially in autumn. In plain rodents heart indices slowly rise from July to November (from 5.1 ± 0.13 to $5.6 \pm 0.10\%$ in male and from 5.4 ± 0.12 to $5.8 \pm 0.11\%$ in female

bank voles; from 6.5 ± 0.12 to $7.0 \pm 0.22\%$ in male and from 6.7 ± 0.14 to $7.5 \pm 0.18\%$ in female northern red-backed voles) and stabilize at the higher level reached.

In voles of the upper belts the index is stable during July, August and perhaps early September. In October a sharp increase is noticed (from 5.9 ± 0.11 to $7.1 \pm 0.15\%$ in male and from 6.7 ± 0.20 to $7.5 \pm 0.17\%$ in female bank voles; from 7.9 ± 0.24 to $8.6 \pm 0.13\%$ in male and from 7.3 ± 0.18 to $8.3 \pm 0.17\%$ in female northern red-backed voles). The high level is retained during the entire winter. As this is accompanied by a sharp increase in hemoglobin content and in relative kidney weight, one can speak of a strongly pronounced physiological reorganization connected with winter approach in the mountains. Seasonal changes of the relative weight of heart and kidney of *C. rufocanus* are shown

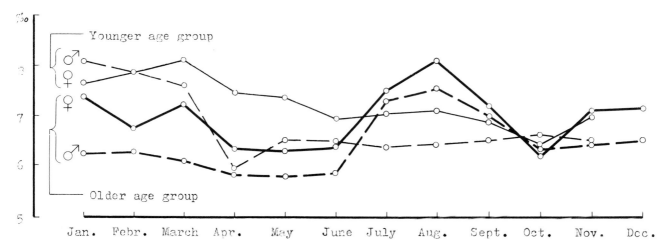

Fig. 4.—Seasonal changes of the relative weight of kidney in *Clethrionomys rufocanus.*

in Figs. 3 and 4, respectively. Relative heart weight changes in *C. rufocanus* are manifested peculiarly, although they are within the scheme (Fig. 3). In nature, these rodents exhibit a sharp decrease in relative heart weight in August and September (from 4.9 ± 0.26 to 4.2 ± 0.17% in males and from 5.0 ± 0.19 to 4.6 ± 0.11% in females), followed by a sharp increase in October to November (up to 5.6 ± 0.15% in males and 5.6 ± 0.19% in females). In December, the exponent somewhat decreases again and is stable in winter and early spring. Comparisons of *C. rufocanus* taken from the South and Polar Urals revealed parallel changes in their relative heart weights (voles from the Polar Urals have larger hearts), which suggests that similar reactions to seasonal changes of the environment. It is interesting to note that the most pronounced seasonal changes in the relative body weight of altitudinal *C. rufocanus* occur in the summer period, while in winter the exponents are stable. The same regularity is recorded for other internal indices which suggests considerable stable living conditions under the snow.

The last chapter concerns dynamics in the vole population age structure. Our investigations in the Ural Mountains have shown that regarding litter size, mountainous and plain populations of rodent and insectivorous species do not differ. In subarctic populations, this index is higher. For example, average litter size of the northern red-backed vole in the South Urals is 6.0 for plain, 6.1 for mountainous, and 9.8 for the Jamal tundra.

Early sex maturation is typical for mountainous populations. Their reproduction period is calendar (but not phenological), coincident with the repro-

ductive period of the same species on the plain. In different regions of the Ural Mountains, reproduction of the field and northern red-backed voles begins at the same time—late April to early May. In the upper belts, this period is shifted. Pregnant females are caught in large numbers only during the second half of May to early June, when the mountains are still snow-covered. Overwintered mountainous northern red-backed voles start breeding at average weights of 22.8 g (range, 19.3 to 25 0 g for females) and 24.3 g (range, 22.8 to 27.0 g for males). As in the Transurals, they produce two litters per year for the most part. Only single specimens have shown three litters per year. Thus, only two of seventeen old females caught in the Zigalg Range early in August could be considered fruitful as that. In July, pregnant females of the first litter are found in traps. In August they can produce one more litter. Towards late August, reproduction of the northern red-backed vole in mountainous and plain regions ceases entirely. Analysis of the population age structure made on the material obtained from the upper belts of the Denezhkin Kamen and Kosvinsky Kamen Mountains has shown that in autumn young animals of the first year accounted for some 95% of the population. It was determined by Schwarz (1980) that specificity of subarctic population age structure lies in high mortality of offspring of the first litter. In mountainous populations of the northern red-backed vole, a great number of the young born in spring survive to autumn and play an important role in population reproduction. As an illustration let us consider a *C. glareolus* population. We determined age groups by calculating molar tooth development rate and thymus size aside from weight and body size estimations. The results demonstrated that age group abundances in mountainous populations were governed by the number of animals that survived the first generations which meet winter at maturity and readily reproduced in spring. Thus, in late winter a *C. glareolus* population in the Ural Mountains consists of specimens from the first spring generation (20–25%), voles of the second and third generations (70–75%) and a small number of old animals. Animals of the third generation do not reach maturation towards winter (males and females). From the second generation only single specimens mature towards autumn.

LITERATURE CITED

HART, J. S. 1956. Seasonal changes in insulation of the fur. Canadian J. Zool., 34:53–57.

HART, J. S., and O. HEROUX. 1953. A comparison of some seasonal and temperature-induced changes in *Peromyscus*: Cold resistance, metabolism and pelage insulation. Canadian J. Zool., 31:528–534.

KOSTELECKA-MYRCHA, A., M. GEBCZYNSKI, and A. MYRCHA. 1970. Some morphological and physiological parameters of mountain and lowland populations of the bank vole. Acta Theriol., 15:1–12.

SCHWARZ, S. S. 1980. Ecological regularities of evolution. Nauka, Moscow, 277 pp.

Address: Institute of Plant and Animal Ecology, 8 Marta Street 202, 620008 Sverdlovsk, U.S.S.R.

STABILITY OF OVERWINTERING POPULATIONS OF *CLETHRIONOMYS* AND *MICROTUS* AT KILPISJÄRVI, FINNISH LAPLAND

JUSSI VIITALA

ABSTRACT

Winter survival was estimated on the basis of live trapping done in autumn and in spring at Kilpisjärvi, Finnish Lappland. *Clethrionomys rutilus, C. rufocanus,* and *Microtus agrestis* were the most numerous rodent species on the study area. The winter survival was higher than that in summer, except in *C. rutilus.* In this species territorial behaviour stabilizes the population density of breeding animal on very low level. Two overwintering strategies were found. One was associated with territorial behaviour of females through the winter. The other was characterized by high dispersal of both sexes. Possibly this was associated with mid-winter aggregations. Different species may adopt different strategies, and even one species may adopt different strategies in different situations.

INTRODUCTION

During my previous studies I have observed distinct differences in social mechanisms of breeding populations between *Clethrionomys* and *Microtus* (Viitrala, 1977, 1980). *Clethrionomys rufocanus* was characterized by great stability and distinct territoriality of females. *Microtus agrestis,* on the other hand, was characterized by instability of male portion as well as considerable emigration of mature females. Territoriality was less distinct especially in early July.

There are only few studies of note on social mechanisms of winter populations of microtines. West (1977) found mid-winter aggregations in *C. rutilus* in Alaska in one winter, and Merritt and Merritt (1978) reported communal nesting in *Peromyscus maniculatus,* but territorial behaviour in *C. gapperi* in Colorado. *Microtus xanthognathus* lives in groups of 5 to 10 individuals through the winter (Wolff and Lidicker, 1980).

On the basis of my live trapping studies conducted at Kilpisjärvi, Finnish Lappland (69°3'N, 20°49'E), I was able to make indirect suggestions on overwintering strategies of three species of small rodents: *Clethrionomys rutilus* (Pall.), *C. rufocanus* (Sund.) and *Microtus agrestis* (L.).

MATERIAL AND METHODS

The material of the present study was collected during a live trapping study directed to investigate the social ecology of *C. rutilus,* a subordinate species in the vole community of Kilpisjärvi area (Viitala, 1980). Even the territories of females may exceed 0.5 ha. Therefore as large a trapping area as possible was chosen. A 6.7 ha trap station grid was marked in subalpine birch wood (Fig. 1). The stations were with 10-m intervals, but the traps were set on every second line and row with 20-m intervals. The traps were then moved so that every trap station had one for two days—normally four inspections of the traps. Ugglan special traps (Hansson, 1967) were used.

One trapping period lasted 8 days. The traps were set on for 8 to 10 hr a day. Thus the back doors of the traps were open for 14 to 16 hr a day. The animals were then free to move in and out of them.

During the continuous daylight of the summer, the animals exhibited polyphasic activity rhythm, and the trapping was done during day time. After mid-August the animals turned night-active, and the trapping was done at night.

The animals were marked individually with toe-clipping. In every capture the species, identity, sex and sexual status were determined by visual inspection. The weights of the animals were determined at least twice during every trapping period.

The 627 individuals used in the present study have been presented in Table 1.

RESULTS

The spring populations of *C. rufocanus* and *M. agrestis* were about equal in size with the autumn ones but much smaller in *C. rutilus.* The spring density of *C. rutilus* was the same in 1973 and 1974, although the autumn density had been higher in 1973 than in 1972 (Table 1).

The winter survival seems to be high in Kilpisjärvi area. That for marked animals varied from

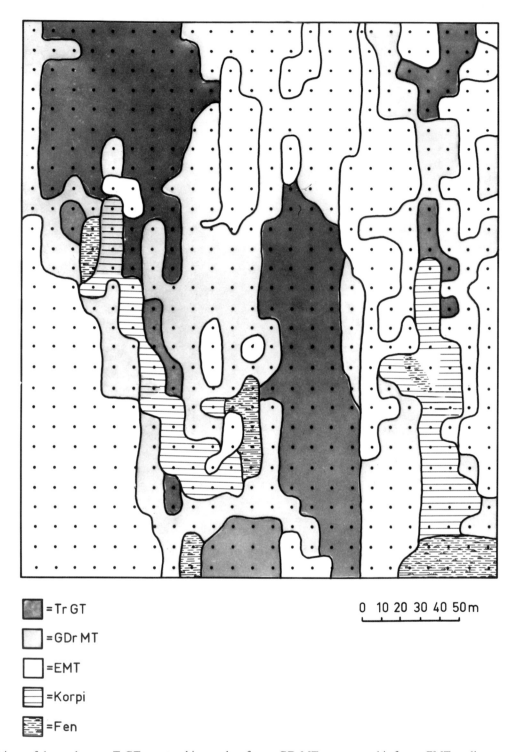

Fig. 1.—Habitats of the study area. TrGT = eutrophic meadow forest, GDrMT = mesotrophic forest, EMT = oligo-mesotrophic heath forest, Korpi = paludified eutrophic Korpi-wood, Fen = treeless fens. Trap stations are indicated by dots.

Table 1.—*Autumn (captured 28.8–22.9) and spring populations (captured before June 24) of* C. rutilus, C. rufocanus *and* M. agrestis *at Kilpisjärvi in 1972–1974. Survivors = animals marked in previous summer or autumn that survived up to the following spring. Visitors (animals captured only once) have been excluded from breeding population. Mature = m and immature = i.*

	C. rutilus					C. rufocanus					M. agrestis					Totals
	mδ	iδ	m♀	i♀	Totals	mδ	iδ	m♀	i♀	Totals	mδ	iδ	m♀	i♀	Totals	
Autumn 72	10	26	17	27	80	—	—	—	—	—	—	—	—	—	—	80
Survivors	2	2	1	2	7	—	—	—	—	—	—	—	—	—	—	7
Spring 73	13		12		25	10		5		15	2		4		6	46
Autumn 73	8	35	12	51	106	28	47	55	68	198	12	8	33	13	66	370
Survivors	1	0	0	8	9	8	8	32	44	92	3	0	10	2	15	116
Spring 74	14		12		26	51		117		168	14		47		61	255
Totals	42	58	52	68	220	81	39	145	24	289	25	8	74	11	118	627*

* This is not the direct sum of totals. The survivors have been excluded from the spring populations. In addition, one *C. rutilus* male survived from autumn 1972 to spring 1974 and has been added to the totals only once.

66%/month in *C. rutilus* males to 95%/month in *C. rufocanus* females (Table 2).

Another aspect of survival is documented by the percentage of spring population that had been marked in previous summer or autumn (Table 3). In this examination, I found two groups. One was formed by *C. rufocanus* and *C. rutilus* in 1974. They differed very significantly from *M. agrestis* and *C. rutilus* in 1973. The former was characterized by a high proportion of marked females and very significantly lower proportion of marked males. The latter was characterized by an equally low number of marked males and females. The proportion of marked females differed very significantly from the former group.

A general phenomenon was that old postbreeding (that had been mature) males survived better through the winter than did the younger immature males. This was not the case in females.

C. rutilus had increased rapidly in summer 1972, and it was almost alone in the study area in following winter. During summer 1973 both *C. rufocanus* and *M. agrestis* increased rapidly, too, and had high population densities in winter 1973–1974. Thus interspecific competition may have affected the survival figures in that winter.

DISCUSSION

The examination of the spring and autumn population sizes alone would suggest extremely high survival for *C. rufocanus* and *M. agrestis* and low survival for *C. rutilus*. The stable number of breeding individuals of *C. rutilus* documents the *Clethrionomys*-type density regulation and great need of space for individual territories of this species (Viitala, 1977, 1980). In autumn 1972, there were several individuals that had been captured on marginal

traps only. They probably lived mostly outside the study plot. Thus the breeding population of the study area was restricted to 12 females and about an equal number of males regardless of the autumn density. The surplus animals were forced to emigrate. The first emigration may take place in late autumn when even the young animals spread on individual territories (Viitala, 1977).

The comparison of autumn and spring popula-

Table 2.—*The mean monthly winter survival percentages of* C. rutilus, C. rufocanus, *and* M. agrestis *from autumn 1972 to spring 1974.*

	C. rutilus		C. rufocanus		M. agrestis	
Years	δ	♀	δ	♀	δ	♀
1972–1973	79	74	—	—	—	—
1973–1974	66	80	87	95	81	86

Table 3.—*Percentage of spring population that has been marked in previous summer and autumn in* C. rutilus, C. rufocanus, *and* M. agrestis *in spring 1973 and 1974.*

	C. rutilus		C. rufocanus		M. agrestis	
Years	δ	♀	δ	♀	δ	♀
1973	30	25	—	—	—	—
1974	7	67	31	65	26	21

tions reveals very high winter survival for *C. rufocanus* and *M. agrestis* as does the high survival of marked *C. rufocanus* females. Winter survival seems generally to be higher than that in summer as documented (for example, Petrusewicz et al., 1971) in an island population of *C. glareolus* in Poland.

The high spring population of *M. agrestis* and the low survival of marked animals reveal low mortality but high dispersal during winter. High dispersal was evident in *C. rutilus* in winter 1972–1973, too. This was exactly what one would expect, if the animals would form winter aggregations as observed by West (1977) in *C. rutilus* in Alaska.

The great proportion of previously marked females in *C. rufocanus* and *C. rutilus* in spring 1974 and the obvious site tenacity of the animals (cf.

Viitala, 1977) suggest that these animals did not aggregate during winter but behaved territorially. This is what Randolph (1977) observed in *Apodemus sylvaticus* in England. Despite the female stability, there does not exist marked inbreeding, however, because of the high dispersal rate of the males.

The sample of *C. rutilus* is small due to its large territory size. However it suggests that the females behave territorially through some winters but not in others. Thus we see that different species may adopt different wintering strategies and that even the same species may adopt different strategies in different situations. For *C. rutilus* one obvious changing factor was the severity of interspecific competition.

ACKNOWLEDGMENTS

I am grateful to Mr. Raimo Kujansuu and Mrs. Sinikka Hakala, M.A., for valuable assistance during the field work. My special thanks are due to the Carnegie Museum of Natural History and Joseph F. Merritt of Powdermill Nature Reserve for organizing this colloquium.

This study has been supported by grants by the Finnish State Board of Natural Sciences to the research group headed by the late Prof. Olavi Kalela and for most of the study by the Eemil Aaltonen Foundation.

Mr. Andres Perendi, M.A., has kindly checked the English language.

LITERATURE CITED

HANSSON, L. 1967. Index line catches as a basis of population studies on small mammals. Oikos, 18: 261–276.

MERRITT, J. F., and J. M. MERRITT. 1978. Seasonal home ranges and activity of small mammals of Colorado subalpine forest. Acta Theriol., 23:195–202.

PETRUSEWICZ, K., G. BUJALSKA, R. ANDRZEJEWSKI, and J. GLIWICZ. 1971. Productivity processes in an island population of *Cleithrionomys glareolus*. Ann. Zool. Fennici, 8:127–132.

RANDOLPH, S. E. 1977. Changing spatial relationships in a population of *Apodemus sylvaticus* with the onset of breeding. J. Anim. Ecol., 46:653–676.

VIITALA, J. 1977. Social organization in cyclic subarctic populations of the voles *Clethrionomys rufocanus* (Sund.) and *Microtus agrestis* (L.). Ann. Zool. Fennici 14:53–93.

———. 1980. Myyrien sosiologiasta Kilpisjärvellä. (Sociology of *Microtus* and *Clethrionomys* in Kilpisjärvi). (In Finnish, with English summary). Luonnon Tutkija 84:31–34.

WEST, S. D. 1977. Midwinter aggregations in the northern red-backed vole, *Clethrionomys rutilus*. Can. J. Zool. 58:1404–1409.

WOLFF, J. O., and W. Z. LIDICKER, JR. 1980. Population ecology of the taiga vole (*Microtus xanthognathus*) in interior Alaska. Can. J. Zool. 58:1800–1820.

Address: Department of Biology, University of Jyväskylä, SF-40100 Jyväskylä 10, Finland.

ABUNDANCE AND SURVIVAL OF *CLETHRIONOMYS RUTILUS* IN RELATION TO SNOW COVER IN A FORESTED HABITAT NEAR COLLEGE, ALASKA

PAUL WHITNEY AND DALE FEIST

ABSTRACT

Small mammals were trapped on two forested grids in the interior of Alaska, near Fairbanks from 1970 to 1981. A *Clethrionomys rutilus* population on the grids did not undergo a 3 to 4 year cycle. Peak yearly density in the forested area was more variable than on a near-by grassland area. Variation in peak yearly density on the forested area was not similar to variation in peak yearly density on another forested area north of Fairbanks trapped by West (1982). Survival of *C. rutilus* on the Fairbanks forested area is judged to be higher than on a near-by grassland area during years when information was available for both areas. Higher survival on the forested area may be due to snow cover remaining longer in forested areas compared to grassland areas.

INTRODUCTION

Systematic live trapping of small mammals in the interior of Alaska near Fairbanks (Fig. 1) from 1968 to 1971 indicated that peak annual density of *Clethrionomys rutilus* was similar each year of the study (Whitney, 1976). Based on this information it was assumed that *C. rutilus* living in a grassland habitat did not undergo a three to four year cycle similar to those of other microtine rodents. West (1982) live-trapped small mammals in the interior of Alaska north of Fairbanks from 1972 to 1976. His data from a forested area also indicated *C. rutilus* undergoes a simple sequence of annual fluctuation in density, however peak annual density was not equal

each year. He hypothesized that fruit production strongly influences vole population density each year and perhaps the amplitude of the annual cycle of abundance.

The purpose of this report is to provide additional information for *C. rutilus* which were live-trapped in a forested habitat in the interior of Alaska from 1970 to 1981. This information will be used 1) to assess the annual fluctuations in density in relation to snow cover and 2) to compare density and survival to a nearby grassland area (Whitney, 1976) and a similar forested area (West, 1982) north of Fairbanks, near Livengood.

CLIMATE AND STUDY AREA

The interior basin of Alaska where Fairbanks is located is generally described as having a continental climate where extremes in nearly all climatic elements occur. The mean annual temperature is −3.4°C; July, the warmest month, has a 15.5°C mean temperature; and January, the coldest month, has a −23.9°C mean temperature. Snow is present on the ground for approximately six months. Early snowfalls in late September are common. Snow falling after early October usually persists for the entire winter (Weather Bureau, 1981).

The regional vegetation of interior Alaska consists of stands of white spruce (*Picea glauca*) on both well-drained soils and south-facing slopes, with black spruce (*P. mariana*), larch (*Larix laracina*), and bogs on poorly drained lowland soils and on north-facing slopes. Fire has been an important environmental factor; consequently, successional stands of elephant grass (*Calamagrostis canadensis*), quaking aspen (*Populus tremuloides*) and Alaska paper birch (*Betula papyrifera*) are common throughout the area (Viereck, 1970).

METHODS

Northern red-backed voles (*Clethrionomys rutilus*) were live-trapped on a study site near Fairbanks, Alaska. Two live-trapping grids C and D were set in a north-facing forested area as part of an experimental study of the environmental impact of a heated pipeline (Fig. 1). Grid C was located in a closed birch forest that merged to the north with a black spruce forest, where Grid D was located. These grids consisted of 64 traps (8 × 8, 10 m interval). Both grids were trapped on a two to four week interval

from mid-1970 to mid-1972 and were trapped less regularly between mid-1972 and 1981 (Fig. 2). The live trapping, mark and recapture procedures were similar to those described by Whitney (1976).

Density estimates are based on minimum numbers known to be alive. Minimum numbers presented in this report have been analyzed using a calendar-of-catch method and the assumption that missed animals did not leave the grid. Survival rates were

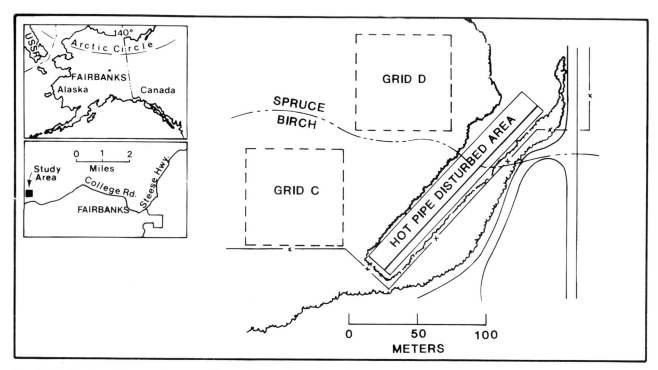

Fig. 1.—Grids C and D are adjacent to the experimental hot pipe disturbed area near College, Alaska. Both grids are in a forested habitat on a north-facing slope.

also estimated using the calendar-of-catch listing. Survival rates were calculated by counting the number of animals surviving a two week period (see Whitney, 1976).

The effective trapped area of the grids is estimated as 0.7 ha for the grids in the forested habitat. Four different methods were used to evaluate the boundary strip, which was finally assumed to be 10 m on all sides. The analyses and the rationale for the edge size are presented by Whitney (1973: 64), who shows similar density estimates on a larger 2.9 ha grid which was snap-trapped.

Snow depth was recorded on an adjacent area (approximately 100 m off the grids) by the Department of Commerce—National Weather Service. Snow density was measured during the 1969–1970 and the 1970–1971 winters by Whitney (1976). Snow data

is analyzed using the ideas presented by Marchand (1982). He indicates that a snow pack can be described by thickness and density and that as the depth of new snow increases diurnal temperature fluctuations in the subnivean environment are completely dampened. Such a dampening is most notable after an accumulation of 20 to 30 cm when the snow density is 0.1 g/cm³ (as is the case for new snow in the Fairbanks area, Whitney, 1976). Comparisons between snow and vole density and survival will assess the data when snow thickness reaches (if at all) 30 cm and the beginning of temperature stability of the subnivean environment, as well as the time during the spring when the snow melting reduces the depth to less than 30 cm.

RESULTS

ANNUAL FLUCTUATION

C. rutilus density on the forested area (Grids C and D) was similar during all years studied (Fig. 2). Notable exceptions were the yearly high density in 1971 on Grid C, which was twice the high density in 1972, and Grid D density in 1972, which was three times greater than density on Grid C. It should be noted that trapping during the latter years of the study was not always conducted during August or September, when the yearly high density is expected (exempli gratia, 1974, 1975, 1977, and 1981). In-

formation from these years was usually collected during July. July density estimates (or mid-winter estimates) for the latter years of the study do not indicate that density was likely to be higher or lower than most of the previous years.

SURVIVAL

Survival data for *C. rutilus* on the forested area (Grids C and D) showed periods of low survival in the summer as well as the winter (Fig. 3). Grid D was not trapped between November 1970 and March

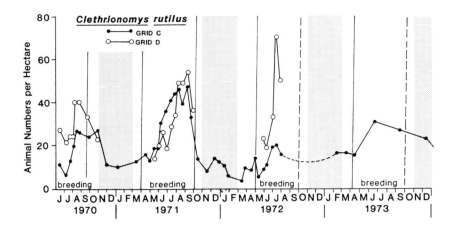

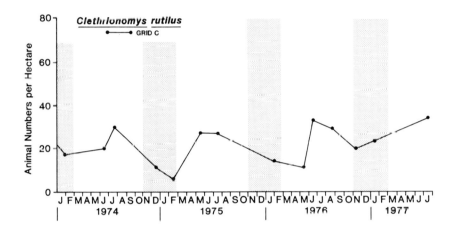

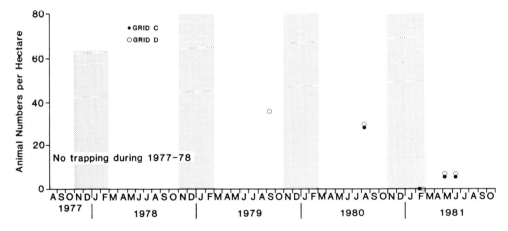

Fig. 2.—Population changes in *Clethrionomys rutilus* on College Grids C and D. Winter months are shaded, and vertical lines mark the length of breeding periods.

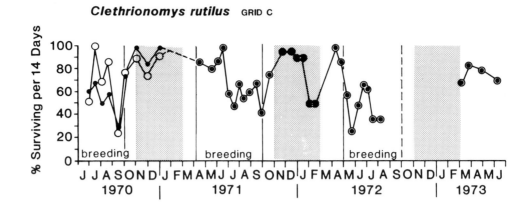

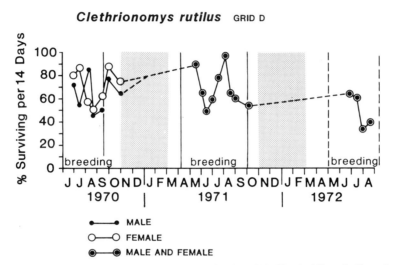

MALE
FEMALE
MALE AND FEMALE

Fig. 3.—*Clethrionomys rutilus* surviving per 14 days on Grids C and D. Vertical lines indicate breeding periods, and shaded areas are winter months.

1971, yet nine out of 16 *Clethrionomys* survived this five month period. When converted to survival on a two week basis, overwinter survival is approximately 95%. The only consistent tendency for low survival was during the summer prior to the end of the breeding season (August–September, Fig. 3). As an example, the lowest survival on Grid D was during July and August 1972, which was the time of the highest density during all years studied.

SNOW

Snow depth (Fig. 4) over the years varied from a record minimum in 1969–1970 to a record maximum in 1970–1971, with more average snow falls in the other winters. The insulative value of the snow varied from poor during the year of low snow

fall to good for years when 30 cm (enough to dampen fluctuations in subnivean temperatures, see Marchand, 1982) of snow fell early in the year and maintained sufficient snow depth to offset slight increases in snow density as the winter progressed. The date when at least 30 cm of snow was on the ground (similar to Pruitt's, 1970, hiemal threshold) varied over the years. For example, 30 cm was present by mid-November during 1970 and 1974, but 30 cm did not occur until mid-February during 1977. Snow depth did not even reach 30 cm during the 1969–1970 winter (Fig. 4). The date when less than 30 cm was present during the spring of each year was less variable and occurred during mid to late April (with the exception of 1972 when 50 cm of snow was present on 1 May, Fig. 4).

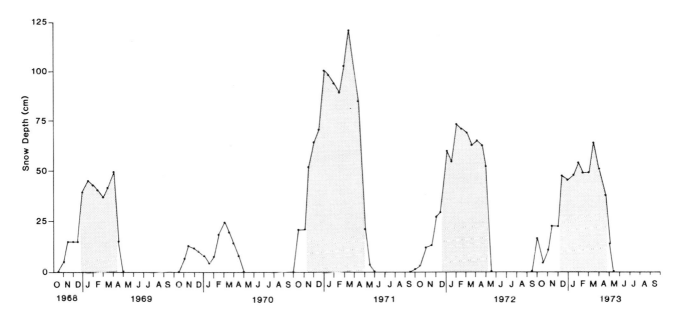

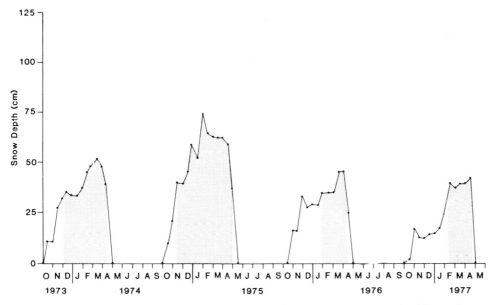

Fig. 4.—Snow fall adjacent to the College Grids C and D is plotted every two weeks. Times when snow fall accumulated to 30 cm or greater are shaded. Shaded periods represent those times when subnivean temperatures are fully dampened (Marchand, 1981).

DISCUSSION

Population density data presented in this report suggest that *Clethrionomys rutilus* does not undergo a 3 to 4 year cycle, a result similar to Whitney (1976) and West (1982). These data also support arguments by West (1982) that the lack of a cycle is not so much related to similar numbers each year as it is to the low densities that occur each year. While yearly variability in density was minimal in the grassland area near Fairbanks (Whitney, 1976), peak yearly density on the forested grids (C and D) was more variable. Variability in peak yearly density was also observed by West (1979) on his forested (control) plot. An interesting point of comparison between density information in this report and West's (1982) is that 1972 was the year with the highest density on the College grids (Fig. 2, this study), but was the year with the lowest density on the control plot trapped by West (1982). Conversely, the time of highest density on his control plot (fall and winter 1975–1976) was a time of low density (see Jan. 1976, Fig. 2) on the College grids.

Accepting West's (1979) idea that variability in fruit production (id est, seed crop) is related to variability in peak annual density, one may assume that fruit production on the College grids was different from fruit production on his control plot. Such a possibility could be tested in future years. Alternatively, fruit may never be abundant on the Grid C and D areas; thus, variability in other food items or other factors would have to be considered.

West (1982) indicated that spring conditions were very difficult for *C. rutilus* and might limit the contribution of individuals from an early spring cohort. Snow depth and density data on the College area (this study) indicate that it is unlikely that good survival of an early cohort was responsible for the high numbers on Grid D, for there was no evidence of breeding under the snow. Perhaps an important question to ask is: why do *C. rutilus* breed under the snow in some years and not others? For example, breeding was observed under the snow only once on the grassland plots trapped by Whitney (1976) and was not observed on the College Grids C and D. To summarize, it appears that the spring melt is a very severe time for small mammals, and without exception, yearly population growth does not start prior to July of each of the years studied by Whitney (1976), West (1982) and covered by this report.

Another purpose of this report is to examine survival between a south-facing grassland habitat and a north-facing forested habitat, and to relate possible differences to variability in depth and density of the snow pack. Survival of *C. rutilus* on the north-facing forested grids is judged to be higher than on the grassland grids (Whitney, 1976) during years when information was available. This result is based on the few times survival dropped below 50% on the forested area (Fig. 3) compared to the relatively high number of times survival dipped below 50% on the grassland grids (Whitney, 1976). This difference is most apparent during the 1970–1971 winter when survival on the grassland grids was low or zero for many weeks (Whitney, 1976), while survival on the forested grids was high (id est, 95% per 2 weeks) during the same period. The snow during this period of low survival on the grassland grids was deep, but the spring thaw period was judged to be severe, resulting in low *C. rutilus* survival. The spring thaw period on the forested area was different in that the snow remained approximately three weeks longer, and when the snow did disappear, ambient temperatures did not drop below freezing. Subnivean temperatures were not available during this period on the forested area, but it seems likely that *C. rutilus* on the north-facing forested area was exposed to fewer periods below freezing during the spring melt period. The tendency for snow to remain longer in forested areas and on north-facing slopes is likely to provide better subnivean temperature conditions, and may explain why *C. rutilus* density and survival on the north-facing forested areas can be higher than on near-by south-facing habitats.

An analysis of yearly accumulation of snow and subsequent dampening of the subnivean temperature indicates considerable variability in the first time (each winter) when 30 cm of snow is on the ground (Fig. 4). This variability could not be consistently related to *C. rutilus* density or survival. For example, the accumulation of 30 cm of snow early in the year in 1974 was not followed by high density on the Grids C or D, but 1970, when snow accumulated early in the year, was followed by high density on Grid D.

Two periods of low survival could be correlated with two extreme snow conditions, one during the record low snowfall year (1969–1970) and the other during the melt period of the record high snowfall (1970–1971). Other periods of low survival (possibly emigration) were consistently associated with the end of breeding season of all years studied on the forested Grids C and D as well as the grassland Grids A, B and E (Whitney, 1976). This trend was not noticeable on West's (1982) control plot near Livengood.

SUMMARY

To summarize, there was no consistent relationship between snow depth and density and vole abundance or survival on the forested grids studied.

Comparisons of trends in abundance and survival between study plots and years in interior Alaska indicate substantial variability. Until such variability within a limited geographical area is better understood, workers should not be discouraged by lack of obvious relationships between *C. rutilus* data in Alaska and other countries where *C. rutilus* is also being studied.

ACKNOWLEDGMENTS

This research was supported by the National Science Foundation under Grant GV 29342 to the University of Alaska. It was performed under joint sponsorship of the International Biological Program and the Office of Polar Programs and was directed under the auspices of the U.S. Tundra Biome. Also, parts of the research were supported by grants GM 10402 and ES 00689 from the National Institute of Health. We would like to thank Wayne Couture, Ray Kendel, Thomas Lahey and Randy Pitney for able field assistance. Funds for report preparation and graphics were provided by Beak Consultants Incorporated.

LITERATURE CITED

MARCHAND, P. 1982. An index for evaluating the temperature stability of a subnivean environment. J. Wildlife Mgmt. 46: 518–520.

PRUITT, W. O., JR. 1970. Some ecological aspects of snow. Pp. 83–99, *in* Ecology of the subarctic regions. Proc. Helsinki Symp. United Nations Scientific and Cultural Organization, Paris, France.

VIERECK, L. A. 1970. Forest succession and soil development adjacent to the Chena River in interior Alaska. Arctic and Alpine Res., 2:1–26.

WEATHER BUREAU. 1981. Daily, monthly and annual record of climatological observations. U.S. Department of Commerce—Environmental Sciences Services Administration—Weather Bureau, College Observatory, College, Alaska.

WEST, S. D. 1979. Habitat responses of microtine rodents to central Alaskan forest succession. Unpublished Ph.D. dissert., Univ. California, Berkeley, 115 pp.

———. 1982. The dynamics of colonization and abundance in central Alaskan populations of the northern red-backed vole, *Clethrionomys rutilus*. J. Mamm., 63:128–143.

WHITNEY, P. 1973. Population biology and energetics of three species of small mammals in the taiga of interior Alaska. Unpublished Ph.D. dissert., Univ. Alaska, Fairbanks, 254 pp.

———. 1976. Population ecology of two sympatric species of subarctic microtine rodents. Ecol. Monogr., 46:85–104.

Address (Whitney): Beak Consultants Incorporated, 317 S.W. Alder Street, Portland, Oregon 97204.

Address (Feist): Institute of Arctic Biology, University of Alaska, Fairbanks, Alaska 99701.

PLANT PRODUCTION AND ITS RELATION TO CLIMATIC CONDITIONS AND SMALL RODENT DENSITY IN KILPISJÄRVI REGION (69°05'N, 20°40'E), FINNISH LAPLAND

Seppo Eurola, Hannu Kyllönen, and Kari Laine

ABSTRACT

Small rodents spend the wintertime chiefly in chionophilous plant communities (snowbeds and *Phyllodoce caerulea—Vaccinium myrtillus* heaths). Their snow cover in January–April is 80 cm or more, and the influence of the air temperature is insignificant. The snowbeds and Norwegian lemmings occur chiefly in the upper part of the fjelds (in the middle alpine belt and in the upper part of the low alpine belt), the other rodents and *Phyllodoce—Vaccinium myrtillus* heaths in the lower part (in the lower part of the low alpine belt and in forests). No drastic lemming invasion has occurred since 1971, and the bryophyte biomass has therefore been increasing continuously in the middle alpine belt. The other rodents eat vascular plants. The flowering, seed and biomass production cycle determined by the effective temperature sum and the occurrence of the rodents are significantly correlated. On the basis of the results, two hypotheses are presented: 1) the nutrient cycle hypothesis, that is, the nutrients occur in soil and in plants by turns; the rodents act as mediators in this cycle, 2) the storage of energy in the form of organic matter in plants; when the energy balance exceeds a certain level, this leads to heavy flowering and good seed production and to good plant biomass quality.

INTRODUCTION

The paper studies the correlation of small rodents and vegetation in the Kilpisjärvi region. Finnish Lapland (69°05'N, 20°45'E) (Fig. 1) during the years 1971–1981. The native rodents met in the area are grey-sided vole (*Clethrionomys rufocanus*), northern red-backed vole (*C. rutilus*), field vole (*Microtus agrestis*), root vole (*M. oeconomus*), and Norwegian lemming (*Lemmus lemmus*). The vegetation studies are concentrated into plant biomass (total plus different fractions, for example, shoots, leaves, rhizomes), net production, flowering, berry and seed crops. The biomass and net production studies are made on 11 different habitats, the other results are based on 14 vascular plants. Only some case examples are given in this short paper. Abiotic factors considered are snow cover, temperature, and some nutrients (N-tot, P-tot, N-NO$_3$, P-PO$_4$, K, Na, and Ca). The amount of the nutrient analysis is not yet significant enough and gives only rough estimates.

The vegetation of the Kilpisjärvi region belongs partly to the subalpine birch forest—often called the mountain section of the northern oroboreal zone (Ahti et al., 1968)—partly to barren fjeld (Fig. 2). The forest limit reaches appr. 600 m a.s.l (Fig. 2II).

The average temperatures in the birch belt are −2.4°C (year-round), +10.4°C (July) and −13.9°C (February) (Federley, 1972). The minimum, average and maximum values for the precipitation are 177-373-519 mm/a (Federley, 1972). The temperature conditions of the air are ecologically arctic,

that of the soil alpine (Eurola, 1974), because of the lack of the permafrost. The length of the growing season (| 5°C . . . +5°C) is appr. 95 days in the birch forest belt (480 m a.s.l.) and <80 days in the low alpine belt (738 m a.s.l.). The correspondong values for the 0°C . . . 0°C periods are appr. 150 and 120 days (cf. Hiltunen, 1980).

By dividing the vegetation roughly into the groups of A) the meadow forests, B) rich birch fens, C) blueberry-rich forest types and heaths, D) crowberry-rich forest types and heaths, E) dwarf birch thickets, F) wind-exposed types, G) short-sedge-grass-forb meadows (snowbeds), H) *Salix herbacea* (incl. *S. polaris* and pure moss-dominated) snowbeds, I) fjeld mires and J) grass heaths (Fig. 2III), four belts are distinguished (Fig. 2II). 1) The subalpine birch forest (northern oroboreal after Ahti et al., 1968) represents the lowest one. In its lower part the forest types and mires are often nutrient-rich, in the upper part poorer (Fig. 2III:A + B). 2) The belt of the dwarf birch (+ *Salix* spp., *Juniperus*) thickets and moss-rich *Phyllodoce caerulea-Vaccinium myrtillus* (blueberry) heaths extends up near to 700 m a.s.l. 3) The lichen-rich blueberry heaths are seen yet in the next belt up to 830–850 m a.s.l., where also short-sedge meadows with different short sedges, grasses and herbs are met frequently. 4) In the topmost belt *Salix herbacea*—rich and pure moss-dominated snowbeds, grass heaths with *Festuca ovina* and *Juncus tridifus,* and *Empetrum hermaphrodi-*

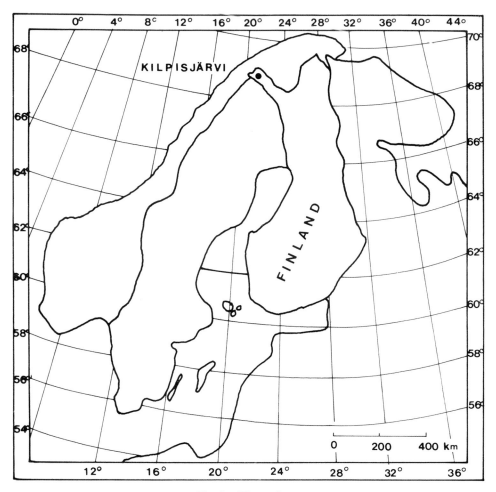

Fig. 1.—The study area.

tum—Cassiope tetragona heaths (vicarious to the blueberry heaths) are typical. Generally taken, different *Empertrum* heaths are met frequently in every belt (Fig. 2III:D). The topmost area (4) belongs into the middle alpine belt (middle oroarctic after Ahti et al., 1968), the areas 2 and 3 into the low alpine one (see for example Eurola, 1974; after Ahti et al., 1968 the belt 2 is hemioroarctic). Different snowbeds represent the habitats of the Norwegian lemmings, the voles populate subalpine forests and thickets. Thus, taking into consideration the vegetation and/ or animals, the fjelds high enough can be divided into the lower (the belts 1 and 2) and upper (the belts 3 and especially 4) parts (cf. also Eurola et al., 1982).

The vertical occurrence of the microtines reflects the difference of the plant biomass between the lower and higher parts of the fjelds. In the lower part,

the living plant biomass is 500 to 2,000 g m^{-2} and net production 130 to 550 g m^{-2}. The vegetation is chiefly composed of fjeld and dwarf birches, willows, dwarf shrubs and high herbs. In the upper part, the living biomass is 200 to 500 g m^{-2} and net production 60 to 100 g m^{-2}. Bryophytes (biomass appr. 150 to 200 g m^{-2}), short sedges and tiny grasslike plants dominate there. Thus the upper part of the fjeld is a natural food niche for lemmings—which during wintertime consume chiefly bryophytes—without any peculiar food competition with voles.

Part of this study has been reported at the Herbivory at Northern Latitudes conference held in Kevo, Finland, on 14–18 September 1981 (Laine and Henttonen, 1983; see also Eurola et al., 1980; Eurola et al., 1982).

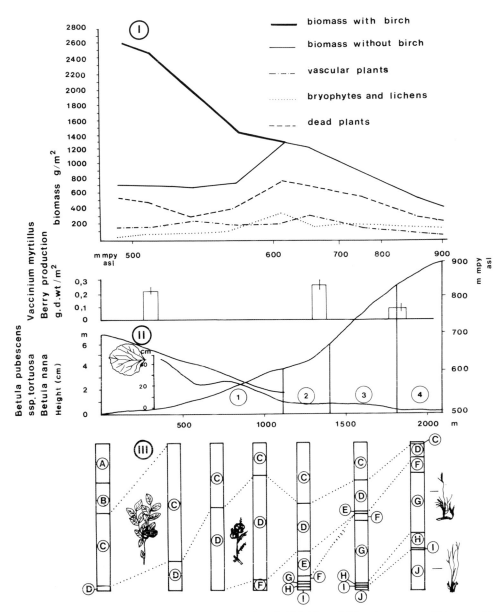

Fig. 2.—The belts, vegetation types and above-ground biomass of the fjeld Jehkats. I. The above-ground biomass. The birch is evaluated after Östbye (1975). The other values are based on the studies of Kyllönen (1980). II. The belts with the heights of *Betula pubescens* sp. *tortuosa* and *B. nana*. The berry crops of *Vaccinium myrtillus* in the years 1972–1978 are also given. 1) The subalpine (northern oroboreal) fjeld birch belt, 2) the lower part of the low alpine belt: the willow-juniper-dwarf birch thickets (the main part of hemioroarctic belt of Ahti et al., 1968), 3) the upper part of the low alpine belt: dwarf shrub heaths and short-sedge meadows (low oroarctic belt), 4) the middle alpine (middle oroarctic) belt of (*Salix herbacea-S. polaris*) snowbeds and *Juncus tridifus-*(*Festuca ovina*) grass heaths. III. The percentage occurrence of the vegetation type groups at different altitudes. A) The meadow forests, B) rich birch fens, C) *Myrtillus-*rich forest types and heaths, D) *Empetrum-*rich forest types and heaths, E) *Betula nana-*(*Juniperus-Salix*) thickets, F) wind exposed types, G) short sedge-grass-forb meadows, H) *Salix herbacea-*(*S. polaris*) snowbeds, I) spring influenced fjeld mires, and J) *Juncus trifidus-*(*Festuca ovina*) grass heaths (Eurola et al., 1982).

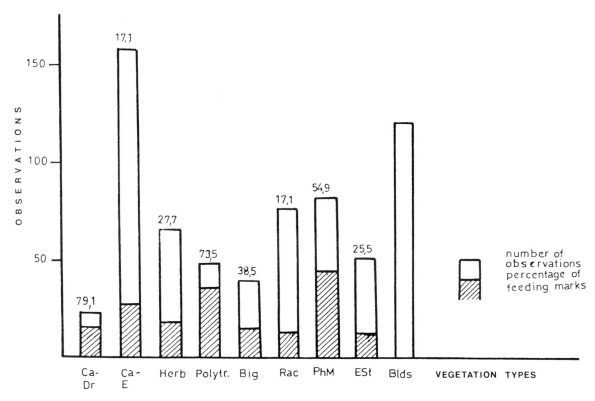

Fig. 3.—The percentage occurrence of feeding marks in some vegetation types. Vegetation symbols: see Fig. 4.

WINTER HABITATS OF THE RODENTS

The amount of feeding marks found after the snow melting reflects the wintertime habitats of the rodents (Fig. 3). Both absolute and percentage occurrence concentrate on habitats with deep snow cover (see Fig. 4): *Phyllodoce caerulea-Vaccinium myrtillus* (PhM) and *Cassiope tetragona-Dryas octopetala* (Dryas) heaths plus dwarf willow + moss-rich snowbeds (Herb). The last ones represent a late (at the end of June—in July)-very late (in August) snow-free vegetation (Gjaerevoll, 1956; Hiltunen, 1980). One can see the low percentage of feeding marks in the *Empetrum*-rich vegetation which occurs in the thinly snow-covered, early snow-free, and therefore microclimatically thermal-continental habitats (Hiltunen, 1980).

RODENT CYCLES AND THE ANNUAL VARIATION OF SOME PLANT INDICATORS

Annual variations in rodent number are known to be particularly strong in the arctic and subarctic areas (Kalela, 1957, 1962; Tast and Kalela, 1971; Lahti et al., 1976; Laine and Henttonen, 1983). In principle, the cycles of different microtines in northern Fennoscandia are synchronous (Henttonen et al., 1977; Hansson, 1979, Laine and Henttonen, 1983). In Fig. 5 is presented only the total density of all microtines in the Kilpisjärvi region.

During the period 1972–1981 (Fig. 5), the peak years of the microtine occurrence were 1974 and 1978 (plus this year 1982). Thus the winter success was good during the winters 1973–1974, 1977–1978 and 1981–1982. Opposite to these periods, the amount of rodents was very small in the years 1971, 1976 and 1979. The amount of voles was greatest in the years 1974 and 1978 (plus 1982) and thus in the lower part of the fjelds. Lemmings occurred frequently in the year 1970 when they wandered also downwards. In the years 1974 and 1978, the lemmings had no such drastic increase in density.

Different plant indicators also vary annually (Fig.

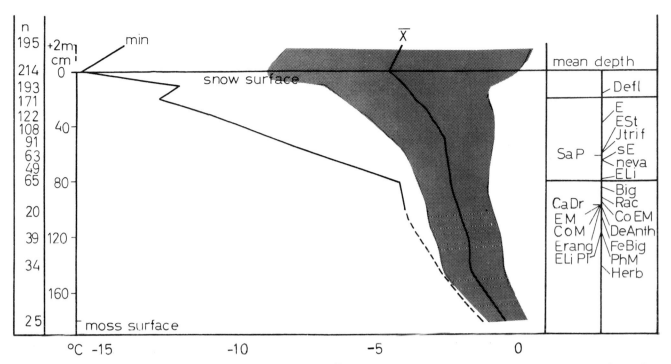

Fig. 4.—Temperatures under the snow, minimum (min) and mean (X̄) values with standard deviation (shadowed area). The figure also gives the mean snow depth in the vegetation types studied and the number of measurements (n) made in different snow depths. Vegetation symbols: Defl = wind exposed heath; sE, ELi, EM, CoEM, CoM = different birch forest types, where E = *Empetrum*, Li = lichen, Co = *Cornus suecica* and M = *Vaccinium myrtillus*; Erang and Neva = fen types; E, ESt and CaE = *Empetrum*-rich open heath types (Ca = *Cassiope tetragona*, St = *Stereocaulon*); PhM = *Vaccinium myrtilus*-rich open heath type (Ph = *Phyllodoce caerulea*, SaP = *Betula nana*-thicket rich in *Pleurozium schreberi* and *Hylocomium splendens*; CaDr = eutrophic heath (Dr = *Dryas octopetala*); Big, Rac, DeAnth, FeBig = different sedge-, grass- and herb-rich snowbeds (Big = *Carex bigelowii*, Rac = *Ranunculus acris*, De = *Deschampsia flexuosa*, Anth = *Anthoxanthum odoratum*, Fe = *Festuca ovina*); Herb = *Salix herbacca*-rich snowbed; Polytr. (Fig. 3) = bryophyte-dominated snowbed; Blds (Fig. 3) = boulder fields, rock, etc., Jtrif. = grass heath. (After Eurola et al., 1980.)

6 and 7). In this paper they are grouped in two: 1) the pure biomass plus the weight of rhizomes, and 2) sexual reproduction (flowers, seeds, berries).

The peak years of the herbs in the group 1 were 1973, 1978 and 1980. In the upper belts there was no peak in the year 1978. It appeared in the next year and joined together the 1980 year's peak. Also the bryophyte fluctuation follows that of the herbs in the lower belts. In the upper belts the bryophyte biomass increased up to the year 1978. Thus the lemming population was not high during the winter 1973–1974. The peak years in the group 2 are 1973, 1977 and 1980 (Fig. 7).

The peak year of the plant indicators (group 1 and/or group 2) occur a year before the microtine invasion. This means a good wintertime success of the rodents and thus a good nutrient status (quantity and/or quality) in plants after the summer of the peak year (see also Tast, 1982). During and after the microtine peaks follows a more or less clear decrease of the plant indicators caused by the rodents themselves (the years 1974, 1979 and 1982). A cold summer can yet lengthen this sinking, as happened in the year 1975. This summer (year 1975) had no influence in bryophytes. However, both these events are not reasons for the regular cycles of the plant indicators and microtines; it must be found in the plants themselves, the inner condition of which, in this paper, is divided in two: 1) the nutrient status and 2) the accumulation of energy in plants.

NUTRIENTS IN PLANTS AND SOIL

The soil and plant samples were taken in the late summer, dried immediately, and then analyzed later in winter. At least a weak tendency into the nutrient circulation between the plants and the humus layer

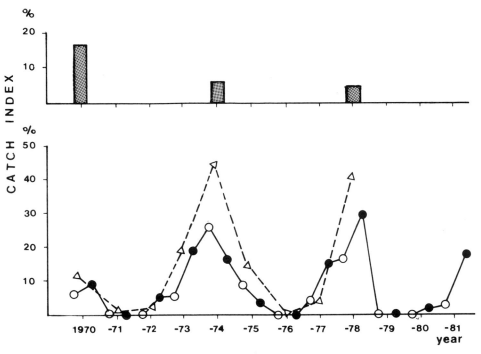

Fig. 5.—Fluctuation of the microtines (% of the total individuals caught per trap night) in the subalpine birch forest (=belt 1 in Fig. 2; continuous line, ● = early summer, ○ = late summer), in the lowermost part of the alpine belts (=belt 2 in Fig. 2; broken line) and in the upper part of the low alpine belt (=belt 3 in Fig. 2; grey columns) in the Kilpisjärvi area in the years 1970–1981.

is seen (Fig. 8), for example, when there are more nutrients in plants, there are not so many in soil and vice versa. In 1973, there was much nitrogen and phosphorous in the humus and, during the next two–three years, also in plants. The nitrogen cycle is more regular and slower than the other ones. Sodium has no cycle. A rodent peak occurs after good nitrogen balance in soil together with high potassium content in plants. All this gives rise to a hypothesis that the rodents act as mediators in the plant/soil nutrient cycle (cf. Pieper, 1964; Schultz, 1969).

ENERGY BALANCE OF PLANTS

Hustich and Elfving (1944) showed that the growth of Scotch pine in thickness is correlated with the temperature of the current growing season, particularly with that of July. Kyllönen (1980) has noticed the same for the herb biomass in the Kilpisjärvi region; exceptions are the summers during or after the peak years of microtines because of their herbivory. Laine (1981) suggested that other plant indicators (flowering, reproduction of seeds, berries and rhizomes) have no clear correlation with the warmest month or with the current growing season. On the other hand, Hustich (1948) and Sarvas (1962, 1966) showed that a warm previous growing season is needed for a rich flowering, and Sarvas (1962) also indicated that a total effective temperature during the current summer is important for flowering.

Sørensen (1941) had shown the responsibility of the arctic plants for flowering (flower buds are ready) if the temperature conditions are suitable. But what are these suitable conditions because rich flowering can occur also in a not very warm summer (for example, 1977, cf. Fig. 6)? Our idea is (Laine and Henttonen, 1983) that the temperature of the whole cycle is needed for the rich occurrence of the non-biomass indicators. In the Kilpisjärvi region 1,500–1,600° d.d. is enough, thereafter the plants can use their energy more than in average for sexual and asexual reproduction. This is an energy balance (or storage) hypothesis. A good energy balance coincides also with a good nutrient status, and both of them form together the so called "power" condition in plants.

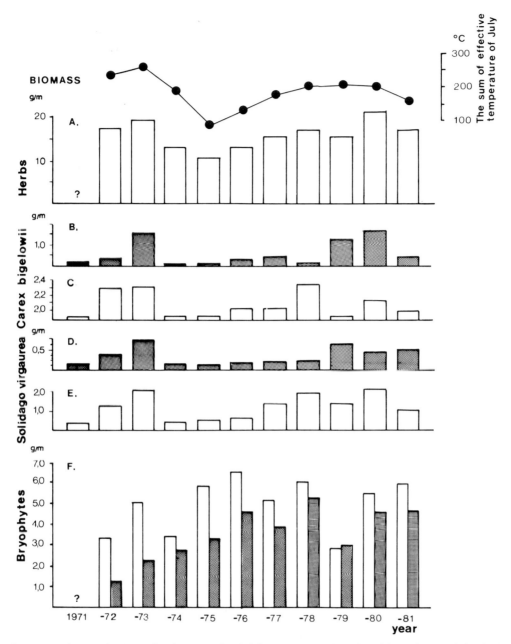

Fig. 6.—The biomasses of herbs A, *Carex bigelowii* (B, C), *Solidago virgaurea* (D, E) and bryophytes (E) in the lower part (white columns) and in the upper part (grey columns) of the low alpine belt in Kilpisjärvi in the year 1971–1981. The effective temperatures of July (>5°C) are seen above the Fig. 6A.

DISCUSSION

The corresponding correlation between the flowering frequency of two plant species, *Eriophorum angustifolium* and *Solidago virgaurea,* and microtine density and winter success was reported by Tast and Kalela (1971) and Tast (1982). Based partly on these results, they presented a hypothesis on production biology, according to which both the quantity and quality of food may effect the growth of small rodent populations and winter success (see also Laine and Henttonen, 1983).

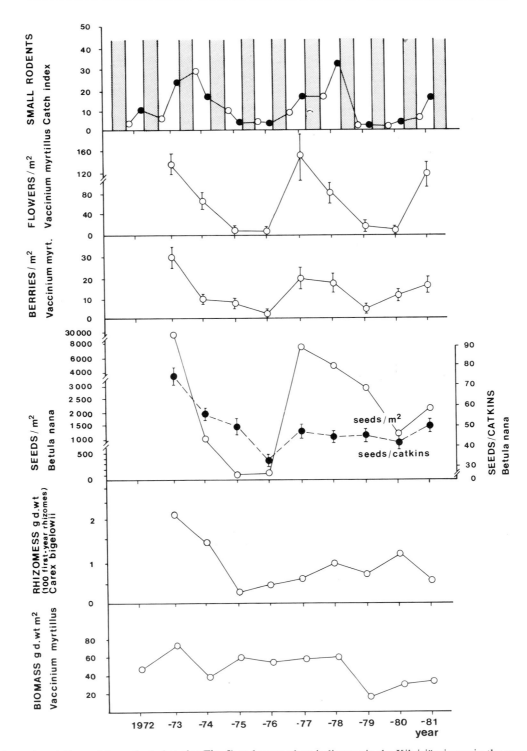

Fig. 7.—Annual variations of the rodents (see also Fig. 5) and some plant indicators in the Kilpisjärvi area in the years 1972–1981.

The results obtained here support this. The variation of plant production like flowering, seed and berry crops, and partly biomass and net production is clearly tended to the same direction as the density of small rodents in Kilpisjärvi area. The correlation can be observed with 14 plant species in all altitudes, even in higher belts where small mammals don't occur.

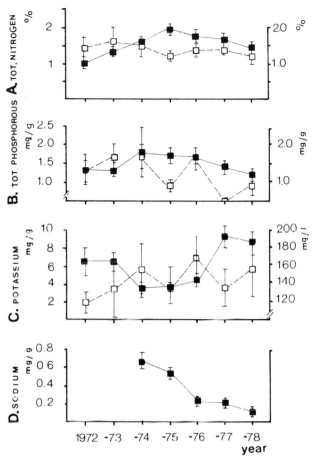

Fig. 8.—The amounts (n = 5) of total nitrogen (A), total phosporous (B), potassium (C) and sodium (D) in plants (left scale, continuous line) and in humus layer (right scale, broken line).

It has been proposed some times in literature that the direct consumption by small rodents causes the variation of flowering frequency (Tikhomirov, 1959; Pruitt, 1968; Batzli, 1975). According to our understanding, the consumption is not a primary reason but more a factor caused by variation. This can be accounted for as follows: 1) the variation of flowering reported by Tast and Kalela (1971) was similar in a closed study area than in open ones; 2) the variation can also be observed in higher belts where rodents don't occur; 3) the variation can also be observed in such plant species and factors which can't be affected directly by small rodents (for example, seed production of birches, the size of first-year-rhizomes).

So, what reasons cause these variations in plants? The answer is not so simple. In alpine and arctic plants, two types of explanations to variation of flowering and seed production have been presented. Firstly, the previous summer temperature has been observed to affect decisively. Secondly, the current summer temperature has been proposed to be significant. According to our observation, the highest correlations were observed with combined temperature sums of two consecutive summers.

But we must bear in mind that plant flowering physiology is still partly unknown. It is well known that the energy demand of the flowering and seed production is great, and in northern latitudes in arctic and alpine areas, the short growth period of one or two summers is maybe not enough for flowering. So maybe the accumulation of energy and nutrients requires a longer period, and after the threshold of energy and nutrients has been exceeded, the flowering and seed production can be observed. In the literature the often-mentioned expressions "the ability to wait for favourable conditions" and "the invigoration stage" can be understood as a time for energy resource gathering (Sørensen, 1941; Söyrinki, 1938, 1939; Holway and Ward, 1963, 1965; Wielgolaski, 1974; Bell and Bliss, 1977). This accumulation of energy and nutrients would also be an answer to the question made by Bulmer (1974), inquiring, "Are there any cyclic variable factors in the environment, which can cause the cyclic variation of a rodent population?"

Based on preliminary results, it seems that we can also find variations in the amount of nutrients. These variations seems to have a connection to the fluctuations of plants and rodents. According to our opinion, the turned variation can be observed in nutrients in soil and plants, and the rodents act as mediators in this cycle. Because of the small amount of samples, all the results are not yet statistically significant, but it is now just at the hypothetical stage.

LITERATURE CITED

AHTI, T., L. HÄME T-AHTI, and J. JALAS. 1968. Vegetation zones and their sections in north-western Europe. Ann. Bot. Fennici, 5:169–211.

BATZLI, G. O. 1975. The role of small mammals in arctic ecosystems. Pp. 243–269, in Small mammals, their productivity and population dynamics (F. B. Golley, K. Petrucewicz, and L. Ryszkowski, eds.), IBP, Cambridge Univ. Press, Cambridge, 5.

BELL, K. L., and L. C. BLISS. 1977. Overwinter phenology of plants in a polar semi-desert. Artic, 30:118–121.

BULMER, M. G. 1974. A statistical analysis of the 10-year cycle in Canada. J. Anim. Ecol., 43:701–718.

EUROLA, S. 1974. The plant ecology of Northern Kioöen, arctic or alpine? Aquilo Ser. Bot., 13:10–22.

EUROLA, S., H. KYLLÖNEN, and K. LAINE. 1980. Snow conditions and some vegetation types in the Kilpisjärvi region. Luonnon Tutkija, 84:43–48. (In Finnish with English summary.)

———. 1982. The nature of the fjeld Jehkats (Kilpisjärvi, NW Finnish Lapland, 69°01′N and 20°50′E) I. The belts, vegetation types and above-ground biomass. Kilpisjärvi Notes, 6:7–13. (In Finnish with English summary.)

FEDERLEY, B. 1972. The invertebrate fauna of the Kilpisjärvi area, Finnish Lapland. I Introduction: the area, its investigation and the plant cover. Acta Soc. Fauna Flora Fennica, 80:1–36.

GJAEREVOLL, O. 1956. The plant communities of the Scandinavian alpine snow-beds. Kgl. Norske Videnskaps Selsk. Skrifter, 1956(1):1–405.

HANSSON, L. 1979. Food as a limiting factor for small rodent numbers. Test of two hypotheses. Oekologia, 37:297–314.

HENTTONEN, H., A. KAIKUSALO, J. TAST, and J. VIITALA. 1977. Interspecific competition between small rodents in subarctic and boreal ecosystems. Oikos, 29:581–590.

HILTUNEN, R. 1980. Temperature and snow conditions on snow beds and wind-exposed places on Pikku-Malla, NW Finnish Lapland. Luonnon Tutkija, 84:11–14. (In Finnish with English summary.)

HOLWAY, J. G., and R. T. WARD. 1963. Snow and meltwater effects in an alpine area of Colorado. Amer. Midland Nat., 60:189–197.

———. 1965. Phenology of alpine plants in northern Colorado. Ecology 46:73–83.

HUSTICH, I. 1948. The Scotch pine in northernmost Finland. Acta Bot. Fennici, 42:1–75.

HUSTICH, I., and G. ELFVING. 1944. Die Radialzuwachsvariationen der Waldgrenzkiefer. Comme. Biol. Soc. Sci. Fenn., 9(8):1–18.

KALELA, O. 1957. Regulation of reproduction rate in subarctic population of the vole Clethrionomys rufocanus (Sund.) Ann. Acad. Scient. Fenn. A, 4(34):1–60.

———. 1962. On the fluctuations in the numbers of arctic and boreal small rodents as a problem of production biology. Ann. Acad. Sci. Fenn. A, 4(66):1–38.

KYLLÖNEN, H. 1980. Tunturikasvillisuuden biomassasta ja tuotoksesta sekä niihin vaikuttavista tekijöistä Kilpisjärvellä vuosina 1971–78. Unpublished manuscript, Dept. Bot., Univ. Oulu, 107 pp.

LAHTI, S., J. TAST, and H. UOTILA. 1976. Fluctuations in small rodent populations in the Kilpisjärvi area in 1950–75. Luonnon Tutkija, 80:97–107. (In Finnish with English summary.)

LAINE, K. 1981. Kukinnan, marja- ja siementuoton sekä eräiden kasvinosien vuotuisista vaihteluista ja niiden suhteesta pikkunisäkkäiden kannanvaihteluihin Enontekiön Kilpisjärvellä vuosina 1972–78. Unpublished manuscript, Dept. Bot., Univ. Oulu, 77 pp.

LAINE, K., and H. HENTTONEN. 1983. The role of plant production in microtine cycles in northern Fennoscandia. Oikos, 40:407–418.

ØSTBYE, E. 1975. Hardangervidda, Norway. Pp. 225–264, in Structure and function of tundra ecosystems (T. Rosswall and O. W. Heal, eds.), Ecol. Bull. (Stockholm), 20.

PIEPER, R. D. 1964. Production and chemical composition of arctic tundra vegetation and their relation to the lemming cycles. Unpublished Ph.D. dissert., Univ. California, Berkeley.

PRUITT, W. O., JR. 1968. Synchronous biomass fluctuations of some mammals. Mammalia, 32:172–191.

SARVAS, R. 1962. Investigations on the flowering and seed crop of Pinus silvestris. Commun. Inst. For. Fenn., 53:1–198.

———. 1966. Temperature sum as a restricting factor in the development of forest in the Sub-Arctic. Symp. Ecol. Sub-Arctic Reg. Helsinki Mimeogr., 5 pp.

SCHULTZ, A. M. 1969. A study of an ecosystem: the arctic tundra. Pp. 77–93, in The ecosystem concept in natural resource management (G. M. Van Dyne, ed.), Academic Press, New York.

SØRENSEN, T. 1941. Temperature relations and plant phenology of the northeast Greenland flowering plants. Medd. om Grønland, 125:1–305.

SÖYRINKI, N. 1938. Studien über die generative und vegetative Vehrmehrung der Samenpflanzen in der alpinen Vegetation Petsamo-Lapplands. I. Allgemeiner Teil. Ann. Bot. Soc. Vanamo, 11:1–311.

———. 1939. Studien über die generative und vegetative Vermehrung der Samenpflanzen in der alpinen Vegetation Petsamo-Lapplands. II. Spezieller Teil. Ann. Bot. Soc. Vanamo, 14:1–406.

TAST, J. 1982. Winter success of root voles, Microtus oeconomus, in relation to population density and food conditions at Kilpisjärvi, Finnish Lapland. This volume.

TAST, J., and O. KALELA. 1971. Comparisons between rodent cycles and plant production in Finnish Lapland. Ann. Acad. Sci. Fenn. A, 4 (186):1–14.

TIKHOMIROV, B. A. 1959. Interrelationships of the animal world and the vegetation cover of the tundra. (Russ. with English summary). Acad. Sci., USSR, Komarov Bot. Inst., Moscov–Leningrad, 104 pp.

WIELGOLASKI, F. E. 1974. Phenological studies in tundra. Pp. 209–214, in Phenology and seasonality modeling. (H. Lieth, ed.), Ecological Studies, Springer-Verlag, Berlin-Heidelberg-New York.

Address: Department of Botany, University of Oulu, Linnanmaa, SF-90570 Oulu, Finland.

METABOLIC AND THERMOGENIC ADJUSTMENTS IN WINTER ACCLIMATIZATION OF SUBARCTIC ALASKAN RED-BACKED VOLES

DALE D. FEIST

ABSTRACT

Physiological and biochemical aspects of seasonal acclimatization have been studied in red-backed voles (*Clethrionomys rutilus dawsoni*) in the subarctic taiga forest of mixed birch and spruce at Fairbanks, Alaska. Measurements of oxygen consumption by voles at low temperatures show that winter voles achieve a much greater maximum metabolic capacity (M_{max}) than summer voles. In tests of the calorigenic response to injected norepinephrine (NE), winter voles show a significantly greater NE stimulated nonshivering thermogenesis (NE-NST) than summer voles. The seasonal winter peak in NE-NST coincides with peaks found for M_{max}, brown fat mass, myoglobin concentration, and heart mass. Fall and winter voles exhibit increased activity of adrenal tyrosine hydroxylase and therefore a greater capacity to synthesize adrenal catecholamine hormones (NE and epinephrine) which could help mediate cardiovascular and metabolic responses to cold. Studies of seasonal energy and water metabolism in free-living voles by the doubly labelled water method show that average daily metabolic rate (ADMR) is lowest in spring and highest (and similar) in summer and winter. Body water turnover rate is positively correlated with ADMR in all seasons except winter when the water turnover rate is reduced about 33% from summer. The significance of these features of seasonal acclimatization are discussed. Further studies are proposed to help clarify the processes and mechanisms of metabolic acclimatization in small mammals.

INTRODUCTION

Small mammals which live in cold-dominated high latitude environments such as subarctic Alaska must acclimatize to winter by metabolic as well as insulative adjustments. The northern red-backed vole (*Clethrionomys rutilus*) is a microtine rodent which is widely distributed in the subarctic regions of North America and Siberia. In the taiga forest of interior Alaska, red-backed voles remain active throughout the winter (that is, there is no voluntary torpor or hibernation) and may venture above the snow even in the coldest months, although most of their winter activity is confined to the subnivean spaces which may be warmer than air above the snow. Several workers have reported on the population dynamics (Pruitt, 1968; Whitney, 1973, 1976, 1977; West, 1974, 1979, 1982), bioenergetics (Grodzinski, 1971; Whitney, 1977) and behavior (West, 1977) of Alaskan red-backed voles. In the 1960's physiological and morphological studies (Morrison et al., 1966; Sealander, 1966, 1967, 1969, 1972; Sealander and Bikerstaff, 1967) showed that in addition to increased pelage density, red-backed voles increased muscle myoglobin, white fat and brown fat mass and decreased in mean body mass during acclimatization to winter. These results indicated that metabolic changes are important for survival in winter, and they provided the basis for further studies of seasonal metabolic and thermogenic adjustments in these small mammals.

The objectives of this paper are 1) to review more recent studies of metabolic seasonal acclimatization in Alaskan red-backed voles along with pertinent findings of earlier work and 2) to suggest areas for future investigations of physiological aspects of survival of small mammals in the subarctic winter.

When we began studies of metabolism and regulatory thermogenesis in red-backed voles in 1972, we asked: 1) do voles exhibit seasonal changes in metabolic capacity (M_{max})?; if so, 2) do these changes in M_{max} involve enhanced nonshivering thermogenesis and to what extent?; and, 3) what physiological and biochemical mechanisms and processes act to support seasonal changes in metabolic capacities? To begin answering these questions we trapped red-backed voles in the taiga forest of mixed birch and spruce (*Betula papyrifera, Picea glauca, P. mariana*) in interior Alaska at Fairbanks (65°N, 148°W). Trapping was carried out as described in detail previously (Rosenmann et al., 1975; Feist and Rosenmann, 1976) with Sherman live traps on grid areas previously established for population ecology studies (Whitney, 1973) near research laboratories on the University of Alaska campus. Captured animals were transported to the laboratory about 200 m distant and used in experiments the same morning of capture.

Air temperatures recorded near our trapping areas range from greater than 25°C in June or July to lower than −40°C in winter (Weather Bureau, 1973; Whitney, 1976). The mean annual temperature in Fairbanks is $-3.4°C$; July, the warmest month, has a mean temperature of $15°C$ and January, the coldest month, a mean of $-24°C$. Annual snowfall averages 151 cm with an average accumulation of 74 cm. Snow falling after early October usually persists for the entire winter until early April (Whitney, 1976). Subnivean temperatures in January may be as low as $-30°C$ as in 1970 when there was little snow cover, or may stablize at about $-5°C$ with adequate insulative snow cover (Whitney, 1976).

SEASONAL CHANGES IN MAXIMUM METABOLIC RATE AND NONSHIVERING THERMOGENESIS

In initial studies we subjected voles at different seasons to cold to elicit M_{max}. As shown in Fig. 1, we found a remarkable doubling of M_{max} from summer to winter (Rosenmann et al., 1975). This would allow winter voles to maintain body temperature down to $-75°C$ and summer adults to $-40°C$. The advantages of this increased M_{max} in winter can be expressed in other ways. For example, the metabolic reserve (M_{max}—resting metabolic rate) is about the same (10 ml $O_2 \cdot g^{-1} h^{-1}$) for a winter vole at $-30°C$ and a summer vole at $15°C$ (Rosenmann et al., 1975). Thus winter voles can continue normal activities without reaching their limit despite low ambient temperatures and associated demands for temperature regulation.

Next we tested for the magnitude of norepinephrine (NE) stimulated nonshivering thermogenesis. After injecting voles at different seasons with the same dose of NE, we found that the capacity for nonshivering thermogenesis also increased from summer to winter (Fig. 1, M_{NE}) essentially in proportion to the increase in M_{max} (Feist and Rosenmann, 1976). This is particularly important since an increase in M_{max} which involves nonshivering thermogenesis (rather than shivering) will provide more efficient thermoregulation during other energy demanding activities in the winter.

When we repeated studies of M_{max} and nonshivering thermogenesis in subsequent years, it was apparent that the magnitude of change from summer to winter may vary from year to year (Feist and Morrison, 1981). For example, M_{max} and nonshivering thermogenesis were significnatly greater in the winter of 1973–1974 than in the winter of 1976– 1977. Body weights of animals tested were similar for both years. We have no measure of subnivean temperatures, but there were no apparent correlations with ambient temperature or snow depth in those years. Year-to-year differences in thermogenic capacity may be related to factors such as changing food supply, leading to variations in availability of metabolic substrates (for example, glucose and fatty acids) (Wang, 1978, 1980) and ambient temperature of exposure during early growth and development in spring, summer and fall (Lynch et al., 1976; Lacy et al., 1978). Wang (1978, 1980) found that rats fasted overnight exhibited a lower M_{max} in the cold than normally fed rats. He also found that overnight fasted rats given a mixture of substrates (intragastrically) 1 hr before testing exhibited the highest M_{max}. He concluded that substrate availability limits thermogenesis in severe cold. Whether this applies to red-backed voles and accounts for year-to-year differences in M_{max} remains to be shown. The Lynchs and their co-workers (Lynch et al., 1976; Lacy et al., 1978) found that house mice exposed to cold ($5°C$) during the first month of life exhibited a greater brown fat mass at 2 months of age than did mice exposed to cold during the second month of life. They also found that rearing mice at $5°C$ from birth to 3 months of age resulted in permanent increases in nonshivering thermogenesis and brown fat weight. Experiments on red-backed voles to determine the influence of cold exposure during early life on the thermogenic capacity in later life could help explain year-to-year differences in thermogenic capacity.

MECHANISMS FOR INCREASING OR SUPPORTING MAXIMUM METABOLISM AND NONSHIVERING THERMOGENESIS

One of the principal mechanisms for increased thermogenic capacity in small mammals, an increase in brown fat (BF) mass (Smith and Horwitz, 1969; Chaffee and Roberts, 1971; Foster and Frydman, 1978), has been shown to occur on a seasonal basis in red-backed voles. In Fig. 1, Sealander's data (Sealander, 1972) show a greater than 2-fold increase in the relative mass of interscapular BF from summer to winter with the peak in mass of this thermogenic tissue coinciding with peaks for M_{max} and nonshivering thermogenesis.

With regard to central nervous system adjust-

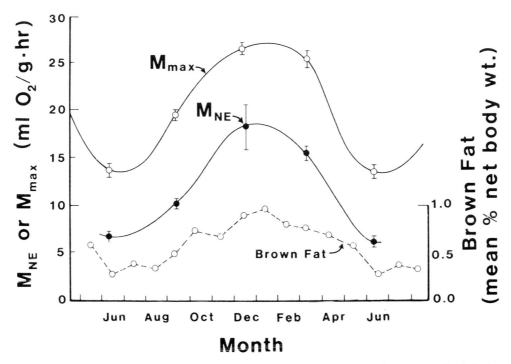

Fig. 1.—Seasonal changes in maximum metabolism (M_{max}; Rosenmann et al., 1975), metabolic response after injection of norepinephrine (M_{NE}; Feist and Rosenmann, 1976) and interscapular brown fat (Sealander, 1972) in Alaskan red-backed voles (*Clethrionomys rutilus*).

ments versus peripheral tissue adjustments for thermogenesis in acclimatization, one can pose the question: does acclimatization to winter involve an increased sympathetic stimulation to BF (that is, a new set point) at any particular temperature below thermoneutrality? That is, if a summer vole and a winter vole were exposed to a temperature below thermoneutrality (such as 0°C, which may elicit a similar total metabolic rate in each vole but would involve more heat production by BF in the winter vole) would the greater nonshivering thermogenesis from BF in the winter vole be due, at least in part, to a greater sympathetic nerve stimulation to BF? Recent experiments on NE turnover in BF indicate that warm acclimated (22°C) and cold acclimated (5°C for 2 mo) red-backed voles exposed acutely to 0°C exhibit a similar intensity of sympathetic stimulation to BF (Feist, 1980). These results, along with those indicating that NE stimulated nonshivering thermogenesis is greater in the cold-acclimated voles (Feist and Rosenmann, 1976) and that nonshivering thermogenesis is based primarily in BF (Foster and Frydman, 1978), suggest that in red-backed voles 1) sympathetic nerve activity in BF depends primarily on the ambient temperature of exposure and this relationship was not altered by a change in acclimation state and 2) enhanced nonshivering ther-

mogenesis after acclimation to cold is based principally in alterations of BF adrenergic receptors and/or related cellular metabolic capabilities. Much more work needs to be done on laboratory-acclimated voles to verify these suggestions and on wild voles to demonstrate their application to acclimatization to winter. Recent results from our laboratory suggest that the enhanced thermogenic capacity in winter may, in part, be due to an increase in the density of β-adrenergic receptors for NE in brown fat (Feist, 1983).

In addition to increases in thermogenic tissue, certain cardiorespiratory and neuro-endocrine changes have been found in winter-acclimatized voles which may be important in supporting the increased M_{max} and nonshivering thermogenesis. Morrison et al. (1966) found an annual cycle of muscle myoglobin that coincides with that of M_{max} and nonshivering thermogenesis. The increased myoglobin may facilitate O_2 transfer and storage and may increase the efficiency of shivering heat production by muscles. Sealander (1966) reported a greater relative heart weight in winter voles, and more recently we have found both a greater relative and absolute heart mass in winter voles. For example, summer voles (23 g) had hearts of 128 mg (=5.7 mg/g body wt.) while winter voles (18 g) had

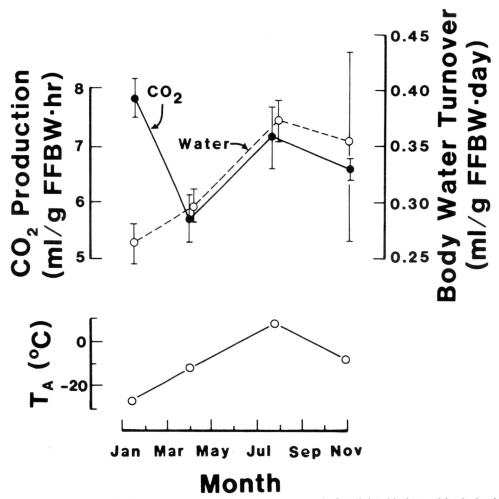

Fig. 2.—Seasonal changes in carbon dioxide production and body water turnover in free-living Alaskan red-backed voles. T_A is ambient temperature. Data are derived from Holleman et al., 1982.

hearts of 143 mg (=8.0 mg/g). A greater volume of the heart and/or greater strength could increase the cardiac output and support the increased metabolic capacity in winter. Since the sympatho-adrenal system stimulates BF and affects cardio-vascular and metabolic responses to cold (Himms-Hagen, 1975; Maickel et al., 1967; Mejsnar and Jansky, 1976), we examined the capacity of adrenals to synthesize NE and epinephrine (E) at different seasons by assaying tyrosine hydroxylase (TH) and phenylethanolamine-N-methyltransferase activities (Feist and Feist, 1978). We found an elevated activity of TH, the rate limiting enzyme, in fall and winter. Thus, winter voles show an increased capability of the adrenals to produce hormones that can affect heat production. However, the precise roles of NE and E in influencing tissue metabolism (White and Feist, 1980) and cardiovascular activity remain to be elucidated in seasonally acclimatized red-back voles.

ENERGY CONSERVING ADJUSTMENTS

An apparent energy conserving mechanism which has been given considerable attention in other papers of this volume is the phenomenon of body weight reduction from summer to winter which occurs in a number of northern latitude small mammals (Rosenmann et al., 1975). As shown by Sealander (1966) and Whitney (1973, 1976), populations of Alaskan red-backed voles undergo an annual cycle of body weight from a mean of about 25 g in early June to about 15 g in December, due to cessation of growth and weight loss of individual voles and mortality of older, larger voles. Fuller (1969), who also measured weight loss in this species, suggested that this reduction in size would be advantageous in reducing food requirements in winter. Our findings on seasonal metabolism (Rosenmann et al., 1975) confirm this view. Comparison of calculated resting daily energy requirements for summer and winter voles at temperatures ranging from +15°C to −30°C shows about a 30% advantage for winter. For example, during exposure to −20°C, theoretically, summer voles (28 g) would have a metabolic rate of about 138 kJd^{-1} while winter voles (15 g) would have a rate of about 100 kJd^{-1}. Much more work needs to be done on changes in body composition involved in the weight loss and on neuroendocrine mechanisms which may control the cessation of growth and loss of body weight.

In a recent study of energy metabolism and body water turnover in free living red-backed voles (Holleman et al., 1982), we were concerned that the energy cost of ingesting frozen food and water might limit intake under conditions of extreme cold in winter and might lead to metabolic adjustments such as reduced metabolic rate and/or reduced body water turnover. After application of the doubly labeled water method (that is, injection of $^3H_2O^{18}$ as described in Holleman et al., 1982) to free-living voles at different seasons, we found (Fig. 2) that the metabolic rate (as CO_2 production) was highest in winter and summer voles and lowest in spring. Calculations of average daily metabolic rate (ADMR) from CO_2 production indicated that ADMR was highest and similar in winter and summer (58.5 kJd^{-1} for 14 g winter voles and 61.8 kJd^{-1} for 16 g summer voles). The high ADMR in our free-living winter voles confirms the prediction of Whitney (1977) based on conventional respirometry. The similar ADMR in our 16 g summer voles (with that in winter voles) contrasts with Whitney's (1977) lower predicted ADMR for larger (23 to 29 g) summer voles. Differences in ADMR among summer voles of different size (as well as age cohort and sex) might be expected and needs much more investigation in free-living voles from the standpoint of total ADMR and also the partitioning of energy into various physiological and behavioral activities in winter versus summer (see also discussion in Whitney, 1977).

In the free-living red-backed voles, we (Holleman et al., 1982) also found (Fig. 2) that body water turnover was correlated with ambient temperature in all seasons and positively correlated with metabolic rate (CO_2 production) in all seasons except winter. In winter when metabolic rate was high, body water turnover was significantly reduced by

Table 1.—*Summary of metabolic and thermogenic adjustments in winter acclimatization of red-backed voles.*

Adjustment	Reference
Increased:	
1. Maximum metabolism	Rosenmann et al., 1975
2. Nonshivering thermogenesis	Feist and Rosenmann, 1976
3. Brown fat mass	Sealander, 1972
4. White fat mass	Sealander, 1972
5. Myoglobin in muscle	Morrison et al., 1966
6. Heart weight	Sealander, 1966; this paper
7. Capacity of adrenals to synthesize catecholamine hormones	Feist and Feist, 1978
8. Andrenergic receptors in brown fat	Feist, 1983
Decreased:	
1. Body mass	Sealander, 1966; Whitney, 1976
2. Body water turnover	Holleman et al., 1982

30% from that in summer. This dissociation between metabolic rate and body water turnover contrasts with the generally assumed correlation between these processes (Yousef et al., 1974; Deavers and Hudson, 1977).

From estimates of body water turnover and production rate of metabolic water, we calculated the rate of ingestion of ice or cold water and then estimated the energy cost of warming to normal body temperature (Holleman et al., 1982). We also calculated the potential energy savings of consuming less ice in winter due to the 30% reduction in body water turnover. The total energy cost for a winter vole at an ambient temperature of −23°C to warm ingested ice plus dry herbage was 2.5 to 3.0% of ADMR. This contrasts with previous estimates of 8 to 30% ADMR for small rodents ingesting ice between −10 to −20°C (Whitney, 1977; Chappell, 1980). The higher percentages were calculated from estimated ingestion rates that were considerably higher than the water turnover rates measured in our study. The energy savings due to reduced water turnover in winter was about 1.7% of ADMR. Thus, reduced body water turnover is a feature of acclimatization to winter, but the energy savings seems insignificant. The importance of body water conservation in the winter acclimatization processes remains to be clarified. The reduced water turnover in winter may be related to: 1) ingestion of drier herbage and/or herbage of low protein and mineral content (Cameron and Luick, 1972) or 2) reduced respiratory water loss due to more effective functioning of a nasal exchange mechanism (Schmidt-Nielsen et al., 1970). Both of these possibilities deserve further investigation.

SUMMARY AND CONCLUSIONS

Table 1 summarizes seasonal adjustments of red-backed voles which have been discussed in this paper. The average daily metabolism appears to be similar in summer and winter red-backed voles of similar size, but a substantial portion of daily metabolism in winter may be expended on thermogenesis. The capacity for thermogenesis is markedly enhanced during acclimatization to winter. The enhanced heat production involves and is supported by changes in the quantity and quality of various tissues. In spite of the continued growth of certain tissues (e.g., brown fat, heart), other tissues apparently cease growth and voles may decrease in body mass by mid-winter. Currently, we have only a superficial understanding of these aspects of metabolic seasonal acclimatization.

Studies to help clarify the details of the neural and endocrine mediation of metabolic acclimatization should be done on: 1) the relative importance of hormones such as norepinephrine, thyroxin and melatonin in mediating changes in quantity and quality of brown fat for enhanced nonshivering thermogenesis in winter; 2) the relationship between the intensity of sympathetic nerve stimulation to brown fat versus ambient temperature below thermoneutrality in different acclimatization states (that is, summer versus winter); and 3) the number and affinity of hormone receptors for norepinephrine on the cell membrane of brown fat cells and possible changes for enhanced nonshivering thermogenesis during seasonal acclimatization.

Further studies of the physiological and biochemical processes underlying metabolic acclimatization should consider: 1) brown fat cell biochemical mechanisms for heat production (see Cannon et al., 1978) in summer and winter; 2) the quantity and quality of lipid and carbohydrate stores and the effect of cold, norepinephrine and epinephrine on the mobilization and oxidation of free fatty acids and glucose in summer and winter; and 3) changes in body composition during winter weight loss and the neuroendocrine mechanisms which mediate cessation of growth and loss of body weight from fall to winter.

Investigations of the external (environmental) and internal factors which limit the processes and mechanisms of acclimatization should assess: 1) the effects of exposure to low temperature during early growth and development of young born in spring, summer, and fall on the magnitude of maximum metabolism and nonshivering thermogenesis in winter; and 2) the nutritional and energetic composition of seasonal diets and the effects of changing food supply and plane of nutrition on maximum metabolism, nonshivering thermogenesis and body water turnover.

Finally, to better understand the bioenergetics of seasonally acclimatized voles, as a result, in part, of the integration of various metabolic aspects of acclimatization, studies should be done on the average daily metabolic rate in voles of different size, age cohort, sex and breeding status in summer and winter in order to partition the energy requirements for various physiological and behavioral activities (for example, thermoregulation, feeding, breeding, lactation).

ACKNOWLEDGMENTS

Portions of the research reported herein were supported by National Institutes of Health research grants GM10402 from the Institute of General Medical Sciences, ES00689 from the Institute of Environmental Health Sciences and AM26864 from the Instiute of Arthritis, Metabolism and Digestive Diseases.

LITERATURE CITED

CAMERON, R. D., and J. R. LUICK. 1972. Seasonal changes in total body water, extracellular fluid and blood volume in grazing reindeer. Canadian J. Zool., 50:107–116.

CANNON, B., J. NEDERGAARD, L. ROWERT, U. SUNDIN, and J. SVARTENGREN. 1978. The biochemical mechanism of thermogenesis in brown adipose tissue. Pp. 567–594, in Strategies in cold (L. C. H. Wang and J. W. Hudson, eds.), Academic Press, New York, 715 pp.

CHAFFEE, R. R. J., and J. C. ROBERTS. 1971. Temperature acclimation in birds and mammals. Ann. Rev. Physiol., 33:155–202.

CHAPPELL, M. A. 1980. Thermal energetics and thermoregulatory costs of small arctic mammals. J. Mamm., 61:278–291.

DEAVERS, D. R., and J. W. HUDSON. 1977. Effect of cold exposure on water requirements of three species of small mammals. J. Appl. Physiol., 43:121–125.

FEIST, D. D. 1980. Norepinephrine turnover in brown fat, skeletal muscle and spleen of cold-exposed and cold-acclimated Alaskan red-backed voles. J. Therm. Biol., 5:89–94.

———. 1983. Increased β-andrenergic receptors in brown fat of winter acclimatized Alaskan voles. Amer. J. Physiol., 245: R357–R363.

FEIST, D. D., and C. F. FEIST. 1978. Catecholamine-synthesizing enzymes in adrenals of seasonally acclimatized voles. J. Appl. Physiol.: Respirat. Environ. Exercise Physiol., 44:59–62.

FEIST, D. D., and P. R. MORRISON. 1981. Seasonal changes in metabolic capacity and norepinephrine thermogenesis in the Alaskan red-backed vole: environmental cues and annual differences. Comp. Biochem. Physiol., 69A:697–700.

FEIST, D. D., and M. ROSENMANN. 1976. Norepinephrine thermogenesis in seasonally acclimatized and cold acclimated red-backed voles in Alaska. Canadian J. Physiol. Pharmacol., 54:146–153.

FOSTER, D. O., and M. L. FRYDMAN. 1978. Non-shivering thermogenesis in the rat. II. Measurements of blood flow with microspheres point to brown adipose tissue as the dominant site of the calorigenesis induced by noradrenaline. Can. J. Physiol. Pharmac., 56:110–122.

FULLER, W. A. 1969. Changes in numbers of three species of small rodents near Great Slave Lake, NWT, Canada, 1964–1967, and their significance for general population theory. Ann. Zool., Fennicae, 6:113–144.

GRODZINSKI, W. 1971. Energy flow through populations of small mammals in the Alaskan taiga forest. Acta Theriol., 16:231–275.

HIMMS-HAGEN, J. 1975. Role of the adrenal medulla in adaptation to cold, In Handbook of physiology. Adrenal gland. Amer. Physiol. Soc., Washington, D.C., sect. 7, vol. 6, 38:637–665.

HOLLEMAN, D. F., R. G. WHITE, and D. D. FEIST. 1982. Seasonal energy and water metabolism in free-living Alaskan voles. J. Mamm. 63:293–296.

LACY, R. C., C. B. LYNCH, and G. R. LYNCH. 1978. Developmental and adult acclimation effects of ambient temperature on temperature regulation of mice selected for high and low levels of nest-building. J. Comp. Physiol. B, 123:185–192.

LYNCH, G. R., C. B. LYNCH, M. DUBE, and G. ALLEN. 1976. Early cold exposure: effects of behavioral and physiological thermoregulation in the house mouse, Mus musculus. Physiol. Zool., 49:191–199.

MAICKEL, R. P., N. MATUSSEK, D. N. STERN, and B. B. BRODIE. 1967. The sympathetic nervous system as a homeostatic mechanism. I. Absolute need for sympathetic nervous function in body temperature maintenance of cold-exposed rats. J. Pharmacol. Exptl. Therap., 157:102–110.

MEJSNAR, J., and L. JANSKÝ. 1976. Mode of catecholamine action during organ regulation of non-shivering thermogenesis. Pp. 225–242, in Regulation of depressed metabolism and thermogenesis (L. Janský and X. J. Musacchia, eds.), Thomas, Springfield, Illinois, 276 pp.

MORRISON, P., M. ROSENMANN, and J. A. SEALANDER. 1966. Seasonal variation of myoglobin in the northern red-backed vole. Amer. J. Physiol., 211:1305–1308.

PRUITT, W. O., JR. 1968. Synchronous biomass fluctuations of some northern mammals. Mammalia, 32:172–191.

ROSENMANN, M., P. MORRISON, and D. D. FEIST. 1975. Seasonal

changes in the metabolic capacity of red-backed voles. Physiol. Zool., 48:303–310.

SCHMIDT-NIELSEN, K., F. R. HAINSWORTH, and D. E. MURRISH. 1970. Counter-current heat exchange in the respiratory passages: effect on water and heat balance. Resp. Physiol., 9:263–276.

SEALANDER, J. A. 1966. Seasonal variations in hemoglobin and hematocrit values in the northern red-backed mouse, Clethrionomys rutilus dawsoni (Merriam), in interior Alaska. Canadian J. Zool., 44:213–244.

———. 1967. Reproductive status and adrenal size in the northern red-backed vole in relation to season. Internat. J. Biometeor., 11:213–220.

———. 1969. Effect of season on plasma and urinary proteins of the northern red-backed vole, Clethrionomys rutilus. Physiol. Zool., 42:275–287.

———. 1972. Circum-annual changes in age, pelage characteristics and adipose tissue in the northern red-backed vole in interior Alaska. Acta Theriol., 17:1–24.

SEALANDER, J. A., and L. K. BICKERSTAFF. 1967. Seasonal changes in reticulocyte number in relative weights of the spleen, thymus, and kidneys in the northern red-backed mouse. Canadian J. Zool., 45:253–260.

SMITH, R. E., and B. HORWITZ. 1969. Brown fat and thermogenesis. Physiol. Rev., 49:330–425.

WANG, L. C. H. 1978. Factors limiting maximum cold-induced heat production. Life Sci., 23:2089–2098.

———. 1980. Modulation of maximum thermogenesis by feeding in the white rat. J. Appl. Physiol., 49:975–978.

WEATHER BUREAU. 1973. Local climatological data, annual summary with comparative data, Fairbanks, Alaska. U.S. Government Printing Office, Washington, D.C., 4 pp.

WEST, S. D. 1974. Post-burn population response of the northern red-backed vole, Clethrionomys rutilus, in interior Alaska. Unpublished M.S. thesis, Univ. Alaska, Fairbanks, 66 pp.

———. 1977. Midwinter aggregation in the northern red-backed vole, Clethrionomys rutilus. Canadian J. Zool., 55:1404–1409.

———. 1979. Habitat responses of microtine rodents to central Alaskan forest succession. Unpublished Ph.D. dissert., Univ. California, Berkeley, 115 pp.

———. 1982. The dynamics of colonization and abundance in central Alaskan populations of the northern red-backed vole, Clethrionomys rutilus. J. Mamm., 63:128–143.

WHITE, R. G., and D. D. FEIST. 1980. Stearate oxidation in the Alaskan red-backed vole: effects of cold and norepinephrine. The Physiologist, 23:84.

WHITNEY, P. H. 1973. Population biology and energetics of three species of small mammals in the taiga of interior Alaska. Unpublished Ph.D. thesis, Univ. Alaska, Fairbanks, 254 pp.

———. 1976. Population ecology of two sympatric species of subarctic microtine rodents. Ecol. Monogr., 46:85–104.

———. 1977. Seasonal maintenance and net production of two sympatric species of subarctic microtine rodents. Ecology, 58:314–325.

YOUSEF, M. K., H. D. JOHNSON, W. G. BRADLEY, and S. M. SEIF. 1974. Tritiated water turnover rate in rodents: desert and mountain. Physiol. Zool., 47:153–162.

Address: Institute of Arctic Biology, University of Alaska, Fairbanks, Alaska 99701.

WINTERING STRATEGY OF VOLES AND SHREWS IN FINLAND

Heikki Hyvärinen

ABSTRACT

Wintering seems to be more clearly guided by endogenous annual cycles in shrews than in voles. During the autumn and early winter the body weight of shrews in Finland decreases by about 35% and the body length by about 7%, with the obvious consequence of a reduced need for food consumption. The body size of wintering populations of voles is also smaller than summer populations. It is assumed that Bergman's rule is not the consequence of evolution caused only by thermoregulatory needs. Reasons for bigger size in the north can be found in other factors,

such as better ability for successful reproduction in short productive period during the summer.

In wintering shrews the metabolism of lipids is obviously more elevated than that of carbohydrates, in contrast to voles, whose carbohydrate metabolism seems to be more important for energy production during the winter. Almost the whole endocrine system, except adrenal medulla of the common shrew, is inactivated during the winter. Catecholamines are assumed to have a very important role in wintering metabolism of both shrews and voles.

INTRODUCTION

The problem of wintering of small mammals has been studied widely in northern regions. Adaptation mechanisms associated with wintering are often species-specific, but common features can also be seen. The aim of the present work is to compare the wintering strategy of some vole species and of the common shrew in Finland from the ecophysiological point of view. This paper is partly a review and partly based on original material.

MATERIALS AND METHODS

The voles and shrews in the original material were captured and killed, using snaptraps at regular monthly intervals in the years 1966–1973 within the administrative district of Oulu (65°N), Finland. For the different purposes the animals were handled in the appropriate way. For most purposes the animals were frozen and stored in sealed polythene bags in deepfreeze containers at −35°C prior to laboratory treatment. Storage time was less than two months. (For further information see Hyvärinen, 1969a and 1969b and Hyvärinen and Heikura, 1971.) The liver glycogen content was measured using the method of van der Vies (1954). After drying the livers 24 hrs in 105°C, the lipid content of the liver was measured by extracting the lipids using the Soxhlet apparatus with hot alcohol (extraction period at least 17 hr) and by weighing the residue after evaporation of alcohol (see Sidney and Samuels, 1946). Lipid content was measured in a total of 195 common shrew specimens.

The lipase-esterase activity was determined from liver or muscle homogenate according to Seligman and Nachlas (1962), using as substrate 2-naphtyl laurate at pH 7.6. The incubation time was one hour and the incubation temperature 37.5°C. The activity of the enzyme is expressed in milligrams naphthol released per hour from the liver/g body weight, or μg naphthol released in hour per mg tissue.

Phosphorylase activity was measured from liver or muscle homogenate in pH 6.1 citrate buffer in the presence of AMP and fluoride according to the method of Hers and van Hoof (1966). Incubation time was 30 min and temperature 37.5°C.

Glucose-6-phosphatase activity was measured from liver homogenate in pH 6.5 citrate buffer according to the method of Hers and van Hoof (1966). Incubation time was 30 min and

incubation temperature 37.5°C. Both phosphorylase and glucose-6-phosphatase activities are expressed in μg liberated phosphorus (P) in 10 min from liver/g body weight. The material for enzyme activity measurements is from the year 1969–1970.

The left adrenal gland of each animal was fixed in Bouin and then treated by the customary wax-embedding method, cut into 5 μ sections and stained with the trichrome method of Wallard and Houette (Romeis, 1948). The material for the histological study consists of 332 (143 ♀ and 189 ♂) specimens of the common shrew, and 231 (112 ♀ and 119 ♂) specimens of the bank vole. The largest section of the whole gland was located by the aid of an ocular micrometer, and the length and breadth of this section was measured. The greatest length and breadth of the adrenal medulla was similarly measured.

The volume of the adrenal gland and the medulla was calculated according to the formula 4/3 a²b, where a is the major semiaxis (½ length) and b the minor semiaxis (½ breadth). The thickness of the zones of the adrenal cortex in section where the medulla was largest was measured with the aid of an ocular micrometer. Four measurements were made of each zone on different sides of each gland, and the mean value was recorded as the thickness of the zone.

The animals used for catecholamine determinations were deep-frozen with dry ice in the field, transported to the laboratory and preserved deep-frozen until handling. This material consists of 53 specimens of the common shrew and 46 specimens of the bank vole from the year 1970–1971.

Catecholamines in the adrenal glands were isolated using the method of Sourkes and Murphy (1961) and measured using a fluorescence spectrophotometer (Perkin Elmer Model 203). Two

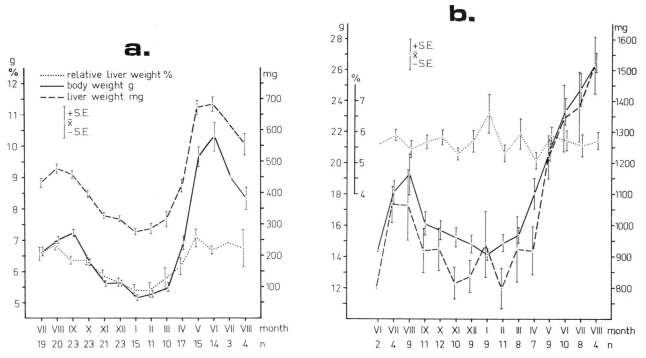

Fig. 1.—Seasonal changes in the body weight (g), liver weight (mg) and relative liver weight (%) of the common shrew (a) and bank vole (b). At the beginning of the curves, the animals are young nonwintered animals and at the end of the curves, old wintered animals.

sets of readings at different wavelengths were used: 1) exciter 405 nm, analyser 495 nm; 2) exciter 435 nm, analyser 540 nm. The calculations for noradrenalin and adrenalin and for total catecholamine content were determined according to the procedure

of Perkin-Elmer Corporation (1968). The results are expressed as µg per both adrenal glands and as µg in both adrenal glands per g body weight.

RESULTS

MORPHOLOGICAL CHANGES

It has been known for a long time that the size of the common shrew (*Sorex araneus*) decreases for winter (Adams, 1912; Dehnel, 1949). In Finland this decreasing is about 35% in body weight (about 15% greater than in Poland) and about 7% in body length (Siivonen, 1954; Hyvärinen and Heikura, 1971). The shortening of body length is caused by the flattening of the invertebral discs and also by the resorption of cartilage in the invertebral discs (Hyvärinen, 1969a). Observed decreasing in the height of the skull (Dehnel, 1949) is caused by the resorption of the parietal and occipital bones at the edges of the *sutura sagittalis* and *sutura lambdoides* (Pucek, 1955; Hyvärinen, 1968, 1969a).

The decreasing of the weight of most internal organs is relatively greater than that of the body weight. Only the relative weight of the heart increases in the winter, when the relative weight of the other organs

measured is smaller during winter than in summer (Pucek, 1965). This is unlike what happens in acclimation experiments of small mammals to cold (see Chaffee et al., 1969).

The body size of the wintering population of all the vole species living in Finland is smaller than that of summer populations (Kalela, 1957; Hyvärinen, 1969b; Hyvärinen and Heikura, 1971; Myllymäki, 1972; Tast, 1972; and others). This is well known in other northern areas, too (see Wasilewski, 1952; Zejda, 1965; Fuller et al., 1969; Merritt and Merritt, 1978). In the skeleton of the voles, there cannot be seen, however, similar resorption phenomena in winter as in the common shrew. The decrease in body weight is known to be caused mainly by the generally smaller size of wintering population and also by the weight loss of individuals in hard conditions (see Heikura, 1977).

In the bank vole (*Clethrionomys glareolus*), which

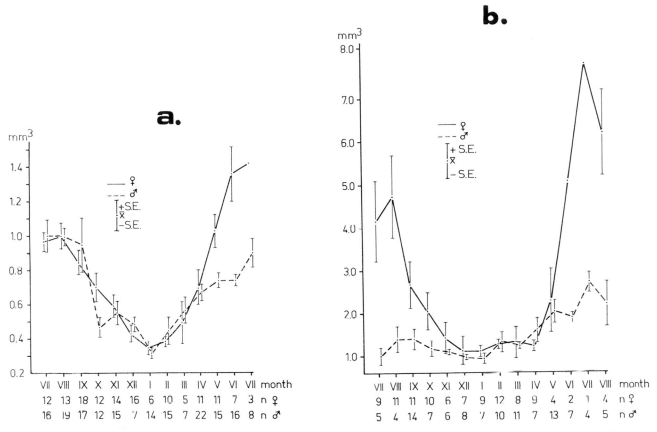

Fig. 2.—Seasonal changes in the volume of adrenal gland of the common shrew (a) and the bank vole (b). See the legend of Fig. 1.

is perhaps the most extensively studied Finnish vole species, the prevention of the growth during the winter is clearly shown (Hyvärinen, 1969b; Hyvärinen and Heikura, 1971). Similar decrease in relative weight of the internal organs as in *Sorex araneus* cannot be seen in *Clethrionomys glareolus* (Hyvärinen, 1967; Fig. 1a, b). In winter the relative liver weight is about the same in *Clethrionomys glareolus* as in *Sorex araneus,* but during the summer months higher in the shrew than in the vole. The relative weight of kidney is about 50% higher in *Sorex araneus* as compared to *Clethrionomys glareolus* (Hyvärinen, 1967), and the relative weight of the interscapular brown fat over 100% higher than in *Clethrionomys glareolus* or in field vole (*Microtus agrestis*) (Pasanen, 1971).

CHANGES IN ENDOCRINE GLANDS

Pituitary Gland

According to histological work, the pituitary gland of *Sorex araneus* is inactive in winter months (Hy-

värinen, 1967, 1969a). Both the proportion of the acidophilic cells (somatotropin secreting cells) and PAS-positive cells (mainly gonadotropins and thyrotropin secreting cells) is very small from December to March. From April to May the volume of the pituitary gland is increasing several hundred percent, especially causing the increase in the number of the gonadotropic cells (Hyvärinen, 1969a). Similar studies have not been published in voles.

Thyroid and Parathyroid Gland

The thyroid gland of *Sorex araneus* is very active in autumn up to the middle of December (autumnal thermal overturn), but clearly inactivates for winter months (Hyvärinen, 1969b). The parathyroid gland is also most active in autumn and early winter and inactivates for midwinter. The active period is connected with resorption of bone during the decreasing of body size (Hyvärinen, 1969a). It has also been shown in the field vole (*Microtus agrestis*) that the activity of the thyroid gland is smallest in winter

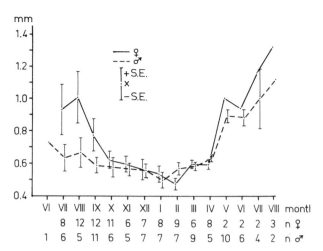

Fig. 3.—Seasonal changes in the thickness of zona fasciculata of the bank vole. See the legend of Fig. 1.

(Rigaudiere, 1969). Similar kind of structural inactivation of the thyroid gland has also been shown in bigger mammals, for example in wintering deer species (see Hoffman and Robinson, 1966).

Adrenal Gland

In *Sorex araneus* the seasonal changes in adrenal gland histology have been studied by Hyvärinen (1969c) and Siuda (1964) and in *Clethrionomys glareolus* by Sorto (1966; unpublished thesis). Sea-

sonal weight changes of adrenal gland of *Clethrionomys glareolus* have been studied by Schvarts (1975). The seasonal changes in adrenal gland volume in both species are very similar according to the earlier results and in the light of present material.

In the common shrew the decrease in adrenal volume in winter (Fig. 2a) is caused by the decrease of zona fasciculata and zona reticularis thickness in both sexes (Hyvärinen, 1969c). Only in spring and summer the thick zona reticularis of adult female shrews causes the difference between the sexes in adrenal gland volume.

In *Clethrionomys glareolus,* however, the clear difference in adrenal gland volume between males and females (Fig. 2b) is mainly caused by the thickness change in the zona reticularis of females, although also the thickness of zona fasciculata is decreasing in winter and clearly increasing in spring (Fig. 3) both in males and females.

The volume of adrenal gland medulla did not decrease significantly for the winter in the male common shrew and bank voles (Fig. 4a, b) but in females the volume of medulla decreased a little for the winter in both species although individual differences were comparatively great.

The increase in medulla volume in wintered adult male bank voles is proportional to body weight changes (see Fig. 1b). In male common shrews, a clear adrenal gland medulla volume increase can be

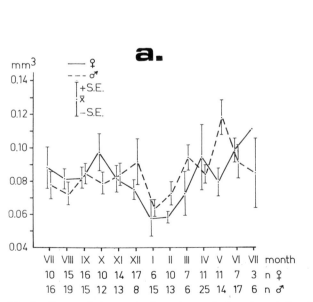

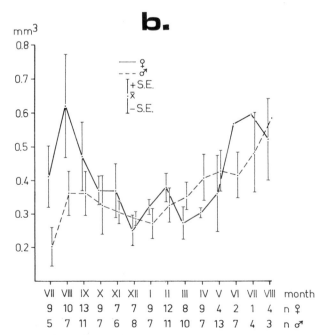

Fig. 4.—Seasonal changes in the volume of adrenal medulla of the common shrew (a) and the bank vole (b). See the legend of Fig. 1.

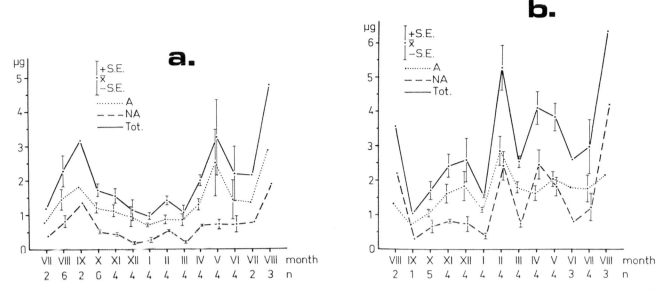

Fig. 5.—Seasonal changes in adrenalin, noradrenalin and total catecholamine contents on adrenal glands of the common shrew (a) and the bank vole (b). See the legend of Fig. 1.

seen (Fig. 4a) in May. Similar increase can also be seen in adrenalin content of the adrenal gland in May (Fig. 5a). Because of the decrease in cortex thickness, the proportion of medulla to the whole adrenal gland is higher in winter than during other seasons (Fig. 6a, b). In the common shrew the catecholamine content tended to be smaller during the winter as compared to other seasons, but no adrenal gland medullary depletion can be seen. In the bank vole no season bound changes in adrenal gland catecholamine content can be seen (Fig. 5b). Catecholamine storages in adrenal gland were surprisingly high in winter in both species (Fig. 7).

METABOLISM AND INDICATORS OF METABOLISM

Pearson (1962) studied the cold acclimation and seasonal acclimatization of *Clethrionomys glareolus* and *Clethrionomys rufocanus* at the University of Helsinki. He observed that in seasonal acclimatization the animals had a lower metabolic rate in winter in all temperatures measured than the animals in summer. Cold acclimated animals had, however, higher metabolic rate in all ambient temperatures than the controls. Similar results have been obtained since then in many other small mammals, for example in *Sorex araneus* (Gebczynski, 1965).

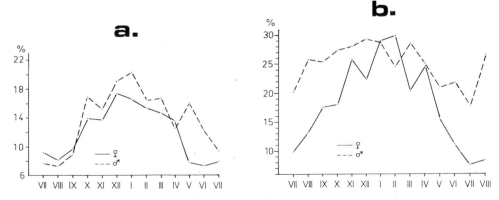

Fig. 6.—Seasonal changes in the relative proportion of adrenal medulla to the whole adrenal gland in the common shrew (a) and in the bank vole (b). See the legend of Fig. 1.

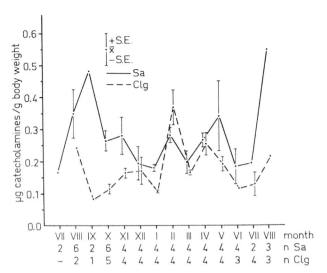

Fig. 7.—Seasonal changes in the relative content (μg catechol-amines/g body weight) of catecholamines in the common shrew (Sa) and in bank vole (Clg). See the legend of Fig. 1.

Regardless of different results (see Rosenmann et al., 1975), it is clear that the energy requirement per individual in a given temperature is lower in winter than in summer, at least when the smaller body size of wintering animals is considered.

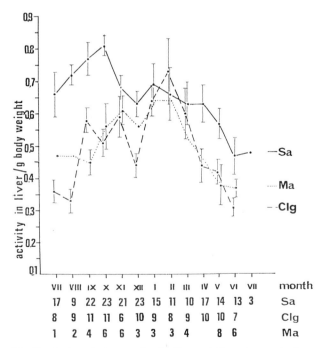

Fig. 8.—Seasonal changes in the liver relative glucose-6-phos-phatase activity (activity in liver/g body weight) in the common shrew (Sa) field vole (ma) and bank vole (Clg). At the beginning of the curves, the animals are young, nonwintered animals and at the end, old wintered animals.

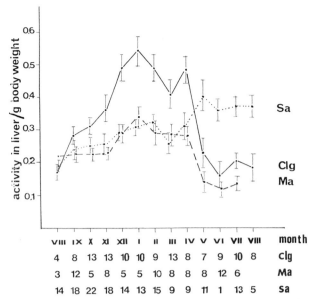

Fig. 9.—Seasonal changes in liver relative glycogen phosphory-lase activity in the common shrew (Sa), field vole (Ma) and the bank vole (Clg). See the legend of Fig. 1.

The activity curves of some liver or muscle enzymes of carbohydrate and lipid metabolism in *Sorex araneus* are quite different as compared with the activity curves of the same enzymes in *Clethrinon-omys glareolus* and *Microtus agrestis*. Relative glu-cose-6-phosphatase activity in liver (total activity in liver/g body weight) is highest in young, nonwin-tered shrews in summer and autumn and decreases for the winter months, and is lowest in adult win-tered animals (Fig. 8). In both vole species glucose-6-phosphatase activity is highest during the winter. Glucose-6-phosphatase is known as gluconeogenic enzyme; its activity correlates with the rate of glu-coneogenesis (Pontremoli and Grazi, 1968). Ac-cording to this gluconeogenesis, or in this case the release of glucose from liver to blood, is most effi-cient in voles during midwinter and in shrews in autumn at the time of autumnal thermal overturn. The relative activity of glycogen phosphorylase, similarly as glucose-6-phosphatase, is highest in voles during the winter months (Fig. 9). In the common shrew, however, its activity is clearly highest in win-tered adults. Especially high glycogen phosphorylase activity was in the liver of *Clethrionomys glareolus* during the winter.

Liver relative lipase-esterase activity was very high in wintering *Sorex araneus* (Fig. 10), and the activity was much higher than that of *Clethrionomys gla-reolus*. In *Clethrionomys glareolus* the highest ac-tivity was measured in the autumn and early winter.

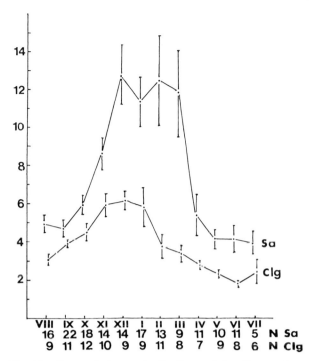

Fig. 10.—Seasonal changes in the liver relative lipase-esterase activity (mg 2-naphthol liberated from 2-naphthol laurate/liver/g body weight/h) in the common shrew (Sa) and the bank vole (Clg). See the legend of Fig. 1.

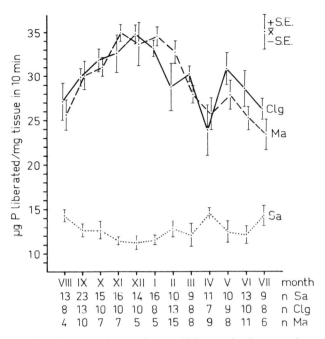

Fig. 11.—Seasonal changes in the thigh muscle glycogen phosphorylase activity in the common shrew (Sa), field vole (Ma) and the bank vole (Clg). See the legend of Fig. 1.

The muscle glycogen phosphorylase activity was much higher in the voles than in the common shrew (Fig. 11). In the voles the highest activity was found in winter, but in the shrew no clear seasonal variation was found. The muscle lipase-esterase activity was clearly higher in the common shrew than in voles. The activity was highest in winter (Fig. 12).

In the common shrew, liver lipid (Fig. 13) and glycogen (Fig. 14) contents have been measured in different seasons. It can be seen that lipid content is normally highest during the winter, although very great differences between the years can be found. Liver glycogen content was, however, lowest in wintering animals and highest in old wintered adults as was glycogen phosphorylase activity.

DISCUSSION

The successful wintering in an active state of such small mammals as shrews has astonished many researchers. The metabolism of shrews is very high (Morrison et al., 1953; Buckner, 1964; Gebczynski, 1965). Although the subnivean temperature is not very harsh, the energy consumption of the shrew during the winter is high. The insulation on shrews is only slightly better in winter than in summer (Gebczynski and Olszewski, 1963). The curious resorption phenomenon of the skeleton leading to smaller size of an individual is possibly developed for diminishing the need of food of an individual (see Mezhzherin and Melnikova, 1966). The endogenous nature of this Dehnel's phenomenon can be seen in the experiments of Pucek (1964). The skull height of the common shrew also decreases in

good nutritive conditions in the laboratory. In general, wintering seems to be more clearly guided by endogenous annual cycles in shrews than in voles.

The Dehnel's phenomenon as measured in the decrease of body weight of shrews is greater in Finland than in Poland. The weight of wintering *Sorex araneus* in Finland is about one gram less than in Poland (see Siivonen, 1954; Pucek, 1965). This difference must be connected with the more severe winter in Finland than in Poland.

The inactivation of the thyroid gland in *Sorex araneus* happens at about the same time as the animals have reached 5.5 g weight (see Hyvärinen, 1969a, 1969b). At about the same time, resorption phenomena also cease and the parathyroid gland is inactivated. It seems that after reaching the 5.0–5.5

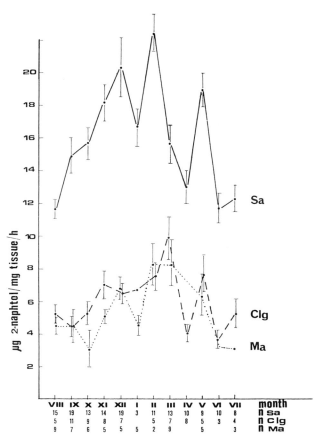

Fig. 12.—Seasonal changes in thigh muscle lipase-esterase activity of the common shrew (Sa), field vole (Ma) and bank vole (Clg). See the legend of Fig. 1.

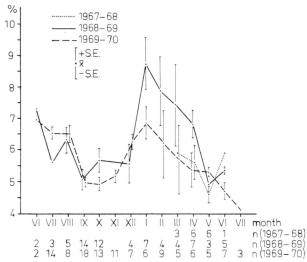

Fig. 13.—Seasonal changes in liver lipid content of the common shrew in three different years. See the legend of Fig. 1.

g weight it is time to interrupt the catabolism. The shrews change after attaining this weight from catabolic autumn metabolic steady state (autumnal thermal overturn) to winter "surviving metabolism." At this time the snow thickness is sufficient to insulate the environment of the animal against fluctuating ambient temperature. In the voles the smaller body size of wintering animals must be originally caused by the same reason as in shrews. In the voles the change from one metabolic steady state to another is not so clear as in *Sorex araneus*, although the metabolic steady state during wintering has similarities with that of the shrew.

Cytochrome c-content in different organs of *Sorex araneus*, *Clethrionomys glareolus* and *Microtus agrestis* should be an indicator of the metabolic rate of the tissue (Hyvärinen and Pasanen, 1973). Cytochrome c-content of the liver, kidney and muscle of the common shrew is about twice that of *Clethrionomys glareolus* and *Microtus agrestis*. In the

brown fat, however, cytochrome c-content is similar in all three species and clearly higher than in the other tissues studied (Hyvärinen and Pasanen, 1973). The so-called autumnal thermal overturn (cf. Merritt and Merritt, 1978) can be seen very clearly in brown fat cytochrome c-content. During winter months cytochrome c-content is smaller than in autumn (Hyvärinen and Pasanen, 1973). In the voles similar metabolic activity can be found only in the brown adipose tissue. The high metabolic capacity of all tissues of the shrew must be important for maintaining the homeothermy in cold. However, the great relative weight (double that of voles) of interscapular brown fat (Pasanen, 1971) indicates

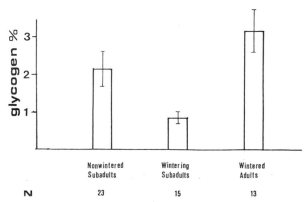

Fig. 14.—Liver glycogen content of the common shrew in young nonwintered subadults (VII–IX), wintering subadults (XII–III) and in wintered adults (V–VII).

that the meaning of brown fat in heat production is also greater in *Sorex araneus* than in voles.

Wintering seems to be more clearly guided by endogenous annual cycles in shrews than in voles. One clear difference between the common shrew and the voles in the midwinter metabolism is in the relative efficiency of carbohydrate and lipid metabolism. According to the results presented here, it is very obvious that the carbohydrate metabolism in voles is relatively more important during wintering than in the shrew, in which the lipid metabolism is especially important. It is very probable that the muscles of the shrew use fatty acids very effectively as an energy source, as it has been found to be the case for example in well trained man (Havel, 1971). The ability of the liver of the shrew to release fatty acids for use is very good, according to the results presented. The content of lipids in food of shrews is high during winter partly because of the high lipid content of wintering insects (see Naumov, 1963).

According to the histological observations, when the most important endocrine glands are inactive in winter, it is obvious that the metabolic acclimatization to winter is brought by quite different hormonal mechanisms than those revealed by study of cold acclimated animals in the laboratory (see Hart, 1963). The relatively great size of adrenal medulla both in *Clethrionomys glareolus* and *Sorex araneus* and the comparatively good stores of catecholamines indicate that catecholamines and especially the noradrenalin are in very important roles in wintering metabolism of small mammals (see Feist and Rosenmann, 1976).

Bergmann (1847) discovered that the dimensions of the bodies of warm-blooded animals increase toward the north. According to the number of results concerning the smaller size of wintering small mammal populations than summer populations, it must be said that in small mammals the so-called Bergmann's rule is not the consequence of evolution caused by thermoregulatory needs. Reasons for bigger size in the north could be found in other factors, such as better ability for successful reproduction in a short productive period during the summer.

LITERATURE CITED

ADAMS, L. E. 1912. The duration of life of the common and lesser shrew, with some notes on their habits. Manchester Mem., 56:1–10.

BERGMANN, C. 1847. Über die Verhältnisse der Värmeökonomie der Tiere zu ihrer Grösse. Göttinger Studien, 3:595–708.

BUCKNER, C. H. 1964. Metabolism, food capacity and feeding behavior in four species of shrews. Canadian J. Zool., 42:259–279.

CHAFFEE, R. R. J., W. C. KAUFMAN, C. H. KRATOCHVIL, M. W. SORENSON, C. H. CONAWAY, and C. C. MIDDLETON. 1969. Comparative chemical thermoregulation in cold- and heat-acclimated rodents, insectivores, protoprimates, and primates. Fedn Proc. Fedn Amer. Socs Exp. Biol., 28:1029–1034.

DEHNEL, A. 1949. Studies on the genus *Sorex* L. Ann. Univ. M. Curie-Sklodowska, Sect. C, 4:17–102.

FEIST, D. D., and M. ROSENMANN. 1976. Norepinephrine thermogenesis in seasonally acclimatized and cold acclimated red-backed voles in Alaska. Canadian J. Physiol. Pharmacol., 54:146–153.

FULLER, W., L. L. STEBBINS, and G. R. DYKE. 1969. Overwintering of small mammals near Great Slave Lake, northern Canada. Arctic, 22:34–55.

GEBCZYNSKI, M. 1965. Seasonal and age changes in the metabolism and activity of *Sorex araneus* Linnaeus 1758. Acta Theriol., 10:303–331.

GEBCZYNSKI, M., and J. L. OLSZEWSKI. 1963. Katathermometric measurements of insulating properties of the fur in small mammals. Acta Theriol., 10:303–331.

HART, J. S. 1963. Physiological responses to cold in non-hibernating homeotherms. Pp. 373–406, *in* Temperature: its measurement and control in science and industry (Z. Herzfeld, ed.), Academic Press, New York, 683 pp.

HAVEL, J. R. 1971. Influence of intensity and duration of exercise on supply and use of fuel. Pp. 315–325, *in* Muscle metabolism during exercise (B. Pernow and B. Saltin, eds.), Plenum Press, New York, 560 pp.

HEIKURA, K. 1977. Effects of climatic factors on the field vole, *Microtus agrestis*. Oikos, 29:607–615.

HERS, H. G., and F. VAN HOOF. 1966. Enzymes of glycogen degradation in biopsy material. Pp. 525–532, *in* Methods in enzymology VIII (S. P. Colowick and N. O. Kaplan, eds.), Academic Press, New York, 759 pp.

HOFFMAN, R. A., and P. F. ROBINSON. 1966. Changes in some endocrine glands of white tailed deer as affected by season, sex and age. J. Mamm., 47:266–280.

HYVÄRINEN, H. 1967. Variation of the size and cell types of the anterior lobe of the pituitary during the life cycle of the common shrew (*Sorex araneus* L.). Aquilo, Ser. Zool., 5:35–40.

———. 1968. On the seasonal variation of the activity of alkaline phosphatase in the kidney of the bank vole (*Clethrionomys glareolus* Schr.) and the common shrew (*Sorex araneus* L.). Aquilo Ser. Zool., 5:7–11.

———. 1969a. On the seasonal changes in the skeleton of the common shrew (*Sorex araneus* L.) and their physiological background. Aquilo Ser. Zool., 7:1–32.

———. 1969b. Seasonal changes in the activity of the thyroid gland and the wintering problem of the common shrew (*Sorex araneus* L.). Aquilo Ser. Zool., 8:30–35.

———. 1969c. Seasonal changes in the morphology of the adrenal cortex and medulla of the common shrew (*Sorex araneus* L.) in Finland. Aquilo Ser. Zool., 9:55–64.

HYVÄRINEN, H., and HEIKURA. 1971. Effects of age and seasonal rhythm on the growth patterns of some small mammals in Finland and in Kirkenes, Norway. J. Zool., London, 165: 545–556.

HYVÄRINEN, H., and S. PASANEN. 1973. Seasonal changes in the cytochrome c-content of some tissues in three small mammals active in winter. J. Zool., London, 170:63–67.

KALELA, O. 1957. Regulation of reproduction rate in subarctic populations of the vole, *Clethrionomys rufocanus* (Sund.). Ann. Acad. Sci. Fenn., A, IV, 34:1–60.

MERRITT, J. F., and J. M. MERRITT. 1978. Population ecology and energy relationships of *Clethrionomys gapperi* in a Colorado subalpine forest. J. Mamm., 59:576–598.

MEZHZHERIN, V. A., and G. L. MELNIKOVA. 1966. Adaptive importance of seasonal changes in some morphophysiological indices in shrews. Acta Theriol., 25:503–521.

MORRISON, P. R., F. A. RYSER, and A. R. DAWE. 1953. Physiological observations on a small shrew, *Sorex cinereus*. Fed. Proc., 12:100–101.

MYLLYMÄKI, A. 1972. Peltomyyrä. Pp. 381–395, *in* Suomen Nisäkkäät I (L. Siivonen, ed.), Otava, Helsinki, 474 pp.

NAUMOV, N. P. 1963. The ecology of animals. N. D. Levine (ed.), Univ. Illinois Press, Urbana, 560 pp.

PASANEN, S. 1971. Seasonal variation in interscapular brown fat in three species of small mammals wintering in an active state. Aquilo Ser. Zool., 11:1–32.

PEARSON, O. P. 1962. Activity patterns, energy metabolism, and growth rate of the voles *Clethrionomys rufocanus* and *C. glareolus* in Finland. Ann. Zool. Soc. "Vanamo," 24:1–58.

PERKIN-ELMER CORPORATION. 1968. Fluorescence clinical chemistry procedures. Norwalk, Connecticut, 90 pp.

PONTREMOLI, S., and E. GRAZI. 1968. Gluconeogenesis. Pp. 259–295, *in* Carbohydrate metabolism and its disoders I (F. Dickens, P. J. Randle, and W. J. Whelan, eds.), Academic Press, London, 576 pp.

PUCEK, Z. 1955. Untersunchungen über die Veränderlichkeit des Schädels in Lebenszyklus von *Sorex araneus araneus* L. Ann. Univ. M. Curie-Sklodowska, Sect. C, 9:113–211.

———. 1964. Morphological changes in shrews kept in captivity. Acta Theriol., 8:137–166.

———. 1965. Seasonal and age changes in the weight of internal organs of shrews. Acta Theriol., 10:369–438.

RIGAUDIERE, N. 1969. Les variations saissonnieres du metabolisme de base et de la thyroide chez les microtines. Arch. Sci. Physiol., 23:215–244.

ROMEIS, B. 1948. Mikroskopische Technik. Oldenbourg, München, 695 pp.

ROSENMANN, M., P. MORRISON, and D. FEIST. 1975. Seasonal changes in the metabolic capacity of red-backed voles. Physiol. Zool. 48:303–310.

SELIGMAN, A. M., and M. M. NACHLAS. 1962. Lipase. Pp. 776–778, *in* Methoden der enzuymatischen Analyse, Weinheim, 980 pp.

SCHVARTS, S. S. 1975. Morpho-physiological characteristics as indices of population process. Pp. 129–152, *in* Small mammals: their productivity and population dynamics (F. B. Golley, K. Petrusewitcz, and L. Ryskowski, eds.), Cambridge Univ. Press, Cambridge, 451 pp.

SIDNEY, R., and L. T. SAMUELS. 1946. Previous diet and the role of the kidney in the metabolism of the eviscerated rat. Amer. J. Physiol., 146:358–365.

SIIVONEN, L. 1954. Über die Grössenvariationen der Säugetiere und die *Sorex macropygmaeus* Mill.-Frage in Fennoskandien. Ann. Acad. Sci. Fenn., Series A, IV, 21:1–24.

SIUDA, S. 1964. Morphology of adrenal cortex of *Sorex araneus* Linnaeus, 1758, during the life cycle. Acta Theriol., 8:115–124.

SORTO, A. 1966. Metsämyyrän lisämunuaisen ja haiman vuodenaikaisesta vaihtelusta. Phil. Cand. thesis, Univ. Oulu, Finland, 25 pp.

SOURKES, T. L., and G. F. MURPHY. 1961. Determination of catecholamines and catecholaminoacids by differential spectrophotofluorimetry. Pp. 147–152, *in* Methods in medical research, 9 (J. H. Quastler, ed.), Chicago, 460 pp.

TAST, J. 1972. Annual variations in the weights of root voles, *Microtus oeconomus*, in relation to their food conditions. Ann. Zool. Fennici, 9:116–119.

VAN DER VIES, J. 1954. Two methods for the determination of glycogen in liver. Biochem. J., 57:410–416.

WASILEWSKI, W. 1952. Badania nad morfologia *Clethrionomys glareolus* Schreb. Ann. Univ. Mariae Curie-Sklodowska, Sect. C, 7:119–211.

ZEJDA, J. 1965. Das Gewicht, das Alter und die Geschlecht aktivität bei der Fötelmaus (*Clethrionomys glareolus* Schreb.). Z. Säugetierk., 30:1–9.

Address: Department of Biology, University of Joensuu, Box 111, 80100 Joensuu 10, Finland.

WINTER TISSUE CHANGES AND REGULATORY MECHANISMS IN NONHIBERNATING SMALL MAMMALS: A SURVEY AND EVALUATION OF ADAPTIVE AND NON-ADAPTIVE FEATURES

W. B. QUAY

ABSTRACT

Winter changes in the tissues of nonhibernating small mammals at intermediate and high latitudes are important in understanding the environmental relations and adaptive strategies and vulnerabilities of these animals. Nonadaptive winter tissue changes are conceptually of two types: (1) pathological changes following tissue injury, as by freezing or other circumstances; and (2) features that are tied to other more primally adaptive changes. Adaptive winter tissue changes include many that are visible in integumentary, cardiovascular, adipose, muscular, hemal and hemopoietic systems. Other winter tissue changes are often complex and are as yet difficult to interpret. Paramount among these are those within the central nervous system. Original research findings are presented concerning two kinds of changes tied secondarily to tissue adaptations: (1) decreased megakaryocyte numbers in bone marrow, probably related to a postulated decrease in platelet production in winter; and (2) a reversed pattern of bone resorption and deposition in the dermal bones of the skull, probably related to decreased intracranial (brain) volume. Winter changes in endocrine and neuroendocrine tissues suggest different adaptive and seasonal strategies in different major taxa. Thus, within Rodentia, pineal size and probable significance in photoperiodic physiological timing of adaptive changes increase with latitude. In Insectivora, no latitudinal pineal cline is evident, and the significance of the pineal in adaptive strategies remains obscure.

INTRODUCTION

Within the broad sweep of winter tissue responses and adaptations of small mammals, only a few have been studied. These represent but the isolated peaks in a figurative, and currently submerged, mountain range. General evidence for this state of affairs comes from two topic areas chiefly: (1) the magnitude and diversity of observationally evident seasonal changes in feral species and populations; and (2) the complexity and plasticity of changes seen in laboratory populations and strains of small rodents in association with changes in particular environmental variables.

The specific features and regulatory mechanisms important in winter responses and adaptations can be studied in a wide variety of ways: anatomically, behaviorally, physiologically and biochemically. We will center our attention here to microanatomical or histological lines of evidence. These are more often easily apparent and readily interpreted than are isolated physiological or biochemical observations. However, true biological understanding requires a synthesis from observations and data from all modes of investigation.

Features and mechanisms of primary concern to us can be considered in accord with either of two rather arbitrary and artificial conceptual subdivisions. The first of these presumes ability to distinguish between adaptive and non-adaptive tissue responses. "Non-adaptive" responses include pathological changes following injury as from freezing, stress, starvation and other or related circumstances, but they might be considered to include as well tissue changes that are in some, often as yet obscure, way a consequence of primary adaptive responses or mechanisms involved in these responses. The second kind of conceptual subdivision divides winter tissue changes and responses into those that are peripheral in location versus those that are in nervous and neuroendocrine control systems. The latter are in many ways the most interesting and challenging. This is because they govern both the temporal biorhythmic programming of seasonal adaptations and the animal's internal evaluation and integration of environmental inputs for appropriate peripheral responses to both acute and chronic environmental stimuli.

This report combines a survey of tissue changes related to winter conditions, with a summary presentation of some original research in two topic areas. These are seasonal tissue changes in bone and bone marrow within the skulls of feral rodents, and a comparison of the pineal's probable contributions in neuroendocrine control mechanisms. These last concerns may lie at the core of seasonal programming of biological adaptations at geographic intermediate and high latitudes. With the objective of putting these observations into a meaningful biological perspective, I will briefly survey and interrelate examples first of non-adaptive winter tissue changes or features, and then what appear to be adaptive winter tissue changes and features. Finally, I will survey briefly some of the evidence for involvements of endocrine, neuroendocrine and nervous systems in the control mechanisms for responses to winter conditions. And lastly, within the framework of these mechanisms, the role of the pineal gland of small rodents and insectivores will be compared and evaluated.

NON-ADAPTIVE FEATURES

Investigations on the effects of cold and other winter conditions have been concerned predominantly with the positive or adaptive side of animal responses, rather than with what might appear to be the negative or moribund side, where pathology and increased mortality are observed results (Hart, 1953). Unfortunately, most of the published work on cold injury is based on results from man and laboratory rats. Nevertheless, these studies do provide insights and technical approaches for study of similar phenomena in feral species under either natural or partially controlled conditions (Crane et al., 1958; Heroux, 1963).

There are several generalizations from such studies with laboratory small mammals, that have important implications for understanding events in feral species in natural winter conditions: (1) Prolonged exposure to cold or related extreme conditions generally modifies the age of onset and the incidence of diseases that occur naturally in the particular species. (2) However, within a particular species, strain or population, a "cold exposure syndrome" of pathological changes can sometimes be defined (Heroux and Campbell, 1960). (3) Circadian and circannual (seasonal) rhythms occur not only in most metabolic, vascular and regulatory parameters in small mammals (Andrews, 1970; Heusner et al., 1971; Stebbins et al., 1980; Lynch and Wichman, 1981), but also in the thermoregulatory responsiveness to behavioral feedback (de Castro, 1978) and in tissue sensitivity to cold (Isobe et al., 1980). Deep hypothermia itself can have different degrees or kinds of effects on the different circadian rhythms of small mammals (Richter, 1975). (4) Freezing or cold injury to tissues may kill some cells and modify some membrane structures and functional capacities, but such injury also sets into motion complex cellular and tissue adaptive responses that maintain some degree of functional integrity while at the same time foster reparative processes (Heroux, 1959; Ehrlich et al., 1981). Therefore, pathological changes following cold injury can not be viewed solely as "nonadaptive," or only as indicators of failure of acclimation or adaptation. Comparative studies of tissue processes in adaptive and "non-adaptive" circumstances are greatly needed if we are to really understand the physiological strategies that are available to northern small mammals. Such studies are lacking as yet.

ADAPTIVE WINTER TISSUE CHANGES

Integumentary System

Winter changes in integumentary structures and characteristics have been studied usually in relation to one or another of two broad biological topic areas: (1) specialized features or adaptations of particular species, and (2) more generally distributed adaptations relating to insulation and thermoregulation. Examples of specialized integumentary features are the winter "double claws" of collared lemmings (*Dicrostonyx*) and the white pelages of these lemmings and of northern weasels (Hall, 1951) and varying hares. "Double claws" of collared lemmings have been well illustrated in their external morphology and growth (Ognev, 1948). It has been surmised for many years that they facilitate digging of tunnels and galleries in subnivean substrates. They appear nearly in full form in late summer young by 26 days of age, even under constant photoperiod conditions (Quay and Quay, 1956). The programming of the development of these structures under these conditions would appear to have to originate from the mother, but the mechanism and physiological routing are not known. Hormonal mechanisms can more readily be attributed to some seasonal changes in pelage pigmentation. These can include changes in levels of, or responsiveness to, melanocyte stimu-

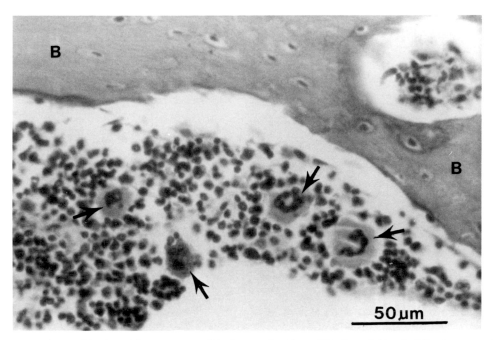

Fig. 1.—Megakaryocytes (arrows) in the red bone marrow of the interparietal bone (B) of an adult male *Microtus pennsylvanicus*, Chester Co., Pennsylvania, July 5, 1980. Acid alum hematoxylin and eosin stains.

lating hormones (Weatherhead and Logan, 1979) for increased pigmentation, and changes in levels of melatonin, a pineal hormone, for decreased pigmentation, as demonstrated in the ermine (*Mustela erminea*) (Rust and Meyer, 1969).

Winter integumentary changes that suggest increased insulation are controversial in regard to their adaptive importance. Winter-caught deer mice (*Peromyscus maniculatus gracilis*) have greater cold resistance than summer mice, and this is accompanied by greater pelage insulation and lower oxygen consumption at 1 to 2°C. However, these changes are not identical to changes produced by cold acclimation in the laboratory (Hart and Heroux, 1953). The interactions and relative contributions of metabolism, insulation and behavior are complex and can vary in relation to subject species, physiological assumptions, methods of study and circadian and circannual phases (Scholander et al., 1950; Hart, 1963; Hart et al., 1965; Chappell, 1980; Aschoff, 1981).

Adipose Tissue

Adipose tissue (fat) is quantitatively the most variable body component of small mammals. It changes markedly in relation to a wide array of factors, including season, nutrition, reproduction, metabolism and stress, in both northern and southern populations (Lynch, 1973; Judd et al., 1978; Rock and Williams, 1979; Millar, 1981). Fat or adipose tissue deposits in mammalian bodies are widespread, but they characteristically include anatomically localized and very specific deposits. These often differ in size and other characteristics in relation to species, developmental stage and such factors as noted above. Two histologic and metabolic categories of adipose tissues are generally recognized: white fat, a metabolically less active nutrient store, and brown fat (BF), an important site of nonshivering thermogenesis (NST) (Himms-Hagen, 1967; Smith and Horowitz, 1969; Jansky, 1973). BF mass increases during cold acclimation (Roberts and Smith, 1967). Cold exposure generally leads to increased oxygen consumption and heat production in BF (Jansky, 1973). This mobilization of BF for NST is mediated by the rich sympathetic innervation of this tissue. Indicative of the increased sympathetic activity of nerve terminals in BF during cold exposure and NST is the increased turnover of norepinephrine (Feist, 1980). This is especially marked in small arctic-subarctic rodents (Feist and Rosenmann, 1976).

Within a more chronic, or less acute, time frame, a number of different hormonal systems have been

Table 1.—*Number of megakaryoctes per parasagittal section of the interparietal bone in adult* Peromyscus leucopus *in winter and summer. All animals were live-trapped at a single site in Ladue, St. Louis County, Missouri, and were immediately dissected and fixed by identical procedures.*

| Dates | Megakaryocytes | | |
	Mean ± SE	N	P*
Dec. 24–27	4.68 ± 0.73	5	<0.01
Aug. 1–5	14.33 ± 2.57	6	

* P = probability based on Student-Fisher *t*.

implicated in regulation of BF in small rodents. These physiological relations can be manifested by changes in BF mass, metabolic level, and specific hormone binding and receptor numbers in the tissue. Among the hormones involved are: growth hormone (Goodman, 1981; Goodman and Coiro, 1981), thyroid hormone(s) (Mory et al., 1981), insulin (Goursot et al., 1981; Castex and Sutter, 1981), adrenal hormones and melatonin (Heldmaier et al., 1981; Sinnamon and Pivorum, 1981; Glass and Lynch, 1981). The last of these, melatonin, a pineal hormone, is of particular interest because it appears to be a component in the physiological linkage of shortening photoperiods and increases in amount and activity of BF during the winter.

Histological studies of the different fat deposits in populations of *Peromyscus* and *Microtus* representing different phases of the annual cycle have led me to two general conclusions. These are not original discoveries, but are merely extensions and correlations in field samples of what has been noted previously, chiefly in laboratory studies. (1) BF is not limited to the interscapular (classical "hibernating gland") region, but is found in many other, and often cryptic sites, such as ventral to the basioccipital bone (Fig. 2). (2) In these regions marked seasonal changes also occur in the cytological indicators of levels of metabolic activity. Thus, in simplest terms, summer BF cells have small nuclei and few but large cytoplasmic lipid droplets, and winter BF cells are smaller overall, but have large nuclei along with reduced average size of cytoplasmic lipid droplets. However, between, and even within, the BF deposits in winter-caught animals there is often great variation in the levels of activity suggested by the quantitative cytological characteristics.

Skeletal Muscle

Skeletal muscle is another of the tissues that contributes to NST in small mammals (Jansky and Hart,

1963). In association with this capacity there have been noted a number of adaptive changes in this tissue either during experimental acclimation to a low temperature, or during winter acclimatization of feral species of northern rodents. Thus, increases have been reported in muscle concentrations of myoglobin (Morrison et al., 1966), and cytochrome c (Depocas, 1966), and in mean capillary density (Wickler, 1981). Mean muscle fiber size or area appears to decrease during either winter acclimatization or cold acclimation. Red and white muscle fibers are affected about equally (Heroux, 1958; Wickler, 1981).

Cardiovascular Tissues

Experimental cold acclimation of small rodents leads to increased cardiac output and increased blood flow to such tissues and organs as brown and white fat, pancreas, kidney, intestine and liver, among others (Jansky and Hart, 1968). In general, thermoregulation is associated with great changes in the partition of blood flow between nutrient and non-nutrient vascular beds (Hales, 1981). Although local sympathetic vasoconstrictor mechanisms are probably involved, much remains unknown about the control of blood vessels and their adaptive changes. Arteriovenous anastomoses (AVAs) form one of the specialized kinds of sites of control of perfusion of vascular beds. Their sympathetic innervation is known the best, but there is evidence for cholinergic and sensory innervations, too (Molyneux, 1981). Studies with humans and other large mammals indicate that vascular responses in the extremities do not depend upon the integrity of the nerve supply, but they are more readily elicited and are greater in magnitude when the sympathetic innervation is intact (Shepherd and Vanhoutte, 1981).

Information on changes within vascular tissues during either cold acclimation or seasonal acclimatization is rare. This is surprising, considering the adaptive importance of these tissues and their further vulnerability in disease processes.

Blood and Blood-Forming Tissues

Blood is a tissue whose cellular components and derivatives are on the one hand (erythrocytes) important for aerobic metabolism and gas exchange, and are on the other hand (leukocytes and platelets) contributors in a wide array of internal defense mechanisms. It is an easily sampled tissue and potentially lends itself to studies monitoring internal events and adaptive success under either experi-

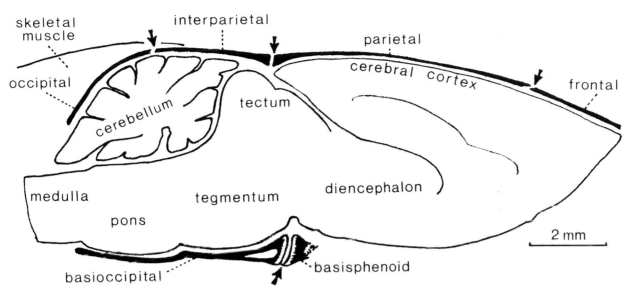

Fig. 2.—Parasagittal section through the brain and some of the adjacent cranial bones of an adult *Microtus pennsylvanicus*, Washtenaw Co., Michigan, Dec. 6, 1952. Major sutures or articulations discussed in the text are indicated by arrows.

mental and artificial conditions, or under natural field conditions. Although such studies have been made with hibernating small mammals, relatively little has been published concerning hemal and hemopoietic changes in non-hibernating small mammals through the course of the annual cycle. This does not deny that hematological and internal fluid volume adjustments have been studied and recognized for many years by studies on mammals subjected to changes in environmental temperatures (Deb and Hart, 1956).

Two major changes are seen in blood cell populations of northern small mammals in winter. The first is an increase in erythrocytes and their immature forms, the reticulocytes (Newson, 1962; Sealander, 1966; Sealander and Bickerstaff, 1967). The second is reduction in numbers of circulating leukocytes, especially in shrews (Wolk, 1981). Satisfactory explanations for this second phenomenon are not available. However, the fact that a peak in immature forms of granulocytic leukocytes occurs in early winter (Wolk, 1981) suggests that an increased turnover in granulocytes may occur at this time.

Personal investigations on seasonal changes in tissues of *Microtus* and *Peromyscus* have revealed a frequently lower concentration or number of megakaryocytes in the bone marrow in winter. This was noted first in samples of *Microtus pennsylvanicus* (Fig. 1), but was substantiated in samples of adult

Peromyscus leucopus from a single locality (Table 1). The late December sample had been exposed to subzero (°C) temperatures for many days, and trapping was subnivean. Since megakaryocytes are the cells of origin for blood platelets, the data suggest that production of platelets is likely to be less in winter than in summer. This in turn implies that the transport and blood-clotting mechanisms in which platelets participate may also be diminished in mid-winter, as compared with summer. More direct and extensive data for such an interpretation are of course needed.

Skeletal Tissues

Examination of seasonal changes in body tissues, and especially bone among the skeletal tissues, can provide important clues concerning the adaptive significance and controlling mechanisms in changes in body weight and composition. Furthermore, within the microstructure of the osseous skeleton, there is evidence of an engraved record of earlier events in seasonal and growth cycles and fluctuations. This has its conceptual basis in the well established plasticity and responsiveness of bone tissue to such influences as diet, hormones, reproductive activity, levels and special attributes of body (muscular) activity, changes in vascular patterns and relative changes in physical pressures or tensions from adjacent structures. The records of the actions of these influences are inscribed locally

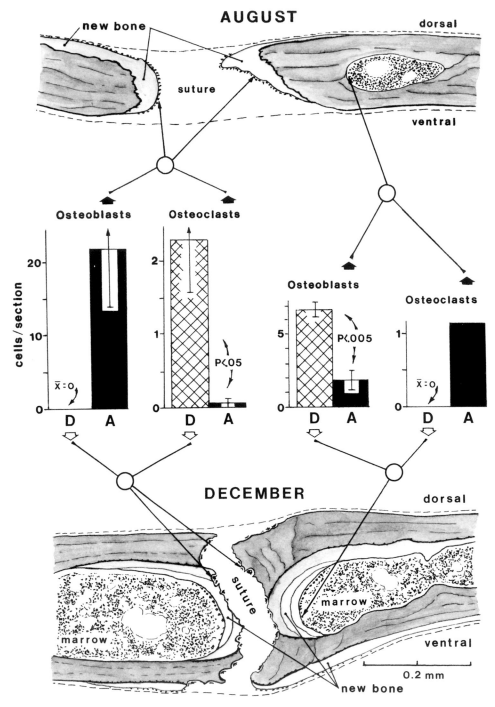

Fig. 3.—Comparisons of bone resorption and deposition in the interparietal-parietal bone region of the skull in relation to season, in *Peromyscus leucopus* at a single site in Ladue, St. Louis Co., Missouri. Parasagittal tissue sections through the parietal (left) and interparietal (right) bones and the connective tissue suture are shown at top (August) and bottom (December), based upon camera lucida outline drawings. Histograms in the center panel summarize quantitative data based upon counts of osteoblasts and osteoclasts in two regions shown, the suture, and the endosteum at the anterior end of the interparietal bone's marrow space. Means ± standard errors are shown where informative. Adult males and females (reproductively inactive) fixed at 08:00–09:40 AM were used (N = 5 [August 1–4] and 6 [December 24–27]).

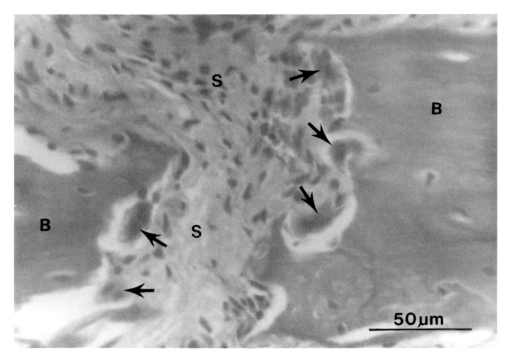

Fig. 4.—Small area of suture (S) between the parietal (left) and interparietal (right) bones (B) of an adult male *Peromyscus leucopus*, Ladue, St. Louis Co., Missouri, December 27, 1980. Numerous osteoclasts are seen (arrows) in excavations (Howship's lacunae) where resorption of bone matrix was occurring. Acid alum hematoxylin and eosin stains.

in bone, through bone deposition by osteoblasts and bone resorption in association with the presence of osteoclasts. Microscopic examination of standard histological slide preparations reveals not only the distribution and numbers of these cells, but can also often distinguish between new and old bone layers (lamellae) by the effects of relative mineralization on staining, and even by some principles and techniques adapted from historical geology! In this way, temporal sequencing of different bone lamellae is often possible.

Winter remodeling of the skull in northern small mammals was first described by Polish mammalogists working with shrews (Dehnel, 1949; Pucek, 1955, 1957). This phenomenon was confirmed in populations of *Sorex araneus* in other regions and by other investigators (Schubarth, 1958; Crowcroft and Ingles, 1959). However, the questions of the taxonomic, geographic and climatographic distributions of winter remodeling of the skull in small mammals remains little studied. Nevertheless, there is evidence for cold or winter conditions being associated with diminished or modified growth, and/or resorption, in other parts of the skeleton, such as tail (Thorington, 1966, 1970), mandible and digital

phalanges (Klevezal and Fedyk, 1978), and trunk vertebrae (Hyvärinen, 1984).

Winter depression of cranial depth in shrews may be a consequence of winter reduction in size of the contained brain (Cabon, 1956; Yaskin, 1980, 1984). The phenomenon appears to be true of some northern small rodents also (Yaskin, 1980, 1984). Other kinds of possible factors, for which however there is only circumstantial evidence, are changes in daily rhythms of skeletal growth (Klevezal and Gebczynski, 1978) and increased blood lactic acid levels with cold exposure and increased energy output (Hart and Heroux, 1954; Ruben and Bennett, 1981).

Personal studies on seasonal changes in the skulls of North American small rodents suggest the following: (1) A seasonal cycle in patterns of bone deposition and resorption in the dermal bones of the cranium can be demonstrated in *Peromyscus leucopus* and *Microtus pennsylvanicus*, even in populations as far south as St. Louis (38–39°N). (2) These changes are in general similar to those described in palearctic species wherein winter depression of cranial depth appears to be associated with winter decrease in brain volume. (3) The nature and localizations of the remodeling processes at the time of

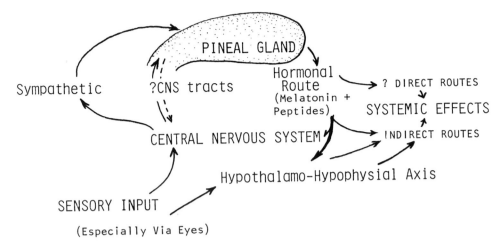

Fig. 5.—Diagrammatic representation of the physiological relations of the mammalian pineal gland. For additional information comprehensive reviews are available (Quay, 1974; Kappers and Pévet, 1979; Reiter, 1981).

death are detectable in part on the basis of localizations and numbers of osteoblasts and osteoclasts. (4) The nature and localizations of earlier remodeling are suggested by differences in mineralization (staining) and lamellar patterns of different parts of the dermal bones.

MATERIALS AND METHODS

To illustrate these and some related points, I will use samples of *Microtus pennsylvanicus* from around the year at Washtenaw County, Michigan, and samples of *Peromyscus leucopus* from two seasons at the St. Louis area, Missouri. The procedures followed that were most satisfactory are as follows: (1) Animals obtained through frequent examination of live-traps are killed within an hour or two by cervical section or pressure. (2) A median slice of the head, excluding skin and rostral, mandibular and palatal regions, is made by cutting with fine tapered scissors passed anteriorly from near the lateral limits of the foramen magnum. The head slice, 4 to 6 mm in thickness, is kept intact, with its bones and meninges in normal anatomical relations. (3) The slice is fixed for two days in standard Bouin's fluid (75 parts sat. aq. picric acid: 25 parts concentrated formalin solution: 5 parts glacial [concentrated] acetic acid) at a ratio of fixative volume to specimen volume of at least 10:1. (4) The fixative is then decanted and replaced with two to three changes of 70 to 80% ethanol. This stops or limits the continued hydrolytic and demineralizing actions of the acidic components and serves as a longterm preservative. The acidity of the fixative and the first alcoholic rinses decalcifies the bone for sectioning, but prolonged immersion in fixative or acidic first rinses hinders staining of cell nuclei and many other structures. (5) The slice is trimmed and dehydrated, cleared, infiltrated and embedded according to standard histological techniques. (6) Serial parasagittal sections of 7 μm thickness are cut with a microtome and affixed to glass slides. (7) These are stained and covered, forming permanent preparations. Two simple staining procedures that I have found particularly useful with such specimens are Ehrlich's acid alum hematoxylin and eosin, and Azure A and orcein. The resulting slides can be used for microscopic study and morphometry of brain regions, skeletal tissues and other structures.

RESULTS

Major seasonal differences are most clearly evident in the regions of the dorsal sutures of the skull (Fig. 2). These differences are consistent with a summer expansion of the dermal bone roof of the skull and a winter depression or subsidence. The basicranial components (Fig. 2), joined by synchondroses rather than sutures, show less change and somewhat more subtle features. Among the dorsal sutures of the skull similar trends are seen in all, but they are perhaps most marked or dramatic in the region of the medial and the parietal-interparietal sutures (Fig. 3). In our description here, only the latter suture will be used as an example. In mid to late summer, poorly mineralized regions of new bone covered with numerous osteoblasts occur on the ends of the dermal bones bordering the sutures and on the dorsal suface (Fig. 3, top panel). There is often an expansion of the connective tissue sutures at this time, along with thickening, lengthening and widening of the bones. In mid to late winter the suture zones are narrow, and active bone resorption occurs at the bone surfaces facing the sutures (Fig. 3, bottom panel; Fig. 4). The integrity of the internal support of these bones is maintained at this time by bone deposition in the endosteum at the bone edges where resorption is taking place externally within the periosteum of the suture. There is also deposition of new bone at some sites along the ventral surface of the bones. As illustrated in Fig. 3, the winter and summer differences are differences primarily in *patterns* of bone deposition and resorption. This conclusion is reinforced by the changes in numbers and locations of osteoblasts and osteo-

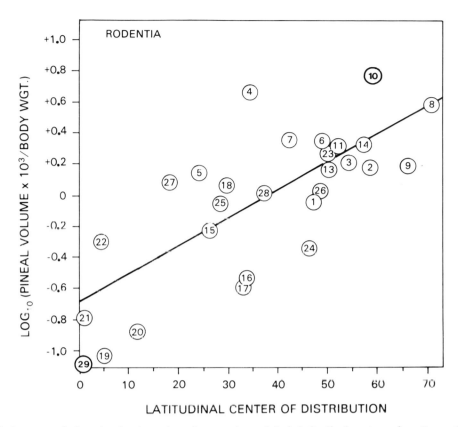

Fig. 6.—Relationship between relative pineal volume in rodent species and their latitudinal centers of north–south geographic distribution. Each circle represents a particular species; the number within each circle is the species number as listed in Table 2. Information on the derivations and correlations of these data is available elsewhere (Quay, 1980*a*). Modifications from the previous presentation of this material consist of the deletion of one unsubstantiated species (#12), the correction of another (#10) (Quay, 1981*b*), and the addition of a third (#29) (Quay, 1981*a*).

clasts (Fig. 3, middle panel). In summary, at either season, osteoblast and osteoclast representations in the sutures are opposite to those in the endosteum bordering the marrow nearby, and similarly, the osteoblast and osteoclast numbers at the two sites have opposite seasonal patterns of change.

DISCUSSION

It is important to appreciate that much variation can occur in these patterns, either at the local microscopic level, as for example with changes in the adjacent blood vessels, or at the individual animal level, as with changes in age, nutrition-related growth rate, or changes in hormone levels or receptors during stress or reproductive activity. A major conclusion still remains: The changes in the skull in winter can not be attributed solely to less bone material through nutritional, hormonal or biochemical mechanisms. Instead, at least in the studied and presumably representative north temperate-zone populations, the winter changes in the skull are adaptive changes in the anatomical *patterns* of bone resorption and deposition. These could well be responses to changes in intracranial volume, especially volume of some brain regions. Morphometry of brain regions in these same specimens should be able to provide one kind of evaluation of this hypothesis.

REGULATORY MECHANISMS

The central regulatory mechanisms for adaptive changes in the winter environment surely involve the nervous and endocrine systems and their interactions. A relatively new approach, the deoxyglu- cose method (Sokoloff, 1981), is available for mapping changes in functional levels of activity within the central nervous system. It is applicable to the problem of localizing changes at times leading to

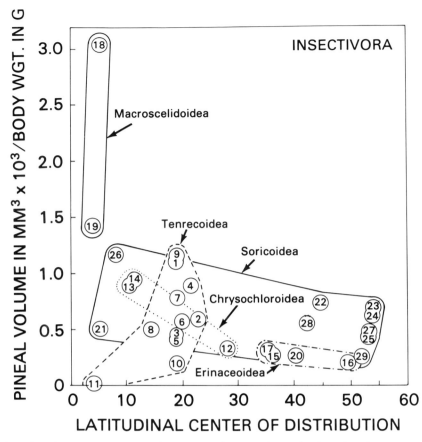

Fig. 7.—Relationship of relative pineal volume to latitudinal centers of geographic distribution of species and superfamilies of Insectivora. Each small numbered circle represents a particular species number as listed in Table 3. Original data are from Legait et al. (1976) and the author's research.

and following cold acclimation and natural seasonal acclimatization. It has been applied only very recently to mapping changes in the brain during hibernation (Kilduff et al., 1982). Contributing to mammalian thermoregulation however, are a great many neural and chemical mediators (Bligh, 1981; Blatteis, 1981). Two major overlapping sites can be recognized where many of these agents or their effects converge, the hypothalamus and the autonomic nervous system, especially its sympathetic (adrenergic) subdivision (Riedel et al., 1981; Boulant, 1981).

Lastly, I wish to point out some comparative aspects of the pineal gland's probable role within seasonal and thermoregulatory adaptations in northern small mammals. Although the marked responsiveness of the pineal's specialized cells, the pinealocytes, to changes in photoperiod was demonstrated early (Quay, 1956), it is only recently that its wide

involvement in seasonal adaptive phenomena has been appreciated (Quay, 1969; Ralph et al., 1979; Reiter, 1980; Hoffman, 1981). Current ideas concerning the mammalian pineal's major physiological relations are summarized in Fig. 5. The major input to the pineal is generally considered to be via its sympathetic innervation, bringing photic and other information from the CNS. But there is increasing evidence for physiological actions within the pineal through more direct neural pathways and involving more complex interactions with photic and other sensory systems. The chief output of the pineal is believed to be hormonal, but much remains to be determined about sites and mechanisms of the actions of this hormonal output, and even about the chemical identity of some of its presumptive hormonal products.

A latitudinal gradient in pineal size and activity in vertebrates has been implied or postulated for

Table 2.—*Rodent species used in Fig. 6. Methods and derivations of data as presented previously (Quay, 1980a).*

Species		Sample size	Body wt. (g)	Pineal vol. (mm³)	Pineal vol. × 10³ Body wt.	Lat. center
No.	Name					
Sciuridae						
1.	*Citellus citellus*	4	291.	0.2753	0.946	47°N
2.	*Marmota marmota*	1	1,900.	2.849	1.499	59°N
3.	*Sciurus vulgaris*	3	319.	0.5331	1.671	54°N
Cricetidae						
Tribe: Hesperomyini						
4.	*Peromyscus leucopus*	34	23.6	0.1090	4.619	34°N
5.	*Sigmodon hispidus*	3	221.	0.3179	1.438	24°N
Tribe Cricetini						
6.	*Cricetus cricetus*	3	325.	0.7296	2.245	49°N
7.	*Mesocricetus auratus*	1	83.	0.1926	2.320	42°N
Subfamily Microtinae						
8.	*Dicrostonyx groenlandicus*	6	53.5	0.2188	4.090	71°N
9.	*Lemmus lemmus*	4	43.5	0.0707	1.625	66°N
10.	*Microtus orcadensis*	1	34.	0.4050	11.912	59°N
11.	*Microtus pennsylvanicus*	9	46.	0.0920	1.999	51°N
13.	*Phenacomys intermedius*	1	28.	0.0435	1.554	50°N
14.	*Synaptomys borealis*	1	30.	0.0656	2.187	57°N
Subfamily Gerbillinae						
15.	*Meriones crassus*	2	167.5	0.1006	0.601	26°N
16.	*Meriones libycus*	3	162.	0.0466	0.288	34°N
17.	*Meriones persicus*	1	130.	0.0331	0.255	33°N
18.	*Meriones shawi*	1	121.	0.1429	1.181	29°N
Muridae						
19.	*Cricetomys gambianus*	1	780.	0.0726	0.093	5.0°S
20.	*Dasymys incomtus*	1	120.	0.0157	0.131	12°S
21.	*Lemniscomys striatus*	2	83.	0.0137	0.165	0.5°N
22.	*Praomys [Rattus] morio*	2	36.5	0.0183	0.501	4°N
Gliridae						
23.	*Eliomys quercinus*	4	82.	0.1592	1.941	50°N
24.	*Glis glis*	86	141.	0.0653	0.463	46°N
25.	*Graphiurus graphiurus*	4	23.	0.0206	0.896	28°S
26.	*Muscardinus avellanarius*	23	21.	0.0226	1.076	48°N
Caviidae						
27.	*Cavia aperea*	1	735.	0.9172	1.248	17°S
Capromyidae						
28.	*Myocastor coypus*	1	3,700.	3.892	1.052	37°S
Bathyergidae						
29.	*Heterocephalus glaber*	1	19.	0.0021	0.112	0.2°N

over a decade and has been associated conceptually with the physiological premium for photoperiodism in arctic species (Folk, 1978). Within major Orders of mammals however, a latitudinal cline in relative pineal size has been demonstrable so far only in the Rodentia (Quay, 1980a; Fig. 6; Table 2). The Insectivora comprise the other group of small mammals whose geographic ranges extend into the far north. They do not show any suggestions of increased relative pineal size with higher latitude (Quay, 1980b; Fig. 7; Table 3). This remains true when one compares the species within the most relevant superfamilies, Erinaceoidea and Soricoidea (Fig. 7). Among rodents are some, such as *Pero-*

Table 3.—*Insectivore species used in Fig. 7. Methods of obtaining the data are described elsewhere (Quay, 1980a, 1980b).*

Species	Pineal volume (P.V.) (mm³)	P.V. × 10³ Body wt.	P.V. × 10² Brain wt.	Latitudinal distribution Center	Highest
Solenodontidae					
1. *Solenodon paradoxus*	0.7350	1.119	14.759	19.	23.
Tenrecidae					
2. *Echinops telfairi*	0.1052	0.584	16.567	23.	25.5
3. *Hemicentetes semispinosus*	0.0920	0.454	10.698	19.	24.5
4. *Limnogale mergulus*	0.0450	0.900	4.545	22.	25.
5. *Microgale cowani*	0.0068	0.447	1.619	19.	25.
6. *Microgale (Nesogale) dobsoni*	0.0223	0.558	4.055	20.	23.
7. *Microgale (Nesogale) talazaci*	0.0439	0.798	5.628	19.	24.5
8. *Oryzorictes talpoides*	0.0216	0.489	3.724	14.	15.5
9. *Setifer setosus*	0.2587	1.150	17.599	19.	25.
10. *Tenrec ecaudatus*	0.2343	0.117	9.261	19.	25.5
Potamogalidae					
11. *Potamogale velox*	0.0016	0.002	0.034	4.	14.
Chrysochloridae					
12. *Chrysochloris asiatica*	0.0146	0.324	2.086	28.	34.5
13. *Chrysochloris leucorhina*	0.0533	0.888	15.229	10.	15.
14. *Chrysochloris (Chlorotalpa) stuhlmanni*	0.0307	0.945	4.264	11.	18.
Erinaceidae					
15. *Aethechinus algirus*	0.1533	0.267	4.746	36.	43.5
16. *Erinaceus europaeus*	0.2460	0.207	6.989	50.	65.
17. *Hemiechinus auritus*	0.0681	0.307	3.547	35.	55.
Macroscelididae					
18. *Elephantulus fuscipes*	0.1600	3.048	12.030	5.	10.
19. *Rhynchocyon stuhlmanni*	0.7050	1.410	11.371	4.	6.
Soricidae					
20. *Blarina brevicauda*	0.0044	0.263	1.275	40.	54.
21. *Crocidura occidentalis*	0.0161	0.503	3.833	5.	12.
22. *Crocidura russula*	0.0073	0.741	4.271	44.5	53.
23. *Neomys fodiens*	0.0114	0.704	3.619	54.	70.
24. *Sorex araneus*	0.0044	0.682	2.110	54.	71.
25. *Sorex minutus*	0.0024	0.438	2.295	53.	70.
26. *Suncus murinus*	0.0312	1.182	8.667	8.	26.5
Talpidae					
27. *Desmana moschata*	0.2037	0.485	7.275	53.	60.
28. *Galemys pyrenaicus*	0.0391	0.559	2.985	42.	43.
29. *Talpa europaea*	0.0212	0.225	2.356	51.5	62.

myscus leucopus, in which the pineal has a mediating role in timing of seasonal reproductive activity, and probably of torpor, in response to changes in photoperiod (Johnston and Zucker, 1980; Heath and Lynch, 1981; Petterborg and Reiter, 1981). There is also intraspecific variability, in part latitudinal, for photoperiodic regulation of seasonal reproduction, with photoperiod-induced regression more extensive in northern populations (Lynch et al., 1980). Also within rodents, the naked mole-rat (*Heterocephalus glaber*) has the relatively smallest pineal, little thermoregulatory ability and non-photoperiodically timed reproductive activity (Quay, 1981a; Moon et al., 1981). It represents the low end and *Dicrostonyx* represents the high end of the latitudinal gradient in the pineal's relative size (Fig. 6). Studies are greatly needed on pineal function of shrews. They may represent both a more generalized mammalian pattern of adaptive mechanisms, at least as compared with rodents, and a better subject for study of possible pineal involvement in seasonal thermoregulation and related phenomena.

LITERATURE CITED

ANDREWS, R. V. 1970. Circadian variations in adrenal secretion of lemmings, voles and mice. Acta Endocrinologica, 65:645–649.

ASCHOFF, J. 1981. Thermal conductance in mammals and birds: its dependence on body size and circadian phase. Comp. Biochem. Physiol. Pt. A, 69:611–619.

BLATTEIS, C. M. 1981. The newer putative central neurotransmitters: roles in thermoregulation, Fed. Proc., 40:2735–2745.

BLIGH, J. 1981. Amino acids as central synaptic transmitters or modulators in mammalian thermoregulation. Fed. Proc., 40:2746–2749.

BOULANT, J. A. 1981. Hypothalamic mechanisms in thermoregulation. Fed. Proc., 40:2843–2850.

CABON, K. 1956. Untersuchungen über die saisonale Veränderlichkeit des Gehirnes bei der kleinen Spitzmaus (*Sorex minutus minutus* L.). Ann. Univ. Mariae Curie-Sklodowska, Sec. C, 10:93–115.

CASTEX, C., and C. J. SUTTER. 1981. Insulin binding and glucose oxidation in edible dormouse (*Glis glis*) adipose tissue: Seasonal variations. Gen. Comp. Endocrinol., 45:273–278.

CHAPPELL, M. A. 1980. Insulation, radiation and convection in small arctic mammals. J. Mamm., 61:268–277.

CRANE, W. A. J., B. L. BAKER, and D. J. INGLE. 1958. Pathologic changes in adrenalectomized and non-adrenalectomized rats exposed to cold. Endocrinology, 62:216–226.

CROWCROFT, P., and J. M. INGLES. 1959. Seasonal changes in the brain-case of the common shrew (*Sorex araneus* L.). Nature, 183:907–908.

DEB, C., and J. S. HART. 1956. Hematological and body fluid adjustments during acclimation to a cold environment. Canadian J. Biochem. Physiol., 34:959–966.

DE CASTRO, J. M. 1978. Diurnal rhythms of behavioral effects on core temperature. Physiol. Behav., 21:883–886.

DEHNEL, A. 1949. Studies on the genus *Sorex* L. Ann. Univ. Mariae Curie-Sklodowska, Sec. C, 4:17–105.

DEPOCAS, F. 1966. Concentration and turnover of cytochrome *c* in skeletal muscles of warm- and cold-acclimated rats. Canadian J. Physiol. Pharmacol., 44:875–880.

EHRLICH, H. P., R. L. TRELSTAD, and J. T. FALLON. 1981. Dermal vascular patterns in response to burn or freeze injury in rats. Exp. Molecular Path., 34:281–289.

FEIST, D. D. 1980. Norepinephrine turnover in brown fat, skeletal muscle and spleen of cold exposed and cold acclimated Alaskan red-backed voles. J. Therm. Biol., 5:89–94.

FEIST, D. D., and M. ROSENMANN. 1976. Norepinephrine thermogenesis in seasonally acclimatized and cold acclimated red-backed voles in Alaska. Canadian J. Physiol. Pharmacol., 54:146–153.

FOLK, G. E., JR. 1978. The pineal and photoperiodism in arctic species. Prog. Reprod. Biol., 4:157–168.

GLASS, J. D., and G. R. LYNCH. 1981. The effect of superficial pinealectomy on reproduction and brown fat in the adult white-footed mouse, *Peromyscus leucopus*. J. Comp. Physiol., 144:145–152.

GOODMAN, H. M. 1981. Separation of early and late responses of adipose tissue to growth hormone. Endocrinology, 109:120–129.

GOODMAN, H. M., and V. COIRO. 1981. Effect of growth hormone on adipose tissue of weanling rats. Endocrinology, 109:2046–2053.

GOURSOT, R., B. PORTHA, C. LEVACHER, and L. PICON. 1981. Effect of early and chronic hypoinsulinism on adipose tissue cellularity in the rat. Diabetologia, 21:418–421.

HALES, J. R. S. 1981. Thermoregulatory implications for partition of the circulation between nutrient and non-nutrient circuits. Pp. 316–328, *in* Progress in microcirculation research (D. Garlick, ed.), Committee in Postgrad. Med. Ed. U.N.S.W., Sydney, Australia.

HALL, E. R. 1951. American weasels. Univ. Kansas Publ., Mus. Nat. Hist., 4:1–466.

HART, J. S. 1953. The relation between thermal history and cold resistance in certain species of rodents. Canadian J. Zool., 31:80–98.

———. 1963. Surface cooling versus metabolic response to cold. Fed. Proc., 22:940–943.

HART, J. S., and O. HEROUX. 1953. A comparison of some seasonal and temperature-induced changes in *Peromyscus*: cold resistance, metabolism, and pelage insulation. Canadian J. Zool., 31:528–534.

———. 1954. Effect of low temperature and work on blood lactic acid in deer mice. Amer. J. Physiol., 176:452–454.

HART, J. S., H. POHL, and J. S. TENER. 1965. Seasonal acclimatization in varying hare (*Lapus americanus*). Canadian J. Zool., 43:731–744.

HEATH, H. W., and G. R. LYNCH. 1981. Effects of 18 weeks of daily melatonin injection on reproduction and temperature regulation in the mouse, *Peromyscus leucopus*. J. Exp. Zool., 216:193–195.

HELDMAIER, G., S. STEINLECHNER, J. RAFAEL, and P. VSIANSKY. 1981. Photoperiodic control and effects of melatonin on nonshivering thermogenesis and brown adipose tissue. Science, 212:917–919.

HEROUX, O. 1958. Weights and composition of muscles of warm- and cold-acclimated rats. Canadian J. Biochem. Physiol., 36:289–293.

———. 1959. Histological evidence for cellular adaptation to non-freezing cold injury. Canadian J. Biochem. Physiol., 37:811–819.

———. 1963. Patterns of morphological, physiological, and endocrinological adjustments under different environmental conditions of cold. Fed. Proc., 22:789–794.

HEROUX, O., and J. S. CAMPBELL. 1960. A study of the pathology and life span of 6°C- and 30°C-acclimated rats. Lab. Invest., 9:305–315.

HEUSNER, A. A., J. C. ROBERTS, and R. E. SMITH. 1971. Circadian patterns of oxygen consumption in *Peromyscus*. J. Appl. Physiol., 30:50–55.

HIMMS-HAGEN, J. 1967. Sympathetic regulation of metabolism. Pharm. Rev., 19:367–461.

HOFFMAN, K. 1981. The role of the pineal gland in the photoperiodic control of seasonal cycles in hamsters. Pp. 237–250, *in* Biological clocks in seasonal reproductive cycles (B. K. and D. E. Follett, eds.), Wright, Bristol.

HYVÄRINEN, H. 1984. Wintering strategy of voles and shrews in Finland. This volume.

ISOBE, Y., S. TAKABA, and K. OHARA. 1980. Diurnal variation of thermal resistance in rats. Can. J. Physiol. Pharmacol., 58:1174–1179.

JANSKY, L. 1973. Non-shivering thermogenesis and its thermoregulatory significance. Biol. Rev., 48:85–132.

JANSKY, L., and J. S. HART. 1963. Participation of skeletal muscle and kidney during nonshivering thermogenesis in cold-acclimated rats. Canadian J. Biochem. Physiol., 41:953–964.

———. 1968. Cardiac output and organ blood flow in warm- and cold-acclimated rats exposed to cold. Can. J. Physiol. Pharmacol., 46:653–659.

JOHNSTON, P. G., and I. ZUCKER. 1980. Antigonadal effects of melatonin in white-footed mice (Peromyscus leucopus). Biol. Reprod., 23:1069–1074.

JUDD, F. W., J. HERRERA, and M. WAGNER. 1978. The relationship between lipid and reproductive cycles of a subtropical population of Peromyscus leucopus. J. Mamm., 59:669–676.

KAPPERS, J. A., and P. PÉVET. 1979. The pineal gland of vertebrates including man. Prog. Brain Res., 52:xvi + 1–562.

KILDUFF, T. S., F. R. SHARP, and H. C. HELLER. 1982. (^{14}C)2-Deoxyglucose uptake in ground squirrel brain during hibernation. J. Neurosci., 2:143–157.

KLEVEZAL, G. A., and A. FEDYK. 1978. Adhesion lines pattern as an indicator of age in voles. Acta Theriol., 23:413–422.

KLEVEZAL, G. A., and M. GEBCZYNSKI. 1978. Diurnal rhythm of the skeleton growth of some rodent species. Acta Theriol., 23:527–539.

LEGAIT, H., R. BAUCHOT, H. STEPHAN, and J.-L. CONTET-AUDONNEAU. 1976. Etude des corrélations liant le volume de l'épiphyse aux poids somatique et encéphalique chez les rongeurs, les insectivores, les chiropteres, les prosimiens et les simiens. Mammalia, 40:327–337.

LYNCH, G. R. 1973. Effect of simultaneous exposure to differences in photoperiod and temperature on the seasonal molt and reproductive system of the white-footed mouse, Peromyscus leucopus. Comp. Biochem. Physiol., 44A:1373–1376.

LYNCH, G. R., J. K. SULLIVAN, and S. L. GENDLER. 1980. Temperature regulation in the mouse, Peromyscus leucopus: effects of various photoperiods, pinealectomy and melatonin administration. Int. J. Biometeorology, 24:49–55.

LYNCH, G. R., and H. A. WICHMAN. 1981. Reproduction and temperature regulation in Peromyscus: effects of chronic short days. Physiol Behav., 26:201–205.

MILLAR, J. S. 1981. Body composition and energy reserves of northern Peromyscus leucopus. J. Mamm., 62:786–794.

MOLYNEUX, G. S. 1981. Neural control of cutaneous arteriovenous anastomoses. Pp. 296–315, in Progress in microcirculation research (D. Garlick, ed.), Committee in Postgraduate Med. Ed. U.N.S.W., Sydney, Australia.

MOON, T. W., T. MUSTAFA, and J. B. JORGENSEN. 1981. Metabolism, tissue metabolites and enzyme activities in the fossorial mole rat, Heterocephalus glaber. Molecular Physiol. 1:179–194.

MORRISON, P., M. ROSENMANN, and J. A. SEALANDER. 1966. Seasonal variation of myoglobin in the northern red-backed vole. Amer. J. Physiol., 211:1305–1308.

MORY, G., D. RICQUIER, P. PESQUIES, and P. HEMON. 1981. Effects of hypothyroidism on the brown adipose tissue of adult rats—comparison with the effects of adaptation to cold. J. Endocrinol., 91:515.

NEWSON, J. 1962. Seasonal differences in reticulocyte count, hemoglobin level and spleen weight in wild voles. Brit. J. Hematol., 8:296–302.

OGNEV, S. I. 1948. Mammals of USSR and adjacent countries (Mammals of eastern Europe and northern Asia). Acad. Sci. USSR, Moscow-Leningrad, vol. VI, 588 pp.

PETTERBORG, L. J., and R. J. REITER. 1981. Effects of photoperiod and subcutaneous melatonin implants on the reproductive status of adult white-footed mice (Peromyscus leucopus). J. Androl., 2:222–224.

PUCEK, Z. 1955. Untersuchungen über die Veränderlichkeit des Schädels im Lebenszyklus von Sorex araneus araneus L. Ann. Univ. Mariae Curie-Sklodowska, Sec. C, 9:163–211.

———. 1957. Histomorphologische Untersuchungen über die Winterdepression des Schädels bei Sorex L. und Neomys Kaup. Ann. Univ. Mariae Curie-Sklodowska, Sec. C, 10:399–428.

QUAY, W. B. 1956. Volumetric and cytologic variation in the pineal body of Peromyscus leucopus (Rodentia) with respect to sex, captivity and day-length. J. Morphol., 98:471–495.

———. 1969. The role of the pineal in environmental adaptation. Pp. 508–550, in Physiology and pathology of adaptation mechanisms: neural—neuroendocrine—humoral (E. Bajusz, ed.), Pergamon, Oxford.

———. 1974. Pineal chemistry in cellular and physiological mechanisms. Charles C. Thomas, Springfield, xvi + 430 pp.

———. 1980a. Greater pineal volume at higher latitudes in Rodentia: exponential relationship and its biological interpretation. Gen. Comp. Endocrinol., 41:340–348.

———. 1980b. Pineal volume and latitudinal and chronobiological characteristics: comparisons of five mammalian orders. Amer. Zool., 20:898.

———. 1981a. Pineal atrophy and other neuroendocrine and circumventricular features of the naked mole-rat, Heterocephalus glaber (Rüppell), a fossorial, equatorial rodent. J. Neural Transmission, 52:107–115.

———. 1981b. Unusual pineal size and morphology in the Orkney vole (Microtus arvalis orcadensis). J. Mamm., 62:622–624.

QUAY, W. B., and J. F. QUAY. 1956. The requirements and biology of the collared lemming, Dicrostonyx torquatus Pallas, 1778, in captivity. Säugetierkundliche Mitt., 4:174–180.

RALPH, C. L., B. T. FIRTH, W. A. GERN, and D. W. OWENS. 1979. The pineal complex and thermoregulation. Biol. Rev., 54:41–72.

REITER, R. J. 1980. The pineal and its hormones in the control of reproduction in mammals. Endocrine Rev., 1:109–131.

———. 1981. The pineal gland. vols. I–III, CRC Press, Boca Raton, Florida.

RICHTER, C. P. 1975. Deep hypothermia and its effect on the 24-hour clock of rats and hamsters. Johns Hopkins Med. J., 136:1–10.

RIEDEL, W., P. K. DORWARD, and P. I. KORNER. 1981. Central adrenoceptors modify hypothalamic thermoregulatory pattern of autonomic activity in conscious rabbits. J. Autonom. Nerv. Syst., 3:525–533.

ROBERTS, J. C., and R. E. SMITH. 1967. Time-dependent responses of brown fat in cold-exposed rats. Am. J. Physiol., 212:519–525.

ROCK, P., and O. WILLIAMS. 1979. Changes in lipid content of the montane vole. Acta Theriol., 24:237–247.

RUBEN, J. A., and A. F. BENNETT. 1981. Intense exercise, bone structure and blood calcium levels in vertebrates. Nature, 291:411–413.

RUST, C. C., and R. K. MEYER. 1969. Hair color, molt, and

testis size in male, short-tailed weasels treated with melatonin. Science, 165:921–922.

SCHOLANDER, P. F., R. HOCK, V. WALTERS, and L. IRVING. 1950. Adaptation to cold in arctic and tropical mammals and birds in relation to body temperature, insulation and basal metabolism. Biol Bull. Woods Hole, 99:259–271.

SCHUBARTH, H. 1958. Zur Variabilität von *Sorex araneus araneus* L. Acta Theriol., 2:175–202.

SEALANDER, J. A. 1966. Seasonal variations in hemoglobin and hematocrit values in the northern red-backed mouse, *Clethrionomys rutilus dawsoni* (Merriam), in interior Alaska. Canadian J. Zool., 44:213–224.

SEALANDER, J. A., and L. K. BICKERSTAFF. 1967. Seasonal changes in reticulocyte number and in relative weights of the spleen, thymus, and kidneys in the northern red-backed mouse. Canadian J. Zool., 45:253–260.

SHEPHERD, J. T., and P. M. VANHOUTTE. 1981. Cold vasoconstriction and cold vasodilation. Pp. 263–271, *in* Vasodilation (P. M. Vanhoutte and I. Leusen, eds.), Raven Press, New York.

SINNAMON, W. B., and E. B. PIVORUM. 1981. Melatonin induces hypertrophy of brown adipose tissue in *Spermophilus tridecemlineatus*. Cryobiol., 18:603–607.

SMITH, R. E., and B. HOROWITZ. 1969. Brown fat and thermogenesis. Physiol Rev., 49:330–425.

SOKOLOFF, L. 1981. The deoxyglucose method for the measurement of local glucose utilization and the mapping of local functional activity in the central nervous system. Int. Rev. Neurobiol., 22:287–333.

STEBBINS, L. L., R. ORICH, and J. NAGY. 1980. Metabolic rates of *Peromyscus maniculatus* in winter, spring and summer. Acta Theriol., 25:99–104.

THORINGTON, R. W. 1966. The biology of rodent tails: a study of form and function. Arctic Aeromed. Lab., TR-65-8:1–137.

———. 1970. Lability of tail length of the white-footed mouse, *Peromyscus leucopus noveboracensis*. J. Mamm., 51:52–59.

WEATHERHEAD, B., and A. LOGAN. 1979. Seasonal variations in the response of hair follicle melanocytes to melanocyte-stimulating hormone. J. Endocrinol., 81:167P–168P.

WICKLER, S. J. 1981. Capillary supply of skeletal muscles from acclimatized white-footed mice *Peromyscus*. Amer. J. Physiol., 241:R357–R361.

WOLK, E. 1981. Seasonal and age changes in leukocyte indices in shrews. Acta Theriol., 26:219–229.

YASKIN, V. A. 1980. Seasonal changes in brain morphology, basic morphological indicators and behavior of the red-backed vole. Pp. 152–159, *in* Adaptations of animals for winter conditions, Inst. Evol. Morphol. Animal Ecol., Acad. Sci. USSR, Moscow.

———. 1984. Seasonal changes in brain morphology. This volume.

Address: 2003 Ida Street, Napa, California 94558.

STRATEGIES FOR, AND ENVIRONMENTAL CUEING MECHANISMS OF, SEASONAL CHANGES IN THERMOREGULATORY PARAMETERS OF SMALL MAMMALS

Bruce A. Wunder

ABSTRACT

Many small, non-hibernating mammals from environments with cold winters show seasonal changes in several thermoregulatory parameters. A few species of *Peromyscus* and *Phodopus sungorus* show increased capacity for nonshivering thermogenesis (NST) and increased incidence of daily torpor in winter. It was originally thought that such changes were stimulated by seasonal temperature changes. However, more recently it has been found that photoperiod may also be involved.

Since hibernation and daily torpor are physiologically similar and since some hibernators are cued by photoperiod, the cueing of daily torpor by photoperiod is to be expected. However, no species of microtine rodent has ever shown torpor, yet many show increases in basal metabolism (BMR) and NST during winter. Thus, I studied cueing of these parameters in *Microtus ochrogaster*. Animals held at a constant temperature of 23°C all year with a seasonally varying photoperiod showed winter increases in NST capacity, but no change in BMR. However, animals held outside on seasonally and daily varying temperature and photoperiod showed not only winter peaks in NST but also BMR.

Thus, the roles of temperature and photoperiod as cueing mechanisms for thermoregulation are discussed. Also, the significance of these changes for seasonal energy balance and allocation of energy to other processes, such as reproduction, are considered.

INTRODUCTION

Winter poses a variety of stresses for small mammals in North Temperate and Arctic latitudes. Two of the most important of these are reduced food availability and increased cold, both of which affect energy balance. In response, small mammals show a variety of adaptations which can be grouped into two broad categories: 1) Avoidance (mechanisms for reducing the magnitude of the stress) and 2) Resistance (mechanisms for combating the stress). Specific mechanisms for effecting such adaptations are listed in Table 1, and it should be noted that an animal can utilize a variety of combinations of these factors in adjusting to winter.

Given the topic of this symposium, I will confine my discussion to small mammals which are active during winter. With that constraint there are limitations in the extent to which small mammals can utilize the various tactics listed in Table 1. Hibernation is excluded and migration to warmer climate is practical only for bats. Increased insulation through fat and fur is of limited value because small mammals cannot change these attributes greatly and still function (Hart, 1971). Thus, the primary adaptations one might consider for small mammals active in winter are: 1) microclimate, 2) occasional short-term torpidity, and 3) increased capacity for thermogenesis.

Since others in the symposium have discussed the subnivean environment and effects of nests on the thermal environment of small mammals, I will restrict my comments primarily to the use of increased thermogenesis and torpidity in adaptation to winter.

In this regard several species have been reported to show changes in thermogenic capacity at different seasons. Red-backed voles (*Clethrionomys rutilus*—Rosenmann et al., 1975), white-footed mice (*Peromyscus leucopus*—Lynch, 1973; Wickler, 1980), deer mice (*Peromyscus maniculatus*—Stebbins et al., 1980), Djungarian hamsters (*Phodopus sungorus*—Heldmaier et al., 1981) and prairie voles (*Microtus ochrogaster*—Wunder et al., 1977) all show increased thermogenesis during winter relative to summer values.

It was initially thought that such changes were due to colder temperature exposures in winter because it is well known that similar responses occur in the lab due to cold acclimation (see Wunder, 1979, for review). However, if an animal in the field must be cold exposed in order to develop a response to that cold, it may be caught at certain times of year without much thermogenesis during severe cold snaps which occur quickly, since it usually takes two weeks to develop maximal responses to cold (Hart, 1971). Thus, it seems reasonable that other, anticipatory cues might be used by these small forms to prepare for winter cold.

One obvious such cue is photoperiod. Indeed, recently Lynch et al. (1978) have found that photoperiod changes alone will affect incidence of spontaneous torpor and capacity for NST in white-footed

Table 1.—*Mechanisms for adjusting to winter cold.*

A. Avoidance
   1. Migration
   2. Microclimate Usage
   3. Torpidity
   4. Increased Insulation
B. Resistance
   1. Increased Thermogenesis
      a. Basal Metabolic Rate
      b. Nonshivering Thermogenesis
      c. Shivering
      d. Activity

Table 2.—*Body mass* of prairie voles during the year.*

Sample period	Animals held inside	Animals held outside
October	45.8 ± 6.7 (20)	45.9 ± 5.8 (15)
November	44.3 ± 7.6 (13)	44.8 ± 5.1 (15)
January	42.2 ± 6.4 (11)	44.8 ± 6.9 (14)
February	42.8 ± 7.2 (10)	47.1 ± 7.5 (11)
March–April	44.9 ± 9.7 (9)	50.2 ± 6.6 (10)
May	43.8 ± 7.7 (9)	48.9 ± 9.5 (8)

* Values are the mean ± 1 std. deviation. Sample size is given in parenthesis.

mice. Also, Heldmaier and Steinlechner (1981) have reported photoperiod effects on basal metabolic rate (BMR), non-shivering thermogenesis (NST; Jansky, 1973), body weight, and incidence of torpor in Djungarian hamsters.

NST is important in arousal of species which show torpor (or hibernation); thus it may be that the stimulus for increased NST is really associated with changes in capacity for torpor rather than thermogenesis in general, since both species discussed above do show daily torpor. Therefore, I undertook a study to determine whether photoperiod stimulates changes in thermogenesis in the prairie vole (*Microtus ochrogaster*), a species which shows seasonal changes in thermogenic capacity (Wunder et al., 1977), but with no known capacity for torpor.

MATERIALS AND METHODS

My experimental design was simple. Voles were trapped near Fort Collins in early to mid-September 1980. They were housed in individual cages (29 by 18 by 12 cm) with wood shavings for litter and lab chow and water available *ad libitum*. One group was kept in a lab room at a constant temperature (23 ± 2°C), but the light timer was changed weekly to simulate natural photoperiod for that week. The other group was kept outside in a screened area with a roof, so that they were exposed to natural photoperiod and daily and seasonal changes in temperature, but were protected from weather extremes such as snow and rain. In both groups animals had cardboard tubes and cotton which they used for making nests. During winter, the animals outside had water bottles removed at night (since they would freeze and break), and ice cubes were available for water (we could not duplicate this inside as the cubes would melt). Body mass was measured weekly, and BMR and NST were measured at six week intervals until May 1981. BMR was measured at 29°C using open flow system respirometry, as in Wunder et al. (1977). NST was measured as the highest consistent oxygen consumption in unanesthetized animals following a subcutaneous injection of noradrenaline. The dose, 1.35 mg/kg body mass, of noradrenaline for maximal NST response was calculated from Heldmaier (1971). As metabolism increased and the animals warmed, they usually spread out at rest on the wire mesh grid in the respirometer.

RESULTS

There were no statistically significant differences in body mass of voles either within a group or between groups throughout the study (Table 2). Animals held inside showed no seasonal changes in BMR; however, they did show seasonal differences in NST (Fig. 1). The NST capacity was significantly higher ($P < 0.05$) in winter than in fall or spring. Thus, photoperiod alone appears to be a stimulus for seasonal changes in NST. Animals held outside showed a significant ($P < 0.05$) increase in BMR by November and maintained that high rate until March to April, at which time they dropped back to the rate shown in early fall (Fig. 1). They also showed winter increases in NST. The NST pattern is similar to that seen in animals held inside, but the values are slightly higher ($P < 0.05$) than animals from inside (Fig. 1). Hence, temperature appears to modulate the seasonal change in NST and to stimulate an increase in BMR.

DISCUSSION

As indicated in the Introduction and as discussed by Fretwell (1972), winter is potentially a stressful time for small mammals, particularly with regard to energy balance since food (energy input) may be less available and cold stress (energy expenditure) is increased. In essence then we can think of many adaptations to winter in small mammals as problems in achieving and maintaining energy balance.

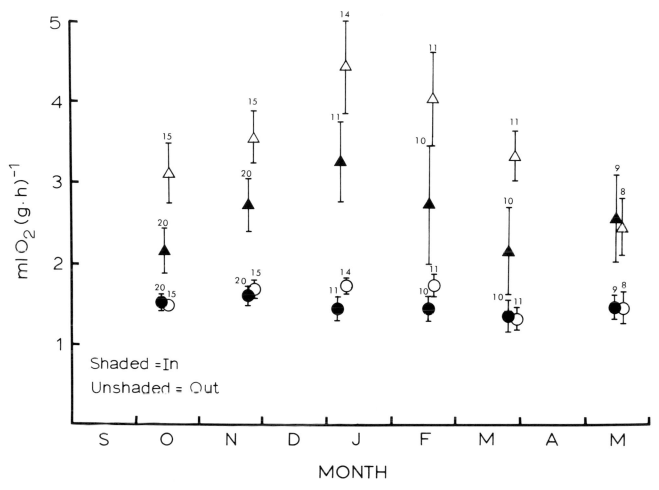

Fig. 1.—Metabolic responses of prairie voles at different months throughout the year. Shaded symbols represent animals held inside, and unshaded symbols represent animals held outside. Circles designate basal metabolic rate, and diamonds designate nonshivering thermogenesis. Vertical lines represent ±2 std. errors of the mean, and sample size is given above each group.

Within the constraints of my discussion which I outlined in the Introduction, torpidity is a tactic for saving energy, and increased thermogenesis, which allows a small mammal to maintain body temperature during cold stress, demands increased energy turnover.

Energy balance in mammals has been recently reviewed in several contexts (Hart, 1971; Grodzinski and Wunder, 1975). More recently, I have discussed the interaction of various factors involved in energy balance, and how they might relate to distribution patterns (Wunder, 1978a). In this latter paper I proposed that small mammals must partition their energy use in a priority fashion, as depicted in Fig. 2. As indicated by the cascade of energy flow lines, I proposed that a primary priority for energy use is thermoregulation. My reasoning was quite simple. If a mammal is not thermoregu-

lating, then it is either hypo- or hyperthermic or is becoming so, and ultimately cannot perform its normal functions; therefore, channeling energy to other functions would not be possible in any case. Secondly, I suggested that feeding is important because that activity provides the energy for all other functions. When those two needs are met, an animal can then funnel energy to other activities. Of course, some small mammals can utilize torpor to save on maintenance energy costs and hence may have more energy to divert to feeding or other activities when normothermic, but some processes may be precluded by bouts of torpor (for example, reproduction?). Before continuing I should also point out that the lines in Fig. 2 represent not only total energy flow per unit time but also turnover rates (Kleiber, 1975). It is especially important to differentiate these concepts, as I have previously indicated (Wunder,

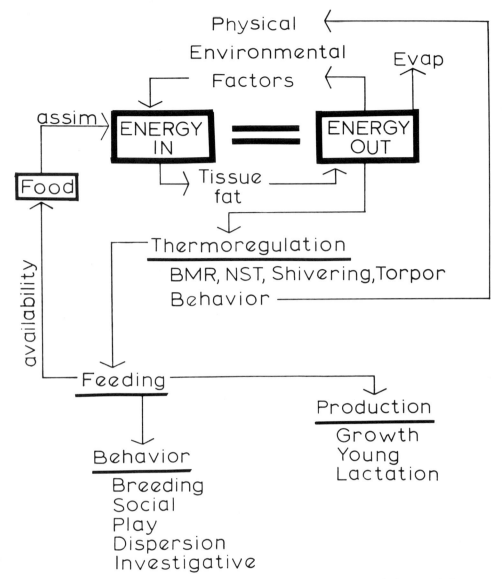

Fig. 2.—A conceptual model of energy balance avenues for small mammals, indicating a priority cascade for energy allocation. Lines represent both total energy flow and rate functions. The behavior listed under thermoregulation is merely thermoregulatory behavior, and feeding is separate from other behavior since it is important as a feedback loop. See text for further explanations (with permission from Wunder, 1978a).

1978a). Turnover rates relate to how fast animals can turn over energy and produce heat to meet their thermoregulatory needs. An animal can freeze to death, even though there may be abundant energy in the environment, if it cannot assimilate that energy and produce heat fast enough to balance heat loss. Secondy, even with a high thermogenic capacity, a small mammal can die if there is not enough total energy (food) available in the environment to meet its needs. Thus in response to winter cold,

small mammals must meet their total flow needs and have high turnover capacity to meet thermoregulatory needs.

We know relatively little about seasonal changes in digestibility or assimilation rates in small mammals (especially given changes in diet), but as mentioned in the Introduction, several species of small mammals change their capacity for thermogenesis seasonally, showing increases in winter. Such increases, many of which involve increases in basal metabolic rate (BMR), should be energetically expensive and occur just at the time of year when energy (as food) may be most scarce. Consequently, it is not surprising to find that many of these species show compensatory adjustments. Red-backed voles (Rosenmann et al., 1975) and prairie voles (Wunder et al., 1977, but see Wunder 1978b) decrease body mass in winter such that, although capacity for thermogenesis is increased, total energy needs are similar in summer and winter or actually are reduced in winter due to the decreased mass which is maintained. White-footed mice (Lynch, 1973), and perhaps deer mice (Fuller, 1969), do not decrease body mass during winter when thermogenesis is increased. However, both species have the capacity for torpor, and the incidence of spontaneous torpor is increased in winter (Lynch et al., 1978; Stebbins, 1971). Thus maintenance costs can be reduced, depending on the amount of time the animals spend in torpor. The Djungarian hamster shows both responses; although thermogenesis is increased in winter, these hamsters show reduced body mass and increased incidence of spontaneous torpor (Heldmaier and Steinlechner, 1981).

Given these sorts of seasonal adjustments, it seems reasonable to ask how capacity for thermogenesis changes, which aspects are modifiable, and what seasonal environmental factors may cue such changes. To increase thermogenesis in response to cold, small mammals could increase basal metabolic rate (BMR), change capacity for non-shivering thermogenesis (NST) (Jansky, 1973), change capacity for shivering or some combination of all of these. Heat production due to shivering has been little studied. However, one study does show that the mass-specific rate of thermogenesis due to shivering in Djungarian hamsters does not change in different seasons or following cold acclimation (Steinlechner, 1980). Furthermore, Steinlechner and I have data (unpublished) from Syrian hamsters showing that this rate does not change following cold exposure in that species. Therefore, total heat production for thermoregulation due to shivering should change only as body mass changes (and will be smaller in smaller animals). Thus, we find that following cold acclimation small mammals usually show increases in BMR and NST (see Webster, 1974, or Wunder, 1979, for review). Given that NST is a relatively new finding in biology (Smith and Horowitz, 1969; Johansson, 1959), there is still some confusion as to how it is measured or defined. Jansky (1973) has

reviewed the subject very thoroughly and has shown how BMR (as a form of NST), thermoregulatory NST, and shivering all interact to effect total thermogenesis for a small mammal (Fig. 3).

Results of this study suggest that for *Microtus ochrogaster* winter increases in BMR are stimulated by temperature as photoperiod alone does not affect BMR. However, in dwarf hamsters, *Phodopus sungorus,* photoperiod alone does stimulate winter increases in BMR (Heldmaier and Steinlechner, 1981). Thus, there are obviously some species differences in the cues used to control this thermogenic parameter. Dwarf hamsters are native to the Siberian steppes, and it may be that this provides a consistently colder environment than the North American Great Plains, and hence, selection for photoperiod alone as a cue for BMR increases. This may also explain why BMR varies from year to year in field caught prairie voles (Wunder, 1978b). Temperatures will be different and since temperature is the major stimulus for BMR changes, the different BMR levels may be tied to the temperature at any particular time.

However, photoperiod does appear to be a more general cue for increases in NST, as it will stimulate winter increases in prairie voles, dwarf hamsters (Heldmaier and Steinlechner, 1981), and white-footed mice (Lynch et al., 1978). Thus, many small mammals may be adapted to "anticipate" the cold of winter as day lengths shorten and the animals increase their capacity for NST. Further, during winter warm spells small mammals would presumably not lose their NST capacity, as would be the case if NST were cued by low temperature alone. Also, given that prairie voles have not been shown to demonstrate torpor in winter, it appears that increases in NST cued by photoperiod may be a general adaptation for increased thermogenesis to combat cold in winter for small mammals, and not simply an attribute associated with forms which show increased torpor in winter and hence need NST for arousal.

As a last point it might be useful to consider how (mechanistically) such changes in thermogenic capacity might be affected. At present it is not clearly known what physiological mechanisms control these changes in thermogenesis (but see Webster, 1974 and Wunder, 1979, for discussion), but from an ecological perspective, it is interesting to consider the possibility that the reproductive hormones may in some way be involved. At this point my postulate

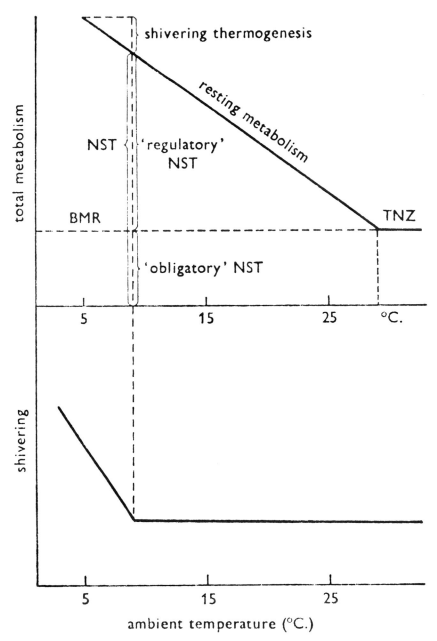

Fig. 3.—Scheme of the thermogenic mechanisms of mammals during cold-exposure. At lowered temperatures, when the capacity of non-shivering thermogenesis (NST) is exceeded, shivering starts to take place. BMR, basal metabolic rate; TNZ, thermoneutral zone (with permission from Jansky, 1973).

is merely an hypothesis based on the following reasoning. We know that much of the increase in thermogenesis shown in winter by small mammals is due to NST (Heldmaier and Steinlechner, 1981; Steinlechner, 1980, for review), and we know that most NST occurs in brown adipose tissue (BAT-Foster and Frydman, 1978). BAT mass, and presumably NST capacity, can be modified by a variety

of factors such as: 1) low temperature (Smith and Horowitz, 1969), 2) thyroxine (LeBlanc and Villemiere, 1970), 3) noradrenaline (LeBlanc and Villemiere, 1970), 4) overeating (Rothwell and Stock, 1979), and 5) melatonin or short photoperiod (Heldmaier and Hoffman, 1974). Reproduction can also be influenced by photoperiod (Sadlier, 1974; Negus and Berger, 1972). However, the responses are counter to one another, that is, short photoperiod stimulates BAT and NST but reduces reproduction, and vice versa. Both functions require high energy demand and turnover from a small mammal (Grodzinski and Wunder, 1975; Wunder, 1978a), and it seems reasonable to postulate that evolutionarily it would be simpler to have one endocrine pathway control both rather than have two or several, so that demand would not be high for both processes at a given time. Indeed, the reproductive hormones of small mammals control functions other than reproduction, functions which may relate to thermoregulation such as: 1) molt (Ling, 1970; Smith, 1970),

2) hibernation (Hall and Goldman, 1980), and 3) patterns of daily activity (Rowsemitt, 1981).

Microtine rodents should be an excellent model to test this hypothesis, since photoperiod can control reproduction (Negus and Berger, 1972), but that control can be overridden by chemicals in the food (Berger et al., 1977; Berger et al., 1981). This, in addition to temperature differences, may indeed, explain why year to year differences in the level of thermogenesis occur in some small forms (Wunder, 1978b; Feist, personal communication for Clethrionomys rutilus).

ACKNOWLEDGMENTS

I thank G. Heldmaier for many discussions which led to this study. Rob Shuman, Paul Arnold and Maureen Miller helped care for animals and gather data. I thank J. Cole for special assistance. This study was supported by an NIH-BRSG grant.

LITERATURE CITED

BERGER, P. J., E. H. SANDERS, P. D. GARDNER, and N. C. NEGUS. 1977. Phenolic plant compounds functioning as reproductive inhibitors in Microtus montanus. Science, 195:575–577.

BERGER, P. J., N. C. NEGUS, E. H. SANDERS, and P. D. GARDNER. 1981. Chemical triggering of reproduction in Microtus montanus. Science, 214:69–70.

FOSTER, D. O., and M. L. FRYDMAN. 1978. Non-shivering thermogenesis in the rat: II. Measurements of blood flow with microspheres point to brown adipose tissue as the dominant site of the calorigenesis induced by noradrenaline. Canadian J. Physiol. Pharmacol., 56:110–122.

FRETWELL, S. D. 1972. Populations in a seasonal environment. Monogr. in Population Biology, 5, Princeton Univ. Press, Princeton, New Jersey, 217 pp.

FULLER, W. A. 1969. Changes in numbers of three species of small rodents near Great Salt Lake, N.W.T., Canada, 1964–1967, and their significance for general population theory. Ann. Zool. Fennicae 6:113–144.

GRODZINSKI, W., and B. A. WUNDER. 1975. Ecological energetics of small mammals. Pp. 173–204, in Small mammals: Their productivity and population dynamics (F. B. Golley, K. Petrusewicz and L. Ryszkowski, eds.), Cambridge Univ. Press, London, xxv + 451 pp.

HALL, V., and B. GOLDMAN. 1980. Effects of gonadal steroid hormones on hibernation in the turkish hamster (Mesocricetus brandti). J. Comp. Physiol., 135:107–114.

HART, J. S. 1971. Rodents. Pp. 1–149, in Comparative Physiology of thermoregulation, Vol. II, mammals (G. C. Whittow, ed.), Academic Press, New York, xi + 410 pp.

HELDMAIER, G. 1971. Zitterfreie Wärmbildung und körpergröe bei Säugetieren. Z. vergl. Physiologie, 73:222–248.

HELDMAIER, G., and K. HOFFMAN. 1974. Melatonin stimulates growth of brown adipose tissue. Nature, 247:224–225.

HELDMAIER, G., and S. STEINLECHNER. 1981. Seasonal control of energy requirements for thermoregulation in the Djungarian hamster (Phodopus sungorus), living in natural photoperiod. J. Comp. Physiol., 142:429–437.

HELDMAIER, G., S. STEINLECHNER, J. RAFAEL, and P. VSIANSKY. 1981. Photoperiodic control and effects of melatonin on nonshivering thermogenesis and brown adipose tissue. Science, 212:917–919.

JANSKY, L. 1973. Non-shivering thermogenesis and its thermoregulatory significance. Biol. Rev., 48:85–132.

JOHANSSON, B. 1959. Brown fat: A review. Metabolism, 8:221–239.

KLEIBER, M. 1975. Metabolic turnover rate: A physiological meaning of the metabolic rate per unit body weight. J. Theor. Biol., 53:199–204.

LeBLANC, J., and A. VILLEMIERE. 1970. Thyroxine and noradrenaline on noradrenaline sensitivity, cold resistance, and brown fat. Amer. J. Physiol., 218:1742–1745.

LING, J. K. 1970. Pelage and moulting in wild mammals with special reference to aquatic forms. Quat. Rev. Biol., 45:16–54.

LYNCH, G. R. 1973. Seasonal changes in thermogenesis, organ weights, and body composition in the white-footed mouse. Oecologia, 13:363–376.

LYNCH, G. R., S. E. WHITE, R. GRUNDEL, and M. S. BERGER. 1978. Effects of photoperiod, melatonin administration and thyroid block on spontaneous daily torpor and temperature regulation in the white-footed mouse, Peromyscus maniculatus. J. Comp. Physiol., 125:157–163.

NEGUS, N. C., and P. J. BERGER. 1972. Environmental factors and reproductive processes in mammalian populations. Pp. 89–98, in Biology of reproduction (J. T. Velardo and B. A. Kasprow, eds.), III Pan American Congress of Anatomy Symposium, Publ. Pan Amer. Assoc. Anatomy, 421 pp.

ROSENMANN, M., P. MORRISON, and D. FEIST. 1975. Seasonal changes in the metabolic capacity of red-backed voles. Physiol. Zool., 48:303–310.

ROTHWELL, N. J., and M. J. STOCK. 1979. A role for brown adipose tissue in diet induced thermogenesis. Nature, 281:31–34.

ROWSEMITT, C. 1981. Seasonal variations in activity rhythms

in male *Microtus montanus*: the role of gonadal function. Abst. of Tech. Papers, 61st Annual Meeting Amer. Society of Mammalogists.

SADLIER, R. M. F. S. 1974. The ecology of the deer mouse, *Peromyscus maniculatus,* in a coastal coniferous forest. II. Reproduction. Canadian J. Zool., 52:119–131.

SMITH, A. 1970. The seasons. Rhythms of life: cycles of changes. Wiedenfeld and Nichalson, London.

SMITH, R. E., and B. A. HOROWITZ. 1969. Brown fat and thermogenesis. Physiol. Revs., 49:330–425.

STEBBINS, L. L. 1971. Seasonal variations in circadian rhythms of deer mice in northwestern Canada. Arctic, 24:124–131.

STEBBINS, L. L., R. ORICH, and J. NAGY. 1980. Metabolic rates of *Peromyscus maniculatus* in Winter, Spring and Summer. Acta Theriol., 25:99–104.

STEINLECHNER, S. 1980. Photoperiodische Kontrolle der Thermogenesekapazität beim Dsungarischen Zwerghamster, *Phodopus sungorus.* Unpublished Ph.D. dissert., J. W. Goethe Univ., Frankfurt, West Germany, 126 pp.

WEBSTER, A. J. F. 1974. Adaptation to cold. Pp. 71–106, *in* Environmental physiology (D. Robertshaw, ed.), Univ. Park Press, Baltimore, Maryland, 326 pp.

WICKLER, S. J. 1980. Maximal thermogenic capacity and body temperature of white-footed mice (Peromyscus) in summer and winter. Physiol. Zool., 53:338–346.

WUNDER, B. A. 1978a. Implications of a conceptual model for the allocation of energy resources by small mammals. Pp. 68–75, *in* Populations of small mammals under natural conditions (D. Snyder, ed.), Pymatuning Lab. of Ecology, Univ. Pittsburgh, Spec. Publ. Ser., 5:1–237.

———. 1978b. Yearly differences in seasonal thermogenic shifts of prairie voles (*Microtus ochrogaster*). J. Thermal Biology, 3:98.

———. 1979. Hormonal Mechanisms. Pp. 143–158, *in* Comparative mechanisms of cold adaptation (L. S. Underwood, L. L. Tiezen, A. B. Callahan, and G. E. Folk, eds.), Academic Press, New York, x + 379 pp.

WUNDER, B. A., D. DOBKIN, and R. GETTINGER. 1977. Shifts of thermogenesis in the prairie vole (*Microtus ochrogaster*). Oecologia, 29:11–26.

Address: Department of Zoology-Entomology, Colorado State University, Fort Collins, Colorado 80523.

CLIMATE EFFECTS ON GROWTH AND REPRODUCTION POTENTIAL IN *SIGMODON HISPIDUS* AND *PEROMYSCUS MANICULATUS*

Warren P. Porter and Polley Ann McClure

ABSTRACT

A combination of biophysical modelling and physiological experimentation has been used to predict the potential levels of seasonal growth and reproduction for individual deer mice and cotton rats living at different latitudes. Microenvironmental effects on maintenance costs and amount of food and water available combine with species specific responses to determine different seasonal potentials for reproduction and growth. We suspect that the different responses of these two species to climatic factors result from their evolutionary experience in tropical vs. north temperate ecosystems.

INTRODUCTION

The subject of this symposium is the winter ecology of small mammals. The question we would like to ask is, "Why do some small mammals reproduce in winter in some areas and not in others?" Our hypothesis is that in some cases, climate is the ultimate factor which determines whether reproduction can occur. It does so through effects on metabolic rate, water loss, food processing capacity and food quality or availability.

We will compare two species at southern versus more northerly latitudes. One species is the cotton rat, *Sigmodon hispidus*. The cotton rat is tropical in origin, ranging from Central America through the southeastern United States. It has recently invaded northern Kansas and the east coast in Virginia. The other species that we will use for comparison is the deer mouse, *Peromyscus maniculatus*. This animal is temperate in its origins and extends from Alaska southward into Baja, California, and from the West Coast to the East Coast of the United States. It has a wide range of reproductive patterns at different latitudes (Millar et al., 1979).

PHYSICAL ENVIRONMENT

Fig. 1 illustrates the physical environment experienced by a small mammal under the snow and in tall deep grass in summertime. These microenvironments are simpler than the physical environments above the surface of snow and grass in that there is very little sunlight, and the infrared radiation is nearly uniform in all directions. Typically the dominant mechanisms of heat exchange consist of infrared radiation, free convective heat exchange, and evaporation. On the basis of our experience and measurements out of doors, both air and radiant temperature are very similar in tall grass microenvironments. Air temperatures near the surface of the ground are approximated well by shade air temperatures two meters above the soil surface. In winter, temperatures near the ground under a substantial snow pack are close to freezing (Schmid, 1984). Recently we have developed the ability to calculate expected values for available microenvironmental variables above and below these protective coverings. We have also developed and tested sophisticated heat transfer models for endotherms that allow us to compute heat losses and, therefore, energy requirements to within about 10% of what can be measured experimentally (Kowalski and Mitchell, 1979; McClure and Porter, 1983). Thus, tools are now available for computing heat and mass transfer for animals in natural environments and the implications of those exchanges for growth and reproduction in different climates at different latitudes.

CLIMATE AND PHYSIOLOGY

Fig. 2 (Porter and Jaeger, 1982) illustrates the interaction of climate with physiology. There are three primary (unboxed) equations represented here, a heat balance and two mass balances. They share common elements which make them interdependent. The central diagonal equation is a heat balance

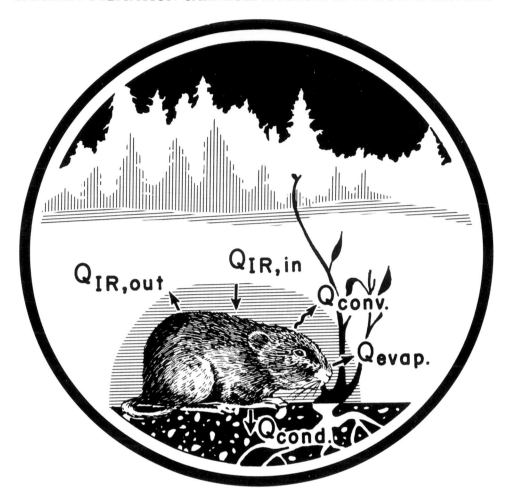

Fig. 1a.—Heat transfer pathways for small mammals under snow.

equation. It consists of heat inputs in the form of solar and infrared radiation. Heat is generated by metabolic processes, and is lost by evaporation, infrared radiation, convection, and conduction. Changes in body temperature indicate heat storage.

The leftmost diagonal equation is a part of the two primary mass balances. The mass of food ingested is equal to the sum of the dry food mass and the mass of water in the food. The top horizontal equation is a dry food mass balance. The dry mass ingested less what is lost due to defecation is equal to what is absorbed. Absorbed dry food mass may be oxidized to release heat, as represented by the slash in the metabolism term. Water and carbon dioxide are products of carbohydrate, lipid, and protein metabolism. Catabolism of proteins also yields nitrogenous waste products. Mass absorbed that is not metabolized and is not lost as water, carbon

dioxide, or nitrogenous waste is available for growth, reproduction, or storage.

The horizontal water balance equation is below the dry food mass equation. The mass of water ingested less that lost by defecation is equal to the mass of water absorbed. Total available water is the sum of absorbed water and water generated by metabolism. Available water may be used for evaporation, resulting in heat loss when water changes its phase, represented by the slash in the evaporation term. Water not lost as evaporation may be used in the production of urine or used in the processes of growth, reproduction, or storage, as represented by the boxed diagonal equations to the right of the heat balance equation. They involve terms from both mass balance equations. Thus, a reduction in the amount of mass of available water, for example, may result in a decline in reproduction or growth

Fig. 1b.—Heat transfer pathways for small mammals in tall grass.

even though sufficient dry mass may be available. This has been demonstrated for the deer mouse (Porter and Busch, 1978).

Finally, mass of oxygen has a stoichiometric relationship to mass of dry food metabolized. The mass of food metabolized may be computed either from the heat balance equation or from measurements of oxygen consumption. Thus, measurements of oxygen consumption can be used to test predictions from heat transfer models.

GROWTH AND REPRODUCTION

The potential for growth and reproduction as a function of climate is illustrated in Fig. 3 (Porter and Jaeger, 1982). One horizontal axis is solar and infrared radiation increasing from back to front. The other axis is air temperature increasing from left to right. The third axis is food required increasing vertically upward. The asymmetric bowl or dish shaped surface defines the maintenance requirements for an endotherm. The surface may be related to metab-

olism versus air temperature curves, commonly seen in the literature by imagining a diagonal trace across the surface from the high upper left portion of the surface where the label "food for maintenance" touches the surface. This is the highest metabolism under the coldest conditions and lowest radiation levels that the animals can tolerate and still survive. Moving diagonally across the surface toward the front right corner involves a traverse down the slope,

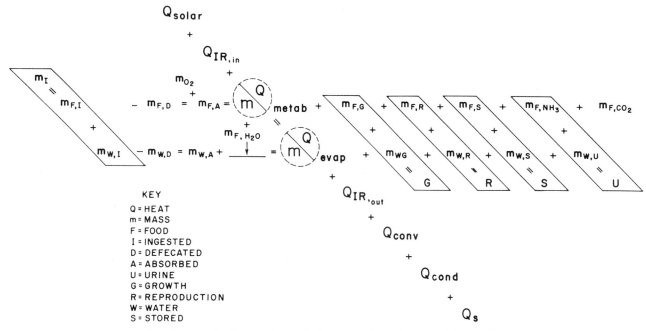

Fig. 2.—Coupled heat and mass balance equations (Porter and Jaeger, 1982).

across the flat thermoneutral zone and up the other side where the animal is beginning to go into heat stress. Deep grass, subnivean spaces and metabolic chambers are typically on the diagonal of this figure. The precipitous edges of the surface indicate a loss of physiological function and approaching death.

The horizontal surface suspended above the floppy dish is the maximum food processing capacity of the animal. It intersects the floppy dish at the highest point of the dish, where it is the coldest that the animal can tolerate for extended periods of time. The space between the upper maximum processing capacity or "lid" and dish represents the food available for growth, reproduction, activity, or internal storage. The space between the dish and lid is greatest over the thermoneutral zone. The further up the dish slope an animal must go because of environmental conditions, the less food there is for growth or reproduction. Thus, there is an environmentally dictated reduction in reproductive potential under colder environmental conditions represented by winter, high latitude, or high altitude. The position of the dish in the horizontal plane may be altered by changes in fur quality or by skin or core temperature. An increase in fur insulation, such as an increase in hair length, will shift the entire dish toward the left rear corner. A reduction in body

temperature or alterations in posture that frequently occur at lower temperatures will reduce the slope of the dish's surface, thereby increasing the space between the dish and lid. The dish also rotates clockwise with decreases in wind speed or increases in body size due to changes in the animal's boundary layer thickness. The magnitude of all such effects are now calculable from the coupled equations in Fig. 2.

The height of the lid may be determined most easily using a lactating female with maximum litter size or lactating in the cold (McClure, unpublished data; Porter and Jaeger, unpublished data).

Our hypothesis of latitude-associated climate effects on growth and reproduction may be seen in Fig. 4 (Porter and McClure, in preparation). The upper figure represents a more northerly environment, such as Kansas, where both cotton rats and deer mice occur. The maintenance cost throughout the year is indicated by the shaded area. Available food far exceeds the maximum food processing capacity illustrated by the uppermost line, F_{max}. The mass of food processed minus the mass required for maintenance is the mass of food available for growth and reproduction in late spring and throughout the summer. We expect excess available food to be shifted from growth and reproduction to fat storage with

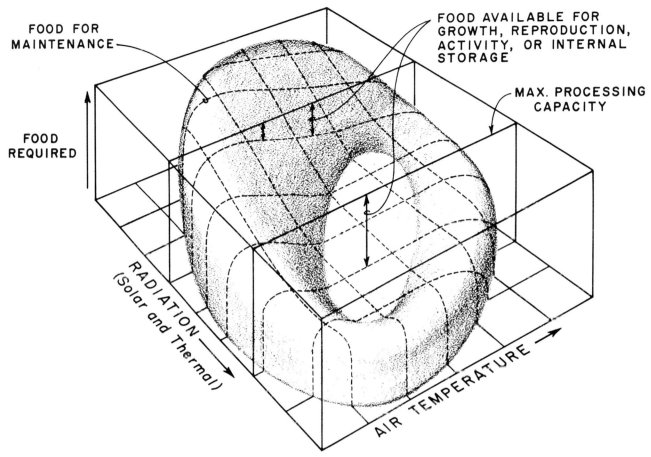

Fig. 3.—Solutions of the coupled equations in Fig. 2 yield the floppy dish and lid illustrating mass of food required for maintenance as a function of environmental variables and food mass available for growth and reproduction (Porter and Jaeger, 1982).

the onset of fall. We expect that there is lower food availability in winter for a cotton rat as illustrated here. In contrast, in a more southerly environment, such as Texas, food availability changes seasonally with a peak in springtime and again in the fall, when rains are more abundant. Maintenance costs may increase slightly in summer. We propose that food absorbed declines in midsummer due to a decrease in food availability and food quality. A consequence of this environmental effect on food quality and availability is twofold. First, the quantity of food available for reproduction may not be as great in a more southerly environment. Second, the potential reproductive periods are distributed over a broader time scale. We might expect that reproductive patterns in more southerly environments would consist of smaller litter sizes and more litters during the year. In contrast, in a more northerly environment, we

might expect to see fewer litters, but larger litters produced and larger animals in the litters. We would expect that the climatic differences and the types of reproduction that should occur would result in selection of larger animals in more northerly climates that could process more food when it is available and be able to produce larger litters and larger individual offspring.

We will now compare calculations of expected reproduction potential for cotton rats and deer mice in the Lawrence, Kansas area and the eastern Texas area, near Houston. Cotton rats and deer mice are commonly found in deep grass habitats along roadsides, fence rows, and regions of thick overgrowth of ground layer vegetation.

To make calculations of growth or reproductive potential, we need to know the position of the lid, metabolism as a function of the physical environ-

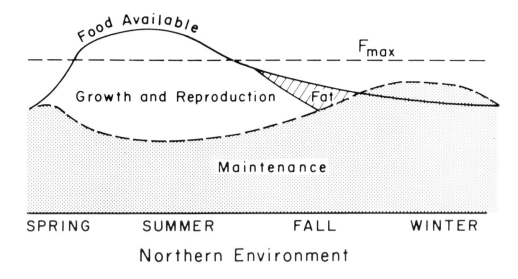

Northern Environment

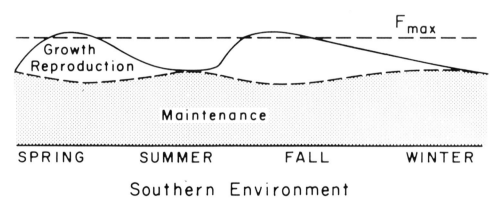

Southern Environment

Fig. 4.—Hypothesized time dependent climatic changes at different latitudes interacting with the coupled equations of Fig. 2 to produce time dependent changes in potential growth and reproduction (Porter and McClure, 1982).

mental variables, and how the microclimate varies at different latitudes throughout the year. The maximum position of a lid is estimated to approximately equal the maximum food consumption during the last three days of lactation. In the case of cotton rats, we know that there is a maximum because when they are put in the cold, the lactating females eat less food and lose body mass, and the young lose body mass and begin to die (McClure, unpublished data). Metabolism as a function of temperature was computed from Scheck's (1982) data for metabolism for Kansas and Texas *Sigmodon hispidus*. Metabolic data for *Peromyscus maniculatus bairdii* was

from Conley (1983). Microclimate air temperature values came from ten year averages of monthly U.S. climate data at Lawrence, Kansas (38°57′N latitude) and Houston, Texas (29°46′N latitude).

Fig. 5 shows calculated grams of dry food per day consumed versus day of the year for cotton rats in Kansas and Texas. The symbols A and R stand for active and resting metabolism, respectively. We estimate that activity metabolism is two times resting metabolism. This is consistent with data from R. W. Pauls (1981) and data of J. C. Randolph (1980) using doubly labelled water on free living animals and caged animals. The maximum food processing ca-

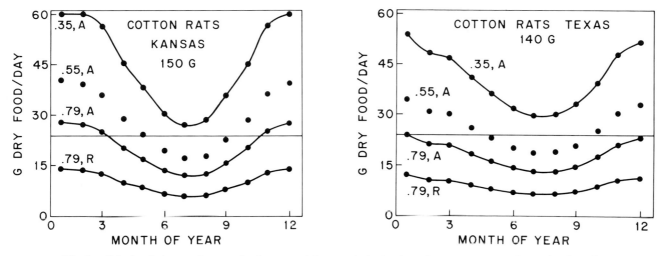

Fig. 5.—Calculated changes in reproductive potential at two latitudes for solitary cotton rats, *Sigmodon hispidus*.

pacity is represented by the horizontal line at 24 grams. This is based on an assimilation effiency of 0.79 for a diet of Purina Lab Chow, which was determined in Dr. McClure's lab. On the basis of a lid and dish using 79% assimilation efficiency, we would predict that cotton rats in Kansas should begin reproduction in April and cease reproduction in late September under average conditions. In Texas, cotton rats should be able to breed year round in the average year. Two additional dish activity metabolism curves are computed for assimilation efficiencies of 55% and 35%, respectively. These efficiencies were measured for animals eating natural food from Texas (Wrazen, 1981). We are estimating the amount of food increase that will occur when assimilation

efficiencies decline from 79% to 55% or 35% assimilation. We are currently collecting data on food consumption as assimilation efficiency declines. Data are needed from field studies to assess the time dependent changes in digestive efficiencies. We might expect that in Texas the decline in food quality in midsummer should result in a maintenance requirement nearly equal to the food absorption capacity of our animals. In Kansas, we do not expect declines in food quality, since midsummer is a moist time of year when food quality should remain high.

Fig. 6 shows our calculations for food consumption by deer mice in the same localities. Here the calculations again suggest reproduction beginning in late April and ending in late September in Kansas,

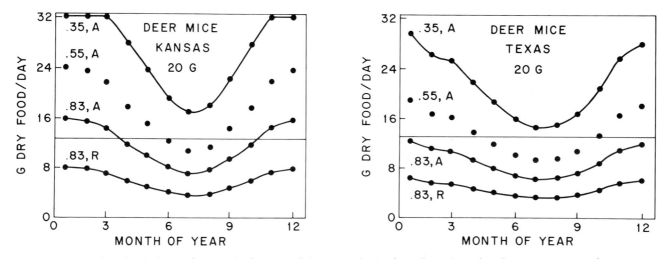

Fig. 6.—Calculated changes in reproductive potential at two latitudes for solitary deermice, *Peromyscus maniculatus*.

and reproduction occurring in the winter in Texas. The assimilation efficiency on lab chow and the maximum consumption of approximately 13 g of food per day for lactating females are based on data collected by Porter and Jaeger (unpublished data). The assimilation percentage of 83% is taken from data by Sadlier et al. (1973). Our calculations and those of the manufacturer of the lab chow agree with his figure within 1%.

Our calculated dates of onset and cessation of reproduction at these two latitudes are consistent with field observations for cotton rats (Cameron, 1977; McClenaghan and Gaines, 1978) and deer mice in Texas (Bronson, 1983; Table 2). Deer mice in Kansas have been reported to breed occasionally in winter (Svendson, 1964). Our calculations are for solitary animals. Solitary deer mice should not be able to reproduce in winter in Kansas. It may be that groups of deer mice huddling together create warm enough microenvironments to reduce maintenance costs and permit some reproduction. Some deer mice in Wisconsin huddle in groups in winter, and others do not (Porter, unpublished observation). It is interesting to note that cotton rats are solitary animals and do not build well insulated nests. They normally do not go underground, although in Kansas there is some evidence they do so in winter (Baar et al., 1975).

CONCLUSION

The original question prompting this research was whether the hypothesis that climate limits the seasonal breeding patterns of small mammals at different latitudes was a reasonable one that merited further investigation. Our calculations were based on: 1) physiological measurements of metabolic rate at a range of environmental temperatures, 2) estimated microclimate temperatures based on weather station data, 3) measurements of maximum food processing capacity at a range of temperatures and one level of food quality, 4) measurements of assimilation efficiency at one level of diet quality and estimates at several other levels, 5) estimates of the level of diet quality experienced by animals in the field, and 6) the assumptions that animals exist in nature year-round as thermally solitary individuals in the microclimate beneath a tall grass cover. While we do not regard these results as definitive, they do indicate that our hypothesis should be retained and subjected to more complete analysis.

Our confidence in the results of such calculations is limited by our confidence in the measurements, estimates, and assumptions on which they are based. It is apparent that further work is needed in several areas. We need to know how food quality and quantity vary seasonally at different latitudes and how this variation affects assimilation efficiency of the rats. Some of this work is underway (Cameron, Wrazen, and Randolph, personal communication). We are currently measuring maximum food processing capacity of *Sigmodon* for a range of diet quality. We also need information about seasonal and latitudinal effects on social behavior, especially huddling.

Actual measurements of microclimate variables in the field are needed to check our estimates. We plan to expand our calculations to include the complete heat balance of the animals rather than just metabolic heat production as done here. The final step, of course, will be field determinations of metabolic rates.

Climate could act on animal reproduction as a proximate control by directly limiting the amount of chemical energy available for productive processes as described above. Alternatively, it could have the same effect by acting as an ultimate factor selecting for another proximate mechanism, such as response to photoperiod, that limits breeding to the appropriate season. Recent evidence supports the second alternative, at least for *Sigmodon*. Desjardins (personal communication) has discovered that male cotton rats from Kansas populations experience testicular regression under winter photoperiods, but those from Texas populations do not. This indicates perhaps that in Kansas, climate has exerted reasonably strong selection against animals breeding in winter. We hypothesize that the selection has been intense because populations in northern environments are very recent. Also supporting this idea is the finding of substantial intra-population variation in response to photoperiod among Kansas animals (Desjardins, personal communication).

While it is clear that much work remains to be done, the initial results presented here show how effective a combination of biophysical and physiological studies can be in screening and clarifying

1984 PORTER AND McCLURE—GROWTH AND REPRODUCTION 181

hypotheses at the ecological and evolutionary levels. Our calculations indicate the likelihood that selection has favored a limitation on winter breeding in northern climates because of the compression of the amount of chemical energy available for reproduction at cold temperatures.

LITERATURE CITED

BAAR, S. L., E. D. FLEHARTY, and M. F. ARTMAN. 1975. Utilization of deep burrows and nests by cotton rats in west-central Kansas. Southwestern Nat., 19:437–453.

BRONSON, F. H. 1983. Chemical communication in house mice and deermice: functional roles in reproduction of wild populations. Pp. 198–238, in Advances in the study of mammalian behavior (J. F. Eisenberg and D. G. Kleiman, eds.), Spec. Publ., Amer. Soc. Mamm., 7:xvi + 1–753.

CAMERON, G. N. 1977. Experimental species removal: Demographic responses by Sigmodon hispidus and Reithrodontomys fulvescens. J. Mamm., 58:488–506.

CONLEY, K. E. 1983. A heat transfer analysis of thermoregulatory heat loss in the deer mouse, Peromyscus maniculatus. Unpublished Ph.D. dissert., Univ. Wisconsin, Madison.

KOWALSKI, G. J., AND J. W. MITCHELL. 1979. An analytical and experimental investigation of the heat transfer mechanisms within fibrous media. Trans. ASME Paper, 79-WA/HT-40: 1–7.

McCLENAGHAN, L. R., JR., AND M. S. GAINES. 1978. Reproduction in marginal populations of the hispid cotton rat (Sigmodon hispidus) in northwestern Kansas. Occas. Papers Mus. Nat. Hist., Univ. Kansas, 74:1–16.

McCLURE, P. A., AND W. P. PORTER. 1983. Development of fur insulation in neonatal cotton rats (Sigmodon hispidus). Physiol. Zool., 56:18–32.

MILLAR, J. S., F. B. WILLE, and S. L. IVERSON. 1979. Breeding by Peromyscus in seasonal environments. Canadian J. Zool., 57:719–727.

PAULS, R. W. 1981. Energetics of the red squirrel: A laboratory study of the effects of temperature, seasonal acclimation, use of the nest and exercise. J. Therm. Biol., 6:79–86.

PORTER, W. P., and R. L. BUSCH. 1978. Fractional factorial designs applied to growth and reproductive success in deer mice. Science, 202:907–910.

PORTER, W. P., and J. JAEGER. 1982. Impact of toxicants, disease and climate on growth and reproduction using Peromyscus maniculatus. Pp. 126–147, in Environmental biology state-of-the-art seminar (P. A. Archibald, ed.), Office of Exploratory Res. Publ. EPA-600/9-82-007.

RANDOLPH, J. C. 1980. Daily energy metabolism of two rodents (Peromyscus leucopus and Tamias striatus) in their natural environment. Physiol. Zool., 53:70–81.

SADLIER, R. M. F. S., K. D. CASPERSON, and J. HARLING. 1973. Intake and requirements of energy and protein for the breeding of wild deer mice, Peromyscus maniculatus. J. Reprod. Fertility, Supp., 19:237–252.

SCHECK, S. H. 1982. A comparison of thermoregulation and evaporative water loss in the hispid cotton rat, Sigmodon hispidus texianus, from northern Kansas and south-central Texas. Ecology, 63:361–369.

SCHMID, W. D. 1984. Materials and methods of subnivean sampling. This volume.

SVENDSON, G. 1964. Comparative reproduction and development in two species of mice in the genus Peromyscus. Trans. Kans. Acad. Sci., 67:527–538.

WRAZEN, J. A. 1981. Feeding patterns of cotton rats (Sigmodon hispidus): seasonal, experimental, and physicochemical factors in the selection of natural foods. Unpublished Ph.D. dissert., Indiana Univ., Bloomington.

Address (Porter): Department of Zoology, University of Wisconsin, Madison, Wisconsin 53706.

Address (McClure): Department of Biology, Indiana University, Bloomington, Indiana 47401.

SEASONAL CHANGES IN BRAIN MORPHOLOGY IN SMALL MAMMALS

VLADIMIR A. YASKIN

ABSTRACT

Small mammals were trapped in the western part of West Sibirian Lowland. Mean brain weight in voles (*Clethrionomys glareolus, C. rutilus, Microtus oeconomus,* and *M. gregalis*) and shrews (*Sorex araneus* and *S. minutus*) were maximum in summer and declined during autumn and winter. This decline was caused both by changes in population structure and mainly by brain weight decrease in separate individuals. During the period of "growth jump" in spring, the brain weight of small mammals began to increase again.

Winter reduction of brain weight was determined mainly by water losses, and by the decrease in the dry rest weight as well. The most intensive decline in weight was observed for the telencephalon (neocortex especially, 22% in bank voles and 35% in common shrews). Weights of such brain parts as bulbus olfactorius and myelencephalon did not change significantly during winter. Weight reduction of the telencephalon was accompanied by some cytomorphological changes in its architectonic.

INTRODUCTION

Adaptations of animals are based on the constant maintenance of their energetic balance. Of main importance is the regulatory function of the central nervous system, which determines both behavior and activity patterns and physiological processes connected with regulation of heat production, metabolic rate and some processes within tissues. The central regulation of adaptive changes has not been discovered in detail. Therefore, studies on morphological changes in the brain connected with the cycling pattern of an animal's vital activity are undoubtedly very urgent.

Dehnel (1949) was the first to describe seasonal changes of the skull height in *Sorex araneus*. Decrease of the brain weight in shrews (*Sorex minutus*) was discovered by Cabon (1956). Several other authors described such a phenomenon in *Sorex araneus* (Bielak and Pucek, 1960; M. Pucek, 1965; Z. Pucek, 1965). Seasonal changes of the brain weight were also observed in small rodents (Yaskin, 1976, 1980). The functional importance of seasonal changes of the brain size in small mammals still remains obscure.

MATERIAL AND METHODS

This material was collected over a period of four full annual cycles (1975-1979) in the flood-lands biotope of the Pyshma River (West Siberia). Six species of small mammals, dominant in the forest ecosystem of this region, were used for study. The main object of our investigation was the bank vole (*Clethrionomys glareolus*). Five other species were also studied—*Clethrionomys rutilus, Microtus oeconomus, M. gregalis* (Rodentia), *Sorex araneus* and *Sorex minutus* (Insectivora). A total of nearly 3,000 individuals were caught.

Voles were divided into two groups, depending on their time of birth: 1) spring voles—appeared from the end of May to July, began to reproduce early, and disappeared before autumn; 2) autumn voles—appeared from the end of July–beginning of August to September, did not reach sexual maturity in the year of their birth, and constituted the group of mature, overwintered animals the next spring.

Shrews did not mature in the year of their birth, and all of them corresponded to autumn voles.

All captured animals were measured and weighed. The brain weight was determined after its separation from the spinal cord at the level of the lower border of attachment to the pyramids. After being weighed to the nearest 1 mg, the brain was put into 10% formalin. The brain weights and some morphophysiological indices of 785 *C. glareolus*, 70 *C. rutilus*, 687 *M. oeconomus*, 572 *M. gregalis*, 217 *S. araneus*, and 40 *S. minutus* from natural populations were determined. For the purpose of studying brain growth during the first stages of postnatal ontogenesis, *C. glareolus* and *M. gregalis* were reared in laboratory conditions.

A method of determination of weight of brain parts was used for assessment of the peculiarities in the brain structure (Latimer, 1950; Dmitrieva, 1969). To convert the weight of brain parts into natural weight, a conversion index was calculated using the weight of fresh and fixed brains.

The brains of five bank voles caught in summer and five caught in winter were processed hystologically, using 20 mk sections. In all, about 1,000 stained sections were made.

183

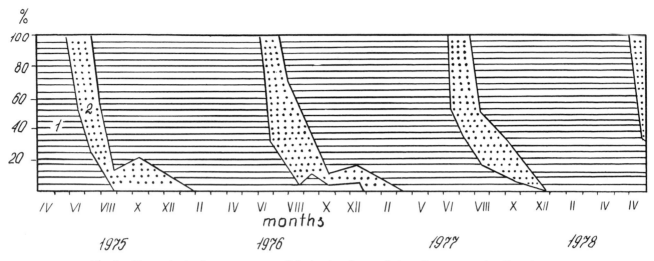

Fig. 1.—Dynamics in the age structure of the bank vole population. 1) autumn voles; 2) spring voles.

RESULTS

The period of our study included practically the whole population cycle of the bank vole. A regular decline in bank vole numbers was observed during autumn and winter of every year of the study period; however, such a decrease was not too pronounced. From October to March the frequency of captures

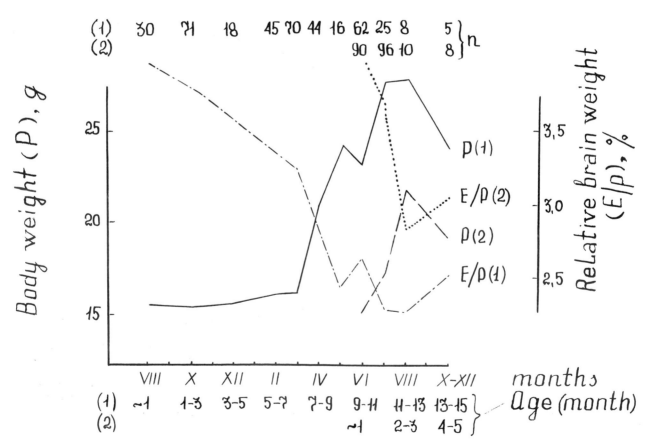

Fig. 2.—Seasonal and age changes in body weight and relative brain weight in the bank vole. 1) autumn voles; 2) spring voles.

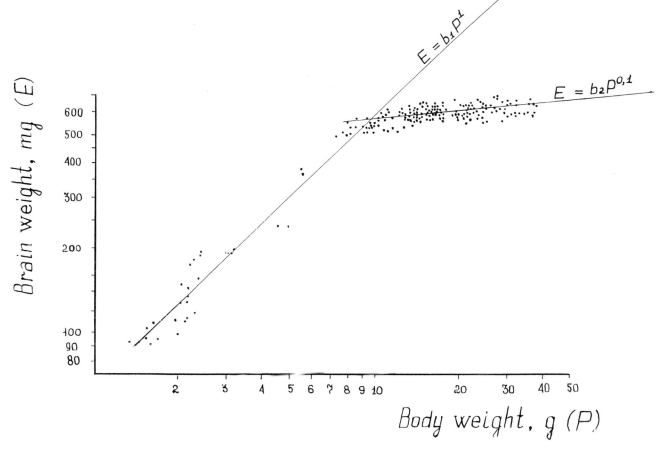

Fig. 3.—The relationship between brain weight and body weight of the bank vole in postnatal ontogeny.

decreased less than half as compared with summer months.

The age composition of the population of bank voles changed regularly as the season progressed (Fig. 1). Before the end of May–beginning of June only overwintered autumn individuals were captured. In the midsummer, 40–70% of the population was represented by yearlings born in spring, and the major part of them, as well as overwintered individuals, disappeared in autumn. After the beginning of October, the population became rather homogenous in its age structure with autumn voles making up 80 to 100% of it.

Patterns of development of autumn and spring voles are significantly different (Schwarz et al., 1964). Our data also indicated these differences. The weight (Fig. 2) and body length in spring voles were appreciably higher than in autumn voles of the same age (of 1.5–2 months). Spring voles matured early, grew rapidly, and their body weight reached the maximum value at the age of approximately 2 months. A break in the growth and development of autumn voles was observed in autumn and winter periods. Their growth was resumed in the beginning of April.

The intensive brain growth occurred during the early postnatal ontogenesis (Fig. 3), before voles passed on to independent existence (body weight near 10 g). In the period of active growth of animals after leaving the nest, the weight of the brain increased relatively little. In yearling and overwintered individuals captured in summer months, the value of an allometric exponent of relationship between brain and body weight was close to 0.1.

Brain weight to body weight ratios regularly declined in voles as their age and body weight increased (Fig. 2). Therefore, the pattern of changes in the relative brain weight of voles resembled a mirror reflection of the dynamics of their body weight. As an index of cephalization which is less dependent on the changes in body weight, the following ratio is applied: brain weight in relation to

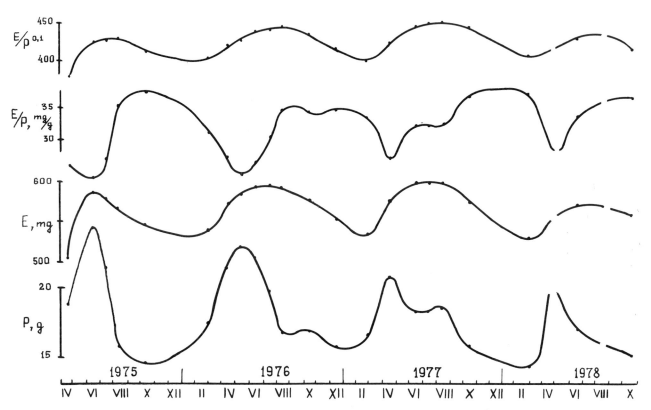

Fig. 4.—Seasonal changes in average body weight (P), brain weight (E) and relative brain weight indices (E/P; E/P$^{0.1}$) in the bank vole.

body weight raised to the power equal to the value of allometric exponent of the relation of brain weight to body weight. So, if necessary, one may use the ratio E/P$^{0.1}$ (where E = brain weight and P = body weight).

According to results of our study, the mean and maximum weights of brain and body in the population of voles reached their top values during spring and summer months and decreased in autumn (Fig. 4). In the late autumn and in winter (from October to March), changes in body and brain weights did not always coincide by direction. In winter the mean body weight declined, was stable, or even increased to some extent in different years, while the mean weight of the brain inevitably decreased in every case. The index (E/P$^{0.1}$) appeared to change regularly and simultaneously with the changes in absolute mean brain weight. This means that cephalization of voles during winter would be lower than during summer; that is, in animals of equal body weight, the brain would weigh less in winter than in summer.

In winter, minimum values of the brain weight in bank voles decreased simultaneously with its mean values, as shown in Fig. 5 where data collected during several years of study are summarized. Winter

samples contained a significant number of individuals (14.4%) with the brain weight less than 500 mg (it was the lower limit for more than 600 specimens collected from May to October 1975–1978). A similar picture was observed in each of four species of voles and in shrews. Every year the number of specimens of voles captured in winter, whose brain weight was less than minimum brain weight of those captured in summer, was about 25%. It should be noted that the appearance of the young voles in populations ceased no later than in September, and the decrease in numbers for the period from October to April did not exceed a value of 50% (that is, intensive dispersal played a minimal role). One may conclude, on the basis of these data, that the decrease of the brain weight in voles would not be connected exclusively with changes in the population structure. Such a decrease apparently occurred in particular individuals as a response to the winter conditions.

Brain growth of voles born in different seasons was significantly different (Fig. 6). The brain weight of spring voles reached its maximum value at the age of 1.5 to 2 months. This figure agrees with some results described in the literature for brain growth of rodents reared under laboratory conditions. A

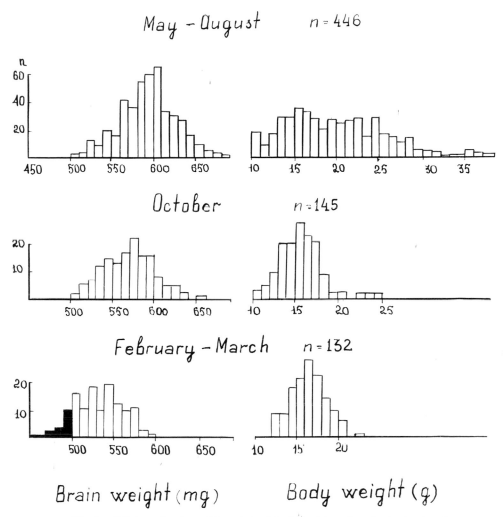

Fig. 5.—Winter reduction of the minimal brain weight values in the bank vole.

different trend was observed in autumn voles which had appreciably lower brain weights at the same age as spring voles. In winter the brain weight of autumn voles decreased by 8 to 13%, and the brain growth was not restored until spring. Therefore, the development of the brain and its growth ceased at the age of 10 to 12 months. A slight decrease of the mean brain weight during autumn was also observed in spring voles (Fig. 6).

We followed the dynamics of the mean body weight and some internal organs of the autumn bank voles during their vital cycle. It should be noted that it was the brain weight that suffered the most pronounced winter regression, as compared with other organs—the liver, the adrenal gland, the kidney, the heart (this organ increased in winter).

Seasonal and age changes of the brain weight in autumn voles and shrews had the same direction during their life (Table 1). The winter regression of the brain weight was more pronounced in insectivores. During the period of "growth rate jump" in spring, the brain weight in voles exceeded its autumn level. On the contrary, the brain weight in overwintered shrews was lower than in young animals before its winter regression. Our results on shrews agree with data on seasonal changes of the brain weight of these animals obtained previously (Bielak and Pucek, 1960; M. Pucek, 1965).

It is known that the water content of an organism tends to decrease with age. There is also plenty of available information on the lowering water content of tissues in animals of different species during a winter period (M. Pucek, 1965; Sawicka-Kapusta, 1974). Our data indicated that the dry weight of the brain in spring bank voles regularly increased with age, and the water content decreased. In autumn

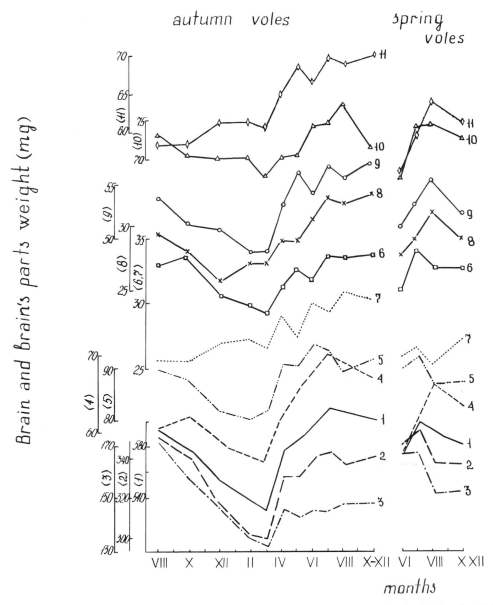

Fig. 6.—Seasonal and age changes in the absolute weight of brain and its parts in autumn and spring bank voles. 1) brain; 2) forebrain; 3) neocortex; 4) hippocampus; 5) paleocortex; 6) striatum; 7) bulbus olfactorius; 8) mesencephalon; 9) diencephalon; 10) cerebellum; 11) myelencephalon.

voles, the dry brain weight and water content decreased during autumn and winter after they reached an age of 1 month. In voles of 6 to 8 months old in March, the mean dry weight of the brain was 3.9% lower ($P < 0.001$) than in animals of the age of 1 to 3 months in October. The water weight was 9.0% lower ($P < 0.001$). During the period of rapid growth in spring, the dry brain weight in voles of 7 to 9 months old intensively increased (as much as 9.9% ($P < 0.001$) from March to June). The water weight

increased by 13.9% ($P < 0.001$) during this period. From winter to summer, the water content in brain weight increased by 0.8% ($P < 0.001$).

The present results indicate that the decrease in the brain weight of bank voles during winter is caused both by lowering of the weight of water and a decline in the dry brain weight. The former cause is more pronounced.

Although a considerable number of works has been dedicated to studies on age-dependent variability of brain parts in animals, the problem of seasonal changes in the proportions of the brain of mammals is still obscure.

The pattern of seasonal changes in the weight of different parts of the brain in bank voles was significantly diverse (Fig. 6). The most pronounced winter regression of the brain mass was observed in the noecortex of autumn voles (-22.1%, $P < 0.001$). The basal part of the forebrain (corpus striatum) also suffered significant weight losses (-13.0%, $P < 0.001$), as it is closely related to the cortex by its morphology and functions. We did not observe any noticeable change in the weight of such parts as olfactory bulbs and myelencephalon. During a period of rapid increase in the growth rate of body and brain (at the age 7 to 9 months) in spring, the rate of weight increase was different in different brain parts. The most considerable increase was observed in the hippocampus ($+22.9\%$, $P < 0.001$). Seasonal changes in weight of brain parts in the common shrew (Fig. 7) were similar with such changes in the bank vole, but they were more pronounced; for example, the weight of the neocortex decreased in winter by 35%.

In so far as seasonal variability in the weight of different parts of the brain in voles had a different regularity, seasonal changes in the proportions were also appreciable. During winter the relative weight of the forebrain declined; the most intensive decrease was observed in the neocortex and striatum. The relative weight of olfactory bulbs, myelencephalon and cerebellum significantly increased. Increase of the relative weight of hippocampus during the period from winter to summer was utmost.

The proportions of the brain in young autumn voles and overwintered voles were considerably different, while differences in the weight of the whole brain were relatively small. In overwintered voles, the relative weight of forebrain, neocortex and striatum was lower than in young voles born in autumn (in the period of high rate of growth during spring, the weight of these parts did not reach the autumn level), and the weights of olfactory bulbs, myelencephalon and hippocampus were higher.

The value of sexual differences in the weight of the brain and its particular parts also changed seasonally. The weight of the brain in males of bank voles was significantly higher than in females only during winter months (the body weight of both sexes was equal). Among mature voles in summer, the weight of the brain was equal in both sexes while the body weight was higher in females. Usually the relative weight of the brain was somewhat higher in

Table 1.—*Seasonal and age changes (in percent) of the brain weight in autumn voles and shrews (1, winter regression; 2, spring "jump of growth"; 3, differences between yearlings (recruits) before winter and overwintered animals; im, immature; m, mature; * P < 0.05; ** P < 0.01; in other cases P < 0.001).*

Species	Autumn (im)→ Winter (im) 1	Winter (im)→ Summer (m) 2	Autumn (im)→ Summer (m) 3
Clethrionomys glareolus	-8.7	$+13.0$	$+3.2*$
Clethrionomys rutilus	-10.3	$+19.6$	$+7.2**$
Microtus gregalis	-12.7	$+26.5$	$+10.4$
Microtus oeconomus	-10.8	$+26.7$	$+13.0$
Sorex araneus	-23.6	$+11.8$	-14.6
Sorex minutus	-31.7	$+21.0$	-17.4

males. The ratio of the weight of neocortex to the weight of the whole cortex was higher in females (with the exception of winter months). On the contrary, the absolute and relative weights of the other part of the cortex—hippocampus—was significantly higher in males. The latter was the part where differences between sexes were most pronounced (as much as 6%). One could observe also some sexual differences in relative weight of striatum, mesencephalon and cerebellum.

The results of quantitative study of hystological sections of the brain have shown that considerable changes in physiology and morphology of rodents during winter were accompanied not only by depression of the growth and decrease of the brain size, but also by certain cytomorphological alterations in the architectonic of some of its parts. The relative area of the cellular elements in the parts of the forebrain under study appreciably decreased from October to February in autumn bank voles, while the number of cells per one unit of area of the section increased. The mean value of the area of neuron section measured in the V layer of the 17th field of the cortex significantly declined. On the contrary, the mean value of these indices of the myelencephalon had no visible seasonal differences, and were not different in voles captured in October and February.

In addition, certain changes of brain weight and its structure in bank voles correlated with effects of drought and at the background of population numbers changes were observed (Yaskin, 1981). In the year of drought (in late summer), a decrease in body and brain weight in young bank voles occurred. In addition, changes in brain proportions were analogous to such changes which occurred in winter. The mean brain weight of voles (*C. glareolus* and *M. gregalis*) was high at the phase of density increase, reaching its maximum values in the peak

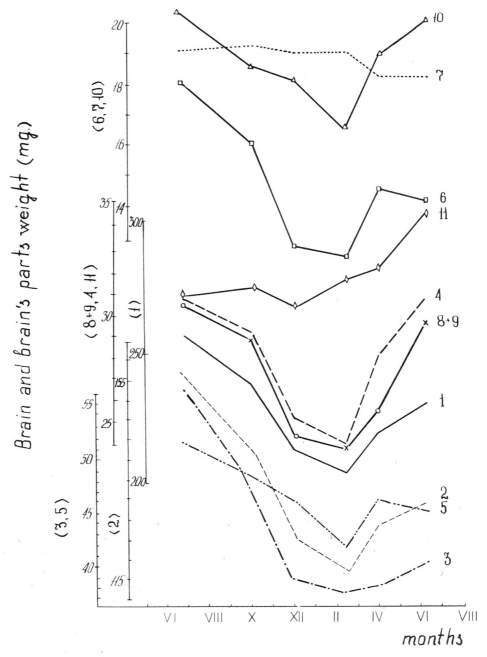

Fig. 7.—Seasonal and age changes in the absolute weight of brain and its parts in common shrew. Designations as in Fig. 6.

year, and significantly decreased in autumn of the peak year (*M. gregalis*) or in the following year (*C. glareolus*). An analysis of annual variability of the relative weight of different brain parts, accompanied with density changes, demonstrate that annual changes were most pronounced in the mean relative weight of the hippocampus.

Therefore, our results revealed the broad morphological variability on such a highly specialized and relatively stable organ as the brain. This variability is connected with the influence of seasonal factors, extreme conditions of the drought, and factors depending on changes in population numbers.

DISCUSSION

On the basis of our data, the winter decrease in the brain weight and changes in the main morphophysiological indices should be considered as the general phenomenon for a large group of mammalian species. Formerly this phenomenon was evident only for Insectivora as the "Dehnel's effect." The mechanisms of such a phenomenon are sufficiently complicated. It may be assumed that small mammals as organisms function less intensively during the winter.

It should be noted that the intensity of the brain metabolism, that is, its energetic requirements, are rather high. The brain requires about 20% of the total energy utilized by an animal, while its weight forms only 2%. The cortex has the greatest energetic requirement among other brain parts. Consequently, the evolutionary fixed winter decrease of the brain size, which occurred mainly at the expense of its proximal parts, may be of positive value for small mammals, as the seasonal changes of morphophysiological peculiarities are directed to the reduction of energetic expenditures of an animal.

It is that the seasonal and annual changes in examined indices of the brain in voles are closely connected with the animal's morphophysiological peculiarities, their way of living and also with natural selection. The alterations in metabolism and hormonal background, defining the growth rate of juveniles of different generations, as well as some profound changes in the brain development and the relative growth of its parts are probably of special importance. A certain role in observed seasonal changes of the brain can be played by changes of such factors as the light, the availability and quality of food, characterized by clear seasonal rhythms under natural conditions.

LITERATURE CITED

BIELAK, T. and Z. PUCEK. 1960. Seasonal changes in the brain weight of the common shrew (*Sorez araneus araneus* Linnaeus, 1758). Acta Theriol., 3:297–300.

CABON, K. 1956. Unetrsuchungen über diesaisonale Veränderlichkeit des Gehirnes bei der kleinen Spitz maus (*Sorex minutus minutus* L.). Ann. Univ. M. Curie–Sklodowska, Sect. C., 10:93–115.

DEHNEL, A. 1949. Studies on the genus *Sorex* L.—Ann. Univ. M. Curie–Sklodowska, Sect. C., 4:17–102.

DMITRIEVA, N. I. 1969. Growth of brain and spinal cord in postnatal ontogeny in white rat. Pp. 132–144, *in* Brain development in animals, Nauka, Leningrad. (In Russian.)

LATIMER, H. B. 1950. The weights of the brain and its parts and the weight and length of the spinal cord in the adult male guinea pig. J. Compar. Neurol., 93:37–51.

PUCEK, M. 1965. Water contents and seasonal changes of the brain weight in shrews. Acta Theriol. 10:353–367.

PUCEK, Z. 1965. Seasonal and age changes in the weight of the internal organs of shrews. Acta Theriol. 10:369–438.

SAWICKA-KAPUSTA, K. 1974. Changes in the gross composition and energy value of the bank vole during their postnatal development. Acta theriol., 19:27–54.

SCHWARZ, S. S., A. V. POKROVSKI, V. G. ISTCHENKO, V. G. OLENJEV, N. A. OVTSCHINNIKOVA, and O. A. PJASTOLOVA. 1964. Biological peculiarities of seasonal generations of rodents with special reference to the problem of senescence in mammals. Acta Theriol., 8:11–43.

YASKIN, V. A. 1976. Seasonal variability of brain weight in *Microtus oeconomus*. P. 48 *in* Informal materials of Institute of Animal and Plant Ecology, Sverdlovsk. (In Russian.)

———. 1980. Seasonal changes of brain morphology, main morphological indices and behaviour in bank voles.—Pp. 152–159, *in* Animal adaptations to winter conditions, Nauka, Moskwa. (In Russian.)

———. 1981. Reaction of the bank voles to winter conditions, drought and changes in population density. Soviet Ecology, 12:36–44. (In Russian.)

Address: Institute of Plant and Animal Ecology, 8 Marta Street 202, 620008 Sverdlovsk, U.S.S.R.

AUTUMN MASS DYNAMICS OF RED-BACKED VOLES (*CLETHRIONOMYS GAPPERI*) IN COLORADO IN RELATION TO PHOTOPERIOD CUES AND TEMPERATURE

DOUGLAS C. URE

ABSTRACT

Red-backed voles, *Clethrionomys gapperi,* lose mass in autumn in response to seasonal photoperiods. Cold temperature exposure is not necessary for normal mass loss. Voles exposed to declining photoperiods and 21°C, and declining photoperiods coupled with seasonal cold exposure lose mass at the same rates. Animals on long days track temperature, losing mass during cold periods and gaining during warm. Digestive efficiences are independent of consumption levels, daylength and cold exposure.

Resting metabolic rates are the same in all animals regardless of temperature and photoperiod. Non-shivering thermogenic capacity increases only when an animal is exposed to both short day photoperiods and low ambient temperatures. The data are not exclusive, but the best candidate for a mechanism of autumn mass reduction remains hypophagy. However, nutritional aspects of reduced consumption under winter stress remain unknown.

INTRODUCTION

Energy balance is essential to the continued well-being of any animal. It is a particularly serious problem for herbivorous small mammals in temperate and arctic regions. As endotherms, they are faced with relatively high metabolic requirements at the best of times, but during winter months those requirements can rise drastically. Lower average temperatures result in a large animal/environment thermal gradient and a subsequently high rate of heat flow. This is coupled with a cessation in plant productivity and decline in the quality of available food. Small mammals have evolved a number of strategies which enable them to avoid or reduce the stress of survival during winter months.

Most temperate and arctic small mammals increase insulation by developing a thicker and longer pelt as winter approaches (Heldmaier and Steinlechner, 1981) and combine this with regional heterothermy. Because pelt length must not impede movement and because regional heterothermy is affected by the length of the thermal gradient, the effectiveness of these modifications is limited in very small mammals. Effective insulation can be increased and net energy demands reduced through the use of microclimates (Vogt and Lynch, 1982). Nests (Glaser and Lustick, 1975) and subnivean spaces (Kucera and Fuller, 1978) can dampen the thermal fluctuations to which a small mammal is exposed. Energy demands can also be reduced through social interaction. Winter aggregations and huddling have been observed in some northern small mammals (Fedyk, 1971; West, 1977), and these behaviors certainly reduce total thermoregulatory costs (Lynch et al., 1978). One component of the energy budget that increases greatly in winter is the cost of foraging for food items in progressively lower abundance and under snow. Most northern small mammals compensate for reduced winter food availability, in part, through hoarding or caching of food items during months of high productivity (Barry, 1976). Hoarding requires leaving the nest periodically to retrieve food from caches. Thus, along with huddling and microhabitat use, it can only partially compensate for increased winter energy demands.

Escape from thermal stress by migration has evolved in the most mobile of small mammals, the bats, but size restricts mobility in all other small mammals. Daily torpor and hibernation are a means of avoiding cold stress and reducing winter energy demands (Muchlinski and Rybak, 1978) in a number of less mobile small mammals, notably shrews and several families of rodents. Hibernation not only reduces energy demands, but also slows the rate of depletion of body stores (Mrosovsky, 1977). In the family Cricetidae, torpor is common only among members of the subfamily Cricetinae, the hamsters and deer mice. No member of the subfamily Microtinae, the voles, has torpor available as an energy conservation option.

Yet another means of reducing net energy demands remains, albeit a poorly understood one. In the face of only partial success in balancing the winter energy budget by the means already mentioned, a small mammal might reduce it's body mass and thereby reduce the amount of tissue that must be maintained homeothermic. A smaller mass would mean an increase in heat loss to the environment (Aschoff, 1981), but this could be overcome, in part, through increased insulation. Reduced body mass, then, would mean a reduction in net energy demands at the time of year when energy balance would be the most critical.

In recent years, it has been shown that autumn mass loss is not just the result of differential age class mortality in overwintering populations, but is an individual phenomenon in many small mammals. Lynch and Gendler (1980) reported it in laboratory populations of white-footed mice, *Peromyscus leucopus*. It has been seen more commonly in microtines, particularly in wild populations of meadow voles, *Microtus pennsylvanicus* (Iverson and Turner, 1974), prairie voles, *M. oeconomus* (Whitney, 1976), and Townsend's voles, *M. townsendii* (Beacham, 1980). The phenomenon is also found in red-backed voles from Alaska (*Clethrionomys rutilus*) (Whitney, 1976) and Colorado (*C. gapperi*) (Merritt and Merritt, 1978).

If an aspect of an animal's physiological or behavioral repertoire is to be considered a seasonal adaptive strategy, it should be observed to confer a survival advantage to individuals that possess it. In addition, the phenomenon should be under the control of an environmental cue or endogenous circannual rhythm. Autumn mass reduction has potential as such an adaptive strategy. If it does serve to balance energy demands with availability, it would have an obvious positive effect on survivorship. Iverson and Turner (1974) suggested that it could be under photoperiodic control. At the very least, an increase in energy demands due to low ambient temperatures could serve as a signal to decrease energy assimilation.

Red-backed voles inhabit an environment in Colorado where heavy snow cover is relatively predictable and food resources are highly seasonal in distribution. This, combined with their readily observable fall and winter mass reduction, led to their selection for a study of the nature and mechanism of environmental cuing of autumn mass reduction.

METHODS

Adult red-backed voles were live-trapped in spruce/fir, *Picea engelmanii/Abes lasiocarpa,* and lodgepole pine, *Pinus contorta,* forests of northern Colorado and southern Wyoming, between 40° and 42° north latitude. All capture sites were at about 3,000 m elevation, high enough to be subject to heavy winter snow cover and early autumn frosts. Trap lines were run between 8 August and 28 September, 1980. Apples were given as a food and water source while in transit. Voles were housed individually in 11 × 16 × 28 cm plastic trays, given sawdust as bedding, laboratory chow (Wayne Lab Blox) and water ad libitum, and placed on a 16L:8D photoperiod regime.

At the beginning of the experiment, bedding was replaced with a 0.64 cm mesh hardware cloth insert, which raised the voles 1.0 cm above the floor of the tray. A sheet of 0.15 cm mesh window screen below this captured feces. The voles were divided into three experimental groups, each consisting initially of 12 animals. One group (P_nT_n) was placed in a screened outdoor enclosure. A roof excluded snow, and black plastic on two sides excluded light from nearby greenhouses, but the animals were otherwise exposed to natural autumn light and temperature fluctuations. The second group (P_lT_n) was placed in another portion of the same enclosure and screened on all sides with plastic to exclude extraneous light sources. Fluorescent lights provided a 15L:9D photoperiod regime, and temperature fluctuated as in P_nT_n. For the third group (P_nT_c), temperatures were maintained at 21°C, and fluorescent lighting simulated natural autumn photoperiods. All animals were given water and measured amounts of laboratory

chow in ad libitum portions. Ice cubes were given as a nocturnal water source, and water bottles were refilled daily when ambient temperatures fell below 0°C. P_nT_n animals were placed under experimental conditions on 14 October. P_lT_n and P_nT_c animals were started 1 and 3 days later, respectively.

Bedding was provided for all animals from the 5th through the 10th weeks to reduce cold exposure mortality in animals under natural temperatures. Ambient temperatures were recorded weekly. At that time all uneaten food was removed and replaced with new measured portions, feces were collected, and both were dried to constant mass at 65°C. The caloric content of feces from each animal during weeks 1 and 6, and of representative samples of food was determined by combustion in a Parr Adiabatic Calorimeter. Digestive efficiency was calculated as [(consumed energy—fecal energy)/consumed energy] × 100. After 10 weeks, resting metabolic rate at 30°C and non-shivering thermogenic capacity at 15°C were measured with a Beckman model G-2 oxygen analyzer and subcutaneous injection of nor-epinephrine (Heldmaier, 1971).

Differential mortality resulted in four (P_nT_n), six (P_lT_n), and 11 (P_nT_c) voles surviving after 10 weeks. All calculations were based on voles surviving to the end of the experiment. Inter-group changes in body mass were compared with an Analysis of Covariance based on Snedecor and Cochran (1967:432–435). Digestive efficiencies, resting metabolic rates, and non-shivering thermogenic capacities were compared with the Student's t statistic.

RESULTS

Autumn mass reduction in red-backed voles was stimulated by photoperiod (Fig. 1). Both groups of animals on natural photoperiods showed a significant mass reduction ($P < 0.005$) over the 10-week experimental period. The rate of mass loss by voles on autumn photoperiod and temperature (P_nT_n) did not differ significantly from that of voles on autumn photoperiod and constant temperature (P_nT_c) ($P > 0.25$). In the same fashion both groups of animals showed the same ultimate proportion of mass lost at the end of 10 weeks ($P > 0.5$). P_nT_n voles compensated for the additional cost of thermoregulation under low seasonal temperatures by increasing mass specific food consumption, when compared with P_nT_c animals ($P < 0.05$). Thermal stress played no role in the ultimate loss of mass by red-backed voles. In spite of differing thermal regimes, both groups on natural photoperiod showed the same pattern of mass loss, and mass loss patterns showed no correlation with ambient temperatures.

Red-backed voles on long days and seasonal temperatures (P_lT_n) exhibited a body mass dynamics pattern that was distinctly different from either group on natural photoperiod ($P < 0.025$). P_lT_n animals showed no significant average mass change, but did appear to track temperature changes. Body mass increased dramatically when thermal stress was mild enough to permit a positive energy flow.

Digestive efficiencies did not change in the face of photoperiod cues or temperature stress (Table 1). Voles on natural temperatures had a higher mass specific consumption rate than those at 21°C, but extracted the same proportion of energy from food as animals kept at milder temperatures. P_nT_c voles generally ate less per gram body mass than did those exposed to seasonal cold stress (Fig. 2), but there was no detectable difference in mass specific consumption between P_nT_n and P_lT_n animals. The greatest variance in consumption was found in P_lT_n animals, which increased consumption drastically as thermal stress increased.

Resting metabolic rates (RMR) were not affected by photoperiod or by temperature (Table 2). Measurements were consistent with RMR's reported for red-backed voles from laboratory colonies (McManus, 1974) and from field caught Colorado red-backed voles (Merritt and Merritt, 1978). Non-shivering thermogenic (NST) capacity was not exceptionally high in P_nT_c or P_lT_n animals. NST-capacity appeared to be higher in P_nT_n voles than in animals of the other two groups, but not significantly so. This was due to a combination of small sample size (n = 4) and the lack of response on the part of one P_nT_n vole to the nor-epinephrine dosage. When the non-responding animal was omitted from the calculations, the elevation in P_nT_n NST-capacity was significant ($P < 0.005$). A failure to respond fully to a nor-epinephrine injection is not unusual in NST studies and may be the result of an overdose based on Heldmaier's (1971) calculations (B. Wunder, personal communication). In spite of being elevated in P_nT_n animals, NST capacity was still substantially lower than that in winter *C. rutilus* from Alaska (Feist and Rosenmann, 1975).

DISCUSSION

While absolute proof that autumn mass reduction is a strategy for coping with winter stress remains to be collected, it does have an important characteristic of other known winter strategies. Mass reduction is under a predictable, seasonal environmental cue, photoperiod, and takes place regardless of the presence of other less predictable cues. The loss of body mass by red-backed voles in fall is not simply a passive response to a negative energy flow. Mass reduction is witnessed under fall photoperiods even when food is plentiful and thermal demands are minimal. A similar role for photoperiod has been demonstrated in *Peromyscus leucopus* (Lynch and Gendler, 1980) with a critical minimum daylength of 12 to 13 hr, below which mass begins to decline. This study was not designed to seek a critical minimum photoperiod in red-backed voles, but it is evident that a minimum exists at some point below 15 hr of light per day. As long as food is available, ambient temperature plays no role in red-backed vole mass loss. Similar conditions are reported for montane voles, *Microtus montanus*, (Petterborg, 1978) in which cold stress neither stimulates nor regulates autumn mass reduction.

While endogenous circannual mechanisms may have some effect on the susceptibility of red-backed

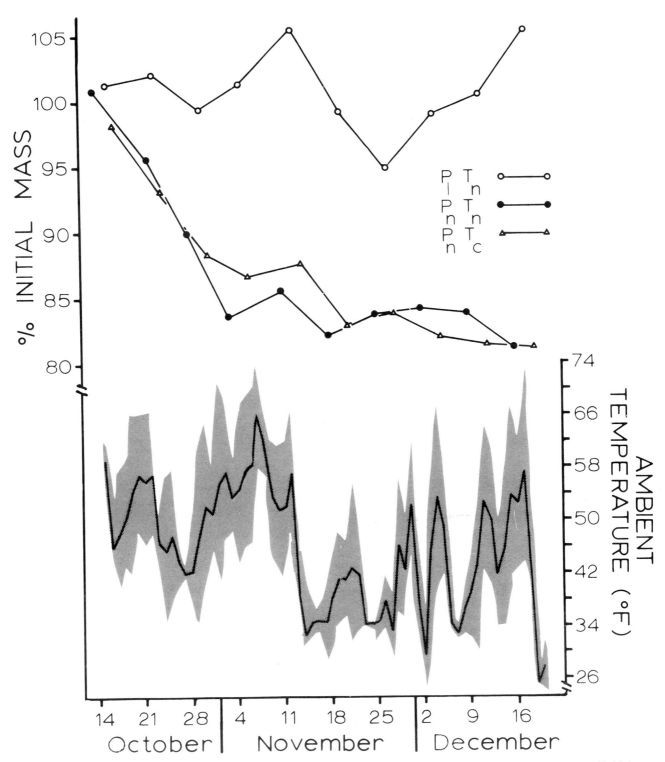

Fig. 1.—Mean body masses of three groups of red-backed voles (*C. gapperi*) over a 10-week period, given as percent of initial mass and compared with mean and range of daily ambient temperatures. Ambient temperatures were recorded in an outdoor screened enclosure. P_nT_n, natural photoperiods and temperatures; P_lT_n 15L:9D and natural temperatures; P_nT_c, natural photoperiods and 23°C.

voles to fall photoperiods, endogenous mechanisms do not themselves bring about mass loss. In spite of identical pre-test exposure to natural environmental conditions for all animals, I found no regular mass loss in animals that did not receive autumn photoperiods. Subsequent work on a second group of voles has indicated that there is a circannual pattern to photoperiod receptivity. This is certainly the case for reproductive condition (Sealander, 1967).

There are several pathways for mass loss that are open to small mammals—1) increase net energy demand while holding consumption constant, 2) decrease assimilation of energy from consumed food, or 3) decrease total consumption.

Energy demands can be increased through activitiy or through metabolic pathways. Increased activity raises metabolic demands and dispells excess energy as heat. Activity also increases thermal conductance (Aschoff, 1981) and has a positive effect on the rate of energy loss to the environment. A negative aspect to dispelling energy through activity is the increased risk of predation inherent in moving about more than is necessary to obtain food. Although I did not measure activity in red-backed voles, I observed no increase in activity of voles on autumn photoperiods. In fact, activity appeared to be negatively correlated with temperature. Barnett et al. (1978) observed a decline in activity with cold exposure in house mice, *Mus musculus*. A similar inhibition in activity as a response to cold ambient temperatures is seen in deer mice, *P. maniculatus* (Sheffield and Andrews, 1980). As in this study, Herman (1977) found a decline in extra-nest and total activity of *C. gapperi* during winter. Thus, activity was probably not the means of mass loss.

Metabolic rate may also be increased to remove excess energy under certain conditions. White rats, *Rattus norvegicus*, on a diet that will induce rapid mass gain can increase non-shivering heat production (Diet Induced Thermogenesis, DIT) to remove excess energy (Rothwell and Stock, 1980, 1981). Under certain conditions small mammals might use this mechanism to reduce assimilation of digested energy without reducing the intake of essential nutrients. In this study it was apparent that, under experimental conditions, voles increase metabolic capacity only when exposed to both fall photoperiod cues and cold stress. Feist and Morrison (1981) obtained similar results with *C. rutulis* in Alaska and proposed that photoperiod primes the system facilitating response to cold stress when it appears. If

Table 1.—*Mean digestive efficiencies of red-backed voles at weeks 1 and 6. Symbols are as in Fig. 1. Digestive efficiencies are given as (kcal digested/kcal consumed)* × 100 ± 1 SE.

Group	Week 1	Week 2
P_nT_n	78.79 ± 3.59	80.37 ± 2.54
P_nT_c	81.32 ± 2.08	79.94 ± 2.06
P_lT_n	82.63 ± 2.58	81.96 ± 2.26

this is always the case, then metabolic pathways are an unlikely means of attaining mass reduction, since they would be dependent on exposure to cold temperatures. As a strategy for reducing energy demands when under increased thermal stress, mass reduction should precede cold exposure rather than be dependent on its appearance (Rosenmann et al., 1975). Nothing is currently known about the presence of DIT in microtines or its role in mass control under natural conditions. Controlled studies must be conducted before DIT can be entirely discounted, but its role in this case was not obvious.

It has also been suggested that small mammals might reduce body mass in autumn by decreasing the digestion or assimilation of energy in the food consumed, thus maintaining a high nutrient intake without having to cope with the excess energy (Cherry and Verner, 1975). While McManus (1974) found this to be the case, the results of this study are at odds with his work. Digestive efficiency in Colorado red-backed voles did not change when ambient temperatures fell and the rate of consumption increased. A lack of response to hyperphagy by digestive processes has also been reported in laboratory rats (Deslauriers et al., 1971), and it appears that the uptake of energy in the digestive tract is dependent on the inherent digestibility of the food item and is not adjustable by the individual.

The most likely means of reducing tissue mass open to red-backed voles is hypophagy. Reduction in consumption under autumn photoperiod and a subsequent loss in body mass has been reported in white-tailed deer, *Odocoileus virginianus*, (Moen, 1978; Severinghaus, 1981; Warren et al., 1981). Stebbins (1978) found caged *P. maniculatus* holding net consumption constant while body mass increased through fall and into January, thus slowing the rate of mass gain. In addition, suppression of feeding is a means of reducing mass to a lower level in hibernators (Mrosovsky and Sherry, 1980). I observed possible signs of a similar reduction in con-

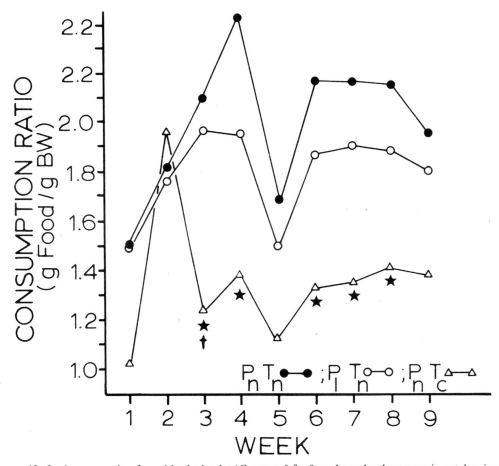

Fig. 2.—Mass specific food consumption for red-backed voles (*C. gapperi*) for 9 weeks under three experimental regimes of photoperiod and temperature (see Fig. 1 for description). The starred values are weeks when P_nT_c mass specific consumption was significantly lower than that of P_nT_n. The daggered value is a week when mass specific consumption of P_nT_c was significantly lower than that of P_lT_n.

sumption in response to photoperiod in red-backed voles. Mass specific consumption increased in P_nT_n relative to P_nT_c voles, but it was not possible to determine if thermal stress accounted entirely for the increase. I made no measurements of pelage change, so the relative contribution of insulation to reduction in energy demands cannot be evaluated. Lynch and Gendler (1980) did show that short days can stimulate winter molt in *P. leucopus*. It is reasonable to assume that this is the case in *C. gapperi* and that insultative changes affected the observed energy demands, as is the case in the Djungarian hamster, *Phodopus sungorus* (Heldmaier and Steinlechner, 1981).

The presence of hypophagy is slightly clearer, however, in the comparison of consumption between P_nT_n and P_lT_n groups. There was no significant detectable difference in mass specific consumption between these groups of animals, but by the end of 10 weeks, P_nT_n animals were substantially the smaller of the two. Mass specific metabolic rates and consumption should have been increased in these smaller animals. The absence of increased food intake implies that hypophagy was probably taking place in response to the differing photoperiod cues. Unfortunately, the extremely small sample sizes

Table 2.—*Mean resting metabolic rate and nor-epinephrine induced non-shivering thermogenic capacity for red-backed voles at the end of 10 weeks. Values given are $ccO_2/g \cdot h \pm 1$ SE. Value in parentheses is NST value including one non-responding animal (see text). Symbols are as in Fig. 1.*

Group	RMR	NST
P_nT_n	3.05 ± 0.54	6.39 ± 0.61[a] (5.24 ± 2.35)
P_nT_c	2.83 ± 0.60	3.42 ± 1.42
P_lT_n	3.00 ± 0.37	3.53 ± 1.08

[a] $P_nT_n > P_lT_n$, $P < 0.005$; $P_nT_n > P_nT_c$, $P < 0.005$.

make it inadvisable to draw definite conclusions about the role of hypophagy in mass reduction from this one test. Still, reduced consumption holds the most promise as the ultimate mechanism for mass reduction in small animals.

One aspect of hypophagy which remains clouded is the affect of nutrient levels on the mechanism of seasonal body mass loss. Maintenance of a constant mass is affected by the quality of the diet in small as well as large mammals. Sinclair et al. (1982) have reported a significant role of forage quality in the mass dynamics of snowshoe hares, *Lepus americanus.* In microtines, the most up-to-date work has been that of Batzli and Jung (1980) and Barkley et al. (1980) on tundra species. They report that nu-

trient levels directly affect forage quality and thereby animal health. It appears that nutrient rather than energy demands limit the activity of tundra microtines. There is also some evidence from domestic species that animal nutrient requirements change with increasing cold stress (see N.R.C., 1981 for a review). Torpor, and subsequent reduced energy and nutrient demands can be induced in the garden dormouse, *Eliomys quercinus,* by feeding a protein deficient diet (Montoya et al., 1979). Unfortunately, no data are available to relate hypophagy and specific nutrient demands to seasonal mass dynamics. Future research in this area will be necessary to our understanding of animal/environmental interactions in seasonal environments.

ACKNOWLEDGMENTS

Space and equipment for this study were provided by Bruce Wunder and the Department of Zoology and Entomology, Colorado State University, Dick Neal helped with the statistical analysis. Special thanks goes to Susan Ure, who helped set traps, typed, and provided much encouragement.

LITERATURE CITED

ASCHOFF, J. 1981. Thermal conductance in mammals and birds. its dependence on body size and circadian phase. Comp. Biochem. Physiol., A, 69:611–619.

BARKLEY, S. A., G. O. BATZLI, and B. D. COLLIER. 1980. Nutritional ecology of microtine rodents: a simulation model of mineral nutrition for brown lemmings. Oikos, 34:103–114.

BARNETT, S. A., W. E. HOCKING, and J. L. WOLFE. 1978. Effects of cold on activity and exploration by wild house mice in a residential maze. J. Comp. Physiol., 123:91–95.

BARRY, W. J. 1976. Environmental effects on food hoarding in deermice (*Peromyscus*). J. Mamm., 57:731–746.

BATZLI, G. O., and H. J. JUNG. 1980. Nutritional ecology of microtine rodents: resource utilization near Atkasook, Alaska. Arctic Alpine Res., 12:483–499.

BECHAM, T. D. 1980. Growth rates of the vole *Microtus townsendii* during a population cycle. Oikos, 35:99–100.

CHERRY, R. H., and L. VERNER. 1975. Seasonal acclimatization to temperature in the prairie vole, *Microtus ochrogaster.* Amer. Midland Nat., 94:354–360.

DESLAURIERS, R., J. ZOAMI, R. S. McCULLOUGH, and O. HEROUX. 1971. Caloric uptake and cold resistance in cold-acclimated rats fed commercial chow of semipurified diet. Canadian J. Physiol. Pharmacol., 49:707–712.

FEDYK, A. 1971. Social thermoregulation in *Apodemus flavicollis* (Melchoir, 1834). Acta. Theriol., 16:221–229.

FEIST, D. D., and P. R. MORRISON. 1981. Seasonal changes in metabolic capacity and norepinephrine thermogenesis in the Alaskan red-backed vole: environmental cues and annual differences. Comp. Biochem. Physiol., A, 69:697–700.

FEIST, D. D., and M. ROSENMANN. 1975. Norepinephrine thermogenesis in seasonally acclimatized and cold acclimated red-backed voles in Alaska. Canadian J. Physiol. Pharmacol., 54:146–153.

GLASER, H., and S. LUSTICK. 1975. Energetics and nesting behavior of the northern white-footed mouse, *Peromyscus leucopus noveboracensis.* Physiol. Zool., 48:105–113.

HELDMAIER, G. 1971. Zitterfreie Wärmbildung und Körpergrösse bei Säugetieren. Z. Vgl. Physiol., 73:222–248.

HELDMAIER, G., and S. STEINLECHNER. 1981. Seasonal control of energy requirements for thermoregulation in the Djungarian hamster (*Phodopus sungorus*), living in a natural photoperiod. J. Comp. Physiol. B., 142:429–438.

HERMAN, T. B. 1977. Activity patterns and movements of subarctic voles. Oikos, 29:434–444.

IVERSON, S. L., and B. N. TURNER. 1974. Winter weight dynamics in *Microtus pennsylvanicus.* Ecology, 55:1030–1041.

KUCERA, E., and W. A. FULLER. 1978. A winter study of small rodents in aspen parkland. J. Mamm., 59:200–204.

LYNCH, G. R., F. D. VOGT, and H. R. SMITH. 1978. Seasonal study of spontaneous daily torpor in the white-footed mouse, *Peromyscus leucopus.* Physiol. Zool., 51:289–299.

LYNCH, G. R., and S. L. GENDLER. 1980. Multiple responses to different photoperiods occur in the mouse, *Peromyscus leucopus.* Oecologia (Berl.), 45:318–321.

McMANUS, J. J. 1974. Bioenergetics and water requirements of the redback vole, *Clethrionomys gapperi.* J. Mamm., 55:30–44.

MERRITT, J. F., and J. M. MERRITT. 1978. Population ecology and energy relationships of *Clethrionomys gapperi* in a Colorado subalpine forest. J. Mamm., 59:576–598.

MOEN, A. N. 1978. Seasonal changes in heart rates, activity, metabolism, and forage intake of white-tailed deer. J. Wildlife Mgmt., 42:715–738.

MONTOYA, R., L. AMBID, and R. AGID. 1979. Torpor induced at any season by suppression of food proteins in a hibernator, the garden dormouse (*Eliomys quercinus* L.). Comp. Biochem. Physiol., A., 62:371–376.

MROSOVSKY, N. 1977. Hibernation and body weight in dormice: a new type of endogenous cycle. Science, 196:902–903.

MROSOVSKY, N., and D. F. SHERRY. 1980. Animal anorexias. Science, 207:837–842.

MUCHLINSKI, A. E., and E. N. RYBAK. 1978. Energy consumption of resting and hibernating meadow jumping mice. J. Mamm., 59:435–437.

N.R.C. 1981. Effect of environment on nutrient requirements of domestic animals. Subcomm. Environ. Stress, Comm. Anim. Nutr., Board Agric. Renewable Res., Commis. Nat. Res., Nat. Res. Coun., National Academy Press, Washington, D.C., 152 pp.

PETTERBORG, L. J. 1978. Effect of photoperiod on body weight in the vole, Microtus montanus. Canadian J. Zool., 56:431–435.

ROSENMANN, M., P. MORRISON, and D. FEIST. 1975. Seasonal changes in the metabolic capacity of red-backed voles. Physiol. Zool., 48:303–310.

ROTHWELL, N. J., and M. J. STOCK. 1980. Similarities between cold- and diet-induced thermogenesis in the rat. Canadian J. Physiol. Pharmacol., 58:842–848.

———. 1981. Regulation of energy balance. Ann. Rev. Nutr., 1:235–256.

SEALANDER, J. A. 1967. Reproductive status and adrenal size in the northern red-backed vole in relation to season. Internat. J. Biometeor., 11:213–220.

SEVERINGHAUS, C. W. 1981. Overwinter weight loss in white-tailed deer in New York. New York Fish Game J., 28:61–67.

SHEFFIELD, M. V., JR., and R. V. ANDREWS. 1980. Interactions of ambient temperature and photoperiod on deer mouse energetics. Comp. Biochem. Physiol., A, 67:103–115.

SINCLAIR, A. R. E., C. J. KREBS, and J. N. M. SMITH. 1982. Diet quality and food limitation in herbivores: the case of the snowshoe hare. Canadian J. Zool., 60:889–897.

SNEDECOR, G. W., and W. G. COCHRAN. 1967. Statistical methods. Sixth edition. Iowa State Univ. Press, Ames, 593 pp.

STEBBINS, L. L. 1978. Some aspects of overwintering in Peromyscus maniculatus. Canadian J. Zool., 56:386–390.

VOGT, F. D., and G. R. LYNCH. 1982. Influence of ambient temperature, nest availability, huddling, and daily torpor on energy expenditure in the white-footed mouse Peromyscus leucopus. Physiol. Zool., 55:56–63.

WARREN, R. J., R. L. KIRKPATRICK, A. OELSCHLAEGER, P. F. SCANLON, and F. C. GWAZDAUSKAS. 1981. Dietary and seasonal influences on nutritional indices of adult male white-tailed deer. J. Wildlife Mgmt. 45:926–936.

WEST, S. D. 1977. Midwinter aggregation in the northern red-backed vole, Clethrionomys rutilus. Canadian J. Zool., 55:1404–1409.

WHITNEY, P. 1976. Population ecology of two sympatric species of subarctic microtine rodents. Ecol. Monogr., 46:85–104.

Address: Department of Zoology and Entomology, Colorado State University, Fort Collins, Colorado 80523.

GROWTH PATTERNS AND SEASONAL THERMOGENESIS OF *CLETHRIONOMYS GAPPERI* INHABITING THE APPALACHIAN AND ROCKY MOUNTAINS OF NORTH AMERICA

JOSEPH F. MERRITT

ABSTRACT

Red-backed voles (*Clethrionomys gapperi*) were live-trapped at two-week intervals from January 1974 to September 1975 in a subalpine forest (3,120 m) of the Front Range of Colorado and from September 1979 to June 1981 in a mixed deciduous forest (450 m) of the Appalachian Mountains of Pennsylvania. Both sites were located at 40°N latitude. On the Rocky Mountain Site (RMS), snow covered the ground for 7½ months, reaching a maximum depth of 270 cm; on the Appalachian Mountain Site (AMS), snowcover was intermittent reaching a maximum depth of 30 cm. The minimum temperature on both sites was −28°C. Subnivean activity was monitored by use of live-traps located within trap chimneys.

Growth dynamics of voles were analyzed year-round by plotting body weight distributions, tracing growth of individuals over consecutive trapping periods and plotting instantaneous growth rates of the population through time. Seasonal metabolic shifts were determined in the laboratory by employing oxygen consumption techniques.

Voles on the RMS shows a 30% reduction in mean body weight from those on the AMS. Winter-caught red-backed voles from the RMS weighed significantly less and also showed a lower metabolic rate than the larger summer-caught voles. *C. gapperi* from the AMS did not show significant weight declines during winter and seasonal metabolic responses were similar. Voles from the RMS showed greater changes in seasonal thermogenesis possibly as an adaptation to lower food availability and greater fluctuation in the microclimatic thermal regime of the RMS as compared to the AMS.

INTRODUCTION

Field studies dealing with the year-round ecology of microtine rodents have revealed a general retardation in growth during mid-winter. In general, a decrease in body weight is seen during autumn and winter with a weight gain in spring concomitant with reproductive effort. In North America such trends are reported for *Microtus pennsylvanicus* (Brown, 1973; Iverson and Turner, 1974); *Microtus oeconomus* (Whitney, 1976); *Microtus xanthognathus* (Wolff and Lidicker, 1980); *Clethrionomys rutilus* (Sealander, 1966; Sealander and Bickerstaff, 1967; Fuller, 1969; Fuller et al., 1969; Whitney, 1976) and *Clethrionomys gapperi* (Fuller, 1969, 1977; Stebbins, 1976; Stinson, 1977; Merritt and Merritt, 1978a; Perrin, 1979). In Europe, mid-winter weight declines are also reported for *M. oeconomus* (Tast, 1972); *Microtus agrestis* (Chitty, 1952); *Clethrionomys glareolus* (Bergstedt, 1965; Zejda, 1967, 1971; Bujlaska and Gliwicz, 1968; Hyvarinen and Heikura, 1971; Wiger, 1979) and *Clethrionomys rufocanus* (Kalela, 1957; Hyvarinen and Heikura, 1971).

It is noteworthy that certain northern microtine rodents do not undergo winter weight declines. The collared lemming (*Dicrostonyx groenlandicus*) for instance, does not conform to the above trends of winter weight decline, but rather shows an accelerated growth under short-day photoperiods (Hasler et al., 1976; Mallory et al., 1981). Possible reasons for this contrasting pattern of growth of *D. groenlandicus* with other microtines studied are discussed by Brooks and Webster (1984).

Researchers working with northern microtines contend that smaller size in winter confers the selective advantage of reducing food needs during this period when food is scarce (Tast, 1972; Stebbins, 1976, and others). Although a decrease in size tends to decrease total caloric needs (McNab, 1971), few studies have actually measured metabolic costs of small mammals inhabiting northern environments during winter months. Notable exceptions are the works of Gorecki (1968), Rosenmann et al. (1975), Wunder et al. (1977) and Merritt and Merritt (1978a) working with *C. glareolus*, *C. rutilus*, *M. ochrogaster*, and *C. gapperi*, respectively.

It is the intent of the present study to examine the means by which montane small mammals cope with selection pressures imposed during long, cold winters. However, in montane regions of North America, most small mammals either hibernate or undergo periodic bouts of torpor, thus complicating a continuous monitoring program. An ideal candidate for study during winter months is the red-backed vole, *C. gapperi*. This vole is characteristic of forests of the Hudsonian and Canadian life zones of North America, and occupies a wide range of habitats, including coniferous, deciduous and mixed forests characterized by an abundant litter of stumps, rotting logs, and exposed roots associated with a rocky substrate—optimal habitats tend to be mesic. Red-backed voles are active year-round; during periods of snowcover, they typically construct elaborate subnivean runway systems.

As is the case with many microtine rodents men-

tioned above, *C. gapperi* is reported to display a decline in growth during midwinter perhaps as a means of reducing energy needs during this cold period of the year when food is in short supply. The objective of the present study was to examine in a comparative fashion, the growth patterns and seasonal thermogenesis of *C. gapperi* in the Appalachian and Rocky Mountains of North America as influenced by food availability and microclimatic thermal regimes.

MATERIALS AND METHODS

STUDY AREAS

Two study sites were chosen to examine the populations dynamics of the red-backed vole. Voles were live-trapped in a Colorado subalpine forest from January 1974 to September 1975 and from September 1979 to June 1981 in a mixed deciduous forest of the Appalachian Mountains of Pennsylvania. Both study sites were located at 40°N latitude.

The Rocky Mountain Site (RMS) was located in Roosevelt National Forest, adjacent to the University of Colorado Mountain Research Station at an elevation of 3,120 m. The study plot was dominated by Engelmann spruce (*Picea englemanni*) and subalpine fir (*Abies lasiocarpa*) and was bisected by a creek. A lower synusia was dominated by myrtle blueberry (*Vaccinium myrtillus*) and abundant herbaceous growth such as *Lupinus argenteus*, *Pedicularis racemosa*, *Arnica cordifolia* and *Epilobium angustifolium*. Streamside habitat supported a more diverse flora. The study area was characterized by a profuse debris of dead trees, herbs, stumps and fallen logs. A more detailed account of the flora of the RMS is given by Merritt and Merritt (1978a). This site was characterized by a continental climate with strong winds frequent due to Pacific air masses. Warm, dry winds (Chinook) are common and a "crestcloud" commonly occurred over the continental divide near the study area. Most of the precipitation on the Eastern Slope of the Front Range is attributable to Gulf air masses, frequent in spring and fall. These air masses are responsible for a rapid build-up of snow on the study area in spring. Snow covered the ground for about 7½ months of the year. Precipitation is usually greater than 75 cm per year, much of it in the form of frequent late winter snow storms and summer convectional storms. Precipitation for the vicinity of the study area was reported as 64.3 cm for the years 1952–1953 (Marr, 1967:127). Temperatures for the area ranged from −25.6 to 23.9°C (mean annual temperature, 1.1°C). The growing season lasted for about 87 days at the elevation of the study site.

The Appalachian Mountain Site (AMS) was located in southeastern Westmoreland County, Pennsylvania (Powdermill Nature Reserve, Carnegie Museum of Natural History) at an elevation of 450 m. The study area occupied a northeast exposure and was bisected by a creek. The upper canopy was formed principally by beech (*Fagus grandifolia*), yellow poplar (*Liriodendron tulipifera*), sugar maple (*Acer saccharum*), cucumber tree (*Magnolia acuminata*), and red oak (*Quercus borealis*). The middle canopy was formed by striped maple (*Acer pennsylvanicum*), spice bush (*Lindera benzoin*), witch-hazel (*Hamamelis virgini-*

ana), and rhododendron (*Rhododendron maximum*). During spring a lush low synusia consisted of many herbs such as *Trillium* spp., *Viola* spp., mayapple (*Podophyllum peltatum*), white clintonia (*Clintonia umbellulata*), dwarf ginseng (*Panax trifolium*), partridgeberry (*Mitchella repens*), Indian cucumber-root (*Medeola virginiana*), and Virginia creeper (*Parthenocissus quinquefolia*) to mention just a few. By midsummer, ferns (for example, *Polystichum acrostichoides*, *Dryopteris* spp.) constituted most of the groundcover. The forest floor was mesic and well drained and characterized by its deposits of residual sandstone boulders with an accumulation of humus and leaf litter between the rocks. Selective logging in early 1900 occurred throughout the study area.

The AMS is located within the physiographic division referred to as the Allegheny Mountains Section of the Appalachian Plateau Province of Pennsylvania (Grimm and Roberts, 1950:6). The area experiences a humid, continental climate marked by warm summers and cold winters. Prevailing winds are from the west. Local weather patterns show strong topographic influence due to the location of the study area between Chestnut and Laurel ridges (maximum elevation 606 m and 848 m respectively). Annual precipitation in the vicinity of the study area ranges from 102 to 148 cm (mean, 122 cm). Temperatures range from a low of −31°C (usually occurring in January) to a high of about 36°C (occurring in July or August). Average annual temperature for the vicinity is about 7°C. Snowfalls are frequent and may be heavy from December through March. Snow may occur from October to April, however, a permanent snowcover normally persists for only about two months of the year. Snowfall amounts in the study region range from 66 to 225 cm (mean snowfall, 139 cm). The growing season lasts from 110 to 120 days in the vicinity of the study area.

FIELD PROCEDURES

The RMS contained a 12 by 12 live-trapping quadrat with stations located at 10 m. Two Sherman live-traps containing synthetic fiber nesting material and sunflower seeds as bait were located at each station. Trapping chimneys were used to monitor subnivean activity (Merritt and Merritt, 1978a; Merritt, 1982). A similar methodology was employed on the AMS except that this study site was a 10 by 10 quadrat and trapping chimneys were of a different design. See Merritt (1982) for discussion of trapping chimney design and methodology employed in monitoring *C. gapperi*. Two trapping periods per month (each consisting of 4 days) were employed to monitor the population year-round on both study sites.

Upon initial capture, voles were toe clipped for identification, at which time toe number, location on grid, weight, sex and reproductive status was recorded. Animals were weighed to the nearest 0.5 g using a Pesola scale and classified as adult, subadult or juvenile according to body weight and sexual maturity. Position of the testes (either abdominal or scrotal) was used to describe reproductive condition of males. Reproductive status of females was assessed by noting condition of the vulva (whether perforate or not perforate) and nipples (whether small, medium or lactating). Upon recapture during the same period, the toe number, location and weight was recorded and the vole released.

Temperatures were recorded year-round on both study sites. On the RMS, ambient temperature was measured continuously by a Bendix hygrothermograph maintained at 30 cm from the ground/snow surface. Subnivean temperatures were recorded by a Dixon thermograph probe which was located in a subnivean runway (Fig. 1). On the AMS a three-point thermograph recorded temperatures below ground surface (sub-surface tunnel within a talus deposit), on ground surface/subnivean level and ambient level (Fig. 2). Snow depths were recorded for both study sites at selected points, whereas on the AMS daily precipitation and snowfall were also recorded.

LABORATORY PROCEDURES

Resting metabolic rate (RMR) for *C. gapperi* was determined in the laboratory at both the RMS and AMS by employing oxygen consumption techniques. Animals were collected in the field during summer (late August) and winter (early February) and used in metabolic tests within 24 hr after capture. Mice were maintained in the laboratory and given food and water *ad libitum*. Oxygen consumption was monitored using a positive pressure "push through" assembly with a Beckman G4 (RMS) and a 755 (AMS) paramagnetic oxygen analyzer. Resting metabolic rate was computed according to the method of Depocas and Hart (1957), expressed in cubic centimeters of oxygen consumed per gram per hour (cc O_2/g/hr) and corrected for standard temperature and pressure. Differences in mean oxygen consumption of summer and winter animals were tested for significance at the 0.05 confidence level by Student's *t*-tests.

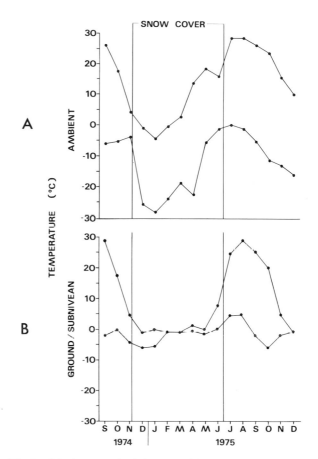

Fig. 1.—Maximum and minimum ambient temperatures recorded on the Rocky Mountain Site from September 1974 to December 1975. Temperatures were recorded at 30 cm above ground or snow (A) and on bare ground surface or in subnivean runway (B).

RESULTS

GROWTH

Growth dynamics of *C. gapperi* were evaluated for 21 months on the RMS and for 22 months on the AMS by employing three different methods: 1) plotting body weight distributions of the entire marked population through time; 2) tracing growth of selected individuals over consecutive trapping periods; 3) plotting instantaneous growth rates (% growth per day) of the population through time.

Rocky Mountain site. — Body weight distributions for red-backed voles from the RMS were depicted as percent of males and females in six age-weight classes over time (Fig. 3). Most voles overwintered at weights of 10 to 14 g, although some juveniles began the winter at weights of 6 to 9 g (January 1974). With the onset of breeding in March, weights of both sexes increased with most voles in the 15 to 19 g and 20 to 24 g classes. Recruitment occurred in June, July and August, at which time about 20% of all males captured consisted of juveniles. Females gained weight in August due to pregnancy—25% weighed 30 to 34 g. In October, the 10 to 14 g weight class appeared and increased to form the majority of the overwintering cohort. This cohort remained

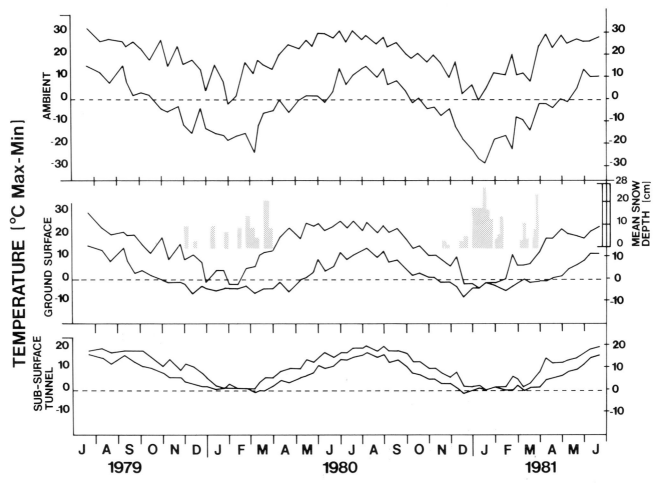

Fig. 2.—Maximum and minimum temperatures recorded on the Appalachian Mountain Site from July 1979 to June 1981. Temperatures are recorded from 1.5 m above ground surface (ambient), on ground surface and in a sub-surface tunnel. Snow depth is shown by stippled bars.

in the 10 to 14 g class until March 1975 at which time increases in size were seen culminating in June with 100% of males occupying the 20 to 24 g class. Males tended to gain weight before females as they became reproductively active. In September, over half the voles were distributed in the 10 to 14 g class as they were in the preceeding year.

Appalachian Mountain site.—The same procedure was employed to analyze body weight distributions on the AMS as for the RMS above. Unlike the RMS eight age-weight classes were defined and monitored for 22 months (Fig. 4). In contrast to *C. gapperi* on the RMS, it was found that voles on the AMS were 30% larger than those from the RMS (mean annual weight for population). From autumn to early winter, most males were distributed in the

22 to 25 g weight class, however, about 50% of September-caught females were in the 26 to 29 g class due primarily to pregnancy. Juvenile females were present in October and November 1979. No juvenile males were captured during this time period, although they did appear in the following year. During 1979–1980 most overwintering females were distributed in the 18 to 21 g class and 22 to 25 g class, while males were distributed in slightly heavier classes. Unlike the RMS in which a heavier autumn weight class entered and overwintered in a lower weight class, it appeared that on the AMS most voles overwintered at a heavier weight than seen in autumn, especially males. Weights increased in March 1980 concommitant with reproductive activity and juveniles appeared in May. During sum-

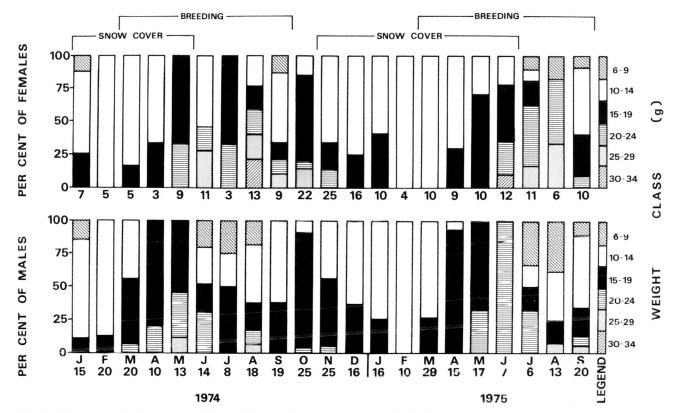

Fig. 3.—Histogram showing body weight distribution of male and female *Clethrionomys gapperi* from the Rocky Mountain Site. Numbers on horizontal axis are monthly sample size.

mer most females were represented by the 18 to 21 g class, while males were distributed more evenly into four different weight classes. The transition from autumn to winter 1980 was similar to that of 1979, however the 14 to 17 g class persisted one month longer in 1980 (December) than in 1979 (November). This was due to a longer reproductive season. In March, the population consisted of about 60% of males in the 26 to 29 g weight class. This increase in weight from February (18% of males in the 26 to 29 g class) was a result of reproductive activity. In April 1981, 25% of females formed the 34 to 37 g class indicating pregnancy—this was 5% higher than female representation in this weight class in 1980. The juvenile weight class was first seen in May 1980 as was the case in 1979.

Rocky Mountain site.—Selected individual growth records are presented in Fig. 5. Mean body weight for October 1974 was 15.4 g for males and 15.1 g for females. These weights declined slowly to a mean low of 13.6 g for males and 13.3 g for females in February 1975. Weights of males increased during

March and reached a high of 17.9 g in May. Females gained weight more slowly and in May averaged only 14.9 g, but in June showed a rapid increase in weight due to pregnancy. A significant increase ($P < 0.05$) in weight occurred between February and May-caught voles of both sexes when compared by Student's t-test. Growth records showed voles weighing more in autumn and spring and less during midwinter—these differences were significant at the 0.05 confidence level (see Merritt and Merritt, 1978*a*:585 for SE for above means).

Appalachian Mountain site.—Individual growth rates are presented in Fig. 6. Mean body weight for October 1980 was 22.2 g (SE = 0.58) for males and 22.8 g (SE = 1.92) for females. These weights increased slightly to an average of 24.9 g (SE = 1.23) for males and 23.3 g (SE = 1.23) for females during winter (February) 1981. Weight of voles increased during March and April due to reproductive effort reaching an average of 30.3 g (SE = 0.42) for males and 30.0 g (SE = 2.12) for females in May 1981. A gradual increase in weight was seen in individuals

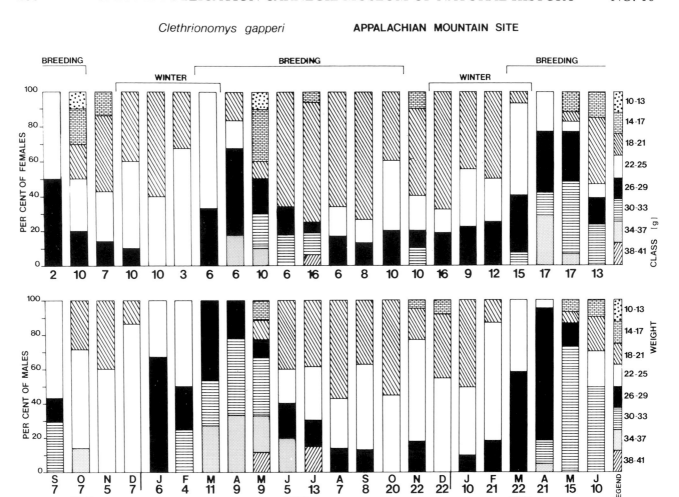

Fig. 4.—Histogram showing body weight distribution of male and female *Clethrionomys gapperi* from the Appalachian Mountain Site. Numbers on the horizontal axis are monthly sample size.

traced from October 1980 to May 1981. Average weight for sexes combined was 22.36 g (N = 10) in October, 23.86 g in February and 30.61 g in May 1981. A significant increase ($P < 0.05$) in weight existed between October and May-captured voles. This contrasts to the Rocky Mountain Site in which a significant decrease in weight was seen from October to February, with an increase beginning in the latter month due to reproductive activity.

Rocky Mountain and Appalachian Mountain sites.—Fig. 7 shows the instantaneous growth rate (% growth per day) of *C. gapperi* on the RMS and AMS from 1974 to 1975 and 1979 to 1981 respectively. Data for each 4-week period were condensed by use of linear regression between mean body weight and growth rate for each time period. A single representative growth rate was calculated for each

regression by adjusting the growth rate to a hypothetical 15 g vole for the RMS and a 25 g vole for the AMS. These weights are representative as the median weight of voles from each study site. As evidenced in Fig. 7, growth rates were highest in spring and summer on both study sites. An exception was seen on the AMS for the period 9 April to 31 May 1980 possibly due to a low sample size. Negative growth was seen in females on the RMS during midwinter 1974 but this changed to high rates of growth in spring. A rapid deterioration of growth culminating in a −0.7% growth per day occurred in autumn 1975. This increased slightly but remained negative during the period from 15 November to 23 February. Growth again reached a positive value during spring 1975 reflecting growth rates of juveniles principally. On the Appalachian

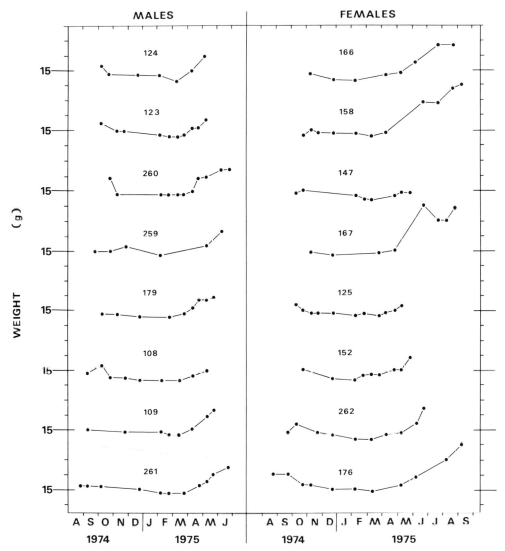

Fig. 5.—Weight records of selected individual male and female *Clethrionomys gapperi* (N = 16) from the Rocky Mountain Site. Divisions on the vertical axis are increments of 5 g. The number of the animal is given above each weight record.

site, positive growth was seen during the autumn period for both years. This was in sharp contrast to the negative growth rates seen for this time of year on the RMS. Rates of growth generally remained positive throughout winter, spring and summer 1980. Two exceptions were seen—females during early winter 1980 and males during spring 1981 showed a zero growth per day. A negative growth rate occurred on the AMS during late November to mid-January 1981 and ranged from -0.25% for females to -0.6% for male voles. This negative growth per day lagged behind a similar decline seen on the RMS by about one month. The autumn cohort on both the RMS and AMS was principally composed of voles born in late summer or spring. This cohort

tended to undergo a weight loss during midwinter on the RMS as contrasted to a gradual yet steady weight increase on the AMS.

METABOLISM

Table 1 summarizes metabolic studies of *C. gapperi* collected during summer and winter on the RMS and AMS. Metabolic data derived from oxygen consumption techniques is expressed in two ways: 1) metabolism per unit weight of animal (cc O_2/g/hr) and referred to as RMR and 2) total caloric needs of an animal per unit time (cal./animal/hr) and referred to as caloric requirement. On the RMS, resting metabolic rate for summer-trapped voles was

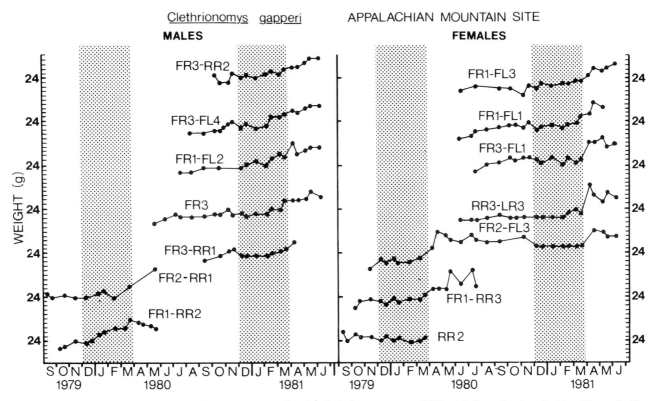

Fig. 6.—Weight records of selected individual male and female *Clethrionomys gapperi* (N = 14) from the Appalachian Mountain Site. Divisions on the vertical axis are increments of 2 g. The number of the animal is given above each weight record.

about 8% higher than that of winter-trapped voles. The caloric requirement of voles during summer was slightly greater than for winter. This difference was due to the heavier summer weights of voles on the RMS. On the AMS summer metabolic rate was also slightly higher than that determined for winter-caught voles. The winter caloric needs of voles on the AMS was higher than the summer requirement again due principally to heavier body weights at this time of year on the AMS. Voles on the RMS showed metabolic rates that were about 32% higher than rates measured for voles on the AMS. Body weights of adult voles tested from the AMS were 54% heavier than adults tested from the RMS. As mentioned earlier, the mean annual weight of voles monitored on the AMS was about 30% greater than the mean annual weight of voles from the RMS. No significant difference was found between body weights, RMR, and caloric requirements of voles from the AMS when compared between winter and summer.

When comparing seasonal changes in body temperature of voles from the RMS, no significant difference was noted. However, on the AMS a significant difference was found when comparing winter body temperature (33.9°C) to summer body temperature (35.9°C). Comparisons of significance were made employing Student's *t*-tests ($P < 0.05$, df = 16). It was felt that the lower body temperature during winter may aid voles in thermogenesis by permitting maintenance of a lower RMR during the winter season.

DISCUSSION

There are many physiological and behavioral adjustments available to small mammals which enable them to cope with harsh climatic extremes, albeit these tactics are not evenly distributed among all taxonomic groups, but some act synergistically to optimize survival. Some important survival "strategies" of small mammals occupying seasonal environments include the following: 1) decrease in body weight during autumn and weight gain in spring (Kalela, 1957; Sealander, 1966; Iverson and Turner,

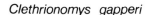

Clethrionomys gapperi

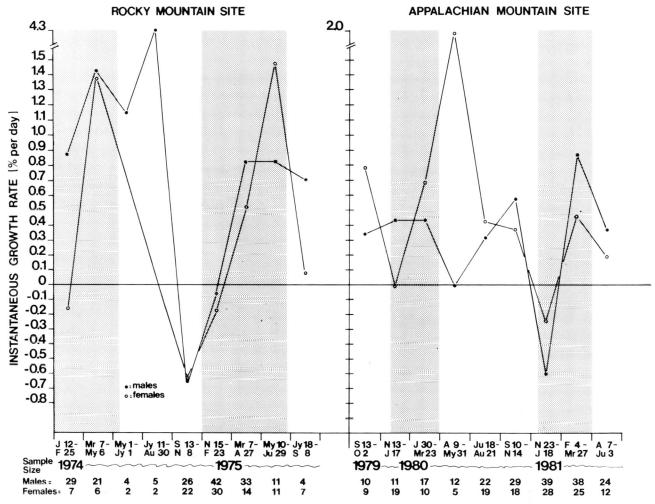

Fig. 7.—Growth rate changes for male and female *C. gapperi* on the RMS and AMS from January 1974 to September 1975 and September 1979 to June 1981 respectively. Instantaneous growth rate (% per day) is shown on the vertical axis. Time period and sample sizes are shown on the horizontal axis. Stipple represents the period of snowcover on the RMS and the calendar dates for winter on the AMS.

1974; Fuller, 1977); 2) lower total rate of metabolism during winter months (Wunder et al., 1977; Merritt and Merritt, 1978*a*; Petterborg, 1978); 3) increased pelage density and length and utilization of energy reserves (brown and white fat) during periods of cold stress (Sealander, 1951, 1972; Khateeb and Johnson, 1971; Heldmaier, 1975; Wunder et al., 1977); 4) conservation of heat during winter months by huddling thermogenesis and/or torpor (Pearson, 1947; Sealander, 1952; Weigert, 1961; Lynch and Gendler, 1980; West, 1977; Wolff and Lidicker, 1981; Lyman et al., 1982); 5) availability of energy-rich food during winter months (Tast, 1966, 1972); 6) winter foraging confined to stable thermal regime within subnivean or subterranean

microhabitat (Pruitt, 1957; Hayward, 1965; Selander, 1966; Kucera and Fuller, 1978; Merritt, 1984; Merritt and Merritt, 1978*a*; Webster and Brooks, 1981).

The present research thrust at Powdermill Nature Reserve concentrates on many of the above "survival strategies" as applied to montane small mammals, specifically *C. gapperi*. In the present paper, only two of the above behavioral and physiological adjustments to winter will be discussed. Seasonal growth patterns and metabolic requirements are compared in two contrasting montane environments in North America—the Appalachian and Rocky Mountains.

Table 1.—*Seasonal energy requirements of* Clethrionomys gapperi *from Rocky Mountain and Appalachian Mountain sites. Resting metabolic rates measured at 29°C. Summer sample, late August; winter sample, early February.*

Location	Season	Body weight (g)	Body temperature (°C)	Sample size	RMR (ccO_2/g/hr)	Caloric requirement[1] (cal/animal/hr)
Rocky Mountain site	Winter	13.3 ± 0.62	37.3 ± 0.49	4	2.76 ± 0.20	193.8 ± 20.0
	Summer	14.1 ± 1.30	37.1 ± 0.61	9	2.98 ± 0.48	225.2 ± 27.4
Appalachian Mountain site	Winter	21.5 ± 3.41	33.9 ± 0.82	6	3.74 ± 0.20	382.4 ± 65.1
	Summer	20.7 ± 2.16	35.9 ± 0.45	12	3.81 ± 0.40	380.5 ± 51.3

[1] Caloric requirement based on RMR.
Values given are means ± 1 SD.

COHORT DYNAMICS

A voluminous literature supports the fact that many northern microtine rodents tend to be smaller in winter than in summer. There is reportedly strong selection favoring small size in winter—it permits a decrease in caloric needs during a time of food shortage. At this point in our discussion, it is important to pinpoint what specific cohort(s) in a given population actually undergoes a weight decline during winter months. A major confusion in viewing weight declines of a given population is concerned with the expression "mean population weight" commonly used by authors. By emphasizing the mean weight of a population we tend to "mask" the contribution of a specific cohort within our population. This point is further illustrated by the work of Fuller et al. (1969) and Iverson and Turner (1974) that reported four processes by which mean population weight of voles can decrease. These processes include: 1) death or emigration of the oldest (largest) animals; 2) cessation of growth in the immature animals; 3) decline in weight of older animals and 4) recruitment of light animals into the population. After evaluating the influence of each of the above factors on *M. pennsylvanicus,* Iverson and Turner (1974:1039) concluded that ". . . the fall and winter decrease in the population mean weight is produced primarily by a combination of recruitment of small animals and weight loss by individuals." Further studies by Zejda (1971), Tast (1972), and Merritt and Merritt (1978a) with *C. glareolus, M. oeconomus,* and *C. gapperi* respectively, indicated that the overwintering cohort is comprised principally of "immature" individuals—those animals born during late summer and fall that have not reached sexual maturity. These voles tend to show a decline in weight from autumn to late winter with an increase in body weight in spring concommitant with reproductive effort. Results of the present study (AMS) show most overwintering voles belong to the spring-born cohort. This group did not exhibit winter weight declines at a significant level when viewed on an

individual basis (Fig. 6). In contrast, on the RMS (Fig. 5) most individuals comprising the overwintering cohort were born in fall (immature during winter) and demonstrated a significant decrease in weight during winter. These voles gained weight in spring as reproductive activity began. Growth trends from the RMS and AMS were based on continuous monitoring of individuals for periods ranging from 10 to 20 months. On the AMS, a female born in late summer and monitored for a period of 20 months, showed an increase in weight during the first winter (immature) and a slight decline in the second winter (adult). However, when viewed as instantaneous growth (% per day), voles from both the RMS and AMS demonstrated retardation in growth but this decline lasted from mid-September to late February on the RMS and only from late November to late January on the AMS. Growth patterns in a population vary, not only as to the cohort affected, but in the length of the period of weight decline in respective ecosystems. The energetic advantage of this "strategy" will be discussed later.

FACTORS INFLUENCING WEIGHT DECLINE IN SMALL MAMMALS

Food Availability and Quality

As mentioned earlier, smaller size during the winter period may confer an energetic advantage in the form of reducing food requirements during this time of food scarcity. However, Iverson and Turner (1974) questioned this reported advantage. They found weight loss in *M. pennsylvanicus* commencing in August, at which time food was abundant, and therefore suggested that food was not a proximate cause of individual weight loss. Merritt and Merritt (1978a) found the winter diet of *C. gapperi* in Colorado to consist principally of seeds. Since endosperm possesses a comparatively high caloric value, they concurred with Iverson and Turner (1974), contending that food, per se, may not be the sole cause of winter weight declines. Iverson and Turner

(1974) and Merritt and Merritt (1978*a*) studied voles living in the subnivean environment during winter; here, the actual act of procurement may pose hardships for acquisition of food by small mammals. During a continuous snowcover, it is possible that voles were prevented from securing adequate quantities of food, due to the morphology of the snowpack at the ground level. Frozen ground or perhaps the presence of a hard crystalline layer on the ground surface may well restrict foraging activities, thus preventing access to a given food.

In the present study, growth declines were more pronounced and voles weighed an average of 30% less on the RMS than those on the AMS. If food availability and nutritive quality were regulating factors to growth trends in winter, as some authors contend, we may expect less food in the form of high energy seed material during winter in the coniferous site than in the deciduous site. This trend could be predicted based upon the fact that gross primary productivity of coniferous forests (RMS) is lower (ca. 3,000 kcal/m²/year) than a temperate deciduous forest such as the AMS (ca. 8,000 kcal/m²/year) (Odum, 1971:51). Although winter food habits of *C. gapperi* from the AMS have not been published at this time, research indicates that the diet of red-backed voles is more varied in eastern deciduous forests than in western coniferous forests—this reflects the greater diversity of available food items in the former versus the latter forest community. See Merritt (1981) for a review of the food habits of *C. gapperi*.

Results of the present study suggest that a metabolic adjustment accompanied the mid-winter growth decline on the RMS. Winter-caught voles weighed significantly less and also showed a lower weight-specific metabolic rate than the larger summer-caught voles. This rate gave winter animals a slight advantage in terms of calories necessary for existence. In contrast, on the AMS both winter and summer voles were approximately equal in weight, but with the winter-caught voles showing a slightly lower weight-specific metabolic rate (Table 1). However, actual seasonal caloric rquirements were similar. It is interesting to note that winter voles did exhibit a slightly lower body temperature than summer animals, thus permitting a lower metabolism. To conclude, seasonal metabolic rate did not show significant differences on the AMS, as did those on the RMS. It is tempting to conclude that voles on the RMS were exhibiting weight declines and lower metabolic rates as an adjustment to low food availability. Indeed, food may act as an important con-

tributor to growth patterns and resulting energy budgets, but must be balanced against the cost of foraging for such food sources. Here, we must measure the impact of encountering different thermal regimes during the foraging period. Red-backed voles tend to forage within the lush humus matt of the forest floor on both study sites. Because of their notably omnivorous diet, it seems unlikely that food quantity would be a sole determinent of thermoregulatory tactics of this small mammal. In order to more fully understand the importance of food quantity and quality to the red-backed vole, detailed studies of net primary production and diet analysis are required.

Microclimatic Regime

Microtine species are active during mid-winter and do not undergo physiological heterothermy. Therefore, they must forage in order to meet their total caloric needs during this period of increased thermal stress. The thermal regime of the foraging zone plays a major role in determining energy budgets of small mammals. As we have seen in Fig. 1, snowcover provided good insulation against fluctuating ambient temperatures during mid-winter on the RMS. The subnivean environment represented the principal foraging area for *C. gapperi* on the RMS. Red-backed voles confined their movements to this zone during winter months (Merritt and Merritt, 1978*b*). In contrast, on the AMS, snowcover was sparce and intermittent. Here, voles foraged principally in subsurface tunnel systems within talus slopes. This zone was, in a sense, analogous to the subnivean environment, in that during mid-winter it supported a stable thermal regime of about 0°C, as did the subnivean environment.

Because of the stable microclimate on both sites, voles were able to avoid the full impact of low temperatures of the macroenvironment. This stable thermal regime permitted lower metabolic rates which, in turn, reduced caloric requirements. Scholander (1955) has indicated that it is not possible for small mammals to greatly increase insulation in order to cope with low temperatures. Instead, they must either avoid cold stress (that is, selection of stable microhabitats) or increase metabolic rates. On both study sites, it seemed apparent that red-backed voles conserved energy by foraging and residing in either subnivean or subsurface runways during the winter period. An additional energy savings may also have been gained by foraging within the thick layer of humus present on both study sites.

As one would expect, this litter layer was thicker and better developed on the AMS.

It is well known that the subnivean environment provides a very stable microclimate for small mammals (Pruitt, 1957; Kucera and Fuller, 1978; Merritt and Merritt, 1978a). Results from the present study indicate that the RMS was characterized by a snow-cover for 7½ months of the year, reaching a depth of 270 cm in late April. If the microenvironmental regime dictates thermoregulatory strategies, then why did voles on the RMS show greater fluctuations in growth patterns and thermogenesis than voles on the snow-free AMS? It is possible that low thermal conditions during autumn on the RMS may have elicited increased physiological adjustments in red-backed voles. Although snowcover provided a stable subnivean environment on the RMS from mid-November to mid-June, there existed a period of about 70 days in autumn (early September to mid-November) in which subzero temperatures occurred at the ground level. At this time, snowcover did not provide a shield against fluctuating ambient temperatures. During this time, the population consisted of primarily juvenile and subadult voles destined to comprise the majority of the overwintering cohort.

The period of autumn freeze has been shown to produce low survival in small mammals inhabiting northern regions. Formozov (1946) and Pruitt (1970) referred to the autumnal thermal overturn as that time of the year when ambient temperatures fall below ground temperatures. When snow thickness is sufficient to insulate soil against fluctuating ambient temperatures a "hiemal threshold" is established, which marks the beginning of winter for small mammals, who then restrict foraging to the subnivean environment. Vaughan (1969) and Stinson (1977) contended that the period of autumn freeze produced high mortality in small montane mammals. In contrast, good survival was found in *C. gapperi* from Great Slave Lake, Canada, the Rocky Mountains, and the Appalachian Mountains (Fuller et al., 1969; Merritt and Merritt, 1978a; Merritt, 1984). In addition to influencing survival of small mammals, this autumn period may be critical to the life of small mammals by presenting two major stimuli—low temperatures and short-day photoperiods. These environmental "signals" may act in concert, as cues to young voles, eliciting cessation of growth and lower metabolic rates. As mentioned earlier, this lower metabolic rate would be adaptive for survival during long winter months when food may be in short supply.

LITERATURE CITED

BERGSTEDT, B. O. 1965. Distribution, reproduction, growth and dynamics of the rodent species *Clethrionomys glareolus* (Schreber), *Apodemus flavicollis* (Melchior) and *Apodemus sylvaticus* (Linn) in southern Sweden. Oikos, 16:132–160.

BROOKS, R. J., and A. B. WEBSTER. 1984. Relationship of seasonal change to changes in age structure and body size in *Microtus pennsylvanicus*. This volume.

BROWN, E. G., III. 1973. Changes in patterns of seasonal growth of *Microtus pennsylvanicus*. Ecology, 54:1103–1110.

BUJLASKA, G., and J. GLIWICZ. 1968. Productivity investigation of an island population of *Clethrionomys glareolus* (Schreber, 1780). III. Individual growth curve. Acta Theriol., 13:427–433.

CHITTY, D. 1952. Mortality among voles (*Microtus agrestis*) at Lake Vyrnwy, Montgomeryshire, in 1936–1939. Phil. Trans. Roy. Soc. London, Ser. B, 236:505–552.

DEPOCAS, F., and J. S. HART. 1957. Use of the Pauling oxygen analyzer for measurement of oxygen consumption of animals in open-circuit systems and in a shortlag, closed-circuit apparatus. J. Appl. Physiol., 10:388–392.

FORMOZOV, A. N. 1946. Snow cover as an integral factor of the environment and its importance in the ecology of mammals and birds. Occas. Publ. Boreal Inst. Univ. Alberta, 1:1–143.

FULLER, W. A. 1969. Changes in numbers of three species of small rodents near Great Slave Lake, N.W.T., Canada, 1964–1967, and their significance for general population theory. Ann. Zool. Fennici, 6:113–144.

———. 1977. Demography of a subarctic population of *Cleth-

rionomys gapperi*: size and growth. Canadian J. Zool., 55:415–425.

FULLER, W. A., L. L. STEBBINS, and G. R. DYKE. 1969. Overwintering of small mammals near Great Slave Lake, northern Canada. Arctic, 22:34–55.

GORECKI, A. 1968. Metabolic rate and energy budget in the bank vole. Acta Theriol., 13:341–365.

GRIMM, W. C., and H. A. ROBERTS. 1950. Mammal survey of southwestern Pennsylvania. Pennsylvania Game Comm., 99 pp.

HASLER, J. F., A. E. BUHL, and E. M. BANKS. 1976. The influence of photoperiod on growth and sexual function in male and female collared lemmings (*Dicrostonyx groenlandicus*). J. Reprod. Fert., 46:323–329.

HAYWARD, J. S. 1965. Microclimate temperatures and its adaptive significance in six geographic races of *Peromyscus*. Canadian J. Zool., 43:341–350.

HELDMAIER, G. 1975. The effect of short daily cold exposures on development of brown adipose tissue in mice. J. Comp. Physiol., 98:161–168.

HYVARINEN, H., and K. HEIKURA. 1971. Effects of age and seasonal rhythm on the growth patterns of some small mammals in Finland and in Kinbenes, Norway. J. Zool., London, 165:545–556.

IVERSON, S. L., and B. N. TURNER. 1974. Winter weight dynamics in *Microtus pennsylvanicus*. Ecology, 55:1030–1041.

KALELA, O. 1957. Regulation of reproduction rate in subarctic populations of the vole *Clethrionomys rufocanus* (Sund.). Ann. Acad. Sci. Fenn., A, IV, 34:1–60.

KHATEEB, A., and E. JOHNSON. 1971. Seasonal changes of pelage in the vole (*M. agrestis*). I. Correlation with changes in the endocrine glands. Gen. Comp. Endocr., 16:217–228.

KUCERA, E., and W. A. FULLER. 1978. A winter study of small rodents in aspen parkland. J. Mamm., 59:200–204.

LYMAN, C. P., J. S. WILLIS, A. MALAN, and L. C. H. WANG. 1982. Hibernation and torpor in mammals and birds. Academic Press, New York, x + 317 pp.

LYNCH, G. R., and S. L. GENDLER. 1980. Multiple responses to different photoperiods occur in the mouse, *Peromyscus leucopus*. Oecologia (Berl.), 45:318–321.

McNAB, B. K. 1971. On the ecological significance of Bergmann's rule. Ecology, 52:845–854.

MALLORY, F. F., J. R. ELLIOTT, and R. J. BROOKS. 1981. Changes in body size in fluctuation populations of the collared lemming: age and photoperiod influences. Canadian J. Zool., 59:174–182.

MARR, J. W. 1967. Ecosystems of the East Slope of the Front Range in Colorado. Univ. Colorado Studies, Ser. Biol., 8:1–134.

MERRITT, J. F. 1981. *Clethrionomys gapperi*. Mammalian Species, 146:1–9.

———. 1982. Red-backed vole. Pp. 196–198, *in* CRC handbook of Census methods for terrestrial vertebrates (D. E. Davis, ed.), CRC Press, Inc. 397 pp.

———. 1983. Influence of snowcover on survival of *Clethrionomys gapperi* inhabiting the Appalachian and Rocky Mountains of North America. Ann. Zool. Fennici, in press.

MERRITT, J. F., and J. M. MERRITT. 1978a. Population ecology and energy relationships of *Clethrionomys gapperi* in a Colorado subalpine forest. J. Mamm., 59:576–598.

———. 1978b. Seasonal home ranges and activity of small mammals of a Colorado subalpine forest. Acta Theriol., 23:195–202.

ODUM, E. P. 1971. Fundamentals of ecology. W. B. Saunders Company. xiv + 574 pp.

PEARSON, O. P. 1947. The rate of metabolism of some small mammals. Ecology, 28:196–223.

PERRIN, M. R. 1979. Seasonal variation in the growth, body composition, and diet of *Clethrionomys gapperi* in spruce forest. Acta Theriol., 24:299–318.

PETTERBORG, L. J. 1978. Effect of photoperiod on body weight in the vole, *Microtus montanus*. J. Zool., 56:431–435.

PRUITT, W. O., JR. 1957. Observations on the bioclimate of some taiga mammals. Arctic, 10:131–138.

———. 1970. Some ecological aspects of snow. Pp. 83–99, *in* Ecology of the subarctic regions. Proc. Helsinki Symp., (UNESCO, Paris), 364 pp.

ROSENMANN, M., P. MORRISON, and D. FEIST. 1975. Seasonal changes in the metabolic capacity of red-backed voles. Physiol. Zool., 48:303–310.

SCHOLANDER, P. E. 1955. Evolution of climatic adaptation in homeotherms. Evolution, 9:15–26.

SEALANDER, J. A. 1951. Survival of *Peromyscus* in relation to environmental temperature and acclimation at high and low temperatures. Amer. Midland Nat., 46:257–311.

———. 1952. The relationship of nest protection and huddling to survival of *Peromyscus* at low temperature. Ecology, 33:63–71.

———. 1966. Seasonal variations in hemoglobin and hematocrit values in the northern red-backed mouse, *Clethrionomys rutilus dawsoni* (Merriam), in interior Alaska. Canadian J. Zool., 44:213–244.

———. 1972. Circum-annual changes in age, pelage characteristics and adipose tissue in the northern red-backed vole in interior Alaska. Acta. Theriol., 17:1–24.

SEALANDER, J. A., and L. K. BICKERSTAFF. 1967. Seasonal changes in reticulocyte number and relative weights of the spleen, thymus, and kidneys in the northern red-backed mouse. Canadian J. Zool., 45:253–260.

STEBBINS, L. L. 1976. Overwintering of the red-backed vole at Edmonton, Alberta, Canada. J. Mamm., 57:554–561.

STINSON, N. S., JR. 1977. Species diversity, resource partitioning and demography of small mammals in a subalpine deciduous forest. Unpublished Ph.D. thesis, Univ. Colorado, Boulder, 238 pp.

TAST, J. 1966. The root vole, *Microtus oeconomus* (Pallas), as an inhabitant of seasonally flooded land. Ann. Zool. Fennici, 3:127–171.

———. 1972. Annual variations in the weights of wintering root voles, *Microtus oeconomus*, in relation to their food conditions. Ann. Zool. Fennici, 9:116–119.

VAUGHAN, T. A. 1969. Reproduction and population densities of a montane small mammal fauna. Pp. 51–74, *in* Contributions in mammalogy (J. K. Jones, Jr., ed.), Misc. Publ. Mus. Nat. Hist., Univ. Kansas, 51:1–428.

WEBSTER, A. B., and R. J. BROOKS. 1981. Social behavior of *Microtus pennsylvanicus* in relation to seasonal changes in demography. J. Mamm., 62:738–751.

WEIGERT, R. G. 1961. Respiratory energy loss and activity pattern in the meadow vole, *Microtus pennsylvanicus*. Ecology, 42:245–253.

WEST, S. D. 1977. Midwinter aggregation in the northern red-backed vole, *Clethrionomys rutilus*. Canadian J. Zool., 55:1404–1409.

WHITNEY, P. 1976. Population ecology of two sympatric species of subarctic microtine rodents. Ecol. Monogr., 46:85–104.

WIGER, R. 1979. Demography of a cyclic population of the bank vole *Clethrionomys glareolus*. Oikos, 33:373–385.

WOLFF, J. O., and W. Z. LIDICKER, JR. 1980. Population ecology of the taiga vole, *Microtus xanthognathus*, in interior Alaska. Canadian J. Zool., 58:1800–1812.

———. 1981. Communal winter nesting and food sharing in taiga voles. Behav. Ecol. Sociobiol., 9:237–240.

WUNDER, B. A., D. S. DOBKIN, and R. D. GETTINGER. 1977. Shifts of thermogenesis in the prairie vole (*Microtus ochrogaster*): strategies for survival in a seasonal environment. Oecologia (Berl.), 29:11–26.

ZEJDA, J. 1967. Mortality of a population of *Clethrionomys glareolus* Schreb. in a bottomland forest in 1964. Zool. Listy, 16:221–238.

———. 1971. Differential growth of three cohorts of the bank vole, *Clethrionomys glareolus* Schreb. 1780. Zool. Listy, 20:229–245.

Address: Powdermill Nature Reserve, Carnegie Museum of Natural History, Rector, Pennsylvania 15677.

REPRODUCTION OF THE MONTANE VOLE, *MICROTUS MONTANUS,* IN SUBNIVEAN POPULATIONS

FREDERICK J. JANNETT, JR.

ABSTRACT

Two populations of *Microtus montanus* (Rodentia: Muridae) were followed by capture-mark-recapture from late summer, 1972, through spring, 1973. Winter subnivean trapping was done at approximately monthly intervals beneath previously established shelters. Populations at the same two areas were trapped from January through spring, 1974.

Extensive winter breeding was displayed by the two populations during the first winter, but not the second, and was characterized demographically by the disappearance of summer-breeding parous females, the continued presence of some males with scrotal testes, and the achievement of puberty by late summer- and fall-born voles. Additionally, winter-born females bred in at least one population. Populations exhibiting late fall and winter breeding were characterized by operational sex ratios approaching unity, whereas non-breeding populations had a paucity of adult males. It has been demonstrated with experimental groups that social stimulation may account for late fall breeding in this species, and other hypothesized proximate causes are not necessary.

Winter breeding in other microtine rodents is reviewed, and different types of winter breeding are recognized herein. Mid-winter breeding in cyclic populations is correlated with the increasing phase of the cycle, as it was in *M. montanus.* In most reports, mid-winter breeding is an extension of fall breeding. This subnivean study indicates that mid-winter breeding in high latitude subnivean populations of lemmings and low latitude populations of voles without permanent snow cover may be the result of the same population dynamics.

INTRODUCTION

One of the most enigmatic phenomena in mammalian biology is the occurrence of extensive mid-winter breeding of microtine rodents beneath a deep mantle of snow, but apparently no such population has heretofore been followed by capture-mark-recapture. During the course of field work on the montane vole in 1972, behavioral observations and longidutinal data on weight and head-body length had been taken on voles in a population since August 10, and by October it was obvious that not only was reproduction continuing in old females, indeed late summer-born females were already breeding. A second study area was therefore begun nearby in mid-October, and these populations were followed through the winter. The voles were removal trapped from the grids in the following spring, and both areas were trapped again in the subsequent winter. The populations which exhibited winter breeding were found during the increase phase of their population cycle, and here I characterize the observed reproduction and sex ratios.

ANIMALS AND METHODS

This work was part of a larger project in which voles were studied in unconfined populations in a high mountain valley (about 2,057 m) in northwestern Wyoming over the period 1971–1977 by capture-mark-recapture and removal trapping. The purpose of the field work was to identify major features of the social system and how they change with density changes as has been reported elsewhere (Jannett, 1978, 1981, 1982).

The two study sites in which voles were trapped during winters were about 1.5 km apart on a large open plain and were on opposite sides of a two-lane paved road. Vegetation was undisturbed for 30–60 m on each side of the road, beyond which cattle were sometimes enclosed by fences. The pastures, which by the fall of each year had been heavily grazed, and the road perhaps inhibited voles from leaving the study areas, but the voles could not be considered to have been constrained. At each of the two sites, the entire field between the road and the fence was covered with a grid of traps at a 6.1 m interval, but in the second year each site had some peripheral traps moved which slightly altered the size of the grid from its previous area (Table 1). However, Grids 6 and 13 were essentially the same area and Grids 7 and 14 were the same area. Both sites had heavy grass cover (*Poa, Bromus,* and *Agropyron*), and few meadow voles (*Microtus pennsylvanicus*), gophers (*Thomomys talpoides*), and ground squirrels (*Spermophilus armatus*). The most common predators were coyotes (*Canis latrans*), shorttail weasels (*Mustela erminea*), longtail weasels (*M. frenata*), sparrow hawks (*Falco sparverius*), Swainson's hawks (*Buteo swainsoni*), red-tailed hawks (*B. jamaicensis*), and great horned owls (*Bubo virginianus*).

Fig. 1 shows the snow pack at Moran, Wyoming, recorded by the U.S. Bureau of Reclamation about 6.4 km from the study sites. The winter study sites differed from that at which the complete data record is available in that the snow pack was slightly

Table 1.—*Basic features of the study grids.*

Grid	Dates	Size of grid (ha)	Traps per station	Number of fall and spring stations	Number of winter stations	Voles per hectare[a]	M. montanus: M. pennsylvanicus trapped-out (n:n)
6 (1972–1973)	August 10 . . . September 28	—	—	—	—		
	October 13–17	0.31	1	104	—	173	
	November 15 . . . 22	0.42	1 or 2	136	(40)[b]		
	January 7–8	0.42	2	—	40		
	February 11–12	0.42	2	—	40		
	March 9–10	0.42	2	—	40		
	April 9–10	0.42	2	—	40		
	April 29, May 2–11	0.42	1	136	(40)[b]		
	May 11–17	0.42	1	136	(40)[b]	151	63:1
7 (1972–1973)	October 18–22	0.47	1	154	—	115	
	November 16–17	0.47	1	154	(51)[b]		
	December 30–January 1	0.47	2	—	51		
	February 4–5	0.47	2	—	51		
	March 6–7	0.47	2	—	51		
	April 6–7	0.47	2	—	51		
	May 21–30	0.47	1	154	(51)[b]	179	85:0
13 (1973–1974)	January 28–29	0.46	2	—	43		
	February 21–22	0.46	2	—	43		
	April 15–16	0.46	2	—	43		
	April 30–May 3	0.46	2	—	43		
	May 11–17	0.46	1	155	(43)[b]	140	64:0[c]
14 (1973–1974)	January 9–12	0.42	2	—	35		
	January 22–25	0.42	2	—	35		
	February 25–March 1	0.42	2	—	35		
	April 23–27	0.42	1	—	(35)[b]		79:1

[a] Calculated on basis of voles known to be present.
[b] The winter covers were out during these trapping periods also.
[c] Five *M. pennsylvanicus* were identified out of 17 voles examined during the winter by dental impressions.

less; it averaged 0.67 to 1 m and was wind-packed. Also, permanent complete snow cover during the 1972–1973 winter at the study sites began 26 November but melted off largely between 17 April and 28 April. Permanent snow pack during the winter 1973–1974 began 1 November and melted off largely between 16 April and 25 April. The coldest maximum and minimum temperatures recorded between 23 December 1972, and 17 April 1973, at about 10 day intervals under a winter trap shelter not used for trapping ranged from −3° to −12°C. Table 2 presents monthly mean temperatures, precipitation, respective departures from normal, and monthly degree day totals at Moran.

To prevent the sun from heating the traps during the fall, each was covered by a clean half-gallon milk carton with one side removed. The carton was placed at the best sign within 0.76 m of the grid point as the trap was placed down and opened. Traps and usually the covers were removed between trap periods. The traps were small nonfolding aluminum box traps (H. B. Sherman, Deland, Florida). Only clean traps were used. All traps were baited with oats, fudge, and apple. About 1.5 g of cotton batting (Stearns and Foster, No. 2 Arlo) for nesting material was hooked to the rear of the trap during most trap periods. Some traps were closed at night in the late fall.

Trapping dates are listed in Table 1. Grid 6 was first trapped (10 August–28 September) not as a grid, but as a study site wherein traps were set at vole sign and known nest sites to ascertain movements of parous females and to identify juveniles (Jannett,

1978). Trap density was high (for example, about 100 traps in about 0.15 ha). The October grid was enlarged in November, during which session wooden trap shelters were added to those stations frequented by those voles on which the most data had already been accumulated. These shelters were of the type described by Iverson and Turner (1969) which consist of two vertical boards connected by two nailed boards on top, between which was placed a removable board. Similarly, winter covers were set out in Grid 7 in November. Winter covers were set out at largely the same stations at these two sites on 11 November 1973, by which time there was approximately 13 cm of snow on the ground. Subnivean trapping was not undertaken until late December or January when there was sufficient snow cover so that "pukak" was not disturbed by my movement on the grid.

Winter trapping differed from fall trapping in that a number 3 "tin" can was fitted to the rear of the trap with the rear door open, and more cotton was added. The space around the rim of the can was sometimes sufficient to allow shrews (*Sorex monticolus*) to escape, but never large enough for voles to escape. Winter trap stations were marked with flags on wires suspended in the snow and moved up after each significant snowfall. On the day before a trapping session, each station in the grid was dug out, a large light-colored plastic bag was filled with snow and placed on top of the station, and the station was filled in again with snow to a total depth of about 0.33 m (including bagged snow). This facilitated opening the station at each trap check,

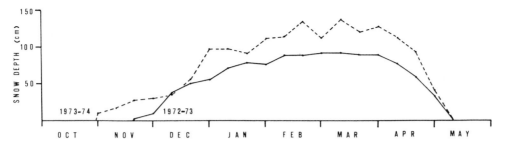

Fig. 1.—Snow pack at Moran, Wyoming, during winters of 1972–1973 and 1973–1974. See text for discussion.

while keeping the station reasonably insulated. The same continuous snowshoe track was used repeatedly in a grid. On the following morning the traps were set out (two per station), and they were checked in late afternoon, midnight, late morning the next day, and were then pulled in late afternoon–early evening. Trap sessions were short so that voles would not be unduly stressed.

Upon capture, a vole was examined and, if it was a female, was palpated. Lactating and advanced pregnant females were usually released immediately. Other voles were brought by automobile about 6 km to a heated laboratory where they were weighed to the nearest 0.1 g on a triple-beam balance, etherized and measured (total, tail, and hindfoot lengths), and sometimes observed in behavioral tests. They were returned to the field at the next trap check. However, voles captured at midnight were usually released immediately, as were most voles which had already been weighed and measured in that trap period. Females were judged parous on the condition of the dugs being suckled or from which there was expressible milk. Males were categorized as having scrotal testes, abdominal testes, or testes in an intermediate state. The intermediate state was categorized as partial

descent, black scrotal pigment present, but cauda epididymides not obvious and testes apparently only about 7 mm long. The use of Chi-square corrected for continuity (χ^2_c) follows Simpson et al. (1960).

Two voles died in traps which were not removed from a station in Grid 14 in January 1974. Also, one female gave birth to a litter in a trap in Grid 7 in February. There was no other trap mortality. Occasionally a pregnant female exhibited placental sign, but there was never loss of a litter by a female between recaptures within a trap period. In the 25 February–1 March session, 33 voles were removed from subnivean Grid 14. Testes biopsies were made on two males in Grid 6. Dental impressions were made from voles which even slightly resembled *M. pennsylvanicus*; wax was inserted into the mouth of an anesthetized vole to obtain an impression. Twenty-seven voles were tested in this manner, most in Grid 13 wherein five *M. pennsylvanicus* were identified, but none of these was still in the grid when voles were removed in May 1974.

A total of 399 *M. montanus* was trapped in the four grids in late summer through winter.

RESULTS

COHORTS OF FEMALES

None of the three breeding females in Grid 6 during the summer was trapped after permanent snow cover. The head-body length of each of two of these during the fall was 124 mm as indicated on Fig. 2 by "X." In Fig. 2, the longer axis for each grid de-

Table 2.—*Climatological data for Moran, Wyoming. Data are from the U.S. Department of Commerce, National Oceanic and Atmospheric Administration, Environmental Data Service.*

Month	Mean temperature (°F) and departure from normal		Degree days[a]		Total precipitation (in) and departure from normal	
	1972–1973	1973–1974	1972–1973	1973–1974	1972–1973	1973–1974
May	43.7 (0.2)	43.3 (−0.2)	653	666	1.56 (−0.29)	.82 (−1.03)
June	52.8 (1.9)	52.0 (1.1)	360	385	1.68 (−0.09)	1.89 (0.12)
July	57.1 (−0.8)	58.8 (0.9)	237	189	1.47 (0.50)	1.24 (0.27)
August	58.6 (2.8)	58.8 (3.0)	195	184	1.04 (−0.26)	.72 (−0.58)
September	46.5 (−1.8)	49.0 (0.7)	550	473	3.05 (1.77)	2.41 (1.13)
October	39.3 (0.5)	40.5 (1.7)	791	754	2.68 (1.23)	1.47 (0.02)
November	26.1 (−0.2)	26.2 (−0.1)	1,165	1,157	.74 (−1.14)	5.21 (3.33)
December	10.5 (−4.9)	18.9 (3.5)	1,685	1,420	3.23 (0.87)	3.39 (1.03)
January	10.8 (0.2)	15.7 (4.3)	1,679	1,525	2.26 (−0.09)	3.32 (0.51)
February	16.0 (1.5)	15.6 (0.0)	1,367	1,378	1.42 (−0.86)	2.17 (0.07)
March	22.9 (2.0)	26.2 (5.1)	1,296	1,204	1.15 (−0.93)	5.52 (3.70)
April	30.0 (−3.0)	34.7 (1.7)	1,043	904	1.74 (0.01)	1.73 (0.01)

[a] Monthly degree day totals are the sums of the negative departures of average daily temperatures from 65°F.

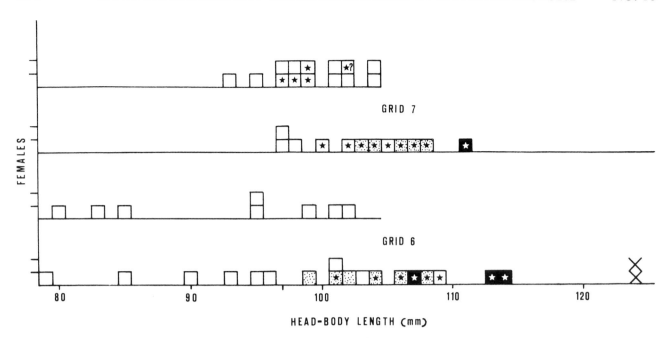

Fig. 2.—Head-body lengths of females trapped in Grids 6 and 7. Long axes represent females trapped in February and March; short axes represent females trapped for the first time in early April. Blackened blocks represent females which bred before permanent snow pack; stippled blocks represent females which were nulliparous when last trapped before snow pack; other blocks represent females not previously trapped. Stars identify females which bred during the winter. See text for further discussion.

picts the head-body lengths of the females measured for February or March (or the average thereof if a female was measured in both months). There were no summer-breeding cohort females in these two winter grids. Of females judged parous in Grid 14, the longest was merely 113 mm. Of parous females in Grid 13, one measured 125 mm, one measured 117 mm, and all others were ≤111 mm. Virtually all females which had bred during the preceding summer in all grids had apparently died by the winter months.

REPRODUCTION IN FEMALES

Voles indicated by stippling in Fig. 2 were known to have been nulliparous when last trapped during the fall (October and/or November), whereas voles indicated by blackened blocks had bred before snowfall. Blocks neither stippled nor blackened represent females not trapped during the summer or fall. Not all subnivean breeding females were brought to the laboratory to be measured. Clearly winter reproduction was a continuation of fall breeding and involved females of different ages. It included primiparous and multiparous females.

Voles represented on the short axes in Fig. 2 are those trapped for the first time in early April, and they are apparently younger females. In Grid 7, four

and perhaps five of these youngest females had already bred. The smallest breeding subnivean female had a head-body length of 97 mm.

Of 23 females ≥97 mm (="potentially breeding") in subnivean Grid 6, each of 9 was pregnant and/or lactating during at least one trap session; additionally, two may have been pregnant. In subnivean Grid 7, of 31 females ≥97 mm, each of 16 was lactating and/or pregnant, two additional females were parous, and one parous and three nulliparous females were perhaps pregnant. In summary, at least 39% and perhaps as high as 48% of the potentially breeding females in Grid 6 bred during the winter; 52 to 68% of the potentially breeding females actually bred in Grid 7, including some winter-born females. There was no significant difference between the grids in the proportion of females (≥97 mm each) which definitely had bred ($\chi^2_c = 0.402$).

Among females known to have been parous or apparently parous on the condition of the teats, only one or two females in these subnivean grids did not breed during the winter; that is, nearly all *parous* females bred.

In contrast, there was no apparently breeding female trapped in either Grid 13 or Grid 14, in each of which all females had imperforate vaginae during the winter months. In Grid 13 there were 25 poten-

Table 3.—*Sex ratios in winter grids. Numbers in parentheses are numbers of individual ♂♂/♀♀.*

Ratio type	Grid 6	Grid 7	Grid 13	Grid 14
	August 10 . . . September 28			
Operational[a]	1:0.43 (7/3)			
Potentially operational[b]	1:1.00 (7/7)			
	October 13–17	October 18–22		
Operational	1:0.86 (7/6)	1:1.50 (6/9)		
Potentially operational	1:1.43 (7/10)	1:2.50 (6/15)		
		[1:2.83 (6/17)][c]		
	February 11–April 10	February 4–April 7	January 28–February 22	January 22–March 1
Operational	1:0.35 (31/11)	1:1.28 (14/18)	1:6.00　(1/6)	1:3.50　(2/7)
	[1:0.45 (31/14)]	[1:1.43 (14/20)]	[1:10.00 (1/10)]	[1:2.00　(4/8)]
Potentially operational	1:0.74 (31/23)	1:2.21 (14/31)	1:25.00 (1/25)	1:16.50 (2/33)
		[1:2.28 (14/32)]		[1:8.25　(4/33)]

[a] Operational sex ratio = reproductively active males: parous and pregnant nulliparous females.
[b] Potentially operational sex ratio = reproductively active males: all females ≥97 mm in head-body length.
[c] Figures in brackets include possibly reproductively active voles.

tial breeders in January and February, including six to 10 judged parous on the condition of the teats. In Grid 14 during these trap periods, there were 33 potential breeders, including seven or eight judged parous. However, in Grid 14 on 1 March, a 91 mm/14.1 g vole and a 98 mm/17.1 g vole were trapped. On the basis of these weights and head-body lengths, breeding appears to have been minimal, but indeed present.

SURVIVAL OF FALL BREEDING MALES

Of the 18 breeding males trapped in Grids 6 and 7 before snow pack, five were still present after spring snow-melt. None of the 13 others were ever trapped during the winter or spring.

REPRODUCTION IN MALES.

Of the five fall-breeding males which were re-trapped in the subsequent spring (in Grids 6 and 7), 4 were trapped during the winter and three had scrotal testes when trapped in February and/or March.

Most of the winter breeding males were late summer- and fall-born individuals (for example, males <10 g each when first trapped in October). Addi-

tionally, eight males, each of which was ≤17 g and ≤100 mm, were trapped for the first time during February or March in Grids 6 and 7; five were re-trapped in early April (subnivean), by which time four exhibited testes descent.

SEX RATIOS

Table 3 lists the operational sex ratios for each population at comparable times. The operational sex ratio is the ratio of receptive males to females at any time (Emlen, 1976). For voles, Jannett (1981) considered in this ratio males exhibiting testes descent and parous females, including those not lactating or pregnant. All voles trapped are considered, regardless of the number of recaptures (that is, regardless of possible residency differences). Table 3 also presents the potentially operational sex ratios wherein I include all females ≥97 mm head-body length, the minimum length at which subnivean females bred. The increasing populations have sex ratios which approach unity or actually favor males. Ratios from the peak populations are highly unbalanced and favor females.

DISCUSSION

Reproduction in small mammals has been reported during the winter months in north temperate regions in various ecological and population contexts.

1) Winter reproduction may be the normal pattern. For example, *Microtus californicus* commonly breeds during the winter in a Mediterranean climate

(Hoffmann, 1958; Krebs, 1966); breeding is not extensive during the dry late summer and early fall months. The muskrat (*Ondatra zibethicus*) commonly breeds during the winter in the southern part of its range (reviewed by Errington, 1963).

2) Breeding in the winter is exceptional and sporadic in *Peromyscus* spp. (for example, Burt, 1948);

Table 4.—*Mid-winter breeding in North American lemmings and voles. Unless otherwise stated, breeding was evidenced by pregnant and/or lactating females.*

Species	Locality	Time of winter reproduction	Extent	Population phase	Was winter breeding an extension of fall breeding?	Was there permanent snow cover?	Reference
Dicrostonyx groenlandicus	Southampton Island	January–April	Extensive	Increase?	?	Yes	Sutton and Hamilton, 1932
Dicrostonyx groenlandicus	Baker Lake, Northwest Territories	November–March	Extensive	Increase	?	Yes	Krebs, 1964
Lemmus sibiricus	Southampton Island	January–March	Extensive	Increase?	?	Yes	Sutton and Hamilton, 1932
Lemmus sibiricus	Baker Lake, Northwest Territories	April–May	Extensive	Increase	Yes?	Yes	Krebs, 1964
Lemmus sibiricus	Barrow, Alaska	Through winter	28% (♀♀ > 100 mm, >25.0 g) December–February	Increase	?	Yes	Mullen, 1968
Synaptomys cooperi	Kansas	Through winter?	Moderate to extensive (~summer rates)	3 successive winters with generally increasing densities	Yes?	No	Gaines et al., 1977
Microtus pennsylvanicus	New York	Through winter	Extensive	Increase	Yes	No	Hamilton, 1941
Microtus pennsylvanicus	Southern Michigan	Until January	Extensive	Increase?	Yes	Almost continuous in January and February	Golley, 1961
Microtus pennsylvanicus	Minnesota	Through winter	12.2% (♀♀ pregnant) January–March	Increase	Yes	Yes	Beer and Macleod, 1961
Microtus pennsylvanicus	Indiana	Mostly through winter	0 to 33% December–February	(one year study)	Yes	No?	Corthum, 1967
Microtus pennsylvanicus	Pennsylvania	Through winter	?	Higher during increase phase	Yes	Some winters	Christian, 1971
Microtus pennsylvanicus	Manitoba	December–January	Very little?	?	?	Yes	Iverson and Turner, 1974
Microtus ochrogaster	Missouri	Through winter	?	?	?	No?	Fisher, 1945
Microtus ochrogaster	Kansas	Through winter except in January	Not extensive	(11 month study)	Yes	No	Jameson, 1947

Table 4.—*Continued.*

Species	Locality	Time of winter reproduction	Extent	Population phase	Was winter breeding an extension of fall breeding?	Was there permanent snow cover?	Reference
Microtus ochrogaster	Kansas	Through winter (as evidenced by patent vaginae)	Reduced from other months	Winter breeding in 2 successive winters	Yes	No	Martin, 1956
Microtus ochrogaster	Kansas	Through winter?	Reduced from other months	No discernable cycle?	Yes?	No	Fitch, 1957
Microtus ochrogaster	Indiana	Through winter	14–37% December–February	(One year study)	Yes	No?	Corthum, 1967
Microtus ochrogaster	Kansas	Through winter, 1971–1972; through winter, 1972–1973	54.1% (♀♀ ≥ 20 g) 1971–1972; 30.4% (♀♀ ≥ 20 g) 1972–1973	Increase and winter following peak	Yes	No	Rose and Gaines, 1978
Microtus townsendii	British Columbia	Through winter	Moderate	Increase? (second winter following a peak)	Yes	No	Krebs et al., 1978

these reports are not from subnivean populations and do not involve large percentages of females in the respective populations.

3) Mid-winter breeding by females in generally low latitude populations of *Clethrionomys* spp. and *Apodemus* spp., species which are less strictly folivores than are lemmings and *Microtus* spp., has been correlated with a particularly good food supply (for example, Smyth, 1966; Hansson, 1982). It has also been shown to be a response to an experimentally increased food supply (Andrzejewski, 1975). The root vole (*Microtus oeconomus*) specializes on first-year rhizomes of *Eriophorum*; a rare occurrence of mid-winter breeding in this species was correlated with peculiar autumn food conditions (Tast and Kaikusalo, 1976).

4) Breeding begun in the late winter or early spring under the winter snow pack in microtine rodents such as *Clethrionomys gapperi* may be the usual prelude to spring breeding, as reported by Merritt and Merritt (1978).

5) Mid-winter reproduction, often under snow pack, has been reported for numerous genera and species of microtine rodents, and its proximate cause(s) is unknown. In reviews of the literature, Sadleir (1969), Everden and Fuller (1972), and Krebs and Myers (1974) suggested that this phenomenon is correlated with increasing population density in a multi-annual "cycle." Table 4 lists records of winter breeding in North American microtine rodents. Additionally, there are reports of winter breeding correlated with increasing population densities in Old World species and these include another genus, *Myopus* (Mysterud, 1966, 1968). Perhaps this is a spurious correlation because so few voles are trapped after population crashes and therefore subnivean breeding cannot be detected. In this study, populations which exhibited winter breeding in Grids 6 and 7 were also increasing. Populations in Grids 13 and 14 had densities very similar to those in the respective areas during the previous year, but had achieved these levels without any contribution of winter breeding. In general, there were high-density populations of *M. montanus* in the area in the fall of 1973 and in 1974.

The most curious case of the generality of "off-season" breeding during population increases comes from observations of *M. californicus*, which normally breeds during the rainy winter: Krebs (1966, and others cited therein) reported that summer (dry season) breeding was sometimes extensive during the population increase phase, and opined that the phenomenon is directly comparable to winter breeding in lemmings.

Mid-winter breeding in subnivean lemmings has generally been documented by sampling, whereas mid-winter breeding in *Microtus* has been documented by longitudinal studies in areas where there is not a regular deep snow pack. A review of the literature also indicates that in those cases where longitudinal data are available for individuals in a population over a winter, most winter breeding was an obvious continuation of the usual summer breeding, as was the case in this subnivean study.

There are reports of mid-winter breeding involving various percentages of females in generally low latitude populations of *Microtus* where snowfall and snow cover are variable and where there is not a regular deep snow pack. For winter breeding in *M. ochrogaster,* Rose and Gaines (1978) concluded that there is no clear correlation of winter breeding with weather or ages of females involved, but Keller and Krebs (1970) found it associated with the increase phase of the population cycle.

Fuller et al. (1975:878) opined that *Microtus* seemed "to be incapable of subnivean breeding except in low latitudes with short and generally mild winters." The present study indicates that *Microtus* is capable of mid-winter subnivean breeding. Therefore, mid-winter breeding in high-latitude subnivean populations of lemmings and low latitude populations of voles without permanent snow cover may be the result of the same population dynamics.

These observations of *M. montanus* may be discussed with respect to three hypotheses regarding the proximate cause of mid-winter breeding in microtine rodents.

Extrinsic factors.—Fuller (1967) suggested that lemming winter breeding occurred during winters when there was sufficient snow cover and quality food. For example, Krebs (1964) reported apparently extensive winter breeding in *Lemmus* and *Dicrostonyx* during 1959–1960, and the fall of 1959 had a lack of freezing rain and a quick buildup of protective snow cover; the fall of 1960 had oscillating freezing-thawing and a lack of good snow cover until mid-December, and there was subsequently no winter breeding detected.

The winter breeding which was observed during my study cannot easily be accounted for by extrinsic factors. Snow-pack was later that year, so voles were more directly subjected to decreasing photoperiod and temperature. There was less protective snow pack in the winter of reproduction than in the following winter. The ambient temperature and precipitation profiles for the summers and falls of the two years were comparable (Table 2). Most significantly with respect to Fuller's hypothesis, mid-winter reproduction was a continuation of breeding by late summer-born and fall-born females, and other populations being studied in 1972 did not exhibit late fall reproduction (Jannett, unpublished).

Genetic polymorphism.—Keller and Krebs (1970) suggested that some genotypes are able to breed in winter while others are not. Johnston and Zucker (1980) reported that exposure to short daylengths resulted in testes regression and inhibition of spermatogenesis in only 70% of male *Peromyscus leucopus* examined in the laboratory. However, Jannett (ms) exposed August-born young of field-trapped *M. montanus* females in 1973 to strange young of both sexes and to males with scrotal testes under natural photoperiod and temperature in an experiment blocked for litter effect. Unstimulated control littermate females did not breed, but 83% of stimulated females did breed. The stimulated adult males maintained scrotal testes whereas unstimulated controls did not. It would seem unnecessary to hypothesize that a genetic polymorphism is responsible for these responses.

Social stimulation.—I suggest that continued fall and mid-winter breeding may be a result of increased social stimulation. There is increasing awareness of social stimulation of reproductive activity in vertebrates. Bronson (1979) reviewed work on the house mouse, *Mus musculus,* and microtine rodent reproduction may be even more dependent on olfactory cues. Male-induced estrus, rather than cyclic estrus, is the common estrus pattern among microtine rodents, including *M. montanus* (Jannett, 1980). Puberty among female microtine rodents may be stimulated by the male (for example, *M. ochrogaster,* Hasler and Nalbandov, 1974; *Dicrostonyx groenlandicus,* Hasler and Banks, 1975). Both resident and transient males could stimulate resident females to breed or to continue breeding by various modes, that is, direct contact and indirect contact (urine, feces, scent secretions). Transient males may also elicit more activity of resident males which in turn may increase stimulation of females directly or indirectly; for example, when a resident male scent marks in response to a transient male, his marks may also stimulate the female. Lastly, females, particularly parous females, may stimulate males. To test this general hypothesis, data are needed on precise reproductive capabilities of males, nesting patterns, measures of subnivean home ranges and territories, indices of sex ratios, contact rates, and the roles of transient versus resident males. Many small mammal studies identify a transient on the basis of an arbitrary maximum period the individual is on a study area, with scant attention to how it affects the population. Its effects may be considerable.

Sex ratios may be an overall index of contact rates between males and females. The pattern of operational sex ratios favoring females in high-density non-breeding populations in Grids 13 and 14 is similar to that of operational sex ratios for late fall high-density populations of this species where breeding ceased early (Jannett, 1981).

Although data for most of the other research areas listed above will prove methodologically or logistically difficult to collect, this overall scheme is attractive in that it is conceptually simple and therefore may apply to various species. More significantly, it does not contradict general patterns in vertebrate populations. The fluctuating numbers of males, relative to females, may be largely reflections of different vagility and mortality rates for the sexes at different population densities, and of the tendency for territorial males to overlap the territories of several females. The following three heuristic stages are presented; the processes involved are continuously changing.

1) At low densities, voles are so spaced out that, regardless of mortality patterns, females contact relatively few males and are little stimulated. Transient males may be uncommon because of low population numbers and because there is sufficient space for most males to establish territories. Furthermore, if mortality of breeding males is much higher than that of females, maternal frequency will decline.

2) In increasing populations, mortality may or may not be greater for males, but each resident male territory overlaps those of several females, and the many "surplus" males are transients. They stimulate females and males. They may also cause pregnancy disruptions (Jannett, 1980). Increased social stimulation may be the proximate cause for the extension of the breeding season, sometimes into the winter. Different ratios may be responsible for different percentages of breeding females between populations and for different percentages in spatial and temporal segments within a population.

3) At very high densities, mortality of males, especially transients, is greater than that of females, there are virtually no transients left, and the operational sex ratios approach 1:5 (Jannett, 1981). Young remain in extended maternal families where puberty is delayed (Jannett, 1978). Breeding stops early in the season.

ACKNOWLEDGMENTS

This work was supported by a Ford Foundation Ecology of Pest Management Traineeship, the New York Cooperative Wildlife Research Unit, and grants-in-aid of research from the Society of the Sigma Xi and the New York Zoological Society.

I thank the New York Zoological Society and the University of Wyoming for their hospitality, the National Park Service for permission to work in Grand Teton National Park, and J. Jannett for assistance with the field work.

LITERATURE CITED

ANDRZEJEWSKI, R. 1975. Supplementary food and the winter dynamics of bank vole populations. Acta Theriol., 20:23–40.

BEER, J. R., and C. F. MACLEOD. 1961. Seasonal reproduction in the meadow vole. J. Mamm., 42:483–489.

BRONSON, F. H. 1979. The reproductive ecology of the house mouse. Quart. Rev. Biol., 54:265–299.

BURT, W. H. 1946. The mammals of Michigan. Univ. Michigan Press, Ann Arbor, 288 pp.

CHRISTIAN, J. J. 1971. Population density and reproductive efficiency. Biol. Reprod., 4:248–294.

CORTHUM, K. W., JR. 1967. Reproduction and duration of placental scars in the prairie vole and the eastern vole. J. Mamm., 48:287–292.

EMLEN, S. T. 1976. Lek organization and mating strategies in the bullfrog. Behav. Ecol. Sociobiol., 1:283–313.

ERRINGTON, P. L. 1963. Muskrat populations. Iowa State Univ. Press, Ames, 665 pp.

EVERDEN, L. N., and W. A. FULLER. 1972. Light alteration by snow and its importance to subnivean rodents. Canadian J. Zool., 50:1023–1032.

FISHER, H. J. 1945. Notes on voles in central Missouri. J. Mamm., 26:435–437.

FITCH, H. S. 1957. Aspects of reproduction and development in the prairie vole (Microtus ochrogaster). Univ. Kansas Publ., Mus. Nat. Hist., 10:129–161.

FULLER, W. A. 1967. Ecologie hivernale des lemmings et fluctuations de leurs populations. La Terre et la Vie, 21:97–115.

FULLER, W. A., A. M. MARTELL, R. F. C. SMITH, and S. W. SPELLER. 1975. High-arctic lemmings, Dicrostonyx groenlandicus. II. Demography. Canadian J. Zool., 53:867–878.

GAINES, M. S., R. K. ROSE, and L. R. MCCLENAGHAN, JR. 1977.

The demography of *Synaptomys cooperi* populations in eastern Kansas. Canadian J. Zool., 55:1584–1594.

GOLLEY, F. B. 1961. Interaction of natality, mortality and movement during one annual cycle in a *Microtus* population. Amer. Midland Nat., 66:152–159.

HAMILTON, W. J., JR. 1941. Reproduction of the field mouse *Microtus pennsylvanicus* (Ord). Cornell Univ. Agr. Exp. Station Mem., 237:1–23.

HANSSON, L. 1984. Winter reproduction of small mammals in relation to food conditions and population dynamics. This volume.

HASLER, J. F., and E. M. BANKS. 1975. The influence of mature males on sexual maturation in female collared lemmings (*Dicrostonyx groenlandicus*). J. Reprod. Fert., 42:583–586.

HASLER, M. J., and A. V. NALBANDOV. 1974. The effect of weanling and adult males on sexual maturation in female voles (*Microtus ochrogaster*). Gen. Comp. Endocrinol., 23:237–238.

HOFFMANN, R. S. 1958. The role of reproduction and mortality in population fluctuations of voles (*Microtus*). Ecol. Monogr., 28:79–109.

IVERSON, S. L., and B. N. TURNER. 1969. Under-snow shelter for small mammal trapping. J. Wildlife Mgmt., 33:722–723.

———. 1974. Winter weight dynamics in *Microtus pennsylvanicus*. Ecology, 55:1030–1041.

JAMESON, E. W., JR. 1947. Natural history of the prairie vole (mammalian genus *Microtus*). Univ. Kansas Publ., Mus. Nat. Hist., 1:125–151.

JANNETT, F. J., JR. 1978. The density-dependant formation of extended maternal families of the montane vole, *Microtus montanus nanus*. Behav. Ecol. Sociobiol., 3:245–263.

———. 1980. Social dynamics of the montane vole, *Microtus montanus*, as a paradigm. Biologist, 62:3–19.

———. 1981. Sex ratios in high-density populations of the montane vole, *Microtus montanus*, and the behavior of territorial males. Behav. Ecol. Sociobiol., 8:297–307.

———. 1982. Nesting patterns of adult voles, *Microtus montanus*, in field populations. J. Mamm., 63:495–498.

JOHNSTON, P. G., and I. ZUCKER. 1980. Photoperiodic regulation of the testes of adult white-footed mice (*Peromyscus leucopus*). Biol. Reprod., 23:859–866.

KELLER, B. L., and C. J. KREBS. 1970. *Microtus* population biology III. Reproductive changes in fluctuating populations of *M. ochrogaster* and *M. pennsylvanicus* in southern Indiana, 1965–67. Ecol. Monogr., 40:263–294.

KREBS, C. J. 1964. The lemming cycle at Baker Lake, Northwest Territories, during 1959–62. Arctic Institute N. Amer., Tech. Paper 15:1–104.

———. 1966. Demographic changes in fluctuating populations of *Microtus californicus*. Ecol. Monogr., 36:239–273.

KREBS, C. J., and J. H. MYERS. 1974. Population cycles in small mammals. Adv. Ecol. Res., 8:267–399.

KREBS, C. J., J. A. REDFIELD, and M. J. TAITT. 1978. A pulsed-removal experiment on the vole *Microtus townsendii*. Canadian J. Zool., 56:2253–2262.

MARTIN, E. P. 1956. A population study of the prairie vole (*Microtus ochrogaster*) in northeastern Kansas. Univ. Kansas Publ., Mus. Nat. Hist., 8:361–416.

MERRITT, J. F., and J. M. MERRITT. 1978. Population ecology and energy relationships of *Clethrionomys gapperi* in a Colorado subalpine forest. J. Mamm., 59:576–598.

MULLEN, D. A. 1968. Reproduction in brown lemmings (*Lemmus trimucronatus*) and its relevance to their cycle of abundance. Univ. California Publ. Zool., 85:1–24.

MYSTERUD, I. 1966. Forplanting hos skoglemen i vinterhalvåret. Fauna, Oslo, 19:79–83. [Not seen; cited by Mysterud, 1968.]

———. 1968. A third case of winter breeding in the wood lemming (*Myopus schisticolor* Lilljeb.). Nytt Mag. Zool., 16:24.

ROSE, R. K., and M. S. GAINES. 1978. The reproductive cycle of *Microtus ochrogaster* in eastern Kansas. Ecol. Monogr., 48:21–42.

SADLEIR, R. M. F. S. 1969. The ecology of reproduction in wild and domestic mammals. Methuen and Co., Ltd., London, 321 pp.

SIMPSON, G. G., A. ROE, and R. C. LEWONTIN. 1960. Quantitative zoology. Harcourt, Brace, and World, New York, 440 pp.

SMYTH, M. 1966. Winter breeding in woodland mice, *Apodemus sylvaticus*, and voles, *Clethrionomys glareolus* and *Microtus agrestis*, near Oxford. J. Anim. Ecol., 35:471–485.

SUTTON, G. M., and W. J. HAMILTON, JR. 1932. The mammals of Southampton Island. Mem. Carnegie Mus., 12 (part 2, sec. 1):1–111.

TAST, J., and A. KAIKUSALO. 1976. Winter breeding of the root vole, *Microtus oeconomus*, in 1972/1973 at Kilpisjärvi, Finnish Lapland. Ann. Zool. Fennici, 13:174–178.

UNITED STATES DEPARTMENT OF COMMERCE. 1972–1974. Climatological data, Wyoming. National Oceanic and Atmospheric Administration, Environmental Data Service.

Address: Section of Ecology and Systematics, Cornell University, Ithaca, New York 14853.

Present address: Department of Biology, The Science Museum of Minnesota, 30 E. 10th Street, St. Paul, Minnesota 55101.

WINTER REPRODUCTION OF SMALL MAMMALS IN RELATION TO FOOD CONDITIONS AND POPULATION DYNAMICS

LENNART HANSSON

ABSTRACT

The incidence of winter reproduction in Scandinavian small mammals was examined from three sources of information: A population study in South Sweden during the winters 1972–1975, another population study in Central Sweden in 1977–1981 and a compilation of all known instances of winter reproduction in whole Scandinavia. The frequency of winter reproduction varied between species, but it appeared in both folivorous and granivorous rodents. From the best studied, widely distributed species, it was most often found in northern Scandinavia in the folivorous *Microtus agrestis* and in southern Scandinavia in the partly granivorous *Clethrionomys glareolus.*

The observed environmental conditions during winter reproduction were compared with theoretical models of reproduction as dependent on the energy balance of the animals. Winter breeding often, but by no means always, followed a rich food supply of seeds or rhizomes produced in previous summer–autumn. In at least the remaining cases, it appeared in especially sparse populations or at population increase. Three alternative hypotheses are put forward to explain winter reproduction under such conditions: 1) genetic shifts according to the Chitty hypothesis, 2) scramble competition for especially nutritious food, and 3) selection for "strong" surviving animals. Furthermore, effects of any of these factors and an additional food supply may be additive.

Winter reproduction seems a regular and important peculiarity of the Norway lemming (*Lemmus lemmus*) population dynamics as in other subarctic lemmings. In most other Scandinavian species, few individuals produced young during winter and litter sizes were often small. The effect of winter reproduction on population dynamics seems thus to be small in the latter species.

INTRODUCTION

It is by now well known that several small rodent species reproduce during winter also in boreal-arctic areas (see references in Table 1), even under thick snow cover. In some winter studies in North Sweden (Larsson et al., 1973; Larsson and Hansson, 1977), however, we found reproduction during only one or two out of several examined winters.

In models of the nutritive condition of field voles (*Microtus agrestis*) and its consequences (Stenseth et al., 1977), it was assumed that only the energy balance affected the reproductive performance of this species. This would imply that winter reproduction could be due to especially good food supply during autumn and early winter or high ambient temperatures at that time.

These predictions will be tested on one series of winter trapping in South Sweden with strongly varying food supply (mainly beechmast), on one series from Central Sweden with varying temperatures and from a compilation of all known observations of winter reproduction in Scandinavia (including the ecologically similar Kola peninsula of the U.S.S.R.). Both field studies cover a potential "vole cycle." However, few workers on small mammals have bothered to work in snowy conditions and, generally, few small mammal individuals reproduce in winter; so the total field data are of limited extent.

METHODS

Winter reproduction is defined as all observations of sexually mature animals or completed reproduction during the time December–March. Spring may begin during March, but reproductive conditions observed in March have to be initiated earlier (Clarke, 1977).

All animals caught were dissected, and females with perforate vulva and visible follicles in their ovaries were considered sexually mature (=adult). In some females, pregnancy and/or lactation were further evidence and will be discussed separately. In males, scrotal testes, tubular cauda epididymis and seminal vesicles more than one cm length were used as criteria of adulthood. Remaining animals were classified as juveniles (still in juvenile pelage), subadults (adult pelage but with undeveloped gonads) and postreproductive animals (with regressed gonads). Pregnant or lactating females, a litter of free-roaming juveniles, and adult females and males were considered as units when computing percentages of reproductive animals (Table 1).

Trapping with the Small Quadrat index method with snap traps (Myllymäki et al., 1977) was performed in the Revinge area (55°37'N, 13°28'E) in South Sweden in February–beginning of March 1972–1975. Both forests (mainly beech) and abandoned fields were sampled. This trapping was performed in periods with little or no snow. Mean standard temperatures (1.5 m above the ground) were obtained for January–February (the two coldest months) from a weather station in Lund, 20 km away. The sizes of the beechmast and acorn crops were evaluated according to a 5-degree scale in nearby areas in Denmark (Holmgaard, in lit.). However, oak is much less common than beech in the Revinge area.

The studies at Uppsala (59°50′N, 17°45′E) were performed as continuous removal trapping in an abandoned field and adjoining parts of a forest (mainly coniferous) during 1977–1981. Small trapping units were employed so only a small part of the surrounding population was removed. Trapping under snow conditions was done in trap boxes according to Larsson and Hansson (1977). Standard temperature measurements were made at an official weather station only some 500 m away from the trapping area. Hansson (1979) found some evidence that winter food con-

ditions for *M. agrestis* were related to the spruce cone crop, being best one winter before peak cone crop. Estimates of this crop were obtained from the Swedish Forestry Survey.

In both areas there was regular long-term index trapping with the Small Quadrat method (Hansson, 1978), and density indices from autumn trappings will be used to illustrate general density variations.

The compilation comprehends all the Scandinavian literature as far as available. All remarks on weather and food conditions were noted. Furthermore, the conditions during a particular year could be compared to extensive data on small mammal density variations (Myrberget, 1973; Lahti et al., 1976; Myllymäki, 1977a; Hansson and Larsson, 1980).

RESULTS

The South Swedish Samples

There were small variations in the winter temperature during the study period (Fig. 1), which can be characterized as generally mild. The fructification of the two main tree species varied considerably, with a peak in beechmast in the autumn of 1974. This should, if at all, have affected reproduction in February–March 1975. A peak in acorn production in autumn 1971 might have influenced reproduction to a lesser extent in February–March 1972. The populations of small mammals did not vary much in size in this area, and their dynamics cannot be described as cyclic. Variations are shown in Fig. 1 for one typical forest species, *Apodemus flavicollis,* and for one corresponding grassland species, *Microtus agrestis*. Both show fairly low densities in 1973–1974, while *A. flavicollis* had a peak in 1972 and both increased in 1975.

In Fig. 2 the animals in the winter catches are separated into non-reproductive (subadult + post-reproductive) animals and adults (plus two juvenile *A. flavicollis*). A first glance shows that there were comparatively more reproductive animals in 1975 in all species except *A. flavicollis*. The latter showed generally more reproductively active animals in winter than did the remaining species. A statistical analysis for *A. flavicollis* with the 1974 and 1975 data pooled showed a significant difference between years ($\chi^2 = 7.29$, df = 2, $P < 0.05$) with few reproductive animals in 1973. For the remaining species there were no obvious differences between the winters 1972, 1973 and 1974 and the proportion of reproductive/non-reproductive animals during these three winters were compared with the conditions in the winter 1975. There were significant differences in *Apodemus sylvaticus* ($\chi^2 = 83.02$, df = 1, $P < 0.001$), *Clethrionomys glareolus* ($\chi^2 = 40.79$, df =

1, $P < 0.001$) and *M. agrestis* ($\chi^2 = 23.65$, df = 1, $P < 0.001$). Shrews (*Sorex araneus* L. and *S. minutus* L.) were generally sexually immature, but in the winter of 1975, two mature *S. araneus* males were caught.

Pregnant and/or lactating females were, with one exception, only obtained in the winter 1975. One pregnant *A. flavicollis* female was caught in winter 1974 and four in 1975. Furthermore, two juvenile *A. flavicollis,* born about a month earlier, were caught in winter 1975. The mean embryo number of five winter litters from *A. flavicollis* was 6.2 (SD = 2.8), which is slightly higher than the summer estimates (see Southern, 1964). Two pregnant *A. sylvaticus* females were caught in the winter 1975 with 4 and 7 embryos, respectively. One pregnant *C. glareolus* from 1975 bore just one embryo. There were no pregnant *M. agrestis* or shrews.

A clear relation between seed crops and reproductive status emerged in the forest species *A. flavicollis, A. sylvaticus* and *C. glareolus*. In *A. flavicollis* the sexual maturation might have been speeded up by both the acorn crop in 1971 and the beechmast crop in 1974, while the other two species seemed to be affected only by the latter crop. Few seeds were available in the winter 1973 which coincided with the lowest proportion of mature *A. flavicollis*. However, more advanced maturation also appeared in the grassland species *M. agrestis* and possibly in the insectivorous *S. araneus* in 1975 than in previous years. It should also be noted that several species showed especially low population densities before the winter of 1975.

The Central Swedish Samples

The winter temperature (here computed for De-

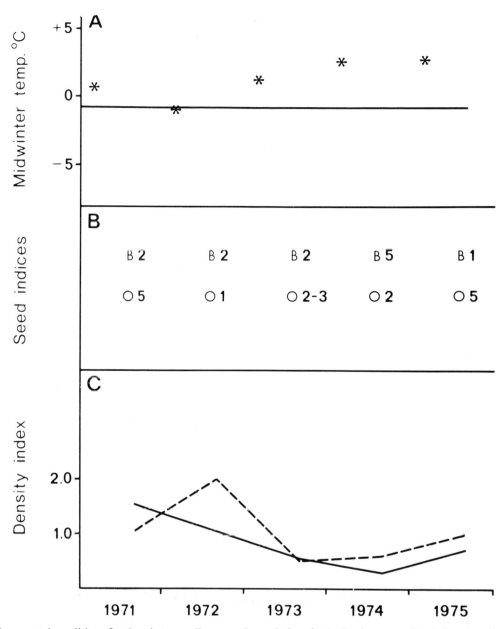

Fig. 1.—Environmental conditions for the winter small mammal populations in the Revinge area, South Sweden. A) Mean midwinter temperatures for January and February during the study period. A 30-year mean is denoted by a horizontal line. B) Seed crops of beechmast (B) and acorns (O) according to a Danish five-degree scale. C) Autumn density indices of one grassland small rodent (*Microtus agrestis,* solid line) and one forest species (*Apodemus flavicollis,* dashed line).

cember–March) was especially low in the winter of 1978–1979 (Fig. 3). However, the March mean temperature was unusually high that winter (−0.3°C), and the mean for December–February was −8.3°C, which is exceptional for this area. The snow cover was of similar total duration during the various winters but was thickest in 1978–1979. The spruce cone crop was large in 1976, at the beginning of a population expansion of *M. agrestis,* but remained low

thereafter. *M. agrestis* populations have shown cyclic behaviour in the Uppsala area since at least 1971 (Hansson and Larsson, 1980), and there was a peak in 1977 and a succeeding decline in 1978. In 1979 the field vole populations started a new increase, but this was broken in an unexpected way in 1980 with an extremely sparse population. 1980 may be considered as an atypical year for the region concerned.

During certain winters some fecund male *A. flav-*

Table 1.—*Reports of winter reproduction in small mammals from the Scandinavian area. Only definite reproduction as established from pregnant or lactating females, nestlings or newly weaned juveniles are included here. Sample size and percentage of reproductive animals, if available, refer to the time of investigation disclosing winter reproduction. The species are arranged in an order from extreme granivores to extreme folivores.*

Species	Locality	Time of reproduction	Samples		Preceding food conditions	Preceding population phase	Reference
			n	% reproductive			
Apodemus flavicollis (Melch).	Revinge, South Sweden	February–March 1974	9	78	Mean	Mean density	Present study
	Revinge, South Sweden	February–March 1975	22	77	Peak beech-mast crop	Low	Present study
	Hestehaven, E. Jutland, Denmark	December 1974–January 1975	?	?	Peak beech-mast crop	Mean density autumn 1974	Jensen (1975)
Apodemus sylvaticus (L.)	Kullaberg, South Sweden	December 1968–February 1969	35	71	Peak beech-mast crop	High density autumn 1968	Hansson, unpublished
	Garpenberg, Central Sweden	March 1973	?	?	?	?	Larsson et al., 1973
	Revinge, South Sweden	February–March 1975	121	60	Peak beech-mast crop	Mean density	Present study
Clethrionomys rutilus (Pallas)	Kirkenes, North Norway	January 1972	5	60	Partly indoors, in barn	After bottom year	Lundquist et al., 1973
Clethrionomys glareolus (Schreb.)	Kullaberg, South Sweden	December 1968–February 1969	49	88	Peak beech-mast crop	High density autumn 1968	Hansson, unpublished
	Hestehaven, E. Jutland, Denmark	December 1974–January 1975	?	?	Peak beech-mast crop	Mean density autumn 1974	Jensen, 1975
	Revinge, South Sweden	February–March 1975	76	28	Peak beech-mast crop	Low-mean density autumn 1974	Present study
	Revinge, South Sweden	December 1976–March 1977	?	?	Much acorn and beech-mast	"Mean" density autumn 1976	Meurling et al., 1981
Microtus agrestis (L.)	Ingels, South Finland	February–March 1961	99	58	(Temporary snowmelt in middle of winter)	After population increase	Myllymäki, 1969, 1977*b*
	Ahtiala, South Finland	March 1969	?	?	(Breeding on clover ley, not in orchards)	After population increase	Myllymäki, 1970
	North Savo, Central Finland	Winters 1965–1978, e.g., 1972, 1977	305	19	(More breeding on fields than in forests)	?	Skarén, 1978
	Robertsfors, North Sweden	February–March 1973	27	78	No indication of food abundance	After population increase	Larsson et al., 1973
	Garpenberg, Central Sweden	March 1973	?	?	No indication of food abundance	After population increase	Larsson et al., 1973

Table 1.—*Continued.*

Species	Locality	Time of reproduction	Samples		Preceding food conditions	Preceding population phase	Reference
			n	% reproductive			
	Stensoffa, South Sweden	March 1973	3	(67)	No indication of food abundance	Mean density	Larsson et al., 1973
	Nordingrå, North Sweden	January–February 1974	41	2	No indication of food abundance	Early peak	Larsson and Hansson, 1977
	Uppsala, Central Sweden	January–March 1979	10	80	No indication of food abundance	After bottom year	Present study
Microtus oeconomus (Pallas)	Lövhögen, Central Sweden	March 1967	1	(100)	No indication of food abundance	After population increase or peak	Höglund, 1970
	Kilpisjärvi, North Finland	January 1973	12	67	After peak rhizome production of *Eriophorum angustifolium*	After population increase	Tast and Kaikusalo, 1976
	Hardangervidda, South Norway	Winters 1965–1979	?	?	No indication of food abundance	?	Stenseth et al., unpublished
Myopus schisticolor (Lillj.)	Nordmarka, South Norway	December–January 1962, 1966	2	(100)	No indication of food abundance	Population increase	Mysterud, 1966
	Brydalen, South Norway	February 1967	1	(100)	No indication of food abundance	Population peak?	Mysterud et al., 1972
Lemmus lemmus L.	Kola peninsula, U.S.S.R.	Winter until snowmelt, 1953–1959	2,747	1–2 litters in winter	?	Most reproduction during increase year	Koshkina and Khalansky, 1962
	Kilpisjärvi, North Finland	Before snowmelt	?	?	Winter habitat with moss	After population increase	Kalela, 1961, Koponen et al., 1961
	Lapland, North Sweden	Before snowmelt 1960	?	?	?	After increase year	Curry-Lindahl, 1962
	Lövhögen, Central Sweden	December 1963	1	(100)	?	After population increase	Höglund, 1964
	Tarfalavagge, North Sweden	March 1974	1	(100)	?	After population peak	Larsson, 1974
	Kilpisjärvi, North Finland	January–April 1978	58	(30?) ≤2 winter litters	"Plenty of subcutaneous fat"	After population peak	Henttonen and Järvinen, 1981

icollis were caught making up 27% out of totally 22 individuals of this species. Two adult male *A. sylvaticus* were also obtained. No fecund *C. glareolus* or shrews appeared, although a considerable number of these animals were caught. Thus, *M. agrestis* was the only species with adult and pregnant females in winter.

The total *M. agrestis* captures and population structures are shown in Fig. 4. In 1977 and 1979 ("peak" years) reproduction ended in September or

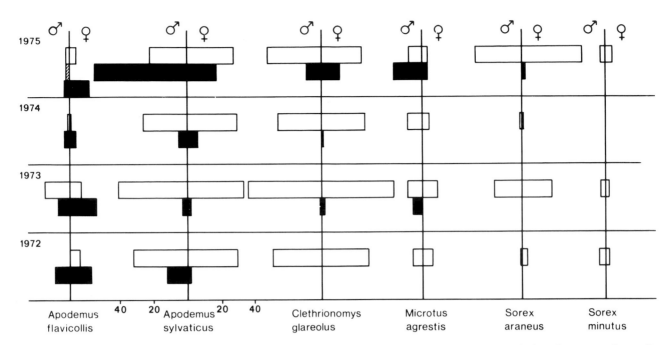

Fig. 2.—Reproductive conditions of six small mammal species in the Revinge area during 1972–1975. Blackened areas equal sexually mature specimens; white areas equal subadult and postreproductive (immature) specimens; cross-hatched areas equal juveniles in juvenile pelage.

earlier, while it continued all the winter in 1978–1979 ("bottom" year). Only four *M. agrestis* were caught in the winter 1980–1981, of which one was an adult male, obtained in the beginning of March. However, also in the winter 1978–1979 few individuals of this species were caught, and none in December. In November one female bore three embryos, and one juvenile was caught as well as two subadults. In January the catch consisted of one pregnant female with three embryos and one subadult animal. In February two fecund *M. agrestis* females were obtained. A juvenile animal had been conceived in January. In March one adult male, two adult females, one juvenile and a few subadults of this species were obtained. One female caught alive at the same place in late summer–autumn 1978 produced litters in the laboratory together with a local male on November 30 (5 young), December 24 (5), January 17 (4) and February (5). Field voles caught in South Sweden and kept in pairs in the same laboratory did not produce any litters this winter.

The winter reproduction of the field voles appeared during what was probably the worst winter of the whole of the Twentieth Century. There were several days with −30°C and little snow in December–early January. Nothing in the changes in the spruce cone crops indicated that food conditions

were good during that winter. However, as in South Sweden, this winter reproduction occurred at the start of a population increase.

REPORTS ON WINTER BREEDING IN SCANDINAVIA

Seven Scandinavian small rodent species have been reported to reproduce in winter (Table 1). Most reports concern *M. agrestis*, which has been studied extensively in several regions of Scandinavia. Most individuals have probably been handled in South Finland but more in South than in North Sweden. In another favourite study object, *C. glareolus*, evidence of winter reproduction appears only from southern Scandinavia, although considerably more animals have been caught in Norway and Central-North Sweden (Myllymäki et al., 1977). However, in folivorous species such as *Microtus agrestis, M. oeconomus, Myopus schisticolor* and *Lemmus lemmus,* almost all winter reproduction was observed in northern areas.

In the mainly granivorous species (*Apodemus* spp., *C. glareolus* and *C. rutilus*), winter reproduction appeared after a heavy acorn or beechmast crop (or other good food supply) during the previous autumn. Tast and Kaikusalo (1976) demonstrated a similar relationship to the production of *Eriopho-*

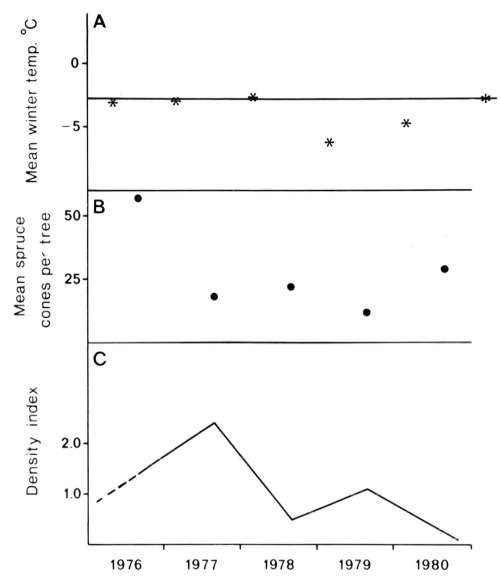

Fig. 3.—Environmental conditions for the *Microtus agrestis* population in the Uppsala area, Central Sweden. A) Mean winter temperatures for December–March during the study period. A 30-year mean is denoted by a horizontal line. B) Mean number of spruce cones per tree in the Uppsala region. C) Autumn density indices of field voles in abandoned fields in the Uppsala region.

rum rhizomes for the folivorous *M. oeconomus.* However, no similar relations were apparent for the remaining observations on the folivorous species.

In these species winter reproduction generally appeared during the increase (or early peak) phase of the population cycles.

DISCUSSION

There were many cases when winter reproduction was preceded by an unusually good food supply. However, such a relationship could not be demonstrated every time, and neither a good food supply nor an otherwise energetically favourable situation seems necessary. Thus the original hypothesis seems to be refuted.

In at least the cases of winter reproduction without any good food supply, the most characteristic feature was low or increasing population den-

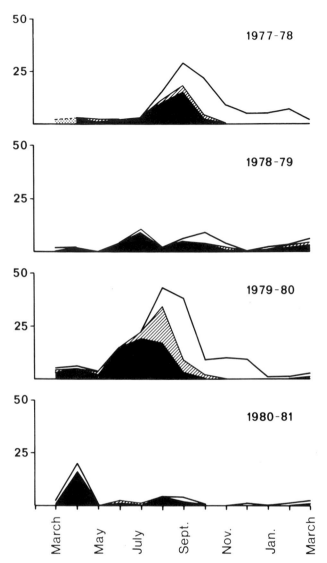

Fig. 4.—Numbers and population structure of *Microtus agrestis* specimens obtained at continuous removal trapping at Uppsala during 1977–1981. Symbols in Fig. 2.

tritious food (or food containing oestrogenic substances) in sparse as opposed to dense populations, bringing the animals to maturation and reproduction during unusual seasons in low or increase phases.

3. Selection for especially strong or "high quality" animals during population "crashes." This quality may be purely physiological. Thus winter reproduction may be common mainly during the low phase and be gradually less frequent as the population increases.

It seems impossible to decide which of these mechanisms, if any, is operative in the present data. Krebs (1979) reported a zero heritability for winter breeding in *Microtus townsendii* (Bachman), which makes an explanation according to alternative 1, and possibly also according to 3, less probable. Future tests of these alternative hypotheses probably have to be done through controlled breeding in laboratories of animals collected at various phases of the population cycles. Kaikusalo (1980) experimented by placing captive voles and lemmings in cages with nutritious food in the snow and obtained winter breeding in all species. However, it has to be considered that effects via food supply and animal quality may be additive. There were no pronounced crashes in the populations of the granivores breeding in winter in South Scandinavia, and for these animals a good food supply seems a sufficient condition for winter breeding.

Winter reproduction has been observed several times in Central European small rodents (Stein, 1963; Zimmerman, 1960; Huminski, 1963; Smyth, 1966, and others), but frequencies do not seem higher there than in Scandinavia. Generally, a fairly small part of the populations was breeding during only certain winters and few juveniles appeared in the samples. However, in North American lemmings (*Dicrostonyx groenlandicus* Traill. and *Lemmus trimucronatus* Richardson) extensive winter breeding was reported before population peaks (Krebs, 1964; MacLean et al., 1974; Batzli et al., 1980), and the effects on the dynamics of the populations are considered important. Conditions seem to be similar in the Norway lemming *L. lemmus*). However, the winter litter size was considerably lower than the summer estimates for this species (Henttonen and Järvinen, 1981). Schwartz (1963) argued that subarctic lemmings had, as an adaptation to their environment, evolved a period of winter reproduction, but otherwise had a moderate fecundity. Other small rodent species expanding ranges from steppes or

sity. Krebs (1964) and others made similar observations on American lemmings and Khlebnikov (1970) on Sibirian voles. Thus, winter reproduction seems also related to the "quality" of the populations and individuals. A change in quality may be due to at least three reasons:

1. A genetically polymorphic population where the animals in the low phase show especially intense reproduction. Such conditions are proposed in the "Chitty hypothesis" (Chitty, 1970; see also Krebs, 1978) and related models (Stenseth, 1978).

2. Less scramble competition for particularly nu-

forest biomes into the subarctic were supposed to retain a summer breeding season but to be characterized by an increased fecundity. Similarly the infrequent winter breeding and usually small litter size in other Scandinavian small rodents seem mainly to indicate variations in animal quality and do not seem to cause profound effects on population dynamics.

ACKNOWLEDGMENTS

The field studies reported were supported by the Swedish Natural Science Research Council and the Swedish Research Council for Forestry and Agriculture, respectively.

LITERATURE CITED

BATZLI, G. O., R. G. WHITE, S. F. MacLEAN, JR., F. A. PITELKA, and B. D. COLLIER. 1980. The herbivore-based trophic system. Pp. 335–410, in An arctic ecosystem: The coastal tundra at Barrow, Alaska (J. Brown, P. C. Miller, L. L. Tieszen, and F. L. Bunell, eds.), Dowden, Hutchinson and Ross, Stroudsburg.

CHITTY, D. 1970. Variation and population density. Symp. Zool. Soc. London, 26:327–334.

CLARKE, J. R. 1977. Long and short term changes in gonadal activity of field voles and bank voles. Oikos, 29:457–467.

CURRY-LINDAHL, K. 1962. The irruption of the Norway lemming in Sweden during 1960. J. Mamm., 43:171–184.

HANSSON, L. 1978. Small mammal abundance in relation to environmental variables in three Swedish forest phases. Stud. Forest. Suecica, 147:1–40.

———. 1979. Food as a limiting factor for small rodent numbers: tests of two hypotheses. Oecologia, 37:297–314.

HANSSON, L., and T.-B. LARSSON. 1980. Small rodent damage in Swedish forestry during 1971–79. Rapp. Inst. Viltekol., 1:1–64.

HENTTONEN, H., and A. JÄRVINEN. 1981. Lemmings in 1978 at Kilpisjärvi: population characteristics of a small peak. Mem. Soc. Fauna Flora Fennica, 57:25–30.

HÖGLUND, N. H. 1964. Fjällämmel ynglande i december. Fauna Flora, 59:145–147.

———. 1970. Vertebratfaunan inom Lövhögen-området. Fauna Flora, 65:161–180.

HUMINSKI, S. 1963. Winter breeding in the field vole, Microtus arvalis (Pall.) in the light of an analysis of the effect of environmental factors on the condition of the male sexual apparatus. Zool. Poloniae, 14:157–201.

JENSEN, T. S. 1975. Population estimations and population dynamics of two Danish forest rodent species. Vid. Medd. Dansk Naturh. For., 138:65–86.

KAIKUSALO, A. 1980. The effect of snow depth and extra food on winter breeding in small rodents. Luonnon Tutkija, 84:56–58.

KALELA, O. 1961. Seasonal change of habitat in the Norwegian lemming, Lemmus lemmus (L.). Ann. Acad. Sci. Fennicae, Ser. A, IV Biol., 55:1–72.

KHLEBNIKOV, A. I. 1970. Winter reproduction of the Northern red-backed vole (Clethrionomys rutilus) in the dark-coniferous taiga of the West Sayan mountains. Zool. Zhurn., 75:801–802.

KOPONEN, T., A. KOKKONEN, and O. KALELA. 1961. On a case

of spring migration in the Norwegian lemming. Ann. Acad. Sci. Fennicae, Ser. A, IV Biol., 52:1–30.

KOSHKINA, T. V., and A. S. KHALANSKY. 1962. On the reproduction of lemming Lemmus lemmus on Kola peninsula. Zool. Zhurn., 41:604–615.

KREBS, C. J. 1964. The lemming cycle at Baker lake, Northwest Territories, during 1959–62. Arct. Inst. North Amer. Techn. Pap., 15:1–104.

———. 1978. A review of the Chitty Hypothesis of population regulation. Canadian J. Zool., 56:2463–2480.

———. 1979. Dispersal, spacing behaviour and genetics in relation to population fluctuations in the vole Microtus townsendii. Fortschr. Zool., 25:61–77.

LAHTI, S., J. TAST, and H. OUTILA. 1976. Fluctuations in small rodent populations in the Kilpisjärvi area in 1950–1975. Luonnon Tutkija 80:97–107.

LARSSON, T.-B. 1974. Vinterreproduktion av fjällammel, Lemmus lemmus (L.) i Kebnekaisefjällen 1974. Fauna Flora, 69:110.

LARSSON, T.-B., and L. HANSSON. 1977. Sampling and dynamics of small rodents under snow cover in northern Sweden. Z. Säugetierk., 42:290–294.

LARSSON, T.-B., L. HANSSON, and E. NYHOLM. 1973. Winter reproduction in small rodents in Sweden. Oikos, 24:475–476.

LUNDQVIST, L., A. NILSSON, and L. HANSSON. 1973. Winter breeding observed in Northern red-backed mouse. Fauna, 26:216–217.

MacLEAN, S. F., JR, B. M. FITZGERALD, and F. A. PITELKA. 1974. Population cycles in arctic lemmings: winter reproduction and predation by weasels. Arct. Alp. Res., 6:1–12.

MEURLING, P., B. ANDERSSON, M. ERIKSSON, T. GUSTAFSSON, E. NYHOLM, and L. WESTLIN. 1981. När smågnagarna blir fler. Forskning och Framsteg, 16(1):32–37.

MYLLYMÄKI, A. 1969. Productivity of a free-living population of the field vole, Microtus agrestis (L.). Pp. 255–265, in Energy flow through small mammal populations (K. Petrusewicz and L. Ryszkowski, eds.), Polish Scientific Publishers, Warsaw, 298 pp.

———. 1970. Population ecology and its application to the control of the field vole, Microtus agrestis (L.). EPPO Publ. Ser. A, 58:27–48.

———. 1977a. Outbreaks and damage by the field vole, Microtus agrestis (L.) since World War II in Europe. EPPO Bull., 7:177–207.

————. 1977*b*. Demographic mechanisms in the fluctuating populations of the field vole *Microtus agrestis*. Oikos, 29: 468–493.

MYLLYMÄKI, A., E. CHRISTIANSEN, and L. HANSSON. 1977. Five-year surveillance of small mammal abundance in Scandinavia. EPPO Bull., 7:385–396.

MYRBERGET, S. 1973. Geographical synchronization of cycles of small rodents in Norway. Oikos, 24:220–224.

MYSTERUD, I. 1966. Winter breeding in the wood lemming (*Myopus schisticolor*) in Norway. Fauna, 19:79–83.

MYSTERUD, I., J. VIITALA, and S. LAHTI. 1972. On winter breeding of the wood lemming (*Myopus schisticolor*). Norw. J. Zool., 20:91–92.

SCHWARTZ, S. S. 1963. Ways of adaptation of terrestrial vertebrates to the conditions of existence in the subarctic. Vol. I. Mammals. Sverdlovsk, 250 pp. (transl. by E. Issakoff/W. A. Fuller).

SKARÉN, U. 1978. Peltomyyrän (*Microtus agrestis* L.) talvilisääntyminen Pohjois-Savossa. Savon Luonto, 10:57–61.

SMYTH, M. 1966. Winter breeding in woodland mice, *Apodemus sylvaticus*, and voles, *Clethrionomys glareolus*, near Oxford. J. Anim. Ecol., 35:471–485.

SOUTHERN, H. N. 1964. Handbook of British mammals. Blackwell, Oxford, 465 pp.

STEIN, G. H. W. 1963. Über Umweltabhängigkeiten bei der Vermehrung der Feldmaus (*Microtus arvalis*). Zool. Jahrb. (Syst.), 8:527–547.

STENSETH, N. C. 1978. Demographic strategies in fluctuating populations of small rodents. Oecologia, 33:149–172.

STENSETH, N. C., L. HANSSON, A. MYLLYMÄKI, M. ANDERSSON, and J. KATILA. 1977. General models for the population dynamics of the field vole *Microtus agrestis* in Central Scandinavia. Oikos, 29:616–642.

TAST, J., and A. KAIKUSALO. 1976. Winter breeding of the root vole, *Microtus oeconomus*, in 1972/1973 at Kilpisjärvi, Finnish Lapland. Ann. Zool. Fennici, 13:174–178.

ZIMMERMAN, K. 1960. Wintervermehrung der Feldmaus (*Microtus arvalis*) bei Postdam-Rehbrücke 1958/59. Z. Säugetierk, 25:94–95.

Address: Department of Wildlife Ecology, Swedish University of Agricultural Sciences, S-750 07 Uppsala, Sweden.

INFLUENCE OF RADIATION ON THE LITTER SIZE OF TWO MICROTINE RODENTS

Jan Zejda and Jaroslav Pelikán

ABSTRACT

Populations of the common vole (*Microtus arvalis*) and the bank vole (*Clethrionomys glareolus*) were monitored for a period of ten years (1968 to 1978) in fields and lowland forest, respectively, in southern Moravia (Czechoslovakia). A total of 1,567 pregnant females of *M. arvalis* and 546 *C. glareolus* were used to evaluate monthly mean litter sizes on a year-round basis. A close positive correlation was found between the sum of circumglobal radiation in open space and under forest canopy and monthly mean litter size of *M. arvalis* and *C. glareolus*. Circumglobal solar radiation was found to be an important factor influencing litter size in microtine rodents.

INTRODUCTION

Variation in litter size in the course of the year has been observed in a number of rodent populations, both confined and free-living. In searching for the factors that are decisive of this variation, greatest attention has been paid to the photoperiod. The dependence of sexual maturation, litter size and breeding intensity upon light intensity has been studied under experimental conditions. Few, if any, data are available from field conditions.

In 1968 through 1978, we participated in a team research into a lowland forest ecosystem, carried out within the International Biological Programme and Man and Biosphere projects. The objective of our participation was a study of the ecology and pro-ductivity of the small mammal community. We had at our disposal data on the primary and secondary production of various taxocoenoses as well as data on the variation in various physical factors. Such complete information on the environment of a species under study is rarely available for a mammalogist.

Therefore, we attempted to analyze the variation in the litter size in dependence on that of the monthly sums of solar radiation. Our model rodent species included the common vole (*Microtus arvalis*), trapped in field habitats, and the bank vole (*Clethrionomys glareolus*), trapped in the lowland forest ecosystems under study.

MATERIALS AND METHODS

Data on solar radiation and other meteorological factors were obtained from an experimental plot in the lowland forest. Particular attention was paid to circumglobal radiation which comprises both incident and reflected radiation. It was measured by means of a pyranograph placed on a meteorological tower over the canopy and by a Bellani's pyranometer placed on the forest floor under the canopy. The data used in our diagrams and tables are monthly sums of circumglobal radiation (in $J \cdot cm^{-2} \cdot month^{-1}$). The evaluation is based on daily measurements made in 1970 through 1974 (Smolík, 1975).

The data on litter size (based on the number of embryos) were collected in 1955 through 1975 in the study area. We examined a total of 1,567 pregnant females of *Microtus arvalis* and 546 pregnant females of *Clethrionomys glareolus*.

All data come from the region of southern Moravia (Czechoslovakia), mostly from the environs of Lednice (48°48′N, 16°48′E). The elevation of the study area varies around 200 m above sealevel, the long-term annual mean air temperature is 9°C, and the annual total precipitation averages 524 mm (Smolík, 1975).

THE STUDY AREA

The lowland forest under study is characterized by several strata of trees and shrubs and a rich herb layer. The growing season in the forest starts in April. The dominant plant species of the herb layer (*Glechoma hederacea* and *Urtica dioica*) begin to develop in the first decade of April and gradually make up 70% of the plant biomass. The major tree species (*Quercus robur*) develops foliage during the last April decade on the average. The development of the herb layer biomass attains its peak in late July. There are 2 ha of leaf blades and 3 ha of assimilation surface per 1 ha of forest.

RESULTS

Character of Circumglobal Radiation

As seen from the data in Table 1, the radiation incident on the study area amounts to 210 kJ·cm⁻²· year⁻¹ on the average. The canopy intercepts four fifths of that amount, so that the forest floor receives only one fifth, that is 42 kJ·cm⁻²·year⁻¹.

The variation in circumglobal radiation is evident from the data in Table 1 as well as from the course of the curves in Fig. 1. In open space, the monthly sum of radiation increases abruptly between January and May; it remains almost unchanged between May and August; and it decreases abruptly between September and December. On the forest floor, the increase in the monthly sum of circumglobal radiation lasts only until April when it attains its peak value. In the months to follow, the monthly sum of radiation decreases again and the minimum falls on December. It is evident that the difference between the radiation over and under the canopy is not only in the attained values, but also in their variation in the course of the year. The abrupt decrease in the sum of radiation on the lowland forest floor between April and August coincides with the development of foliage. From September onwards, the radiation decreases accordingly with the decrease observed in open space (Fig. 1).

Variation in Litter Size

In the course of the year, the litter size shows different variation in *Microtus arvalis* and *Clethrionomys glareolus*. In *Microtus arvalis,* the month-

ly mean litter size increases from the onset of the breeding season until June. In July and August, the monthly mean litter size is still high and insignificantly different from the value attained in June. In September, we recorded a decrease in the mean litter size which continued still in October. The minimum mean litter size was ascertained in January.

In *Clethrionomys glareolus,* the monthly mean litter size increases from the onset of the breeding season until May when it attains its peak. In June, the mean litter size decreases again, the decrease continuing until winter months.

Thus the difference between the above two vole species, as regards the monthly mean litter size and its variation, inheres in the differential course of the curves in June through September (Fig. 2).

Correlation between Circumglobal Radiation and Litter Size in *Microtus arvalis* and *Clethrionomys glareolus*

The graphic representation of the variation in the monthly sums of circumglobal radiation and in the monthly mean litter sizes reveals a conspicuous similarity between the two pairs of curves. The first pair pertains to the circumglobal radiation in open space and the litter size of *Microtus arvalis*. The second pair pertains to the radiation incident on the forest floor and the litter size of *Clethrionomys glareolus*. In the first case, the peak values of radiation are attained one month earlier than the peak litter size (May and June, respectively). In the second case,

Table 1.—*Survey of data on circumglobal radiation and litter size of* Microtus arvalis *and* Clethrionomys glareolus.

| Month | Radiation J·cm⁻²·month⁻¹ | | Litter size | | | | | |
| | Open space | Under canopy | M. arvalis | | | C. glareolus | | |
			n	Mean	SE	n	Mean	SE
Jan.	5,238	1,373	8	3.25	0.453	1	—	—
Feb.	9,282	2,734	1	—	—	6	2.83	0.307
March	16,739	5,928	32	3.84	0.169	5	4.80	0.374
April	21,227	7,004	262	4.76	0.072	53	5.13	0.137
May	27,754	6,113	233	5.53	0.094	141	5.58	0.114
June	27,838	5,200	101	5.93	0.148	103	4.87	0.107
July	27,645	4,283	292	5.91	0.089	83	4.70	0.109
Aug.	27,160	3,123	303	5.81	0.096	94	4.39	0.128
Sept.	19,653	1,880	231	5.37	0.088	40	4.38	0.155
Oct.	15,956	1,490	64	5.05	0.189	13	3.92	0.400
Nov.	7,469	1,578	39	3.72	0.187	4	4.00	0.500
Dec.	5,292	1,097	1	—	—	3		
Year	209,583	41,809	1,567	5.41	0.039	546	4.88	0.055

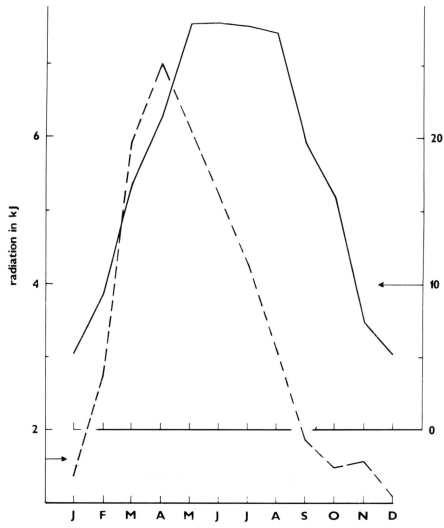

Fig. 1.—Circumglobal radiation in the course of the year as measured over the canopy (solid) and under the canopy (dashed).

the peaks are attained earlier by one month (April and May, respectively).

It is evident from the data in Table 1 and namely from the correlation diagrams in Figs. 3 and 4 that the monthly mean litter sizes are distinctly lower during the first part of the season than during its second part, although their correlation with the amount of radiation should be the same in both these periods. This tendency is evident in both *Microtus arvalis* and *Clethrionomys glareolus*. In our opinion, the higher mean litter sizes in the second part of the season are positively affected by better food supply, namely of seeds. In *Microtus arvalis,* it is the cereal grains and caryopses of meadow grasses; in *Clethrionomys glareolus,* acorns, seeds of forest trees and woodland grasses. It is also possible,

however, that the higher mean litter sizes observed during the second part of the season are also affected by a different population structure during that part of the season.

For this reason, we computed, for both species, the correlation coefficients and regression equations separately for the first and second part of the season. The respective correlation coefficients for *Microtus arvalis* are 0.968 and 0.975; those for *Clethrionomys glareolus* are similar, 0.996 and 0.938. In all four cases, the values of the correlation coefficients are statistically highly significant. The respective regression equations are evident from the diagrams in Figs. 3 and 4.

It is also interesting to compare the annual total radiation with the annual mean litter sizes of the

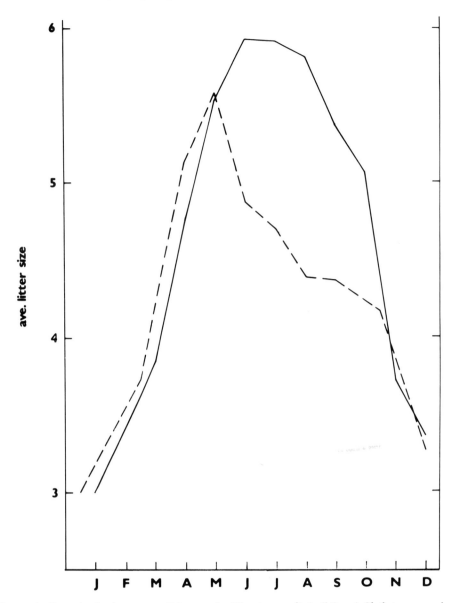

Fig. 2.—Changes in litter size in the course of the year in *Microtus arvalis* (solid) and *Clethrionomys glareolus* (dashed).

two vole species under study. The annual intensity of radiation, available to the *Microtus arvalis* population in the non-woodland environment, is five times as high as that available to the *Clethrionomys glareolus* population on the lowland forest floor. Despite this fact, the mean litter size of *Microtus arvalis* is only 0.53 embryo higher than that of *Clethriono-*

mys glareolus. The difference, however, is statistically highly significant (t = 7.86; $P < 0.001$).

This fact tends to evidence that the litter size is determined genetically and so fixed that radiation, though coming in excessive doses, can modify it to a certain extent but cannot change it markedly.

DISCUSSION

Baker and Ranson (1932) were the first to point out the significance of photoperiod on the breeding

intensity in *Microtus agrestis*. They found a high sensibility of females to photoperiod. A similar re-

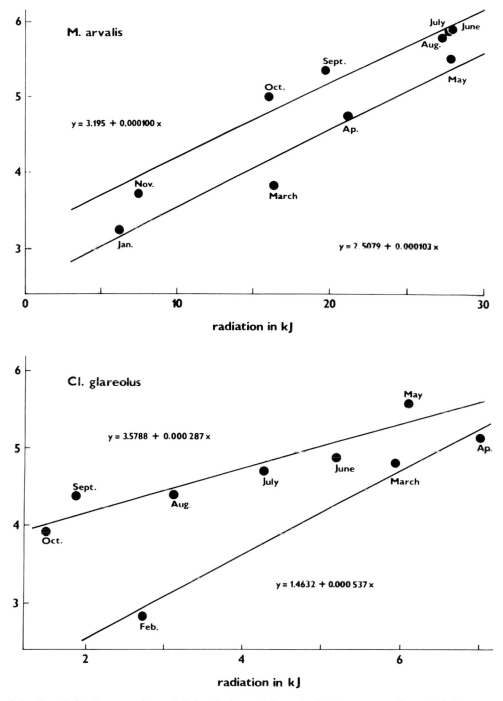

Figs. 3 and 4.—Correlation between circumglobal radiation and litter size in *Microtus arvalis* and *Clethrionomys glareolus*.

sult was obtained in subsequent experiments involving other microtines (Bashenina, 1960; Lecyk, 1962a; Pinter and Negus, 1965). In examining the effects of light, attention was also paid to its wavelength (Lecyk, 1962b; Evernden and Fuller, 1972).

With the example of *Microtus pinetorum*, Geyer and Rogers (1979) demonstrated that reduced light conditions have a deleterious effect on breeding (the number of litters as well as of young was significantly lower). Most of the authors mentioned above, however, have emphasized the stimulating effects of warmth and quality of food upon the intensity of breeding.

In field conditions, the litter size of microtines shows very marked variation in the course of the year (Frank, 1956; Pelikán, 1959; Reichstein, 1964; Zejda, 1966). The correlation between litter size and the periodically increasing and decreasing photoperiod, with temperature variation and with changes in food supply, seems to be generally clear. Comparing two microtine species populating the same geographic region, as we did, concrete examples reveal the complexity of this correlation. Circumglobal radiation and its variation in the course of the year appears to us as a very important factor, as it comprises the total energetic input available to the vole. In the case of *Clethrionomys glareolus*, the amount of available radiation is affected by the density and permeability of the vegetational cover under which it lives. In a sense, our observations resemble those made by Evernden and Fuller (1972) on *Clethrionomys gapperi* when studying the permeability of the snow cover for light of different wavelengths and its significance for rodents living under the snow cover. After the various strata composing the lowland forest have developed full foliage, the photosensitive females of *Clethrionomys glareolus* cannot receive the same amount of radiation as they could before or as is available for the females of *Microtus arvalis* in the open landscape at the same time.

The observed facts also lead to a conclusion concerning the problem of winter reproduction in microtines, apparently of general validity. Although the causes of the winter breeding in rodent populations may be different, the mean litter sizes in winter are the lowest which a particular population is capable of attaining in a particular space throughout the year. We believe that this value is primarily affected by the amount of circumglobal radiation in winter, which may be modified by the permeability of the vegetational cover, the depth of the snow cover, and/or the depth of the layer of litter covering the forest floor.

The observed positive correlation between circumglobal solar radiation and litter size may be one of the causes of geographic variation in the litter size of many small mammal species.

SUMMARY

1) A total of 1,567 pregnant females of *Microtus arvalis* and 546 pregnant females of *Clethrionomys glareolus,* captured in fields and a lowland forest, respectively, in southern Moravia (Czechoslovakia) in various seasons of the year, were used to evaluate the monthly mean litter sizes by the numbers of embryos.

2) In the same region, data were obtained on the variation in sum of circumglobal radiation in the open space and on the lowland forest floor.

3) A close positive correlation was found between the variation of the sum of circumglobal radiation in the open space and that of the monthly mean litter size of *Microtus arvalis*. A similar close positive correlation was found between the sum of radiation under the lowland forest canopy and the monthly mean litter size of *Clethrionomys glareolus.*

4) Circumglobal solar radiation has been found to be a very important factor affecting the variation in the litter size of microtine rodents.

LITERATURE CITED

BAKER, S. R., and R. M. RANSON. 1932. Factors affecting the breeding of the field mouse (*Microtus agrestis*). Part 1: Light. Proc. R. Soc. Ser. B., Biol. Sci., 110:313–321.

BASHENINA, N. V. 1960. The role of the light in the biology of the field vole based on the study of *Microtus arvalis* Pall. Bul. MOIP, 65(5):16–29. (In Russian, with a summary in English.)

EVERNDEN, L. N., and W. A. FULLER. 1972. Light alteration caused by snow and its importance to subnivean rodents. Canadian J. Zool., 50:1023–1032.

FRANK, F. 1956. Beiträge zur Biologie der Feldmaus, *Microtus arvalis* (Pall.) Teil 2: Laboratoriumsergebnisse. Zool. Jb. (Syst.), 84:32–74.

GEYER, L. A., and J. G. ROGERS. 1979. The influence of light intensity on reproduction in pine voles, *Microtus pinetorum.* J. Mamm., 60:839–841.

LECYK, M. 1962a. The effect of the length of daylight on reproduction in the field vole (*Microtus arvalis* Pall.). Zoologica Poloniae, 12:189–221.

———. 1962b. The dependence of breeding in the field vole

(*Microtus arvalis* Pall.) on light intensity and wave-length. Zoologica Poloniae, 12:255–268.

PELIKÁN, J. 1959. Bionomie und Vermehrung der Feldmaus. Pp. 130–179, *in* Hraboš polní (*Microtus arvalis*) (J. Kratochvíl, ed.), Publ. House of CAS, Praha, 359 pp. (In Czech, with a summary in German.)

PINTER, A. J., and N. C. NEGUS. 1965. Effects of nutrition and photoperiod on reproductive physiology of *Microtus montanus*. Amer. J. Physiol., 208:633–638.

REICHSTEIN, H. 1964. Untersuchungen zum Körperwachstum und zum Reproduktionspotential der Feldmaus, *Microtus arvalis* (Pallas, 1779). Zeitschrift Wiss. Zoologie, 170:112–222.

SMOLÍK, Z. 1975. Radiation, climate and microclimate in a lowland forest. Pp. 7–24, *in* Function, productivity and structure of the lowland forest ecosystem. VSZ Brno, 194 pp. (in Czech.)

ZEJDA, J. 1966. Litter size in *Clethrionomys glareolus* Schreber 1780. Zool. Listy Brno, 15:193–206.

Address: Czechoslovak Academy of Sciences, Institute of Vertebrate Zoology, 603 65 Brno, Kvetna 8, Czechoslovakia.

WINTER BREEDING OF MICROTINE RODENTS AT KILPISJÄRVI, FINNISH LAPLAND

Asko Kaikusalo and Johan Tast

ABSTRACT

In five microtine rodents, *Microtus oeconomus, M. agrestis, Clethrionomys rufocanus, C. rutilus,* and *Lemmus lemmus,* reproductive efforts have been recorded several times in winter at Kilpisjärvi, north of the Arctic circle. All of the observations are from the increase phases of the rodent cycles. In four of the species, reproductive activities have led even to propagating of young. The litters in winter were distinctly smaller than in summer. However, their importance to the increase of populations should not be underestimated, especially in the case of *L. lemmus.*

Under natural winter conditions immature pairs of all five species reached sexual maturity and produced young when placed in cages with extra food. Interspecific differences occurred as to need of insulating snow cover. *M. oeconomus* bred even in a cage without any snow. Other species bred only in cages which had above them insulating snow cover. The order in which different species began breeding activities tallied fairly well with the observed order of their onset of breeding in spring. However, *L. lemmus* bred only in a cage with a snow cover of 230 to 320 cm. Also, temperature apparently had an effect on *M. oeconomus,* as breeding began first in the cage with most snow above, and then in the order of the depth of snow.

INTRODUCTION

The regular breeding seasons of the five microtine rodent species living in the Kilpisjärvi area (69°3'N, 20°57'E) are restricted to the short subarctic summer. The first litters are usually delivered in late May or early June. Interspecific variations occur in the onset of breeding. The root vole, *Microtus oeconomus,* is the first, while the field vole, *M. agrestis,* and the grey-sided vole, *Clethrionomys rufocanus,* are the last (Tast, 1966, 1980). The other species within the area are the Norwegian lemming, *Lem-mus lemmus,* and the red vole, *C. rutilus.* Of the species the red vole is a seed-eater by preference, while the others are green-eaters, though in winter the root vole feeds mainly on underground organs of graminids and herbids (Tast, 1966, 1974). The winter food of lemmings consists principally of mosses (Kalela et al., 1961, 1971). The other species eat mainly vascular plants throughout the year. A survey of the food of the species was given by Kalela (1962).

METHODS

Since 1964, winter trapping of small mammals has been performed in the area. In most years, no signs of winter reproduction have been found. However, during four of the winters reproductive efforts were recorded. This paper summarizes these observations.

In 1972/73 experiments with caged animals were made under natural conditions in order to study the influence of extra food and the depth of insulating snow cover on the breeding activities of the five species. The results of these experiments have previously been reported in Finnish (Kaikusalo, 1980).

RESULTS

Winter Reproduction under Natural Conditions

Microtus oeconomus

In January 1973, a parous female root vole and at least four winter-born young voles belonging to three age categories were trapped. Investigations of root voles caught in March and early summer showed that reproduction did not continue after January. The winter reproduction of 1972/73 has been treat-ed in detail already (Tast and Kaikusalo, 1976). In 1980/81, root voles again reproduced in winter. In February 1981, five of eight males obtained were mature, while one was just reaching sexual maturity. His testes were 5 mm long. Testes of mature males exceed 7 mm. Immatures have testes of 2 to 3 mm. Only one of the males was clearly in an immature state. He weighted 30 g, distinctly more than wintering root voles do during normal winters (Tast, 1966, 1972, 1984). One of the males had been

243

mature in the previous summer. He now weighed 55 g. All six females trapped were mature, five being pregnant and one parous. In March, two parous females were caught, one of them being also visibly pregnant. All immatures were winter-born. Their ages ranged from 30 to 60 days. Breeding apparently ceased by April, so there was a time lag of about one month before the onset of summer reproduction.

Root voles bred in the same winter also in forest Lapland at Kolari (67°20′N, 23°45′E). Direct field observations there showed that breeding ended in March (Kaikusalo, in preparation). When the two winters with breeding are compared, it becomes evident that reproductive activities both began and ceased earlier in 1972/73 than in 1980/81.

Microtus agrestis

In the middle of March 1969, a male with testes of 9 mm was obtained. Another male was at the same time reaching sexual maturity. He had 6 mm long testes. In southern Finland winter breeding is not uncommon (Myllymäki, 1970, 1977; Tast and Kaikusalo, 1976).

Clethrionomys rufocanus

Sexually mature males have been trapped in 1972/73 and 1977/78. In January 1973, one male among seven was mature. In February also, one of seven males was mature. The only observation concerning a female grey-sided vole derives from 1973, when in March one among eleven females was pregnant about two months before the ordinary breeding season.

Clethrionomys rutilus

In January 1973, two of twelve males were mature. All twelve females were in immature stage. In March 1973, two of nine females were pregnant. Further, one female was obviously in the beginning of her first pregnancy.

Lemmus lemmus

In most years lemmings are so scarce that no individuals have been obtained in winter trapping. In March 1969, three mature males, a pregnant female and two immatures with ages from 30 to 60 days were caught. A mature male was trapped both in December 1972 and in February 1978. In February 1978, also, a parous female was captured.

Ages of spring wanderers show that in several years breeding has taken place under snow in late winter (Koponen et al., 1961; Koponen, 1964, 1970; Koshkina and Halansky, 1962; Henttonen and Järvinen, 1981). According to placental scars, about 30% of the females had bred under snow in 1960 (Koponen et al., 1961) and in 1978 (Henttonen and Järvinen, 1981). Judging from the pelage phases (Koponen, 1964, 1970), winter breeding has occurred mostly in February through early April.

The breeding season of the species is rather flexible. After April reproduction is suppressed some weeks before snowmelt (Koponen et al., 1961). And after summer breeding, the lemmings change their habitat seasonally and may again set on breeding in suitable habitats in September–October (Kalela et al., 1961). Three mature males were trapped as late as November 1969.

The termination of summer breeding is density-dependent. In years with low or moderate populations, breeding may continue up to mid-August or even later to September, while during high lemming density, as for example in 1970, it may end already in late June (Kalela, 1970; Tast and Kalela, 1971; Henttonen and Järvinen, 1981).

Winter Reproduction and Rodent Cycles

In Fig. 1 the cyclic fluctuations in the numbers of small rodents are compared with the observed winter reproduction. The data used for cycles derive from snap trapping carried out by Kaikusalo during the snowless season. All cases of reproductive efforts in winter coincide with the increase phase of the cycles.

Litter Size in Winter

Reproductive efforts in winter are not so efficient as in summer. In many cases they do not lead to production of offspring, and litters delivered in winter are distinctly smaller than in summer. In Table 1 the litter sizes of root voles in summer and winter are compared. The mean in summer was 7.1 (N = 361; Tast, 1966 and in press). In the winter sample in addition to the six females caught at Kilpisjärvi, five from Kolari trapped in the same winter 1980/81 are included. The mean was 3.5.

There is a clear correlation between the weight of the female and the number of embryos in summer litters (Tast, 1966). Hence, in Table 1 the litters are also grouped according to the weight of the female. The weights of embryos have been subtracted from those of their mothers. The seasonal differences in litter sizes are highly significant also when the weight of the female is taken into consideration.

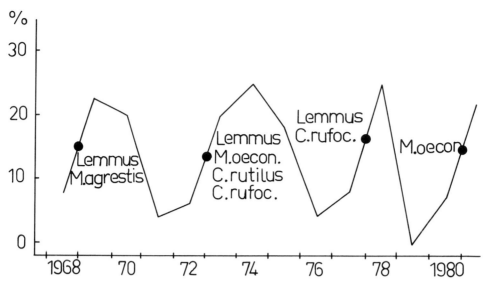

Fig. 1.—Fluctuations in relative numbers of rodents at Kilpisjärvi in 1968 to 1981, according to snap trapping in summer. Scale: individuals per 100 trap nights. Winter breeding is shown by black dots and the names of species in which it was recorded.

The litter sizes of the other species in winter were the following: *C. rufocanus* 3, *C. rutilus* 3 and 4, and *L. lemmus* 5. The averages in summer litters at Kilpisjärvi and its neighborhood are according to Kalela and Oksala (1966): 6.0 (N = 285) for *C. rufocanus*, 6.2 (N = 20) for *C. rutilus,* and 6.1 (N = 364) for *L. lemmus.* In the May sample of 1978, five of the lemming females had delivered litters so recently that the number of young could be determined from the placental scars and corpora lutea, and the mean thus obtained was 3.0 (Henttonen and Järvinen, 1981).

DISCUSSION

Breeding below the snow has been recorded in several small rodent species, but mostly in later winter or early spring only. Even when rodents have been investigated year-round, usually no efforts of reproduction in mid-winter have been recorded (Fuller, 1969; Fuller et al., 1969; Stebbins, 1976; Merritt and Merritt, 1978). However, *Clethrionomys gapperi* commenced breeding in March in Colorado while snow depths ranged from 150 to 230 cm (Merritt and Merritt, 1978).

Of the species studied by us, some scattered observations of actual mid-winter breeding are made. Lundquist et al. (1973) caught a mature male and a pregnant female of *C. rutilus* in a barn in northernmost Norway in January 1972. The litter size, three embryos, fits well with our records. Höglund (1970) trapped in Swedish mountain areas a new-born root vole in March 1967. Obviously, the lemming winter

Table 1.—*Litter size of the root vole in winter and summer. Winter sample A = Kilpisjärvi 1981, and sample B = Kolari 1981.*

No. of embryos	1	2	3	4	5	6	7	8	9	10	11	12	
Winter A	—	1	2	2	1								
B	—	3	2										M = 3.5
Summer	1	1	3	17	38	68	97	69	40	21	5	2	M = 7.1

Litter size according to the weight of the female in the root vole in winter and summer. Among summer litters only those are included which belong to the weight groups represented in the winter sample. The number of females in brackets.

	Winter	Summer
Less than 30 g	5.0 (1)	5.3 (3)
30–39.9 g	3.3 (8)	5.9 (21)
40–49.9 g	3.5 (2)	6.8 (30)

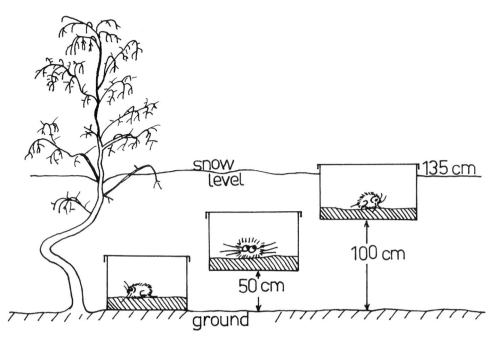

Fig. 2.—The experiment cages in January in the ordinary plot. Turf level hatched.

reproduction occurs regularly before peak years. No doubt, the achievement of a population high is significantly influenced by winter breeding.

It is interesting to find that lemmings living in Alaska are very similar to our species. Thus, in *Lemmus trimucronatus* and *Dicrostonyx groenlandicus* winter breeding occurs regularly, but not in every winter (Krebs, 1964; Pitelka, 1973; MacLean et al., 1974; Batzli et al., 1980). Also, lemmings in Alaska have suppression of breeding both in spring in May and in autumn in September (Batzli et al., 1980). Litter sizes in *Lemmus trimucronatus* differed markedly among seasons. In summer the mean litter size was from six to eight according to the age cohort of the females, while during mid-winter mean litter size declined to three (Pitelka, 1973; Batzli et al., 1980). According to MacLean et al. (1974) population changes of lemmings were related to the intensity of summer and, especially, winter reproduction. Further investigations are needed to determine whether winter reproduction has any significant ef-

fect on the population development of the four species other than *L. lemmus* in our study area. At least in the case of the root vole, this possibility is not excluded.

CAGE EXPERIMENTS

In October 1972, immature male/female pairs of the five rodent species were placed in cages with sizes of 60 by 60 by 80 cm. The rodents used were recently live-trapped in the Kilpisjärvi area. On the bottom of each cage, a 10 cm level of turf was placed

Table 3.—*Monthly mean temperatures at Kilpisjärvi in 1972 and 1973 as compared with the means of 1931–1960, according to the Meteorological Central Office of Finland. The summer of 1972 was exceptionally warm with three months above +10°C instead of the normal one.*

Month	1972	1973	1931–1960
January	−9.5	−9.0	−12.6
February	−11.8	−14.8	−13.1
March	−7.4	−7.9	−10.1
April	−4.3	−5.4	−5.0
May	+1.2	+1.0	+1.5
June	+10.3	+8.3	+6.6
July	+13.2	+13.8	+11.6
August	+10.3	+8.0	+9.6
September	+4.6	+2.6	+4.7
October	−0.8	−4.5	−0.9
November	−9.0	−12.4	−6.3
December	−6.3	−16.5	−10.2

Table 2.—*Depth of snow cover (cm) at ordinary cages (A) and at the lemming cage on the snow bed plot (B).*

			Date			
	Oct. 23	Nov. 28	Jan. 6	Feb. 18	Mar. 18	May 21
A	15	60	120	125	135	90
B			230	270	310	320

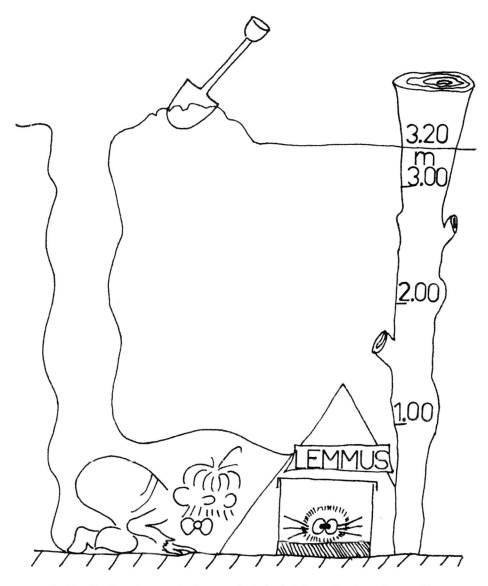

Fig. 3.—The lemming cage in the snow bed plot in May at the time of controlling.

and dwarf shrub above it. The cages were set in the field under natural conditions: 1) on the ground, 2) 50 cm above the ground, and 3) 100 cm above the ground (Fig. 2). The natural snow cover varied during the experiment from 15 to 135 cm (Table 2). The cages 100 cm above the ground were thus all the time partially above the snow level. One cage with a pair of lemmings was set in January 1973 in a snow bed area where the depth of snow cover varied from 230 to 320 cm (Table 2).

In each cage food was given in excess. As principal food, seed of oats were used for all species, except lemmings which were fed on green fodder and em-

bryos of wheat. In the beginning, apples were placed in all cages.

The animals were controlled five times. The experiment began on 23 October, and dates of control were 28 November, 6 January, 18 February, 18 March, and 21 May, when the experiment was ceased. During control, snow was dug away from cages (Fig. 3), the state of rodents was checked, and those young which had reached independent state were removed. Thereafter, food was added and the cage was filled with snow to the original level.

The monthly mean temperatures of 1972 and 1973 at Kilpisjärvi are compared with the means of 1931–

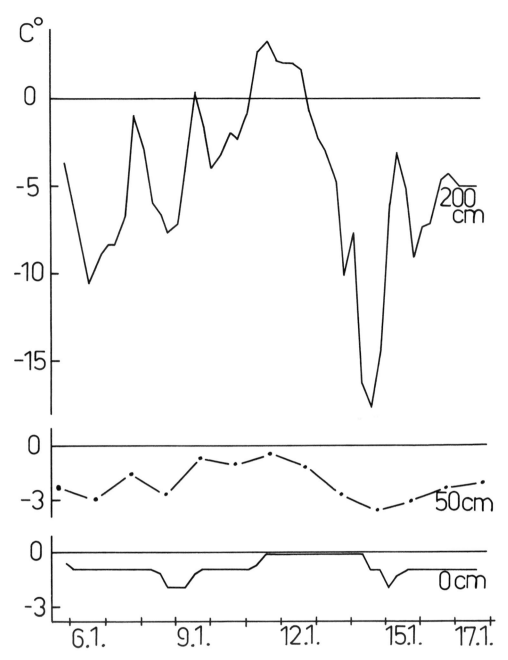

Fig. 4.—Temperature fluctuations from 6 January to 17 January in 1973 at three heights: 200 cm above the ground; 50 cm above the ground; and on the ground.

1960 in Table 3. These figures give an overall view of the temperature of the area. From the point of view of the small rodents living under snow, the influence of insulating snow cover on the temperature is of importance. In Fig. 4, temperatures are shown for a twelve-day period at three heights: 1) 2 m above the ground, 2) on the ground (= 120 cm deep in the snow), and 3) 50 cm above the ground (=70 cm in the snow). Temperatures fluctuated on the ground less than 2°C, while about 20°C in the air. The effect of snow is also seen when the minimum temperatures are compared. Minimum temperatures were reached on the ground 1 to 2 days later than in air.

Table 4.—*Numbers of young in litters delivered in cages with extra food under natural conditions at Kilpisjärvi in the winter of 1972/73. + = the female pregnant at the end of the experiment.*

Species	Height of the cage (cm) above the ground	Month of delivery						
		Nov.	Dec.	Jan.	Feb.	Mar.	Apr.	May
Microtus oeconomus	0	—	4	3	—	4	5	—
	50	—	—	2	5	—	—	+
	100	—	—	—	4	5	—	—
Microtus agrestis	0	—	—	—	3	—	—	4
	50	—	—	—	—	—	—	+
	100	—	—	—	—	—	—	—
Clethrionomys rufocanus	0	—	—	—	4	5	—	—
	100	—	—	—	—	—	—	4
Clethrionomys rutilus	0	—	—	4	3	4		
	100	—	—	—	4	+		
Lemmus lemmus	0	—	—	—	—	—	3	—
	50	—	—	—			—	l
	0 (snow bed)				4	6	6	+

RESULTS OF CAGE EXPERIMENTS

Microtus oeconomus

Results of cage experiments are summarized in Table 4.

The root voles reproduced in all cages. Thus, for the species good food conditions seemed to be enough to release reproductive activities. However, the influence of temperature is seen in the order in which the reproduction in cages began. Breeding activities began almost immediately in the cage on the ground with a snow cover of 15 cm in the beginning of experiment. Thus, most of the cage was above the snow level at that time. The first litter was delivered approximately on 10 December. As pregnancy takes about 20 days, these voles had attained sexual maturity rather soon after the beginning of the experiment. This pair produced four litters during the winter. Breeding obviously ceased in May at the time when breeding in the field just began. The other two pairs produced two litters each in the expected order; that is, the pair with a thicker snow layer above produced first. In both of these pairs reproductive activities also stopped before the ordinary breeding season.

An extra experiment was made with the young of the first litter born in the early December. These young voles were placed in a cage in the shed of Kilpisjärvi Biological Station with shelter only from wind, temperatures being almost the same as in the open. This cage had no insulating snow. Here a female produced two young at the end of February.

Microtus agrestis

Field voles produced young only in the cage on the ground. The female gave birth to three young in February. A second litter was delivered in May a few weeks before ordinary breeding season.

Clethrionomys rufocanus

The species bred only in the cage on the ground, producing two litters, one in February and the other in March. Thereafter, reproduction terminated. In the cage 100 cm above the ground, a litter was produced in May a little earlier than in the field.

Clethrionomys rutilus

Red voles produced offspring both on the ground and 100 cm above the ground. The influence of temperature on the onset of reproductive activities is seen also in this species, as breeding began earlier on the ground. There the female produced three litters in succession in January–March. Obviously she had post partum oestrus. The female living 100 cm above the ground produced her first litter in February and was pregnant for a second time in March, when experiments with this species were ceased.

The red vole thus responded to good food conditions sooner than *M. agrestis* and *C. rufocanus*, being second only to *M. oeconomus*.

Lemmus lemmus

Lemmings did not produce any offspring in the ordinary cages before April, when the female living on the ground gave birth to her first litter. In the cage 50 cm above the ground, the female was pregnant at the end of experiment on 22 May. In the cage placed in the snow bed plot on 6 January, the pair immediately reached sexual maturity and produced three litters in February to April. The female was pregnant for the fourth time at the end of May.

CONCLUSIONS AND DISCUSSION

Generally, the onset of breeding in many mammalian species is released by the combined effects of several external factors, among which day-length is usually considered to be the primary stimulus, while temperature and food regulate the onset in greater detail. However, in subarctic conditions photoperiodism cannot have such a major role as in areas farther south, where most investigations have been made. In our study area, nights are completely light from mid-May to late July, and in the winter, the sun does not rise at all during a period of about two months in December and January. We began cage experiments when days were shortening. In most cages reproductive activities began just at the time of constant darkness. So in the five microtine rodents studied, even the indirect influence of light on their onset of winter breeding is excluded.

The maturation of overwintering rodents was principally a result of food conditions. However, temperature also had an effect as was seen in the order in which reproduction began in cages at different depths of snow.

The order in which reproductive efforts in different species began was in fairly good agreement with observations of their onset of breeding in the field in spring. However, the time lag between species in cage experiments was much longer than in the wild in spring conditions.

The results support the view given earlier (Tast and Kaikusalo, 1976) that the onset of winter reproduction in root voles in 1972/73 was released by unusually good food conditions. The summer of 1972 was exceptionally warm and long (Table 3). The cotton grass, *Eriophorum angustifolium,* is the principal food plant of the root vole in most habitats in Kilpisjärvi area (Tast, 1966, 1972, 1974). The winter food consists mostly of the first-year rhizomes, which are produced in late summer. Usually in the spring, rhizomes reach the earth surface and a few green leaves are developed, but in 1972 this development took place already in the fall. We assume that some chemical changes in food plants functioned as a cue to initiate reproduction. Reference may be made to the investigations on the breeding of *M. montanus,* in which reproductive efforts were induced experimentally by chemical components of their food plants (Negus and Berger, 1977, Negus et al., 1977). Recently the chemical composition of this compound has been revealed (Sanders et al., 1981; Berger et al., 1981). This chemical, 6-methoxybenzoxazolinone, is found in young sprouting plant shoots. Obviously, corresponding changes in *E. angustifolium* shoots in fall 1972 caused the onset of breeding in root voles. Winter breeding in 1972/73 ceased rather soon. It can also be interpreted by food, as no continuation of "spring" development occurred in food plants.

In 1980/81 we have no field observations of probable factors leading to winter breeding. It began later and continued longer than in 1972/73. In both falls there was a time lag between summer and winter reproduction. Also, in Kolari where winter breeding occurred in 1980/81, summer reproduction terminated in due time. In October 1980, 39 root voles were examined, all of them being in an immature state (Kaikusalo, unpublished). In February 1981, more than 80% of root voles were mature. The root vole is usually the species which first begins its reproduction at Kilpisjärvi. Under natural conditions its first litters are delivered under snow (Tast, 1966).

Many of the plants adapted to short growing season and low temperatures are capable of responding to the light that penetrates rather deep through snow. For example, germination of some plant seeds was promoted by exposure of 3 hr to the light beneath 2 m of snow (Richardson and Salisbury, 1977). Leaf greening and the growth and development of different stages of shoot, leaf, and flower has been recorded under snow cover at temperatures near freezing point (Kimball et al., 1973; Kimball and Salisbury, 1974). Even photosynthesis may be possible beneath shallow snow layers. There are many plants that remain green throughout the winter, and in some plants the optimum temperatures for photosynthesis are surprisingly low. In a moss species, *Dicranum elongatum,* studied in northern Finland at Kevo Subarctic Research Station, the optimum for photosynthesis was +5°C (Kallio and Heinonen,

1975), and in this species 70% of photosynthesis activities still remained at 0°C.

The lemming obviously needs almost even temperatures in addition to good food conditions, as it bred only in the cage more than 200 cm under snow. There temperature was almost constant. The wintering sites of lemmings are just the snow bed areas (Kalela et al., 1961, 1971). There snow cover may reach tens of meters, and it can remain even to August or sometimes throughout the whole year. The growth of mosses, their principal winter food item, differs markedly from that of vascular plants. According to Hagerup (1937), their growth in Denmark and the Fair Isles has two annual periods. One of them begins in the middle of winter. In the case of *Polytrichum*, it begins in December and continues about three months. It seems rather probable that corresponding winter growth takes place also in

northern Finland. At low temperatures plants in photosynthesis produce especially lipids (Karunen and Kallio, 1976; Levitt, 1980). In *Dicranum elongatum* a peak of lipids is in March (Karunen and Kallio, 1976; Kallio et al., 1981). Also in *Sphagnum fuscum* derived from southern Finland, a clear winter maximum of lipids was recorded, the peak being in January (Kallio et al., 1981).

Summarizing, it seems evident that the onset of breeding in the five microtine rodents studied is released by good food conditions, but also temperature has an effect on it. Winter reproduction both in the wild and in cages showed that day-length cannot be any primary stimulus. In spring their regular breeding obviously is released by changes in the food plants. In lemming food (mosses), these changes take place in winter frequently resulting in lemming winter reproduction.

LITERATURE CITED

BATZLI, G. O., R. G. WHITE, S. F. MacLEAN, JR., F. A. PITELKA, and D. D. CHRISTIAN. 1980. The herbivore based trophic system. Pp. 335–410, *in* An arctic ecosystem: The coastal tundra at Barrow, Alaska. (J. Brown, P. C. Miller, L. L. Tieszen and F. L. Bunnell, eds.), Dowden, Hutchison and Ross, Inc., Stroudsberg, xxv + 571pp.

BERGER, P. J., N. C. NEGUS, E. H. SANDERS, and P. D. GARDNER. 1981. Chemical triggering of reproduction in *Microtus montanus*. Science, 214:69–70.

FULLER, W. A. 1969. Changes in number of three species of small rodent near Great Slave Lake, N.W.T., Canada, 1964–1967, and their significance for general population theory. Ann. Zool. Fennici, 6:113–144.

FULLER, W. A., L. L. STEBBINS, and G. R. DYKE. 1969. Overwintering of small mammals near Great Slave Lake Northern Canada. Arctic, 22:34–55.

HAGERUP, O. 1937. Zur Periodizität im Laubwechsel der Moose. Kongelige Danske Vidensk. Selsk. Biol. Meddel., 11(9):1–86.

HENTTONEN, H., and A. JÄRVINEN. 1981. Lemmings in 1978 at Kilpisjärvi: population characteristics of a small peak. Memoranda Soc. Fauna Flora Fennica, 57:25–30.

HÖGLUND, N. H. 1970. Vertebratfaunan inom Lövhögsområdet. Fauna o. Flora, 65:161–180.

KAIKUSALO, A. 1980. Häkkikokeita myyrien talvilisääntymisestä Kilpisjärvellä. (The effect of snow depth and extra food on winter breeding in small rodents.) Luonnon Tutkija, 84:56–58.

KALELA, O. 1962. On the fluctuations in the numbers of arctic and boreal small rodents as a problem of production biology. Ann. Acad. Scient. Fennicae A IV, 66:1–38.

———. 1970. Movements of the Norwegian lemming (*Lemmus lemmus*) in a year with extremely large populations. Paper read at Kilpisjärvi Biol. Station during the excursion of the IBP meeting on Secondary Productivity in Small Mammal

Populations, Helsinki, 24–26 August 1970. Paper is published in Kilpisjärvi Notes 6:1–6, 1982.

KALELA, O., in collaboration with T. KOPONEN, E. A. LIND, U. SKARÉN, and J. TAST. 1961. Seasonal change of habitat in the Norwegian lemming, *Lemmus lemmus* (L.). Ann. Acad. Scient. Fennicae A IV, 55:1–72.

KALELA, O., in collaboration with L. KILPELÄINEN, T. KOPONEN, and J. TAST. 1971. Seasonal differences in habitats of the Norwegian lemming, *Lemmus lemmus* (L.), in 1959 and 1960 at Kilpisjärvi, Finnish Lapland. Ann. Acad. Scient. Fennicae A IV, 178:1–22.

KALELA, O., and T. OKSALA. 1966. Sex ratio in the wood lemming, *Myopus schisticolor* (Lilljeb.), in nature and in captivity. Ann. Univ. Turkuensis A II, 37:1–24.

KALLIO, P., and S. HEINONEN. 1975. CO$_2$ exchange and growth of *Rhacomitrium lanuginosum* and *Dicranum elongatum*. *In* Fennoscandian tundra ecosystems 1 (F. E. Wielgolaski, ed.). Ecol. Stud., 16:138–148.

KALLIO, P., P. KARUNEN, H. MIKOLA, K. PIHAKASKI, S. PIHAKASKI, and M. SALIN. 1981. Kasvien pohjoinen stressi ja sopeutuminen (Adaptation of plants to northern conditions). Luonnon Tutkija, 85:98–107.

KARUNEN, P., and P. KALLIO. 1976. Seasonal variation in the total lipid content of subarctic *Dicranum elongatum*. Rep. Kevo Subarctic Res. Stat., 13:63–70.

KIMBALL, S. L., B. D. BENNETT, and F. B. SALISBURY. 1973. The growth and development of montane species at near-freezing temperatures. Ecology, 54:168–173.

KIMBALL, S. L., and F. B. SALISBURY. 1974. Plant development under snow. Bot. Gaz., 135:147–149.

KOPONEN, T. 1964. The sequence of pelages in the Norwegian lemming, *Lemmus lemmus* (L.). Arch. Soc. Zool.-Bot. Fennicae Vanamo, 18:260–278.

———. 1970. Age structure in sedentary and migratory populations of the Norwegian lemming, *Lemmus lemmus* (L.), at Kilpisjärvi in 1960. Ann. Zool. Fennici, 7:141–187.

KOPONEN, T., A. KOKKONEN, and O. KALELA. 1961. On a case of spring migration in the Norwegian lemming. Ann. Acad. Scient. Fennicae A IV, 52:1–30.

KOSHKINA, T. V., and A. S. HALANSKY. 1962. O razmnoženii norvezskogo lemminga (*Lemmus lemmus* L.) na Kol'skom poluostrove. Zool. Zurnal, 41:604–615.

KREBS, C. J. 1964. The lemming cycle at Baker Lake, Northwest Territories, during 1959–62. Arctic Inst. North Amer., Tech. Paper, 15. 104 pp.

LEVITT, J. 1980. Responses of plants to environmental stresses. 1. Chilling, freezing and high temperature stresses. Academic Press, New York and London. 2. edition, 497 pp.

LUNDQUIST, L., A. NILSSON, and L. HANSSON. 1973. Vinterfortplantning hos rödsork. Fauna (Oslo), 26:216–217.

MACLEAN, S. F., JR., B. M. FITZGERALD, and F. A. PITELKA. 1974. Population cycles in arctic lemmings: Winter reproduction and predation by weasels. Arctic and Alpine Res., 6:1–12.

MERRITT, J. F., and J. M. MERRITT. 1978. Population ecology and energy relationships of *Clethrionomys gapperi* in a Colorado subalpine forest. J. Mamm., 59:576–598.

MYLLYMÄKI, A. 1970. Population ecology and its application to the control of the field vole, *Microtus agrestis* (L.). EPPO publ. Ser. A, 58:27–48.

———. 1977. Demographic mechanisms in the fluctuating populations of the field vole, *Microtus agrestis.* Oikos, 29:468–493.

NEGUS, N. C., and P. J. BERGER. 1977. Experimental triggering of reproduction in a natural population of Microtus montanus. Science, 196:1230–1231.

NEGUS, N. C., P. J. BERGER, and L. G. FORSLUND. 1977. Reproductive strategy of *Microtus montanus.* J. Mamm., 58:347–353.

PITELKA, F. A. 1973. Cyclic pattern in lemming populations near Barrow, Alaska. *In* Alaskan Arctic Tundra (M. E. Britton ed.), Arct. Inst. North Amer. Tech. Paper, 25:199–215.

RICHARDSON, S. G., and F. B. SALISBURY. 1977. Plant responses to the light penetrating snow. Ecology, 58:1152–1158.

SANDERS, E. H.. P. D. GARDNER, P. J. BERGER, and N. G. NEGUS. 1981. 6-methoxybenzoxazolinone: A plant derivative that stimulates reproduction in *Microtus montanus.* Science, 214:67–69.

STEBBINS, L. L. 1976. Overwintering of the red-backed vole at Edmonton, Alberta, Canada. J. Mamm., 57:554–561.

TAST, J. 1966. The root vole, *Microtus oeconomus* (Pallas), as an inhabitant of seasonally flooded land. Ann. Zool. Fennici, 3:127–171.

———. 1972. Annual variations in the weights of wintering root voles, *Microtus oeconomus,* in relation to their food conditions. Ann. Zool. Fennici, 9:116–119.

———. 1974. The food and feeding habits of the root vole, *Microtus oeconomus,* in Finnish Lapland. Aquilo, Ser. Zool., 15:25–32.

———. 1980. Breeding season and litter size of the field vole, *Microtus agrestis,* at Kilpisjärvi, Finnish Lapland. Kilpisjärvi Notes, 4:8–11.

———. 1984. Winter success of root voles (*Microtus oeconomus*) in relation to plant production and population density at Kilpisjärvi, Finnish Lapland. This volume.

———. In press. *Microtus oeconomus* (Pallas, 1776)—Nordische Wühlmaus, Sumpfmaus. *In* Handbuch der Säugetiere Europas, Band II (J. Niethammer, and F. Krapp, eds.).

TAST, J., and A. KAIKUSALO. 1976. Winter breeding of the root vole, *Microtus oeconomus,* in 1972/1973 at Kilpisjärvi, Finnish Lapland. Ann. Zool. Fennici, 13:174–178.

TAST, J., and O. KALELA. 1971. Comparisons between rodent cycles and plant production in Finnish Lapland. Ann. Acad. Scient. Fennicae A IV, 186:1–14.

Address: Kilpisjärvi Biological Station, University of Helsinki, P. Rautatiekatu 13, SF-00100 Helsinki 10, Finland.

Present address (Kaikusalo): Ojajoki Field Station, Forest Research Institute, SF-12700 Loppi, Finland.

Present address (Tast): Korvenkatu 44, SF-33300 Tampere 30. Finland.

REPRODUCTION AND SURVIVAL OF *PEROMYSCUS* IN SEASONAL ENVIRONMENTS

JOHN S. MILLAR

ABSTRACT

Mice of the genus *Peromyscus* exhibit a great deal of variation in their annual cycles. Breeding occurs year round in some southern and temperate populations, but is restricted to as little as two months in some northern and montane populations. Examination of life cycle data for nine subspecies of *P. maniculatus* and 15 species of *Peromyscus* indicates that mice in short-season environments breed post-partum more frequently and have their largest litters earlier than mice in long-season environments, but litter size, birth weights, age at weaning and rates of development are not significantly related to length of the breeding season. Life cycle adaptations are insufficient to compensate for limited opportunities for reproduction. Examination of survival data for *P. maniculatus* and *P. leucopus* shows that the annual probability of survival of young of the year and overwintered adults increases with decreasing length of the breeding season (r = −.80 and r = −.49, respectively). These trends are attributable to the fact that average survival during winter (.86 ± .01 per 14 days) is higher than survival during the summer (.81 ± .14 per 14 days for adults and .75 ± .04 per 14 days for young of the year) and increases with decreasing length of the breeding season (r = −.49).

INTRODUCTION

The annual cycle of most small mammals is represented by two biological periods, the breeding and non-breeding seasons. The duration of these periods varies geographically, with northern and alpine populations having shorter breeding seasons than populations in temperate and southern areas. This variation poses interesting questions about the persistence of populations in different environments. How are small mammal populations maintained in environments where opportunities for reproduction are severely limited? What are the differences in life history characteristics among related animals with vastly different opportunities for reproduction?

A number of general studies have reflected on the complexity of seasonal phenomena and provide suggestions about carrying capacity, population growth, habitat use, survival, body size, growth and clutch or litter size in different environments (for example, Fretwell, 1972; Case, 1978, Boyce, 1979; Roff, 1980). Most empirical studies on small mammals, however, have focused on the relationship between litter size and latitude or elevation (for example, Jackson, 1952; Lord, 1960; Dunmire, 1960; Moore, 1961; Barkalow, 1962; Blus, 1966; Smith and McGinnis, 1968; Spencer and Steinhoff, 1968; Innes, 1978).

The purpose of this study was to examine life cycle and survival characteristics of *Peromyscus* in relation to the length of the breeding season.

METHODS

Most data used here were taken from the literature. Subspecies of *P. maniculatus* and *P. leucopus* were recorded when provided; unknown subspecies were either listed as unknown (?) (for example, Sullivan, 1977; Jameson, 1953; Flake, 1974; McKeever, 1964; Merritt and Merritt, 1980) or assigned the subspecies known to occur in the study area (for example, Sullivan, 1977). Data from field studies were tabulated by species or subspecies, year and study area. Since common sampling periods were not used in all studies, all data were adjusted to a common format. The duration of breeding was defined as the period (in weeks) from the first to the last birth each year. This required the conversion of monthly or semi-monthly data into weeks and reducing the duration of breeding by the length of gestation (3 weeks) or lactation (3 weeks) where necessary.

All survival data (except nestling survival) were converted to a consistent time frame of two weeks. Survival data were initially separated by sex, but were eventually combined to provide larger comparable samples. Nestling survival was defined as the proportion of expected young (based on known births of average litter size) that emerged into the trappable population. Juvenile survival and survival of overwintered adults was defined as the average proportion of animals that remained on the study area over two week periods during the breeding season. Winter survival was defined as the average proportion of all animals (all ages combined) that remained on the study area over two week periods during the non-breeding season.

Life cycle data were derived primarily from laboratory studies. These data were tabulated by subspecies or origin of the captive colony. The extent to which laboratory data reflects patterns in the wild cannot be known with certainty, but the general similarity between laboratory and field estimates of litter size (Lackey, 1973; Rickart, 1977; Millar, 1978; Halfpenny, 1980) indicates that differences among laboratory estimates are likely applicable to natural populations. Adult weights represent non-breeding adult weights, litter size represents the mean number of embryos or neonates, birth weights represent the weight within 24 hr of birth, age that eyes open represents an average or approximate date when most young in a litter opened their eyes, nestling growth represents growth from birth to approximately three weeks and post-weaning growth represents growth from approximately 3 to 7 weeks of age. Growth was calculated as g/day and as a geometric growth rate. Age at weaning was recorded only when the estimate was based on the method of King et al. (1963). The time between the first and second litters of the breeding season, based on live-trapping of individual females, provided comparable data on the frequency of breeding in natural populations. Few laboratory studies examined all of these life cycle traits so that composite sets of data were compiled for each subspecies or species. In cases of duplicate data, the average of all recorded estimates was calculated.

All life cycle and survival characteristics were examined in relation to the length of the breeding season for the population or the average breeding season for the subspecies or species.

RESULTS

SEASONAL REPRODUCTION

The duration of breeding in *Peromyscus* is extremely variable within and between populations. A number of populations breed aseasonally. This pattern is most common in southern populations (Davenport, 1964; Caldwell and Gentry, 1965; Robertson, 1975; Wolf and Linzey, 1977; Lackey, 1976; Rickart, 1977) but has also been recorded in Washington (Scheffer, 1924) and Kansas (Svendsen, 1964).

Seasonal breeding occurs over the full latitudinal range of the genus (Appendix I). The shortest recorded breeding season (approximately 8 weeks) is found in *P. m. borealis* at Heart Lake, N.W.T.

The factors precluding reproduction during part of the year are not well known. The duration of reproduction tends to decrease with increasing latitude (Millar et al., 1979) and elevation (Dunmire, 1960; Halfpenny, 1980), indicating that climatic

Table 1.—*Life cycle traits and length of the breeding season among nine subspecies of* Peromyscus maniculatus. *All numbers are means, sample sizes in parenthesis.*

Subspecies	Season (weeks)	Adult weight (g)	Litter size	Birth weight (g)	Age eyes open (days)	Nestling growth		Juvenile growth	
						(g/day)	(geom.)	(g/day)	(geom.)
P. m. borealis	8	21.5	5.5	1.81	14	.34	(.074)	.20	(.015)
	(8)	(6)	(5)	(2)	(2)	(2)		(1)	
P. m. austerus	23	19.1	5.0	—	—	—		—	
	(15)	(3)	(2)						
P. m. bairdii	39	17.3	4.4	1.81	13	.30	(.067)	.21	(.018)
	(3)	(4)	(7)	(3)	(2)	(3)		(3)	
P. m. gracilis	19	17.6	5.0	2.0	17	.34	(.074)	.25	(.023)
	(1)	(1)	(5)	(1)	(1)	(1)		(1)	
P. m. nebrascensis	23	19.7	4.4	1.83	13	.36	(.079)	—	
	(2)	(2)	(3)	(1)	(1)	(2)			
P. m. sonoriensis	13	21.5	4.3	—	—	—		.24	(.023)
	(1)	(1)	(1)					(1)	
P. m. gambelii	40	19	5.0	1.40	14	.34	(.087)	.27	(.023)
	(2)	(1)	(2)	(1)	(1)	(1)		(1)	
P. m. maniculatus	16	18.4	5.3	1.61	14	.31	(.076)	.24	(.022)
	(6)	(2)	(1)	(2)	(1)	(2)		(1)	
P. m. rufinus	18	21.0	4.9	1.78	15	.31	(.076)	.10	(.008)
	(2)	(4)	(3)	(2)	(1)	(2)		(1)	
Correlation with season		−.62	−.35	−.44	−.34	−.16	(.19)	.26	(.33)
		(9)	(9)	(7)	(7)	(7)		(7)	
Correlation with adult weight		.05	−.05	−.14	.27	(.26)	−.42	(−.32)
		(9)	(7)	(7)	(7)		(7)	

Note: Breeding season length from Appendix I. Life cycle data as follows: *P. m. borealis*: Fuller, 1969; Gyug, 1979; May, 1979; Mihok, 1979; Millar, unpublished; Millar and Innes, unpublished; Morrison et al. 1977. *P. m. austerus*: Sullivan, 1979; Fordham, 1971; Sadleir, 1965, 1974. *P. m. bairdii*: Svihla, 1935; Beer and McLeod, 1966; Dice and Bradley, 1942; Drickamer and Vestal, 1973; Eleftheriou and Zarrow, 1961; Glazier, 1979; Howard, 1949; King, 1958; Morrison et al., 1977; Rood, 1966; Svendsen, 1964. *P. m. gracilis*: Rood, 1966; Beer et al., 1957; Coventry, 1937; Drickamer, 1978; Drickamer and Vestal, 1973; Eleftheriou and Zarrow, 1961; King, 1958. *P. m. nebrascensis*: Brown, 1966; Drickamer and Bernstein, 1972; Halfpenny, 1980. *P. m. sonoriensis*: Dice and Bradley, 1942; Dunmire, 1960. *P. m. gambelii*: McCabe and Blanchard, 1950; Scheffer, 1924. *P. m. maniculatus*: Millar, 1979; Millar et al., 1979. *P. m. rufinis*: Anderson, 1972; Dice and Bradley, 1942; Svihla, 1932; Halfpenny, 1980.

Table 2.—*Life cycle characteristics and length of the breeding season among 15 species of* Peromyscus. *All numbers are means, sample sizes in parentheses.*

Subspecies	Season (weeks)	Adult weight (g)	Litter size	Birth weight (g)	Age eyes open (open)	Nestling growth (g/day)	Nestling growth (geom.)	Juvenile growth (g/day)	Juvenile growth (geom.)
P. maniculatus	22 (5)	21.2 (41)	4.6 (48)	1.90 (22)	14.6 (20)	.34 (17)	(.076)	.23 (15)	(.018)
P. leucopus	30 (23)	21.5 (13)	4.5 (20)	1.89 (9)	13.2 (6)	.36 (7)	(.076)	.27 (3)	(.021)
P. truei	27 (1)	27.0 (2)	3.2 (3)	2.32 (2)	—	.39 (1)	(.071)	.37 (1)	(.025)
P. eremicus	all year (1)	20.5 (3)	2.5 (6)	2.38 (2)	15.2 (2)	.26 (1)	(.060)	.20 (1)	(.020)
P. gossypinus	all year (1)	29 (1)	3.7 (1)	2.19 (1)	13 (1)	.50 (1)	(.086)	.38 (1)	(.022)
P. boylii	31 (3)	24.8 (4)	3.2 (4)	—	—	—	—	—	—
P. californicus	30 (2)	39.0 (1)	1.9 (4)	4.61 (2)	14 (1)	.59 (1)	(.065)	.33 (1)	(.020)
P. yucatanicus	all year (1)	28 (1)	2.8 (1)	2.50 (1)	17 (1)	.39 (1)	(.071)	.41 (1)	(.026)
P. polionotus	all year (1)	13.7 (3)	3.7 (6)	2.15 (1)	13.5 (2)	.24 (1)	(.057)	.23 (1)	(.023)
P. crinitus	20 (1)	14.5 (1)	3.3 (3)	2.20 (1)	16 (1)	.42 (1)	(.064)	—	—
P. floridanus	40 (1)	27.3 (1)	2.7 (4)	2.4 (1)	16 (1)	.34 (1)	(.065)	.40 (1)	(.030)
P. oreas	20 (1)	—	6.1 (1)	1.63 (1)	16 (1)	.27 (1)	(.073)	.25 (1)	(.025)
P. mexicanus	all year (1)	56.9 (2)	2.6 (1)	4.40 (1)	19 (1)	.73 (1)	(.073)	.79 (1)	(.027)
P. melanurus	all year (1)	50.5 (1)	2.9 (1)	—	—	—	—	—	—
P. melanocarpus	all year (1)	59 (1)	2.3 (2)	4.5 (1)	22 (1)	.70 (1)	(.071)	.94 (1)	(.023)
Correlation with season		.49 (14)	−.60* (15)	.47 (13)	.24 (12)	.35 (13)	−.12	.43 (12)	.16
Correlation with adult weight			−.56* (14)	.89** (12)	.76** (11)	.92** (12)	.25	.94** (11)	.26

Note: Breeding season length from Appendix I. Life cycle data as follows: *P. maniculatus*: Table 1 plus Vaughan, 1969; Anderson, 1972; Flake, 1974; Jameson, 1953; McKeever, 1964; Merritt and Merritt, 1980; Spencer and Steinhoff, 1968; Linzey, 1970; Svihla, 1932, 1934; Gashwiler, 1979; Drickamer and Bernstein, 1972. *P. leucopus*: Judd et al., 1978; Lackey, 1973; Jackson, 1952; Millar et al., 1979; Millar, 1975, 1978; Anderson, 1972; Drickamer and Vestal, 1973; Svihla, 1932; Brown, 1964; Burt, 1940; Coventry, 1937; Davis, 1956; Long, 1978, 1973; Fleming and Rauscher, 1978; Glazier, 1979; Hansen and Batzli, 1978; May, 1979; Miller and Getz, 1977; Rintamaa, 1974; Rood, 1966; Svendsen, 1964. *P. truei*: Anderson, 1972; McCabe and Blanchard, 1950; Svihla, 1932. *P. eremicus*: Anderson, 1972; Brand and Ryckman, 1968; Davis and Davis, 1947; Drickamer and Vestal, 1973; Glazier, 1979; Rood, 1966; Svihla, 1932. *P. gossypinus*: Pournelle, 1952. *P. boylii*: Robertson, 1975; Anderson, 1972; Brown, 1964; Jameson, 1953. *P. californicus*: Drickamer and Vestal, 1973; McCabe and Blanchard, 1950; Rood, 1966; Svihla, 1932. *P. yucatanicus*: Lackey, 1976. *P. polionotus*: Caldwell and Gentry, 1965; Carmon et al. 1963; Dapson, 1979; Drickamer and Vestal, 1973; Glazier, 1979; Rand and Host, 1942; Rood, 1966. *P. crinetus*: Drickamer and Vestal, 1973; Egoscue, 1964; Rood, 1966. *P. floridanus*: Drickamer and Vestal, 1973; Glazier, 1979; Layne, 1966; Rood, 1966. *P. oreas*: Sheppe, 1963; Svihla, 1936. *P. mexicanus*: Lackey, 1976; Rickart, 1977; Robertson, 1975. *P. melanurus*: Lackey, 1976; Robertson, 1975. *P. melanocarpus*: Rickart, 1977; Robertson, 1975.

conditions may regulate reproduction. Growth of nestlings is known to be influenced by temperature (Hill, 1972) and the initiation of breeding in northern populations coincides with moderate spring temperatures (Sadleir, 1974; Millar and Gyug, 1981), so that low temperatures likely precludes reproduction in some areas. A number of populations show a lull (Burt, 1940; Rintamaa et al., 1976; Drickamer, 1978; Cornish and Bradshaw, 1978) or cessation of breeding (Batzli, 1977) during the warmest summer months, indicating a possible upper temperature limit to breeding. Rainfall and primary productivity may also be important to breeding by *Peromyscus,* as indicated by the initiation of breeding at the beginning of the wet season in southern populations (Robertson, 1975). The variation in the duration of the breeding season within geographic areas (12–40 weeks for *P. m. austerus* and 12–21 weeks for *P. m. maniculatus;* Appendix I) indicates the complexity of factors likely regulating reproduction within pop-

Table 3.—*Age at weaning in relation to length of the breeding season.*

Taxa	Adult weight	Breeding season* (weeks)	Age at weaning (days)	Reference
P. m. bairdii	17.3	39	18.1	King et al., 1963
P. m. gracilis	17.6	19	23.8	King et al., 1963
P. m. maniculatus	18.4	16	21.6	Millar et al., 1979
P. m. borealis (Heart Lake, N.W.T)	21.5	8	21.4	Millar, unpublished
P. m. borealis (Kananaskis, Alta.)	21.5	12	24.9	Millar and Innes, unpublished
P. l. noveboracensis	21.5	24	22.2	Millar et al., 1979
P. m. nebrascensis	19.7	31	18.0	Halfpenny, 1980
P. m. rufinis (mountain)	21.0	20	18.6	Halfpenny, 1980
P. m. rufinis (tundra)	21.0	17	18.9	Halfpenny, 1980

* From Appendix I.

Table 5.—*Time between the first two parturitions by individual females in relation to the length of the breeding season. Sample sizes in parentheses.*

Taxa	Adult weight	Breeding season (weeks)	Days between parturitions	Reference
P. m. borealis	21.5	8	29.3 (3)	Gyug, 1979
P. m. maniculatus	15.4	16	25.8 (19)	Gyug, 1979
P. l. noveboracensis	21.5	24	24.8 (8)	Gyug, 1979
P. m. gambelii	19.0	29	67.3 (6)	McCabe and Blanchard, 1950
P. truei	27.0	27	65.0 (5)	McCabe and Blanchard, 1950
P. californicus	39.0	29	63.7 (4)	McCabe and Blanchard, 1950

ulations. However, whatever the factors controlling reproduction, clear differences are evident among geographic areas.

LIFE CYCLE CHARACTERISTICS

Adult weight, litter size, birth weight, age that eyes open, nestling growth and post-weaning growth for nine subspecies of *P. maniculatus* and 15 species of *Peromyscus* are presented in Tables 1 and 2, respectively. These characteristics show few significant trends in relation to length of the breeding season. Litter size was negatively correlated with length of the breeding season among 15 species of *Peromyscus* ($r = -.60$; Table 2) but not among 9 subspecies of *P. maniculatus* ($r = -.35$; Table 1). Adult

weights, birth weights, age that eyes opened and growth rates did not vary significantly in relation to length of the breeding season in *P. maniculatus* or among species. Age at weaning (Table 3) was also not significantly correlated with length of the breeding season ($r = -.60$; N = 9; $P > .05$) although the correlation coefficient was relatively high.

Other life cycle traits show somewhat clearer trends in relation to length of the breeding season. For example, parity affects on litter size are pronounced in *Peromyscus* (Table 4). Neither first nor maximum litter sizes were related to length of the breeding season ($r = -.48$; N = 11; $P > .05$ and r =

Table 4.—*Litter size among species and subspecies of* Peromyscus. *Largest litters are underlined.*

Taxa	Adult weight	Breeding* season (weeks)	Parity 1	2	3	4	5	6	7	8	9	10	11	Reference
P. m. borealis	21.5	8	4.7	<u>5.9</u>	4.7	5.0	4.5							Millar, unpublished
P. m. gracilis	17.6	19	4.3		<u>5.2</u>		5.2		4.5		4.3		3.7	Drickamer and Vestal, 1973
P. crinitus	14.5	20	2.5		<u>3.0</u>		3.0		3.1					Drickamer and Vestal, 1973
P. l. noveboracensis	21.5	29	4.1		5.2		5.1		<u>5.3</u>		5.0		4.4	Drickamer and Vestal, 1973
P. l. noveboracensis	21.5	29	4.0	4.7	5.1	<u>5.3</u>	5.1	5.3	5.3		5.4			Lackey, 1973
P. californicus	39	40	1.8		1.9		2.0		2.1		2.2		<u>2.3</u>	Drickamer and Vestal, 1973
P. m. bairdii	17.3	39	4.2		<u>5.4</u>		5.2		4.9		4.2		3.8	Drickamer and Vestal, 1973
P. floridanus	27.3	40	2.0		2.3		2.5		2.5		<u>2.7</u>			Drickamer and Vestal, 1973
P. polionotus	13.7	all year	3.0		3.6		<u>4.2</u>		3.7		4.2		4.0	Drickamer and Vestal, 1973
P. l. castaneus	21.9	all year	3.9	4.8	5.2	5.3	6.0	6.0	6.3		<u>6.4</u>			Lackey, 1973
P. eremicus	20.5	all year	2.2	2.5	2.3	2.5	2.8	<u>2.9</u>	2.6	2.6	2.4	2.6	2.2	Davis and Davis, 1947

* From Appendix I.

Table 6.—*Survival from birth to weaning in* P. maniculatus *and* P. leucopus.

Taxa	Study	Breeding* seasons (weeks)	Proportion of young surviving to weaning	Source
P. m. borealis	Heart Lake N.W.T. 1976	8	1.0	Mihok, 1979
P. m. borealis	Heart Lake N.W.T. 1977	8	.39	Mihok, 1979
P. m. borealis	Heart Lake N.W.T. 1978	8	.90	May, 1979
P. m. borealis	Heart Lake N.W.T. 1979	12	.68	Millar and Innes, unpublished
P. m. austerus	Coastal B.C. 1975 forest	23	.32	Sullivan, 1979
P. m. austerus	Coastal B.C. 1976 forest	23	.63	Sullivan, 1979
P. m. austerus	Coastal B.C. 1977 forest	23	.76	Sullivan, 1979
P. m. austerus	Coastal B.C. 1975 burn	23	.78	Sullivan, 1979
P. m. austerus	Coastal B.C. 1976 burn	23	.51	Sullivan, 1979
P. m. austerus	Coastal B.C. 1977 burn	23	.49	Sullivan, 1979
P. m. austerus	Coastal B.C. 1977 clear cut	23	.46	Sullivan, 1979
P. m. austerus	Coastal B.C. 1974	23	.05	Sullivan, 1977
P. m. austerus	Coastal B.C. 1973/1974	23	.71	Fairbairn, 1977
P. m. austerus	Coastal B.C.	23	.36	Fordham, 1971
P. m. austerus	Coastal B.C. 1962	23	.39	Sadleir, 1965
P. m. saturatus	Saturna Island, B.C. 1974	18	.72	Sullivan, 1977
P. m. ?	Samual Island, B.C. 1974	34	.94	Sullivan, 1977
P. m. maniculatus	Pinawa, Manitoba	16	.62	Wille, unpublished
P. m. bairdii	Michigan	33	.56	Howard, 1949
P. l. noveboracensis	Ontario 1975	24	.75	Harland et al., 1979
P. l. noveboracensis	Ontario 1976	24	.60	Harland et al., 1979
P. l. noveboracensis	Illinois 1976	32	.28	Hansen and Batzli, 1978

* Average for the subspecies or geographic location (Appendix I).

−.25; $N = 11$; $P > .05$, respectively), but parity at which the largest litters occurred was positively correlated with length of the breeding season ($r = .72$; $N = 11$; $P < .05$). Individuals in short-season environments have largest litters earlier in their littering sequence than those in long-season environments. Time between litters as indicated by the average time between the first two litters in wild populations was difficult to compare because such data were available from only six populations (Table 5). Days between litters were not significantly correlated with length of the breeding season ($r = .75$; $N = 6$; $P > .05$) but females in three northern populations studied by Gyug (1979) gave birth to second litters 25 to 29 days after their first litters, while females in three temperate populations studied by McCabe and Blanchard (1950) had 64 to 67 days between the first and second litters. These data are insufficient to separate "study affects" from true environmental affects, but the differences between the short-season populations (*P. m. borealis, P. m. maniculatus*) and the long-season populations (*P. m. gambelii*) are considerable.

Although *Peromyscus* show few life cycle differences in relation to length of the breeding season, they do exhibit clear trends in relation to body size among species. Large-bodied species have smaller litters, larger birth weights, and their eyes open later than small-bodied species (Table 2). Absolute growth rates, but not geometric growth rates, are greater in large-bodied species than small-bodied species. None of these characteristics were related to adult size among subspecies of *P. maniculatus* (Table 1), and parity ($r = .57$; $N = 11$; $P > .05$), age at weaning ($r = 13$; $N = 9$) and time between litters ($r = .51$; $N = 6$; $P > .05$) were not significantly correlated with adult size. Taken together, these data indicate that life cycle adaptations are insufficient to compensate for limited opportunities for reproduction. Survival must be greater in short-season environments than in long-season environments.

SURVIVAL

Survival data were readily available for *P. maniculatus* and *P. leucopus,* but not for other species. Nestling survival among all populations of *P. maniculatus* and *P. leucopus* was extremely variable (Table 6), ranging from more than 90% (Mihok, 1979; May, 1979; Sullivan, 1977) to less than 10% (Sullivan, 1977) and averaging .59 ± .05. Even within subspecies or locations, a full range of nestling survival patterns is evident. For example, Mihok (1979) found 100 and 39% nestling survival over two years at Heart Lake, N.W.T. Sullivan (1979)

Table 7.—Survival in P. maniculatus and P. leucopus populations.

Taxa	Year	Breeding season (weeks)	Summer survival Overwintered (2 weeks)	Summer survival Young (2 weeks)	Winter survival (2 weeks)	Reference
P. m. borealis	1976	8	.67	.81	.94	Mihok, 1979
P. m. borealis	1977	8	.77	.74	.95	Mihok, 1979
P. m. borealis	1978	8	.81 (.88 ♀; .76 ♂)	.86 (.79 ♀; .96 ♂)	.94	May, 1979
P. m. borealis	1979	12	.83	.68	.98	Millar and Innes, unpublished
P. m. saturatus	1974	18	.84 (.85 ♀; .84 ♂)	.79 (.98 ♀; .70 ♂)	.83 (.81 ♀; .84 ♂)	Sullivan, 1977
P. m. saturatus					.89 (.88 ♀; .90 ♂)	Sullivan, 1977
P. m. ? (Samual Island)	1974	34	.76 (.82 ♀; .69 ♂)	.58 (.57 ♀; .58 ♂)	.93 (1.0 ♀; .86 ♂)	Sullivan, 1977
P. m. ? (Samual Island)	1975	34			.83 (.87 ♀; .80 ♂)	Sullivan, 1977
P. m. austerus	1974	23	.82 (.88 ♀; .76 ♂)	.75 (.75 ♀; .75 ♂)	.78 (.56 ♀; .89 ♂)	Sullivan, 1977
P. m. austerus	1975	23			.87 (.87 ♀; .87 ♂)	Sullivan, 1977
P. m. austerus	1975 (forest)	23	.92 (.92 ♀; .92 ♂)	.91 (.93 ♀; .89 ♂)	.81 (.74 ♀; .68 ♂)	Sullivan, 1977
P. m. austerus	1976 (forest)	23	.87 (.91 ♀; .83 ♂)	.50	.86 (.85 ♀; .87 ♂)	Sullivan, 1977
P. m. austerus	1977 (forest)	23	.88 (.88 ♀; .88 ♂)	.77 (.67 ♀; .81 ♂)	.81 (.76 ♀; .85 ♂)	Sullivan, 1977
P. m. austerus	1975 (burn)	23	.87 (.89 ♀; .85 ♂)	.86 (.78 ♀; .89 ♂)	.79 (.74 ♀; .81 ♂)	Sullivan, 1977
P. m. austerus	1976 (burn)	23	.78 (.90 ♀; .65 ♂)	.73 (.78 ♀; .62 ♂)	.82 (.84 ♀; .78 ♂)	Sullivan, 1977
P. m. austerus	1977 (burn)	23	.89 (.90 ♀; .89 ♂)	.84 (.87 ♀; .82 ♂)	.83 (.86 ♀; .80 ♂)	Sullivan, 1977
P. m. austerus	1977 (clear cut)	23	.85 (.89 ♀; .82 ♂)	.83 (.79 ♀; .86 ♂)	.84 (.87 ♀; .83 ♂)	Sullivan, 1977
P. m. austerus	1972/1974	23	.72 (♀ only)	—	—	Fairbairn, 1977
P. m. austerus	1962	23		.89	—	Sadleir, 1965
P. m. austerus	1968 grid I	23	—	—	.78	Petticrew and Sadleir, 1974
P. m. austerus	1968 grid II	23	—	—	.84	Petticrew and Sadleir, 1974
P. m. austerus	1968 grid III	23	—	—	.81	Petticrew and Sadleir, 1974
P. m. austerus	1969 grid I	23	—	—	.89	Petticrew and Sadleir, 1974
P. m. austerus	1969 grid II	23	—	—	.82	Petticrew and Sadleir, 1974
P. m. austerus	1969 grid III	23	—	—	.81	Petticrew and Sadleir, 1974
P. m. austerus	1973 grid U	23	.87 (.92 ♀; .77 ♂)	.73 (.83 ♀; .62 ♂)	.91 (.92 ♀; .90 ♂)	Redfield et al., 1977
P. m. austerus	1973 grid F	23	.77 (.82 ♀; .72 ♂)	.52 (.37 ♀; .78 ♂)	.85 (.86 ♀; .84 ♂)	Redfield et al., 1977
P. m. austerus	1974 grid F	23	.80 (.83 ♀; .76 ♂)	.86 (.94 ♀; .73 ♂)	.88 (.92 ♀; .84 ♂)	Redfield et al., 1977
P. m. maniculatus		16	.85	.63	.90	Wille, unpublished
P. m. bairdii		33	.92	.82	.88	Howard, 1949
P. l. noveboracensis		32		.74	—	Hansen and Batzli, 1978
P. l. noveboracensis		32	.85	.72	—	Batzli, 1977
P. l. noveboracensis		24	.69 (.90 ♀; .48 ♂)	.70 (.72 ♀; .69 ♂)	.87	Harland et al., 1979
P. l. noveboracensis		26	.74	.70	—	Lackey, 1973
P. l. noveboracensis		25	.74	.76	—	Miller and Getz, 1977
P. l. noveboracensis		36	—	.80	—	Rintamaa, 1974
P. l. noveboracensis		29	.90	—	.88	Snyder, 1955

found nestling survival of *P. m. austerus* to range from 32 to 78% among years and habitats. These survival patterns were not significantly correlated ($r = -.29$; $N = 22$) with the average length of the breeding season for the location of subspecies among all samples ($r = -.29$; $N = 22$) or among areas ($r = -.12$; $N = 9$).

Survival of post-weaned young during the breeding season was also variable (Table 7). Young females survived better than young males in 7 of 14 cases, indicating no consistent pattern between sexes. Average survival of both sexes combined ranged from <60% per 14 days (Sullivan, 1977, 1979; Redfield et al., 1977) to >90% per 14 days (Sullivan, 1979), and averaged .75 ± .02 per 14 days (Table 7). There was no significant correlation relating juvenile survival and average length of the breeding season among all samples ($r = -.10$; $N = 26$), or among areas ($r = -.007$; $N = 13$).

Adult survival during the breeding season was similar to that of post-weaned young in being variable (range = .67 to .92 per 14 days; Table 7) and not being significantly related to length of the breeding season among samples ($r = .28$; $N = 25$) or among areas ($r = .22$; $N = 12$), but females consistently survived better than males (in 13 of 15 cases) and average survival (.81 ± 0.14) was somewhat better than that of young of the year.

Survival during the non-breeding season ranged from .78 to .98 per 14 days, averaged .86 ± 0.01 per 14 days ($N = 30$), was higher in females than in males in 9 of 16 cases, and was negatively correlated with length of the breeding season among samples ($r = -.49$; $N = 30$; $P < .01$) but not among areas ($r = .61$; $N = 9$). Winter survival was related to length of the breeding season as

$$Y = 76.6 - .37X \qquad (1)$$

among samples where Y = arcsine proportion surviving 14 days and X = length of the breeding season in weeks. In this way, mice with an 8 week breeding season have a survival rate of .92 per 14 days during the non-breeding season while those with a 25 week season survive at a rate of only .85 per 14 days.

Enhanced survival during the non-breeding season permits mice in short seasons to have an annual survival probability that is higher than that of mice with longer breeding seasons. Annual survival of adult animals from the beginning of one breeding season to the next, defined as $[S_b \cdot (B/2)][S_n \cdot (N/2)]$ where S_b = 2 week survival during the breeding season, S_n = 2 week survival during the non-breeding season and B and N is the duration of the breeding and non-breeding seasons in weeks, respectively, varies with length of the breeding season as

$$Y = 18.61 - 0.42X \qquad (2)$$

among samples ($r = .49$; $N = 24$; $P < .05$) where Y = arcsine probability of survival and X = duration of the breeding season in weeks. In this way, an average of 7% of the mice with an 8 week breeding season survive one year while 2% of the mice with a 25 week breeding survive to the next breeding season. Annual survival of young animals, defined as $P\{S_b \cdot [(B - 3)/2]\}[S_n \cdot (N/2)]$ where P = nestling survival, shows a similar pattern. Annual survival of a mouse born at the beginning of the breeding season varies with length of the breeding season as

$$Y = 21.82 - 0.72X \qquad (3)$$

among samples ($r = .80$; $N = 21$; $P < .01$) where Y = arcsine probability of survival and X = length of the breeding season in weeks. In this way, an average of 8% of the mice born into a population with an 8 week breeding season survive to one year of age while only 0.4% of the mice born into a population with a 25 week breeding season survive until the next year.

DISCUSSION

The data presented here indicate that few life cycle characteristics vary significantly in relation to length of the breeding season. Litter size was inversely related to length of the breeding season among species, but this pattern was not evident within *P. maniculatus* or when body size and parity affects were considered. None of the other basic life cycle characteristics (birth weights, growth rates, age at weaning) were significantly related to length of the breeding season. The only factors that might compensate for limited opportunities for reproduction are parity affects, whereby females with limited opportunities for reproduction have their largest litters early in their littering sequence, and frequency of reproduction, whereby post-partum breeding appears characteristic of mice with short breeding seasons. Data presented here is insufficient to prove that post-partum breeding is most common in short-season environments, but May (1979) showed that the proportion of mature females with embryos during the breeding season increases with decreasing length of the breeding season among *Peromyscus* populations. Since time between litters is inversely proportional to the proportion pregnant (assuming random sampling and breeding), the post-partum breeding in short-season environments appears real.

The general pattern that emerges here is that most life cycle characteristics are relatively consistent over wide geographic areas. This view is supported by other studies. Smith and McGinnis (1968) and Millar et al. (1979) reported small differences in litter size of *Peromyscus* with latitude (.07 and .087 offspring per degree latitude, respectively) but no significant latitudinal trends have been found within species (Smith and McGinnis, 1968; Lackey 1978). Studies reporting increasing litter size with elevation (Dunmire, 1960; Spencer and Steinhoff, 1968; Halfpenny, 1980) did not control for parity affects so that much of the apparent increase in litter size at high elevations may be due to age biases in samples (Fleming and Rauscher, 1978; Halfpenny, 1980). Neonate weights of *P. maniculatus* did not vary with latitude (Drickamer and Bernstein, 1972). Growth in the laboratory was better correlated with adult size than latitude (Drickamer and Bernstein, 1972) and growth of wild *Peromyscus* did not vary significantly among three populations with different breeding seasons (Gyug and Millar, 1981).

Taken together, life cycle adaptations are clearly insufficient to compensate for limited opportunities for reproduction. Enhanced survival must be the major factor responsible for the persistence of *Peromyscus* in strongly seasonal environments. This fact has been recognized previously, although the pattern of survival appears different than previously thought. Smith and McGinnis (1968) speculated that juvenile mortality must compensate for high productivity. May (1979) thought that nestling survival might be high in northern populations, but found no significant pattern in relation to length of the breeding season among ten populations. Halfpenny (1980) suggested that both juvenile and adult mortality was enhanced in short-term season environments, although his estimates of mortality were not independent of production. Data presented here indicates that survival is an important factor in the persistence of northern and alpine populations of *Peromyscus*. Survival of nestlings, juveniles and overwintered adults was relatively consistent among populations. However, annual survival of both adults and young (combining all survival categories) was significantly correlated with length of the breeding season. This relationship is explained in part by enhanced winter survival in short-season environments, but length of the season explains much more of the variation in annual survival of young ($r^2 =$ .64) than does winter survival ($r^2 = .24$). The obvious interpretation of these data is that either nestling or juvenile survival during the breeding season must be enhanced in short-season environments, although there is no consistent pattern in the way that this is achieved.

In general, *Peromyscus* populations persist in short-season environments primarily because they commonly breed post-partum and survive better than in long-season environments. The reasons for this life history pattern can only be speculated, but a number of general observations may set the stage for future research.

None of the general life history theories explain the patterns observed here. The theory of r-K selection, originally proposed by MacArthur and Wilson (1967) and Pianka (1970) would predict that climatically harsh or unstable environments should lead to high mortality and favor a compensatory high reproductive rate. This pattern does not apply to *Peromyscus* because those individuals in the harshest environments exhibit low mortality. Even if the sequence is reversed, high survival has not favored K-selected attributes such as small litters, infrequent reproduction, slow development and extended parental care.

Bet-hedging theory (Murphy, 1968; Charnov and Schaffer, 1973; Stearns, 1976) would predict that a high prereproductive/adult survival ratio should favor a high reproductive rate. This concept is difficult to apply to *Peromyscus* because prereproductive time periods are variable; in a northern population the prereproductive period is approximately one year while in a southern population the prereproductive period may be two months for early-born young and many months for late-born young. Despite these complications, some predictions can be formulated. In a northern population (8 week season) early born young have a probability of .08 of breeding the following spring. In a long-season environment, early born young can mature at approximately 8 weeks of age (Clark, 1938; Layne, 1968; Millar et al., 1979). Assuming average nestling survival of .59 over 3 weeks in the nest and .75 per 14 days over the subsequent 5 weeks, a single offspring has a probability of .29 of surviving to maturity. This, along with the fact that relatively more cohorts of young can mature in the summer of their birth in long-season environments indicates that total prereproductive survival is lower in short-season environments than in long-season environments. Since annual adult survival increases with decreasing opportunities for reproduction, the ratio of prereproductive/adult survival must be smallest in short-season environments. Bet-hedging theory would predict northern *Peromyscus* to have K-selected life cycle characteristics. This prediction is not supported by the data presented here. Alternative concepts are needed to explain the life history patterns observed in *Peromyscus*.

Spencer and Steinhoff (1968) speculated that short seasons limit the number of times that a female can reproduce in her lifetime and suggested that short seasons favor large litters despite a high risk to the parent because future opportunities for reproduction are limited. Their specific argument is not supported here because average litter size does not vary in the predicted direction and average parental mortality during the breeding season does not vary in relation to opportunities for reproduction. However, the parity affects observed here lend some support to their general proposal. The large second litters of *P. m. borealis* coincides with the last litter

of the season. Since the energy requirements for reproduction in *Peromyscus* are proportional to litter size (Millar, 1978, 1979; Glazier, 1979), females do appear to devote more resources to their last litters of the season.

Finally, some of the life history patterns may be explained by purely environmental factors, rather than any special evolutionary strategies. Vegetative production in strongly seasonal environments is highly synchronized, resulting in a potential superabundance of food resources during the short growing season. This superabundance of food during the breeding season likely exceeds the abundance of food during any part of a longer, less synchronized growing season. This could permit the high energy demands of overlapped breeding to be more readily met in short-season environments than in long-season environments. Reduced intraspecific competition in populations with low annual productivity and reduced winter activity in northern populations may contribute to enhanced survival. Clearly, detailed studies of survival are needed before the maintenance of *Peromyscus* populations in short-season environments can be fully understood.

ACKNOWLEDGMENTS

This study was supported by the Natural Sciences and Engineering Research Council of Canada.

LITERATURE CITED

ANDERSON, S. 1972. Mammals of Chihuahua. Bull. Amer. Mus. Nat. Hist., 148:149–410.

BARKALOW, F. S., JR. 1962. Latitude related to reproduction in the cottontail rabbit. J. Wildlife Mgmt., 26:32–37.

BATZLI, G. O. 1977. Population dynamics of the white-footed mouse in flood plain and upland forests. Amer. Midland Nat., 97:18–31.

BEER, J. R., and C. F. McLEOD. 1966. Seasonal changes in the prairie deer mouse. Amer. Midland Nat., 76:277–289.

BEER, J. R., C. E. MacLEOD, and L. D. FRENZEL. 1957. Prenatal survival and loss in some cricetid rodents. J. Mamm., 38:392–402.

BLUS, L. J. 1966. Relationship between litter size and latitude in the golden mouse. J. Mamm., 47:546–547.

BOYCE, M. S. 1979. Seasonality and patterns of natural selection for life histories. Amer. Nat., 114:569–583.

BRAND, L. R., and R. E. RYCKMAN. 1968. Laboratory life history of *Peromyscus eremicus* and *Peromyscus interparietalis*. J. Mamm., 49:495–501.

BROWN, L. N. 1964. Reproduction in the brush mouse and white-footed mouse in the central United States. Amer. Midland Nat., 72:226–240.

———. 1966. Reproduction of *Peromyscus maniculatus* in the

Laramie Basin of Wyoming. Amer. Midland Nat., 76:183–189.

BURT, W. H. 1940. Territorial behavior and populations of some small mammals in southern Michigan. Misc. Publ. Mus. Zool., Univ. Michigan, 45:6–58.

CALDWELL, L. D., and J. B. GENTRY. 1965. Natality in *Peromyscus polionotus* populations. Amer. Midland Nat., 74:168–175.

CARMON, J. L., F. B. GOLLEY, and R. G. WILLIAMS. 1963. An analysis of the growth and variability in *Peromyscus polionotus*. Growth, 27:247–254.

CASE, T. J. 1978. On the evolution and adaptive significance of postnatal growth rates in the terrestrial vertebrates. Quart. Rev. Biol., 53:243–282.

CHARNOV, E. L., and W. M. SCHAFFER. 1973. Life history consequences of natural selection: Cole's result revisited. Amer. Nat., 107:791–793.

CLARK, F. H. 1938. Age of sexual maturity in mice of the genus *Peromyscus*. J. Mamm., 19:230–234.

CORNISH, L. M., and W. N. BRADSHAW. 1978. Patterns in twelve reproductive parameters for the white-footed mouse (*Peromyscus leucopus*). J. Mamm., 59:731–739.

COVENTRY, A. F. 1937. Notes on the breeding of some Cricetidae in Ontario. J. Mamm., 18:489–496.

DAPSON, R. W. 1979. Phenologic influences in cohort-specific reproductive strategies in mice (*Peromyscus polionotus*). Ecology, 60:1125–1131.

DAVENPORT, L. B., JR. 1964. Structure of two *Peromyscus polionotus* populations in old-field ecosystems at the AED Savannah River Plant. J. Mamm., 45:95–113.

DAVIS, D. E. 1956. A comparison of natality rates in white-footed mice for four years. J. Mamm., 37:513–516.

DAVIS, D. E., and D. J. DAVIS. 1947. Notes on the reproduction of *Peromyscus eremicus* in a laboratory colony. J. Mamm., 28:181–183.

DICE, L. R., and R. M. BRADLEY. 1942. Growth in the deer mouse, *Peromyscus maniculatus*. J. Mamm., 23:416–427.

DRICKAMER, L. D. 1978. Annual reproduction patterns in populations of two sympatric species of *Peromyscus*. Behavioral Biol., 23:405–408.

DRICKAMER, L. C., and J. BERNSTEIN. 1972. Growth in two subspecies of *Peromyscus maniculatus*. J. Mamm., 53:228–231.

DRICKAMER, L. C., and B. M. VESTAL. 1973. Patterns of reproduction in a laboratory colony of *Peromyscus*. J. Mamm., 54:523–528.

DUNMIRE, W. W. 1960. An altitudinal survey of reproduction in *Peromyscus maniculatus*. Ecology, 41:174–182.

EGOSCUE, H. J. 1964. Ecological notes and laboratory life history of the canyon mouse. J. Mamm., 45:387–396.

ELEFTHERIOU, B. E., and N. X. ZARROW. 1961. A comparison of body weight and thyroid gland activity in two subspecies of *Peromyscus maniculatus* from birth to 70 days of age. Gen. and Comp. Endocrinol., 1:534–540.

FAIRBAIRN, D. J. 1977. Why breed early? A study of reproductive tactics in *Peromyscus*. Canadian J. Zool., 55:862–871.

FLAKE, L. D. 1974. Reproduction of four rodent species in a shortgrass prairie of Colorado. J. Mamm., 55:213–216.

FLEMING, T. H., and R. J. RAUSCHER. 1978. On the evolution of litter size in *Peromyscus leucopus*. Evolution, 32:45–55.

FORDHAM, R. A. 1971. Field populations of deer mice with supplemental food. Ecology, 52:138–146.

FRETWELL, S. D. 1972. Populations in a seasonal environment. Princeton Monogr. Population Biol., 5:1–217.

FULLER, W. A. 1969. Changes in numbers of three species of small rodent near Great Slave Lake, N.W.T., Canada, 1964–1967, and their significance for general population theory. Ann. Zool. Fennici, 6:113–144.

GASHWILER, J. S. 1979. Deer mouse reproduction and its relationship to a tree seed crop. Amer. Midland Nat., 102:95–104.

GLAZIER, D. S. 1979. An energetic and ecological basis for different reproductive rates in five species of *Peromyscus* (Mice). Unpublished Ph.D. thesis, Cornell Univ., Ithaca, New York, 162 pp.

GYUG, L. W. 1979. Reproductive and developmental adjustments to breeding season length in *Peromyscus*. Unpublished M.Sc. thesis, Univ. Western Ontario, London, 84 pp.

GYUG, L. W., and J. S. MILLAR. 1981. Growth of seasonal generations in three natural populations of *Peromyscus*. Canadian J. Zool., 59:510–514.

HALFPENNY, J. C. 1980. Reproductive strategies: intra- and interspecific comparisons within the genus *Peromyscus*. Unpublished Ph.D. thesis, Univ. Colorado, Boulder, 160 pp.

HALL, E. R., and K. R. KELSON. 1959. The mammals of North America. Ronald Press, New York, 1,162 pp.

HANSEN, L., and G. O. BATZLI. 1978. The influence of food availability on the white-footed mouse populations in isolated woodlots. Canadian J. Zool., 56:2530–2541.

HARLAND, R. M., P. J. BLANCHER, and J. S. MILLAR. 1979. Demography of a population of *Peromyscus leucopus*. Canadian J. Zool., 57:323–328.

HILL, R. W. 1972. The amount of maternal care in *Peromyscus leucopus* and its thermal significance for the young. J. Mamm., 53:774–790.

HOWARD, W. E. 1949. Dispersal, amount of inbreeding, and longevity in a local population of prairie deermice on the George Reserve, southern Michican. Contr. Lab. Vert. Biol. Univ. Michigan, 43:1–52.

INNES, D. G. L. 1978. A reexamination of litter size in some North American microtines. Canadian J. Zool., 56:1488–1496.

JACKSON, W. B. 1952. Population of the woodmouse *Peromyscus leucopus* subjected to the application of DDT and Parathion. Ecol. Monogr., 22:259–281.

JAMESON, E. W., JR. 1953. Reproduction of deer mice (*Peromyscus maniculatus* and *P. boylei*) in the Sierra Nevada, California. J. Mamm., 34:44–58.

JUDD, F. W., J. HERRERA, and M. WAGNER. 1978. The relationship between lipid and reproductive cycles of a subtropical population of *Peromyscus leucopus*. J. Mamm., 59:669–676.

KING, J. A. 1958. Maternal behavior and behavioral development in two subspecies of *Peromyscus maniculatus*. J. Mamm., 39:177–190.

KING, J. A., J. C. DESHAIES, and R. WEBSTER. 1963. Age of weaning in two subspecies of deer mice. Science, 139:483–484.

LACKEY, J. A. 1973. Reproduction growth and development in high-latitude and low-latitude populations of *Peromyscus leucopus* (Rodentia). Unpublished Ph.D. thesis, Univ. Michigan, Ann Arbor, 128 pp.

———. 1976. Reproduction, growth, and development in the Yucatan deer mice, *Peromyscus yucatanicus*. J. Mamm., 57:638–655.

———. 1978. Reproduction, growth, and development in high-latitude and low-latitude populations of *Peromyscus leucopus* (Rodentia). J. Mamm., 59:69–83.

LAYNE, J. N. 1966. Postnatal development and growth of *Peromyscus floridanus*. Growth, 30:25–45.

———. 1968. Ontogeny. Pp. 148–253, *in* Biology of *Peromyscus* (Rodentia) (J. A. King, ed.), Spec. Publ. Amer. Soc. Mamm., 2:1–593.

LINZEY, A. V. 1970. Postnatal growth and development of *Peromyscus maniculatus nubiterrae*. J. Mamm., 51:152–155.

LONG, C. A. 1973. Reproduction in the white-footed mouse at the northern limit of its geographic range. Southwestern Nat., 18:11–20.

———. 1978. Populations of small mammals on railroad right-of-way in prairie of central Illinois. Trans. Illinois State Acad. Sci., 61:139–145.

LORD, R. D., JR. 1960. Litter size and latitude in North American mammals. Amer. Midland Nat., 64:488–499.

MACARTHUR, R. H., and E. O. WILSON. 1967. The theory of island biogeography. Princeton Univ. Press, Princeton, New Jersey, 203 pp.

MAY, J. D. 1979. Demographic adjustments to breeding season

length in *Peromyscus*. Unpublished M.Sc. thesis, Univ. Western Ontario, London, 63 pp.

McCABE, T. T., and B. D. BLANCHARD. 1950. Three species of *Peromyscus*. Rood Associates, Santa Barbara, 136 pp.

McKEEVER, S. 1964. Variation in the weight of the adrenal, pituitary and thyroid gland of the white-footed mouse, *Peromyscus maniculatus*. Amer. J. Anat., 14:1–15.

MIHOK, S. 1979. Behavioral structure and demography of subarctic *Clethrionomys gapperi* and *Peromyscus maniculatus*. Canadian J. Zool., 57:1520–1535.

MERRITT, J. F. 1978. *Peromyscus californicus*. Mammalian Species, 85:1–6.

MERRITT, J. F., and J. M. MERRITT. 1980. Population ecology of the deer mouse in the front range of Colorado. Ann. Carnegie Mus., 49:113–130.

MILLAR, J. S. 1975. Tactics of energy partitioning in breeding *Peromyscus*. Canadian J. Zool., 53:967–976.

———. 1978. Energetics of reproduction in *Peromyscus leucopus*: the cost of lactation. Ecology, 59:1055–1061.

———. 1979. Energetics of lactation in *Peromyscus maniculatus*. Canadian J. Zool., 57:1015–1019.

MILLAR, J. S., and L. W. GYUG. 1981. Initiation of breeding by northern *Peromyscus* in relation to temperature. Canadian J. Zool., 59:1094–1098.

MILLAR, J. S., F. B. WILLE, and S. L. IVERSON. 1979. Breeding by *Peromyscus* in seasonal environments. Canadian J. Zool., 57:719–727.

MILLER, D. H., and L. L. GETZ. 1977. Comparison of population dynamics of *Peromyscus* and *Clethrionomys* in New England. J. Mamm., 58:1–16.

MOORE, J. C. 1961. Geographic variation in some reproductive characteristics of diurnal squirrels. Bull. Amer. Mus. Nat. Hist., 122:1–32.

MORRISON, P., R. DIETERICH, and D. PRESTON. 1977. Body growth in sixteen rodent species and subspecies maintained in laboratory colonies. Physiol. Zool., 50:294–310.

MURPHY, G. I. 1968. Pattern in life history and the environment. Amer. Nat., 102:390–404.

PETTICREW, B. G., and R. M. F. S. SADLEIR. 1974. The ecology of the deer mouse *Peromyscus maniculatus* in a coastal coniferous forest. I. Population dynamics. Canadian J. Zool., 52:107–118.

PIANKA, E. R. 1970. On r- and K-selection. Amer. Nat., 104:592–597.

POURNELLE, G. H. 1952. Reproduction and early post-natal development of the cotton mouse, *Peromyscus gossypinus gossypinus*. J. Mamm., 33:1–20.

RAND, A. L., and P. HOST. 1942. Results of the Archbold expeditions. No. 45 Mammal Notes from Highlands Co., Florida. Bull. Amer. Mus. Nat. Hist., 80:1–22.

REDFIELD, J. A., C. J. KREBS, and M. J. TAITT. 1977. Competition between *Peromyscus maniculatus* and *Microtus townsendii* in grasslands of coastal British Columbia. J. Anim. Ecol., 46:607–616.

RICKART, E. A. 1977. Reproduction, growth and development in two species of cloud forest *Peromyscus* from southern Mexico. Occas. Papers Mus. Nat. Hist., Univ. Kansas, 67:1–22.

RINTAMAA, D. L. 1974. Demographic patterns in a population of white-footed mice (*Peromyscus leucopus*). Unpublished Ph.D. thesis, Bowling Green State Univ., 30 pp.

RINTAMAA, D. L., P. A. MAZUR, and S. H. VESSEY. 1976. Reproduction during two annual cycles in a population of *Peromyscus leucopus noveboracensis*. J. Mamm., 57:593–595.

ROBERTSON, P. B. 1975. Reproduction and community structure of rodents over a transect in southern Mexico. Unpublished Ph.D. thesis, Univ. Kansas, Lawrence, 113 pp.

ROFF, D. 1980. Optimizing development time in a seasonal environment: The "ups and downs" of clinal variation. Oecologia, 45:202–208.

ROOD, J. P. 1966. Observations on the reproduction of *Peromyscus* in captivity. Amer. Midland Nat., 76:496–503.

SADLEIR, R. M. F. S. 1965. The relationship between agonistic behavior and population changes in the deer mouse, *Peromyscus maniculatus* (Wagner). J. Animal Ecol., 34:331–352.

———. 1974. The ecology of the deer mouse *Peromyscus maniculatus* in a coastal coniferous forest. II. Reproduction. Canadian J. Zool., 52:119–131.

SCHMIDLY, D. J. 1974a. *Peromyscus attwateri*. Mammalian Species, 48:1–3.

———. 1974b. *Peromyscus pectoralis*. Mammalian Species, 49:1–3.

SCHEFFER, T. H. 1924. Notes on the breeding of *Peromyscus*. J. Mamm., 5:258–260.

SHEPPE, W. 1963. Population structure of the deer mouse, *Peromyscus* in the Pacific Northwest. J. Mamm., 44:180–182.

SMITH, M. H., and J. T. McGINNIS. 1968. Relationships of latitude, altitude and body size to litter size and mean production of offspring in *Peromyscus*. Res. Popul. Ecol., 10:115–126.

SNYDER, D. P. 1955. Survival rates, longevity and population fluctuations in the white-footed mouse, *Peromyscus leucopus*, in southeastern Michigan. Misc. Publ. Mus. Zool., Univ. Michigan, 95:1–33.

SPENCER, A. W., and H. W. STEINHOFF. 1968. An explanation of geographic variation in litter size. J. Mamm., 49:281–286.

STEARNS, S. C. 1976. Life-history tactics: a review of the ideas. Quart. Rev. Biol., 51:3–47.

SULLIVAN, T. P. 1977. Demography and dispersal in island and mainland populations of the deer mouse, *Peromyscus maniculatus*. Ecology, 58:964–978.

———. 1979. Demography of populations of deer mice in coastal forest and clear-cut (logged) habitats. Canadian J. Zool., 57:1636–1648.

SVENDSEN, G. 1964. Comparative reproduction and development in two species of mice in the genus *Peromyscus*. Trans. Kansas Acad. Sci., 67:527–538.

SVIHLA, A. 1932. A comparative life history study of the mice of the genus *Peromyscus*. Misc. Publ. Mus. Zool., Univ. Michigan, 24:1–39.

———. 1934. Development and growth of deermice (*Peromyscus maniculatus artemisiae*). J. Mamm., 15:99–104.

———. 1935. Development and growth of the prairie deer mouse, *Peromyscus maniculatus bairdii*. J. Mamm., 16:109–115.

———. 1936. Development and growth of *Peromyscus maniculatus oreas*. J. Mamm., 17:132–137.

VAUGHAN, T. A. 1969. Reproduction and population densities in a montane small mammal fauna. Mis. Publ. Mus. Nat. Hist., Univ. Kansas, 51:51–74.

WOLF, J. L., and A. V. LINZEY. 1977. *Peromyscus gossypinus*. Mammalian Species, 70:1–5.

Address: Department of Zoology, University of Western Ontario, London, Ontario N6A 5B7, Canada.

APPENDIX I

Seasonal breeding by Peromyscus.

Species	Year or population	Data	Duration (weeks)	Source
P. m. borealis	1965	1st conc. 1 May last preg. 24 June	5	Fuller, 1969
P. m. borealis	1966	1st conc. 29 Apr last preg. 31 July	10	Fuller, 1969
P. m. borealis	1967	1st conc. May last preg. 17 July	7	Fuller, 1969
P. m. borealis	1976	1st litter 27 May last litter 5 August	10	Mihok, 1979
P. m. borealis	1977	1st litter 17 May last litter 17 July	9	Mihok, 1979
P. m. borealis	1978	1st litter 26 May last litter 27 July	9	May, 1979
P. m. borealis	1978	1st preg. 15 May last preg. 10 July	5	Gyug, 1979
P. m. borealis	1979	preg. May through August	12	Miller and Innes, unpublished
P. m. austerus	1974	1st lact. mid-June last lact. early September	12	Sullivan, 1977
P. m. austerus	1973	lact. February through November	40	Fairbairn, 1977
P. m. austerus	1974	lact. March through October	32	Fairbairn, 1977
P. m. austerus	1975 (forest)	spring/summer	23	Sullivan, 1979
P. m. austerus	1975 (burn)	spring/summer	13	Sullivan, 1979
P. m. austerus	1976 (forest)	spring/summer	16	Sullivan, 1979
P. m. austerus	1976 (burn)	spring/summer	16	Sullivan, 1979
P. m. austerus	1977 (forest)	spring/summer	22	Sullivan, 1979
P. m. austerus	1977 (burn)	spring/summer	20	Sullivan, 1979
P. m. austerus	1977 (clearcut)	spring/summer	15	Sullivan, 1979
P. m. austerus	1969	1st conc. 17 March last conc. 20 June	14	Sadleir, 1974
P. m. austerus	1970	1st conc. 27 March last conc. 3 September	23	Sadleir, 1974
P. m. austerus	1973 (grid U)	lact. April through October	27	Redfield et al., 1977
P. m. austerus	1973 (grid F)	lact. February through December	36	Redfield et al., 1977
P. m. austerus	1974 (grid F)	lact. February through November	40	Redfield et al., 1977
P. m. bairdii	1954	preg. March through November	33	Beer and McLeod, 1966
P. m. bairdii	1963	preg. all year	all year	Svendsen, 1964
P. m. bairdii	1940–1942	births late March through early November	33	Howard, 1949
P. m. gracilis	1973–1976	preg. April through August	19	Drickamer, 1978
P. m. nebrascensis	1963	preg./lact. April through August	16	Brown, 1966
P. m. nebrascensis	1977–1979	births 10 April through 14 November	31	Halfpenny, 1980
P. m. sonoriensis	1957 12,000 ft	preg./lact. 4 mo. summer	11	Dunmire, 1960
P. m. sonoriensis	1957 9,800 ft	preg./lact. 5 mo. summer	16	Dunmire, 1960
P. m. saturatus	1974	lact. mid-May through mid-September	18	Sullivan, 1977
P. m. gambelii	1973	preg. all year	all year	Scheffer, 1924
P. m. gambelii	1942	births April through October	29	McCabe and Blanchard, 1950
P. m. maniculatus	1968	1st birth 19 April last birth 15 September	21	Millar et al., 1979

APPENDIX I
Continued.

Species	Year or population	Data	Duration (weeks)	Source
P. m. maniculatus	1970	1st birth 22 April last birth 27 July	14	Millar et al., 1979
P. m. maniculatus	1971	1st birth 2 May last birth 5 September	18	Millar et al., 1979
P. m. maniculatus	1972	1st birth 2 May last birth 27 July	12	Millar et al., 1979
P. m. maniculatus	1973	1st birth 14 April last birth 16 August	18	Millar et al., 1979
P. m. maniculatus	1974	1st birth 30 April last birth 10 August	15	Millar et al., 1979
P. m. rufinus	1977–1979 (mountain)	breed 21 April through 9 September	20	Halfpenny, 1980
P. m. rufinus	1977–1979 (tundra)	breed 1 May through 29 August	17	Halfpenny, 1980
P. m. ? (Samual Island British Columbia)	1974	lact. mid-April through mid December	34	Sullivan, 1977
P. m. ? (California)	1949	preg. April through December	36	Jameson, 1953
P. m. ? (California)	1950	preg. February through August	27	Jameson, 1953
P. m. ? (Colorado)	1969–1970	preg. February through November	40	Flake, 1974
P. m. ? (California)	1958–1961	preg. March through December	41	McKeever, 1964
P. m. ? (Colorado)	1974	breeding April through August (5.5 mo)	19	Merritt and Merritt, 1980
P. m. ? (Colorado)	1975	late April through August (5.5 mo)	16	Merritt and Merritt, 1980
P. m. (Colorado)	1965–1967	preg. May through August	14	Vaughan, 1969
P. l. texanus	1975	preg. all year	all year	Judd et al., 1978
P. l. castaneus	1970–1971	preg. all year	all year	Lackey, 1973
P. l. noveboracensis	1973	lact. February–early June and September through November	13 + 10 = 23	Batzli, 1977
P. l. noveboracensis	1972	1st birth 27 April last birth 1 November	27	Millar et al., 1979
P. l. noveboracensis	1973	1st birth 1 March last birth 7 September	27	Millar et al., 1979
P. l. noveboracensis	1974	1st birth 30 April last birth 2 October	24	Millar et al., 1979
P. l. noveboracensis	1975	1st birth 29 April last birth 17 October	24	Millar et al., 1979
P. l. noveboracensis	1976	1st birth 3 April last birth 21 August	20	Millar et al., 1979
P. l. noveboracensis	1972–1973	1st preg/lact 30 March last preg/lact 19 October	23	Lackey, 1973
P. l. noveboracensis	1969–1970	preg. February through October	36	Cornish and Bradshaw, 1978
P. l. noveboracensis	1935	preg. March through October	32	Burt, 1940
P. l. noveboracensis	1936	preg. March through October	32	Burt, 1940
P. l. noveboracensis	1975 (control)	preg. early April through October	28	Hansen and Batzli, 1978
P. l. noveboracensis	1976 (control)	preg. early March through October	32	Hansen and Batzli, 1978
P. l. noveboracensis	1975	breeding late March through early September	24	Harland et al., 1979
P. l. noveboracensis	1968–1970	preg. March through August	23	Long, 1973

APPENDIX I

Continued.

Species	Year or population	Data	Duration (weeks)	Source
P. l. noveboracensis	1965	preg. March through October	32	Long, 1978
P. l. noveboracensis	1965	preg/lact March through September	24	Miller and Getz, 1977
P. l. noveboracensis	1966	preg/lact April through November	29	Miller and Getz, 1977
P. l. noveboracensis	1973	preg. April through October	27	Rintamaa et al., 1976
P. l. noveboracensis	1974	preg. March through October	32	Rintamaa et al., 1976
P. l. noveboracensis	1963–1964	preg. all year	all year	Svendsen, 1964
P. l. noveboracensis	1961–1962	preg/lact September through May	33	Brown, 1964
P. truei	1942	births early April through mid-October	27	McCabe and Blanchard, 1950
P. eremicus			all year	Hall and Kelson, 1959
P. gossypinus			all year	Wolf and Linzey, 1977
P. boylii (=attwateri)	1961–1962	preg/lact September through June	37	Brown, 1964
P. boylii	1949	preg. April through October	27	Jameson, 1953
P. boylii	1950	preg. April through July	14	Jameson, 1953
		preg. most months	all year	Robertson, 1975
P. californicus	1942	births early April through late October	29	McCabe and Blanchard, 1950
P. californicus			all year	Merritt, 1978
P. yucatanicus			prob. all year	Lackey, 1976
P. polionotus	1955–1961	preg. all months	all year	Caldwell and Gentry, 1965
P. polionotus	1953–1954	preg. all months	all year	Davenport, 1964
P. crinitus		breed late spring/early summer	20(?)	Hall and Kelson, 1959
P. floridanus		preg. June through March	40	Layne, 1966
P. pectoralis			all year	Schmidly, 1974*b*
P. evides		seasonal (end dry season)	?	Robertson, 1975
P. oreas		breeding mid-March through late July	20	Sheppe, 1963
P. mexicanus			all year	Rickart, 1977
P. mexicanus			all year	Robertson, 1975
P. melanurus			all year	Robertson, 1975
P. lepturus		seasonal (end dry season)	?	Robertson, 1975
P. melanocarpus		seasonal (end dry season)	?	Robertson, 1975
P. melanocarpus			all year	Rickart, 1977
P. thomasi		seasonal (end dry season)	?	Robertson, 1975

GROUP NESTING AND ITS ECOLOGICAL AND EVOLUTIONARY SIGNIFICANCE IN OVERWINTERING MICROTINE RODENTS

Dale M. Madison

ABSTRACT

Two different species of rodents, *Microtus pennsylvanicus* and *Microtus pinetorum,* were monitored with radiotelemetry during the breeding and overwintering periods. Personal observations, combined with evidence from the literature, show solitary nesting during the breeding season and communal nesting during the winter to be a widespread pattern among temperate rodents.

Rodents that live communally during the breeding season show few seasonal changes in space use associated with the onset of winter. The costs and benefits of these shifts in space use are discussed in relation to susceptibility to predation and conservation of energy.

INTRODUCTION

One of the most vigorously contested issues in the biological sciences is the question of what factors regulate or limit the density of natural populations of organisms. Microtine rodents have long been used as key organisms in the associated studies, primarily because these small mammals undergo pronounced population fluctuations and occasional population crashes. As evidence of the central importance of microtine rodents in population studies is their repeated appearance as experimental subjects in recent books and symposia (Cohen et al., 1980; Finerty, 1980; Golley et al., 1975; Snyder, 1978; Stoddart, 1979).

The broad spectrum of interest in microtine rodents is not surprising, but what is curious is that so little research effort is being directed at the most vulnerable, and perhaps most important, time in the life of microtine rodents—the over-wintering period. Microtine population crashes often occur during winter, and the intense "battles" of the rodents with the elements, predators, competing species, and even each other, essentially go unrecorded. Investigators typically begin their studies in the late spring with just a handful of survivors. The result of the mid-winter strife (that is, a population decline) is occasionally recorded in terms of reduced numbers of voles trapped, but the natural causes of, and adaptations for, over-winter survival are poorly understood. The importance of understanding what these adaptations are is clear when one realizes that the small number of spring survivors will be the genetic founders of the next, late-season peak in numbers. Essentially, then, how a microtine rodent behaves and what successes it has during the summer may be heavily tailored, if not governed, by adaptations for over-winter survival.

The behavior of over-wintering rodents is perhaps the least understood of the variables that could be measured. For instance, "huddling" or group nesting in the field during the winter has not been studied throughout the winter for any rodent species, yet the general literature suggests that many rodent species commonly "clump" together or become "aggregated" during the winter (see West and Dublin, 1984); and a number of investigators (see below), by studying the energetic benefits of huddling behavior, recognize the potential importance of social grouping as a cold weather adaptation. The social tolerance required of group nesting voles, the apparent importance of group nesting to winter survival, and the variable tolerance levels within species in cycling microtine populations demands integration and resolution. Detailed studies of over-wintering individuals can provide much of the necessary information.

The purpose of the present paper is to define the general modes of spacing behavior that occur for microtine rodents, to present details of the spacing behavior of two species that represent the two major patterns identified for microtine rodents, and to discuss the benefits and costs of group nesting in winter for meadow voles.

DEFINITIONS AND MODES OF SPACING IN MICROTINE RODENTS

Microtine rodents can be roughly categorized as to whether they are social or solitary, based on space use and parental care associations during breeding. If the female is intolerant, or acts independently, of the male after mating, and if the female is the sole parent involved in direct parental care (e.g., nest preparation, protection, grooming, food provisioning), then the species is typically viewed as being "solitary." If the male shows persistence in his attendance of the female long after mating, and if the male assists the female in caring for the young, the species can be said to be "social." In the latter situation in microtine rodents, nest sharing is a useful indicator of paternal care and, hence, of a social species. During the non-breeding season, the species can be considered social if individuals share in nest construction and use. Solitary species would not show cooperative nest building, and individuals would sleep in separate nests.

The social condition above can be further subdivided into communal and noncommunal types (Brown, 1975). The group is communal if its size is fairly constant, if its composition is stable, if members of the group share a common home range and repel non-group members, and if individual recognition (often based on kinship) and stable social relationships characterize the associations within the group. During the breeding season the communal group shows individual associations (often between three or more adults) beyond the obligatory bond between the mother and her offspring. The noncommunal group is best described as an aggregate of individuals who are attracted to a common resource, and who are able to move between such groups with little social resistance. The noncommunal group, therefore, would be more variable in size and composition through time. A "colony" is a noncommunal group made up of pairs of individuals who can join any other colony, space permitting. There is no group bonding or repelling behavior toward non-member conspecifics, nor is there clear evidence of individual recognition beyond the mate and perhaps the nearest neighbors.

When just the social/solitary status of microtine species is considered seasonally, four different patterns are possible. The species can be solitary or social year-round, or the status can change between summer and winter (that is, the breeding and non-breeding periods). Based on direct or suggestive evidence in the literature, only two of the four patterns occur in microtine rodents (Table 1). Microtines are social in winter, but they can be either solitary or social during summer.

SPECIES CASE HISTORIES

The two basic modes of living for microtine rodents are shown for the pine vole (*M. pinetorum*), which is communal year-round, and the meadow vole (*M. pennsylvanicus*), which is solitary during summer, but lives first in communal then in noncommunal groups during winter.

Table 1.—*Seasonal patterns of social and solitary living in microtine rodents. Many categorizations are based on incomplete and anecdotal information, but in all cases some information is available on both winter and summer behavior. See the text for definitions.*

Species	Social behavior		Reference
	Summer	Winter	
C. glareolus	Solitary	Social	Kikkawa, 1964; Mazurkiewicz, 1981
C. rufocanus	Solitary	Social	Kalela, 1957
C. rutilus	Solitary	Social	West, 1977; Whitney, 1976
M. arvalis	Solitary	Social	Chelkowska, 1978; Frank, 1957; Mackin-Rogalski, 1979
M. californicus	Social ?	Social ?	Lidicker, 1980; Stark, 1963
M. montanus	Solitary	Social	Jannett, 1978, 1980; Stark, 1963
M. ochrogaster	Solitary, social	Social	Criddle, 1926; Getz and Carter, 1980; Thomas and Birney, 1979
M. oeconomus	Solitary	Social	Tast, 1966
M. pennsylvanicus	Solitary	Social	Madison, 1980a, 1980b; Webster and Brooks, 1981a, 1981b
M. pinetorum	Social	Social	FitzGerald and Madison, 1981, 1983
M. xanthognathus	Social ?	Social	Wolff, 1980; Wolff and Lidicker, 1981

Microtus pinetorum

The communal groups of pine voles likely defend group territories, as judged by the degree of exclusivity of the groups under crowded living conditions (FitzGerald, 1984; FitzGerald and Madison, 1981, 1983). Fig. 1 shows several communal groups as revealed by trap census and radiotelemetry techniques. The average group is composed of 4.7 voles (2.2 adult males, 1.3 adult females, 0.7 subadults and juveniles). The mating system varies between cooperative polyandry and monogamy. Kinship is high between group members. The home ranges of the voles within each communal group essentially coincide, and group nests are occupied by all members of the group. The size of the group range increases with social complexity.

Microtus pennsylvanicus

Meadow voles occupy non-coinciding home ranges and sleep in separate nests during the breeding season (Madison, 1980a, 1980b; 1984). The breeding females defend territories against other breeding females, but the males commonly overlap among themselves. The sexes overlap considerably, although the nest of the female is rarely, if ever, entered by residential males. During the fall, this dispersed organization of solitary individuals gradually changes to communal groups, which reach their fully developed state by December (Madison et al., 1984). First, related juveniles remain within the mother's home range and sleep together. Then, as the temperatures continue to get colder during the fall, these individuals begin sleeping with their mothers, who continue to breed. One or two breeding males in the area also join these extended family groups. At this point the male(s) no longer move between the "matricenters" of the different groups. By December when breeding has generally ceased, the communal groups consist of juvenile and subadult offspring sleeping with their mothers and an adult male or two. These groups essentially do not overlap with other such groups in the same area. This group structure gradually changes during midwinter as predation reduces the size of each communal group, or destroys groups altogether. When the group size gets too small, individuals move about freely under the snow and join with other voles, who are non-kin, to form noncommunal sleeping groups. Since a certain minimum group size appears to be required for adequate thermal benefit, and since the population continues to decline, the noncommunal groups become fewer and more dispersed as the

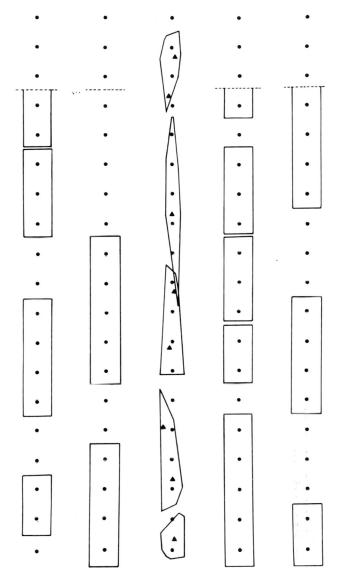

Fig. 1.—Home ranges of groups of pine voles as revealed by radiotelemetry (polygons) and live-trapping (rectangles) in an apple orchard at Modena, New York, during fall 1980. The base of each apple tree (●) and the location of each nest (▲) are indicated. The trees are 5 m apart within rows. The dashed lines show the end of the trapping grid.

winter passes. The trend continues until the spring when the isolated groups disband, and the individual group members disperse and begin their solitary summer existence. Communal nesting between breeding females (sisters?) may occur during cold weather in early spring prior to the spring dispersal (McShea and Madison, 1984). The sequence from solitary living through communal groups to the formation of loose, noncommunal groups is shown in Fig. 2 (see Madison et al., 1984).

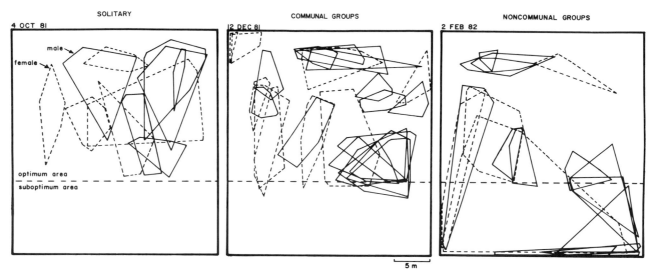

Fig. 2.—Solitary and social groupings of meadow voles radiotracked in field enclosures in Apalachin, New York. Each vole is represented by a polygon enclosing 10 positions collected hourly from 0700 h to 1600 h. The suboptimal area was created by repeated mowing and the collection of cuttings during the summer. Six male and five female adults were present on 4 October; four adults of each sex plus 12 male and five female offspring made up the communal groups; four male and two female adults plus nine male and two female offspring made up the noncommunal groups.

BENEFIT AND EVOLUTION OF GROUP NESTING

Group nesting is a form of social grouping, and the evolution of social groups is thought to occur because of one or more of three basic advantages (Alexander, 1974; Wittenberger, 1981). Each advantage is discussed below with respect to group nesting in *M. pennsylvanicus.* When these explanations are found to be inadequate, a fourth advantage of group formation is proposed—specifically the physical advantages, which include the conservation of heat and moisture.

The first category of benefits of group living is predator avoidance and defense. It involves collective defense, such as mobbing of owls by birds or the defense alliances of baboons against leopards. It also involves the "safety in numbers" hypothesis or the "confusion effect" of fish schools and bird flocks. That meadow voles evolved group nesting in the winter as a defense mechanism against predators seems unlikely. In fact, recent data on weasel predation (Madison et al., 1984) during the winter suggests that group living voles may be more prone to weasel attacks than dispersed voles. Defense alliances seem unlikely because the winter predators of voles (foxes, weasels, raptors) are "competent" in the sense that they are not intimidated by the threat of retaliation or collective action of a group of voles.

Another often cited benefit of group living involves efficiency of food acquisition. This category includes "social carnivory," during which a group of individuals, such as wolves, can bring down a prey that could not be captured by any one group member acting alone. It also involves group foraging with the concomitant attraction of group members toward the individual member that discovers food. In addition, feeding efficiency increases since more time can be spent looking for food in a flock, herd or school than watching out for predators. Considering that visual cues are obscured in the grass runways occupied by meadow voles and that except during fighting the use of vocal signals or alarm calls is nonexistent in meadow voles (Madison, 1980a), food acquisition explanations for group nesting in voles are not very probable.

The third factor given for the evolution of social grouping is habitat scarcity. In this case, some resource in the habitat is very localized, and individuals evolve mechanisms that permit social proximity and aggregation. Such a condition could explain the aggregation of voles during the flooding of low-lying habitat described by Tast (1966), Webster and Brooks (1981b) and West (1977), but this situation does not explain why meadow voles nest in larger groups. Animals can aggregate without having to sleep together, as observed in colonial nesting birds. Besides, group nesting occurs in meadow voles in situations where flooding and habitat limitation do not occur. So what is one left with?

Here I would like to formalize a new category to explain the evolution of social groups, one where voles benefit by each other's presence for reasons independent of predator defense advantages, foraging efficiency and habitat scarcity. I would like to call this category *physical benefits*. Two physical benefits appear to be important, if not mandatory, for winter survival—temperature and moisture conservation.

Perhaps the most important, certainly the most studied, physical benefit of group nesting is the energy conserved during "huddling" (for example Chaplin, 1982; Trune and Slobodchikoff, 1976; see below). In meadow voles, metabolic rates increase with declining environmental temperatures below 25°C (Wiegert, 1961), and the individual metabolic rates are much lower for voles in groups as opposed to voles tested singly (Gebczynski, 1969, 1975; Gebczynska and Gebczynski, 1971; Gorecki, 1968; Hansson and Grodzinski, 1970; Trojan and Wojciechowska, 1968; Wiegert, 1961). The same finding is true for other rodents (Fedyk, 1971; Glaser and Lustick, 1975; Pearson, 1960; Sealander, 1952). Food consumption also decreases per individual in

group living mice (Prychodko, 1958). The huddling of young mice significantly reduces their rate of cooling when the mother leaves the nest (Gebczynski, 1975; Gebczynska and Gebczynski, 1971), but larger nests would also reduce the rate of cooling (MacLean et al., 1974). The benefit of huddling is reduced when mice are tested within the nest and at higher environmental temperatures (Casey, 1981; Gebczynski, 1975; Gebczynska and Gebczynski, 1971). Snow cover would decrease the benefit, as can be inferred from the findings of Cotton and Griffiths (1967) and Sealander (1952); and smaller nests would increase the benefits.

Another benefit of group nesting may be the conservation of moisture. Moisture loss increases at lower temperatures during winter, and this loss combined with the inefficient kidneys noted for voles generally, and for meadow voles in particular, makes water balance an important concern during winter (Church, 1966; Deavers and Hudson, 1979; Getz, 1963, 1968; Heisinger et al., 1973; McManus, 1974; Rhodes and Richmond, 1981). The observation that individual pulmocutaneous water loss is reduced in mice housed in groups (Punzo, 1975) suggests that group nesting may be a behavioral mechanism for conserving moisture. Thus, a high relative humidity in the sleeping compartment may allow recycling of water vapor and an increase in perceived warmth.

COSTS OF GROUP NESTING

Three general disadvantages of social grouping have been recognized by Alexander (1974): increased conspicuousness to prey or predators, increased resource competition, and the greater likelihood of disease and parasite transmission. The *physical costs* of group living should also be added to this list. Since these disadvantages may arise when group nesting occurs in voles, and since group nesting does occur, it is worthwhile to discuss what these disadvantages are for meadow voles and how meadow voles might minimize the negative effects during winter.

Conspicuousness and vulnerability to predators is a likely disadvantage of group nesting in meadow voles. Vole activity under the snow is signalled by the appearance of "breathing holes" in the snow surface. Predators could easily cue in on such signs of abundance (foxes) or even use the holes as entry points into the subnivean space (weasels). Meadow

voles also produce more excretory products in a limited area during group nesting, and these chemical cues could easily be exploited by foxes, weasels and skunks to locate the vicinity of the group nest. Finally, weasels appear to be quite efficient in the mass killing of all the occupants of a given group nest, such that the vulnerability of group nesting voles may be greater than that of solitary nesting voles (Madison et al., 1984; MacLean et al., 1974).

Voles may minimize the above disadvantages by reingesting excretory products produced in or near the nest (Ouellette and Heisinger, 1980) and distributing them widely during periods of activity, by reducing the construction of breathing holes near the nest, and by nesting together only when the benefits are significant (facultative nesting), such as during very cold weather or when little or no snow cover is present. Some of these adjustments appear to exist for meadow voles (Madison, unpublished data).

Resource competition, specifically competition for food, is the most often cited drawback of group nesting or crowded living. Meadow voles in winter live in groups in areas that during the summer were only occupied by single individuals, yet body weights do not decrease appreciably in winter (Madison et al., 1984). One possible reason for these observations is that the benefits of energy conservation by huddling more than offset the disadvantage of having to forage more widely for food. Foraging area increases about four times faster than foraging radius, and thus four voles nesting together would only have to double their average foraging radius in order to have access to the same amount of food as when they were living alone.

The disadvantage of disease or parasite trans-

mission can be partially compensated for by facultative nesting, as described above. Obligatory social nesting, such as by pine voles, occasionally results in the die-off of entire social groups as a result of disease (FitzGerald, personal observations).

The one physical disadvantage of group nesting that is intuitively obvious is the build-up of carbon dioxide in the nest compartment. But this disadvantage may be more apparent than real. Korhonen (1980) released CO_2 into open and closed subnivean tunnel systems, then measured the dissipation rates. He found that the rates were very high and that the measured concentrations of CO_2 were never high enough to have any significant physiological effects on voles. The question of CO_2 build-up is still controversial (MacLean, personal communication).

CONCLUSIONS

Microtine rodents, and probably most small rodent species in temperate and more polar latitudes, are hypothesized to live in social groups during the non-breeding or winter period. Nest sharing is the diagnostic feature of these groups, and energy and moisture conservation are important driving forces behind the evolution of this behavior.

Microtus pennsylvanicus and *Microtus pinetorum* are the most thoroughly studied microtine species in terms of seasonal patterns of space use and be-

havior in the field. They represent the two basic modes of spacing in microtines, specifically the summer solitary and winter social pattern for meadow voles, and the year-round social pattern for pine voles. The former pattern is predicted to be more common among microtine rodents, but the flexibility and facultative nature of microtine spacing behavior is perhaps the most important generality of all.

ACKNOWLEDGMENTS

I thank Randy FitzGerald and Bill McShea for extensive assistance throughout the field studies on pine and meadow voles. The research was supported by the National Science Foundation (DEB-22821) and the U.S. Fish and Wildlife Service (Contract No. 14-16-0009-79-066).

LITERATURE CITED

ALEXANDER, R. D. 1974. The evolution of social behavior. Ann. Rev. Ecol. Syst., 5:325–383.

BROWN, J. L. 1975. The evolution of behavior. W. W. Norton and Co., Inc., New York, 761 pp.

CASEY, T. M. 1981. Nest insulation: Energy savings to brown lemmings using a winter nest. Oecologia, 50:199–204.

CHAPLIN, S. B. 1982. The energetic significance of huddling behavior in common bushtits (*Psaltriparus minimus*). The Auk, 99:424–430.

CHELKOWSKA, H. 1978. Variations in numbers and social factors in a population of field voles. Acta Theriol., 23:213–238.

CHURCH, R. L. 1966. Water exchanges of the California vole, *Microtus californicus*. Physiol. Zool., 39:326–340.

COHEN, M. N., R. S. MALPASS, and H. G. KLEIN (eds.). 1980. Biosocial mechanisms of population regulation. Yale University Press, New Haven, 406 pp.

COTTON, M. J., and D. A. GRIFFITHS. 1967. Observations on temperature conditions in vole nests. J. Zool., London, 153:541–568.

CRIDDLE, S. 1926. The habits of *Microtus minor* in Manitoba. J. Mamm., 7:193–200.

DEAVERS, D. R., and J. W. HUDSON. 1979. Water metabolism and estimated field water budgets in two rodents (*Clethrionomys gapperi* and *Peromyscus leucopus*) and an insectivore (*Blarina brevicauda*) inhabiting the same mesic environment. Physiol. Zool., 52:137–152.

FEDYK, A. 1971. Social thermoregulation in *Apodemus flavicollis* (Melchior, 1834). Acta Theriol., 16:221–229.

FINERTY, J. P. 1980. The population ecology of cycles in small mammals. Yale Univ. Press, New Haven, 234 pp.

FITZGERALD, R. 1984. Temporal dynamics of space use, group composition and social organization in a free-ranging population of pine voles, *Microtus pinetorum*. Unpublished Ph.D. thesis, State Univ. New York at Binghamton, in preparation.

FITZGERALD, R., and D. MADISON. 1981. Spacing, movements and social organization of a free-ranging population of pine voles, *Microtus pinetorum*. Pp. 54–59, *in* Proceedings of the fifth eastern pine and meadow vole symposium (R. E. Byers, ed.), Gettysburg, Pennsylvania, 144 pp.

———. 1983. Temporal dynamics of space use, group composition and social organization in a free-ranging population of pine voles, *Microtus pinetorum*. Behav. Ecol. and Sociobiol., 13:183–187.

FRANK, F. 1957. The causality of microtine cycles in Germany. J. Wildlife Mgmt., 21:113–121.

GEBCZYNSKI, M. 1969. Social regulation of body temperature in the bank vole. Acta Theriol., 14:427–440.

———. 1975. Heat economy and energy cost of growth in the bank vole during the first month of postnatal life. Acta Theriol., 29:379–434.

GEBCZYNSKI, Z., and M. GEBCZYNSKI. 1971. Insulating properties of the nest and social temperature regulation in *Clethrionomys glareolus* (Schreber). Ann. Zool. Fennici, 8:104–108.

GETZ, L. L. 1963. A comparison of the water balance of the prairie and meadow voles. Ecology, 44:202–207.

———. 1968. Influence of water balance and microclimate on the local distribution of the redback vole and white-footed mouse. Ecology, 49:276–286.

GETZ, L. L., and C. S. CARTER. 1980. Social organization in *Microtus ochrogaster* populations. The Biologist, 62:56–69.

GLASER, H., and S. LUSTICK. 1975. Energetics and nesting behavior of the northern white-footed mouse, *Peromyscus leucopus noveboracensis*. Physiol. Zool., 48:105–113.

GOLLEY, F. B., K. PETRUSEWICZ, and L. RYSZKOWSKI (eds.). 1975. Small mammals: their productivity and population dynamics. Cambridge Univ. Press, New York, 451 pp.

GORECKI, A. 1968. Metabolic rate and energy budget in the bank vole. Acta Theoriol., 13:341–365.

HANSSON, L. H., and W. GRODZINSKI. 1970. Bioenergetic parameters of the field vole, *Microtus agrestis* L. Oikos, 21:76–82.

HEISINGER, J. F., T. S. KING, H. W. HALLING, and B. L. FIELDS. 1973. Renal adaptations to macro- and micro-habitats in the family Cricetidae. Comp. Biochem. Physiol., 44A:767–774.

JANNETT, F. J., JR. 1978. The density-dependent formation of extended maternal families of the montane vole, *Microtus montanus nanus*. Behav. Ecol. Sociobiol., 3:245–263.

———. 1980. Social dynamics of the montane vole, *Microtus montanus*, as a paradigm. The Biologist, 62:3–19.

KALELA, O. 1957. Regulation of reproduction rate in subarctic populations of the vole, *Clethrionomys glareolus* and *Apodemus sylvaticus*, in woodland. J. Anim. Ecol., 33:259–299.

KIKKAWA, J. 1964. Movement activity and distribution of the small rodents *Clethrionomys glareolus* and *Apodemus sylvaticus* in woodland. J. Anim. Ecol., 33:259–299.

KORHONEN, K. 1980. Ventilation in the subnivean tunnels of the voles, *Microtus agrestis* and *M. oeconomus*. Ann. Zool. Fennici, 17:1–4.

LIDICKER, W. Z., JR. 1980. The social biology of the California vole. The Biologist, 62:46–55.

MACKIN-ROGALSKI, R. 1979. Elements of the spatial organization of a common vole population. Acta Theriol., 24:171–199.

MACLEAN, S. F., B. M. FITZGERALD, and F. A. PITELKA. 1974. Population cycles in arctic lemmings: winter reproduction and predation by weasels. Arctic and Alpine Res., 6:1–12.

MADISON, D. 1980a. An integrated view of the social biology of *Microtus pennsylvanicus*. The Biologist, 62:20–33.

———. 1980b. Space use and social structure in meadow voles, *Microtus pennsylvanicus*. Behav. Ecol. and Sociobiol., 7:65–71.

MADISON, D. 1984. Activity rhythms and spacing. *In* Biology of New World *Microtus* (R. H. Tamarin, ed.), Spec. Publ., Amer. Soc. Mammal., in press.

MADISON, D., R. FITZGERALD, and W. McSHEA. 1984. Dynamics of social nesting in overwintering meadow voles: possible consequences for population cycling. Behav. Ecol. Sociobiol., in press.

MAZURKIEWICZ, M. 1981. Spatial organization of a bank vole population in years of small or large numbers. Acta Theriol., 26:31–45.

McMANUS, J. J. 1974. Bioenergetics and water requirements of the redback vole, *Clethrionomys gapperi*. J. Mamm., 55:30–44.

McSHEA, W. J., and D. M. MADISON. Submitted. Communal nesting by reproductively active females in a spring population of *Microtus pennsylvanicus*. Canadian J. Zool., 62:344–346.

OUELLETTE, P. E., and J. F. HEISINGER. 1980. Reingestion of feces by *Microtus pennsylvanicus*. J. Mamm., 61:366–368.

PEARSON, O. P. 1960. The oxygen consumption and bioenergetics of harvest mice. Physiol. Zool., 33:152–160.

PRYCHODKO, W. 1958. Effect of aggregation of laboratory mice (*Mus musculus*) on food intake at different temperatures. Ecology, 39:500–503.

PUNZO, F. 1975. The effects of social aggregation on the energy and water metabolism of the cactus mouse, *Peromyscus eremicus eremicus* (Baird, 1958). Saugetierkunde Mitt., 23:143–146.

RHODES, D. H., and M. E. RICHMOND. 1981. Water metabolism in the pine vole, *Pitymys pinetorum*. Pp. 128–130, *in* Proceedings of the fifth eastern pine and meadow vole symposium (R. E. Byers, ed.), Gettysburg, Pennsylvania, 144 pp.

SEALANDER, J. A. 1952. The relationship of nest protection and huddling to survival of *Peromyscus* at low temperature. Ecology, 33:63–71.

SNYDER, D. P. (ed.). 1978. Populations of small animals under natural conditions. The Pymatuning Symposia in Ecology, Univ. Pittsburgh, Pennsylvania, 5:1–237.

STARK, H. E. 1963. Nesting habits of the California vole, *Microtus californicus*, and microclimatic factors affecting its nests. Ecology, 44:663–669.

STODDART, D. M. (ed.). 1979. Ecology of small mammals. John Wiley and Sons, New York, 386 pp.

TAST, J. 1966. The root vole, *Microtus oeconomus* (Pallas), as an inhabitant of seasonally flooded land. Ann. Zool. Fennici, 3:127–171.

THOMAS, J. A., and E. C. BIRNEY. 1979. Parental care and

mating system of the prairie vole, *Microtus ochrogaster*. Behav. Ecol. Sociobiol., 5:171–186.

TROJAN, P., and B. WOJCIECHOWSKA. 1968. The effect of huddling on the resting metabolism rate of the European common vole, *Microtus arvalis* (Pall.). Bull. Acad. Pol. Sci., Cl. II, 16:107–109.

TRUNE, D. R., and C. N. SLOBODCHIKOFF. 1976. Social effects of roosting on the metabolism of the pallid bat (*Antrozous pallidus*). J. Mamm., 57:656–663.

WEBSTER, A. B., and R. J. BROOKS. 1981*a*. Daily movements and short activity periods of free-ranging meadow voles, *Microtus pennsylvanicus*. Oikos, 37:80–87.

———. 1981*b*. Social behavior of *Microtus pennsylvanicus* in relation to seasonal changes in demography. J. Mamm., 62: 738–751.

WEST, S. D. 1977. Mid-winter aggregation in the northern red-backed vole, *Clethrionomys rutilus*. Canadian J. Zool., 55: 1404–1409.

WEST, S. D., and H. T. DUBLIN. 1984. Behavioral strategies of small mammals under winter conditions: solitary or social? This volume.

WHITNEY, P. H. 1976. Population ecology of two sympatric species of subarctic microtine rodents. Ecol. Monogr. 46:85–104.

WIEGERT, R. G. 1961. Respiratory energy loss and activity patterns in the meadow vole, *Microtus pennsylvanicus pennsylvanicus*. Ecology, 42:245–253.

WITTENBERGER, J. F. 1981. Animal social behavior. Duxbury Press, Boston, 722 pp.

WOLFF, J. O. 1980. Social organization of the taiga vole (*Microtus xanthognathus*). The Biologist, 62:34–45.

WOLFF, J. O., and W. Z. LIDICKER, JR. 1981. Communal winter nesting and food sharing in taiga voles. Behav. Ecol. Sociobiol., 9:237–240.

Address: Department of Biological Sciences, State University of New York, Binghamton, New York 13901.

RELATIONSHIP OF SEASONAL CHANGE TO CHANGES IN AGE STRUCTURE AND BODY SIZE IN *MICROTUS PENNSYLVANICUS*

RONALD J. BROOKS and A. BRUCE WEBSTER

ABSTRACT

A population of meadow voles (*Microtus pennsylvanicus*) was studied from July 1977 through April 1978 using both mark-recapture and removal trapping. Mean body weight and body length did not differ among seasons nor between sexes within any season, but range of variation in these measures decreased from summer through autumn and winter. Individuals that bred in summer, lost weight in autumn and disappeared by December. Individuals that did not breed in summer gained weight until early winter, then remained at a stable size until early spring. Mean lens weight increased from a low in summer to a peak in late winter. Body weight and length were highly correlated in all seasons, but a strong correlation of lens weight with these measures in summer and early autumn weakened in late autumn and disappeared in overwintered voles. Rates of dispersal were high in autumn and in winter under deep snow cover. Weight changes in individual voles depend on reproductive activity, age, and seasonal changes in climate and population social structure. The pattern of weight changes of individual voles may be less constrained than has generally been assumed. Both genotypic variation and phenotypic plasticity allow individuals to alter growth patterns in diverse ways to minimize energy expenditure and to maintain fitness.

INTRODUCTION

In microtines, various factors such as age and genotype influence growth rates and body size and proportions (Krebs and Myers, 1974; Mihok and Fuller, 1981), and in northern climates, seasonal variations in environmental parameters such as temperature, photoperiod, food supply, and snow cover also have significant effects on growth (Fuller et al., 1969; Zejda, 1971; Tast, 1972; Iverson and Turner, 1974; Andrzejewski, 1975; Merritt and Merritt, 1978; Petterborg, 1978; Wiger, 1979; Mallory et al., 1981). In spite of this widespread interest in variation in body size measurements, it is still not clear how these influences interact to alter growth rate nor why particular body sizes may be adaptive, although it is generally assumed that body dimensions are determined by natural selection and that they represent optimal solutions to constraints imposed by the environment.

There has been relatively little study of the subnivean activity of microtines (Merritt and Merritt, 1978; Webster and Brooks, 1981), and very few attempts to follow the growth of individuals in the winter environment (Barbehenn, 1955; Brown, 1973; Iverson and Turner, 1974). Therefore, it has been especially difficult to interpret the significance of seasonal changes in body size. Characteristically, in populations of northern microtines, mean body weight undergoes a decline during autumn and winter (Schwarz et al., 1964; Fuller et al., 1969; Hansson, 1971; Hyvarinen and Heikura, 1971; Brown,

1973; Iverson and Turner, 1974; Petterborg, 1978; Wiger, 1979, Wolff and Lidicker, 1980). Several authors have proposed that three processes could account for this decrease: loss of larger individuals from the population; recruitment of small individuals; and loss of weight of individuals (Fuller et al., 1969; Brown, 1973; Iverson and Turner, 1974; Petterborg, 1978). Different authors have tended to treat these processes as three alternatives and, in some cases, have perhaps failed to recognize that a decrease in mean weight can occur without decreases in individual weight. For example, seasonal changes in mean weight or other dimensions may be due to differential growth of individuals from spring—and autumn-born cohorts and therefore, may reflect the different proportions of these cohorts in the population, not individual patterns of growth (Dapson, 1968).

Several studies have suggested that decrease in individual body weight in winter is an adaptive response to limited food supply and demanding climatic conditions, because lower body weights have a reduced energy requirement (Iverson and Turner, 1974; Wunder et al., 1977; Merritt and Merritt, 1978; Wiger, 1979; Wolff and Lidicker, 1980). Such interpretations would predict that these weight declines would be greater in climates with more severely cold winters and this often is the case (Huminski and Krajewski, 1977; Wiger, 1979). On the other hand, it appears that the collared lemming

(*Dicrostonyx groenlandicus*), the most northerly microtine, gains weight when subjected to a short photoperiod (Hasler et al., 1976; Mallory et al., 1981). Furthermore, there are other instances in which individual weight increased or stabilized over winter (Hyvarinen and Heikura, 1971; Zejda, 1971; Tast, 1972; Andrzejewski, 1975).

In a recent study, we used live trapping and radiotelemetry to investigate seasonal changes in behavior and activity of a free-ranging, marked population of meadow voles (*Microtus pennsylvanicus*)

in southern Ontario (Webster and Brooks, 1981). Here, we describe changes in body weight, body length and lens weight in relation to seasonal changes and changes in social behavior and reproductive activity. The three potential sources of weight decrease listed above were examined to determine if they could account for the observed seasonal changes in body weight. A general hypothesis is proposed to account for variation in winter weight dynamics among populations of microtines.

METHODS AND MATERIALS

The study was conducted from June 1977 to April 1978 on a flood plain of the Grand River (43°25′N, 80°20′W) about 75 km southwest of Toronto, Ontario. The study area was covered by a dense mat of vegetation composed mostly of smooth bedstraw (*Galium mollugo*) and bluegrass (*Poa pratensis*). Further description of the study area is in Webster and Brooks (1981).

A 0.2-ha (45 by 45 m) grid was established at the beginning of the study with trap stations placed every 5 m. One Sherman live trap (23.0 by 7.8 by 9.2 cm) was placed under a plywood trap shelter (Iverson and Turner, 1969) at each of the 100 stations. Traps were baited with apples in summer and sunflower seeds in winter and were checked 5 hr after being set. In winter, a paper towel was added to each trap for nest material. Trap mortality was very low.

Traps were set 1 day each week from June to December, 1977 and 2 days per week from January to 15 April 1978 when capture success was lower. All captured voles were marked with individually numbered eartags (National Tag and Band Co., Newport, Kentucky), weighed to the nearest 1.0 g, sexed, and examined to determine reproductive activity and lactation.

In addition, five permanent trap lines were established 200 m from the grid. The lines were 100 m apart and were 100 m long with one Sherman trap and wooden shelter every 5 m. One line was sampled for 2 days every 2 weeks and the trapping sessions were rotated through the five lines in systematic sequence. All animals live trapped in these sessions were sacrificed immediately and stored in 10% buffered formalin for later autopsy. For each

specimen, we recorded sex, body weight, body length, reproductive condition and paired wet eye-lens weight (Myers et al., 1977; Mallory et al., 1981). The cauda epididymes of each male were examined for presence of sperm, and the ovaries and uteri of females were examined for presence of corpora lutea, placental scars and fetuses. Reproductive condition determined from these autopsies was compared to that inferred from the external appearance of the live animal. This allowed us to assess our evaluation of the reproductive state of animals live trapped on the grid.

The 18 biweekly samples from the offgrid lines were combined into five periods, each approximately 7 weeks (Table 1). These periods were designated: period 1 (7 July–31 August); period 2 (1 September–19 October); period 3 (20 October–15 December); period 4 (5 January–20 February); period 5 (21 February–15 April). These periods represented summer, early autumn, late autumn, winter, and late winter–early spring respectively. Reproductive activity ceased by the end of period 2 and did not resume until very late in period 5. Snow cover was continuous from the beginning of period 4 until the last 2 weeks of period 5. During most of this time, snow cover exceeded 0.5 m.

Body weight, body length and lens weight were compared among the five periods using a one-way ANOVA and ANCOVA. As variances were not equal among all comparisons, differences were also compared using "t" tests for unequal variance. Level of significance was set at $P < 0.01$ in these comparisons.

RESULTS

OFFGRID LINES

Between 7 July 1977 and 15 April 1978, 453 voles were trapped on the offgrid lines. Body weight, body length and lens weight were compared among voles trapped in each period. There were no significant differences between sexes within any period in any of these three measures, so data for the two sexes were pooled.

Among the five periods, there were no significant differences in mean body weight or body length (Table 1). However, mean lens weight increased significantly as the seasons progressed, being least in

periods 1 and 2, and greatest in periods 4 and 5 (Table 1).

Changes in the distribution of all three measurements indicated that as the seasons passed from summer through autumn and winter, the range of variation of these measures declined (Table 1; Fig. 1a–f, 2). In summer and early autumn, the population was composed of voles of a wide range of body sizes and lens weights. By late autumn, both large and small animals had disappeared and this trend continued until the winter population was composed entirely of voles weighing close to the

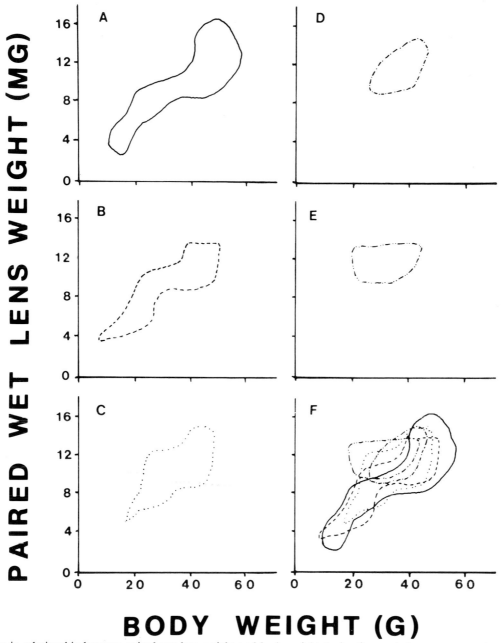

Fig. 1.—Changes in relationship between paired wet lens weight and body weight during five periods. a–e represent periods 1–5 (see text) respectively. f is a compilation of a–e. Each polygon encloses all individuals in the sample. Sample sizes are given in Table 1.

Table 1.—*Changes in means of body weight, body length and paired lens weight in* Microtus pennsylvanicus *trapped during five periods from July 1977 to April 1978.*

	Trapping period				
Measurement	1 7 July–31 Aug. (N = 152)	2 1 Sept.–19 Oct. (N = 57)	3 20 Oct.–15 Dec. (N = 158)	4 5 Jan.–20 Feb. (N = 41)	5 21 Feb.–15 Apr. (N = 45)
Body weight (g)	32.6 ± 11.8†	31.3 ± 10.0	32.6 ± 7.9	34.2 ± 5.2	33.3 ± 6.4
Body length (mm)	103.7 ± 14.0	102.9 ± 12.7	103.3 ± 8.3	106.0 ± 6.2	107.8 ± 5.4
Lens weight (mg)	9.46 ± 0.09*	9.60 ± 0.07**	10.30 ± 0.06††	11.54 ± 0.05	12.24 ± 0.03

† Mean ± SD.
†† N = 124 for lens in period 3.
* Period 1 significantly different from period 3 ($P < 0.01$) and periods 4 and 5 ($P < 0.001$).
** Periods 2 and 3 significantly different from 4 and 5 ($P < 0.001$).

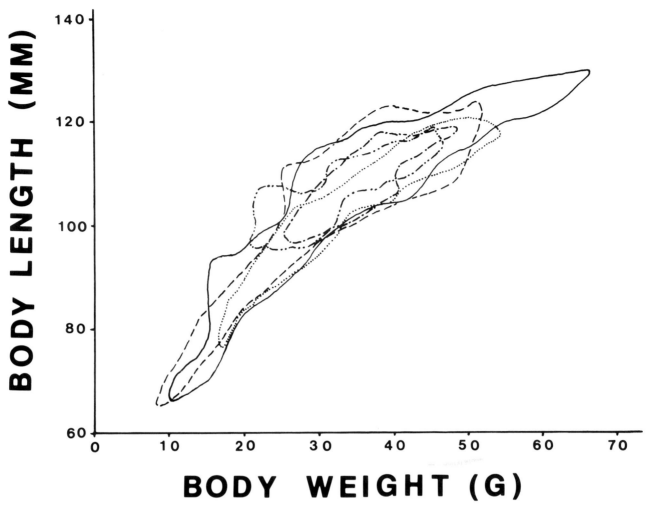

Fig. 2.—Changes in relationship between body length and body weight during periods 1–5. Patterns of lines for each period are the same as those in Fig. 1.

mean of the original weight distribution. In contrast, the range of values for lens weights decreased primarily by the loss of smaller values in the late autumn and winter samples. In winter, the lens weights of most voles fell in the upper range of the original distribution (Fig. 1d, e, f).

Body weight and body length remained strongly correlated with each other throughout the study (Table 2, Fig. 2). Both measures also showed a strong relationship with lens weight in summer and early autumn, but this relationship weakened in the winter sample and in late winter lens weight did not show a significant correlation with either body weight or body length (Table 2).

Table 2.—*Correlations between body weight, body length and paired lens weight in* Microtus pennsylvanicus *trapped during five periods from July 1977 to April 1978.*

| Comparisons | Trapping periods | | | | |
	1 (N = 152)	2 (N = 57)	3 (N = 158)	4 (N = 41)	5 (N = 45)
Body weight versus body length	0.93	0.89	0.92	0.85	0.87
Body weight versus lens weight	0.87	0.78	0.69*	0.51	0.27
Body length versus lens weight	0.87	0.82	0.72*	0.54	0.20

* N = 124.

Table 3.—*Changes in body weights of nonreproductive (<30 g) voles that were first trapped on the mark-recapture grid during period 1.*

Sex	During period 1		Between periods 1 and 2		Between periods 1 and 3		Between periods 1 and 4–5	
	Gain	Lose	Gain	Lose	Gain	Lose	Gain	Lose
Males (N)	28	1	15	1	16	0	14	0
(M)*	+7.24 g		+3.25 g		+7.30 g		+7.90 g	
Females (N)	19	0	6	10	6	3	5	4
(M)	+10.5 g		−1.50 g		+1.78 g		+1.51 g	
Total (N)	47	1	21	11	22	3	19	4
(M)	+8.54 g		+0.88 g		+5.31 g		+5.40 g	

Header: "Number of voles changing weight over time"

* Mean change in weight of individuals.

MARK-RECAPTURE GRID

There were approximately 100 voles of trappable size on the grid when the study commenced in July 1977. Numbers declined to 60 voles by October, but increased due to immigration to over 100 in December. From January to April 1978, the population declined to about 20 animals (Webster and Brooks, 1981). There were no significant differences in body weight between the grid and offgrid samples within any of the five periods ($P > 0.05$ in all cases).

On the grid, 57 voles (35 males : 22 females) that were nonreproductive (weight $\leq$ 30 g) were trapped during period 1 and were recaptured at least once. Forty-eight were recaptured during period 1 and 47 (98%) gained weight between their first and last capture during period 1 (Table 3). Similarly, when these animals were recaptured in later time periods, each individual's last capture weight in period 1 was compared to its last capture weight in each subsequent period. Among males, 45 of 46 (98%) gained weight in such comparisons, whereas females gained weight in only 17 of 34 (50%) comparisons (Table 3).

When weights of these same voles were compared between the last captures in successive periods, 22 of 23 (96%) gained weight between periods 2 and 3, but only 9 of 20 (45%) gained between subsequent periods (Table 4).

Overall, 84 voles (44 males : 40 females) were first captured after the end of period 1 and were recaptured at least once. All of these voles were nonreproductive and all but four of them weighed less than 35 g. As in the previous comparisons, all these voles gained weight between periods 2 and 3 (22 of 22), but only 22 of 45 (49%) comparisons showed an increase in weight between subsequent periods (Table 5).

Table 4.—*Changes in body weights between last captures in successive periods of voles that were nonreproductive (<30 g) when first captured in period 1.*

Sex	Between periods 2 and 3		Between periods 3 and 4		Between periods 4 and 5	
	Gain	Lose	Gain	Lose	Gain	Lose
Males (N)	15	0	5	5	1	1
(M)	+5.41 g		+0.30 g		+1.50 g	
Females (N)	7	1	2	5	1	0
(M)	+3.25 g		−1.43 g		+3.00 g	
Totals (N)	22	1	7	10	2	1
(M)	+4.43 g		−0.41 g		+2.00 g	

Header: "Number of voles changing weight between periods"

Voles that were heavier than 30 g on first capture in period 1 were all reproductively active. Of 35 of these voles recaptured in period 1, 17 (49%) gained weight (mean = −0.65 g, N = 35). Males gained (mean = +0.52 g, N = 19) and females lost weight (mean = −2.00 g, N = 16). Of 25 of these voles recaptured in period 2, only 3 (12%) gained weight (mean = −3.70 g, N = 25). In 15 recapture comparisons between periods 2 and 3, 11 (73%) gained weight (mean = +0.90 g, N = 15). Most weight losses in voles that were more than 30 g occurred shortly before the animals disappeared from the population. None of the voles that were reproductive in period 1 survived beyond period 3.

Of voles that were nonreproductive in period 1, males were significantly less likely to become reproductively active in later periods than were females ($\chi^2 = 7.62$; $P < 0.01$).

The major influx of new voles to the mark-recapture grid occurred between 5 January and 20 February (period 4). During this period, 46 new voles were trapped (24 males : 22 females) and many residents disappeared. Only four new voles, all males, were captured between 21 February and 15 April.

Table 5.—*Changes in body weights between last captures in successive periods of nonreproductive voles that were first captured after period 1.**

Sex	Between periods 2 and 3		Between periods 3 and 4		Between periods 4 and 5	
	Gain	Lose	Gain	Lose	Gain	Lose
Males (N)	9	0	7	2	7	8
(M)	+8.00 g		+1.67 g		+0.33 g	
Females (N)	13	0	4	6	4	7
(M)	+6.23 g		+0.50 g		−0.42 g	
Totals (N)	22	0	11	8	11	15
(M)	+6.90 g		+1.05 g		−0.01 g	

Header: "Number of voles changing weight between periods"

* All but four of these voles were <35 g at first capture.

DISCUSSION

In this study, lens weights were strongly correlated with body weights and lengths during summer and autumn, but these relationships disappeared in winter because lens weights continued to increase at a faster rate than did the other measures. Lens weight appears to increase curvilinearly throughout the normal lifespan of the individual microtine and recent studies have claimed that this increase is independent of the effects of photoperiod, seasonal variations in body growth, sex, litter size and other variables (Spitz, 1974; Gourley and Jannett, 1975; Millar and Iverson, 1976; Huminski and Krajewski, 1977; Hagen et al., 1980; Mallory et al., 1981). Therefore, many people have concluded that lens weight is one of the most reliable measures of age in small rodents (Gourley and Jannett, 1975; Myers et al., 1977; Hagen et al., 1980; Thomas and Bellis, 1980; Mallory et al., 1981). However, most of these investigations were conducted with laboratory populations that experienced uninterrupted growth and unlimited food availability, and there are few data from known-age animals in the field to provide a direct test of whether lens weights grow at similar rates in natural and laboratory populations. There is evidence that lens growth in wild populations is influenced strongly by external perturbations (Schwarz et al., 1964; Myers and Gilbert, 1968; Ostbye and Semb-Johanssen, 1970; Gourley and Jannett, 1975; Millar and Iverson, 1976). In particular, seasonal variations that so clearly affect rates of increase in body weight and body length may also affect lens growth.

In our study, breeding ceased by the end of period 2. Therefore, the mean age of the population should have increased after that time. However, this conclusion depends on an assumption that mortality is independent of age, and in fact, mortality varied with age since larger animals that bred in summer, and were older than most of the population, had a higher rate of mortality in autumn than did voles born in late summer and early autumn (Webster and Brooks, 1981). Despite this, mean lens weight still increased significantly over autumn and winter, probably because lenses continued to grow and because, after period 2, the loss of the older, summer, breeding adults had little effect since these animals represented only a small fraction of the total population.

We can get a reasonable, direct comparison of natural growth rates with laboratory rates by examining the change in mean values of lens weight between periods 3, 4 and 5, when there was no breeding, no significant change in mean body weight and length, and relatively small variation in all three measures. In a laboratory study of growth in known-age *M. pennsylvanicus* from southern Ontario (Brooks, in preparation), paired wet lens weight increased at a rate of 1.0 mg per 25 days in voles with lenses weighing between 10.0 and 12.0 mg. To obtain a rough estimate of the rate of increase in lens weight in our population, we assumed that the mean lens weight in each period occurred at the midpoint of the period. From Table 1, mean lens weight increased 1.9 mg over the 128 days between the midpoints of periods 3 and 5, an increase of 1.0 mg per 67 days. This indicates that in winter the lens grew only 37% as fast in the field as in a laboratory population with unlimited food and in a summer (14: 10 LD) photoperiod. Unfortunately, this comparison could not be made with data from Thomas and Bellis (1980) as they used dry lens weight.

These calculations are only rough approximations, but they indicate there could be a significant discrepancy in the growth of the lens between laboratory and field samples. This discrepancy could lead to an underestimate of the real age of voles from the field. In fact, a comparison of animals from period 5 with predictive curves from the laboratory study indicated that voles in period 5 were, on average, born in early November. This could not be correct as no voles were breeding at that time (Webster and Brooks, 1981). Therefore, it appears that the lens weight in the field samples underestimated age by at least 2 months, and for lens weight to provide more precise estimates more detailed field studies of growth, similar to Zejda (1971) are required. Comparisons of body weight and body length with laboratory data, indicated they were even less accurate as estimators of age, with body weight grossly underestimating age and body length overestimating it. Similar results were presented by Brown (1973).

In our study, the distribution of body size underwent definite change, becoming narrower in late autumn and winter. Several authors have reported this reduced variation in body size in overwintering populations, and some have suggested that this shows there is an optimal body size for winter survival (Hansson, 1971; Iverson and Turner, 1974; Merritt and Merritt, 1978).

In our study, mean weight and body length did not change significantly over the course of the seasons, but many authors have reported that mean and/or individual weights declined in autumn and winter (Fuller, 1969; Fuller et al., 1969; Hansson, 1971; Brown, 1973; Iverson and Turner, 1974; Fuller, 1977; Merritt and Merritt, 1978; Petterborg, 1978). As noted earlier, three processes were proposed as the basis for these declines. They were: (a) individuals lose weight; (b) large animals die or emigrate; (c) smaller animals are recruited to the population. Let us briefly consider the three possibilities in relation to our data. All nonreproductive animals, or animals less than 35 g when first captured, gained weight in summer and autumn. After that, weight changes were slight and in no consistent direction. At the same time, most animals that were reproductive in summer and heavier than 30 g at first capture lost weight in autumn. Most of this weight loss occurred just prior to disappearance of the animals. Only 5 of these animals dropped below 40 g and all did so just before disappearing. Conversely, only four of the animals recruited after period 1 were heavier than 35 g. Therefore it is likely that large animals that disappeared were dying and not dispersing to other areas. In contrast, there was abundant evidence that animals from the nonreproductive group were moving about considerably. Of the 20 residents on the grid in April, 1978, only four were present at the end of October, and 45 new animals appeared between 5 January and 15 February. In addition, we often observed individuals moving among winter nests (Webster and Brooks, 1981). Although few of our immigrant animals weighed over 35 g, most of them were close to the mean body weight. Therefore, in our population, mean body size remained similar from summer through to early spring because of the compensatory effects of the largest animals disappearing, small animals growing and immigrants tending to be close to mean body size when they entered the population.

In their study on *M. pennsylvanicus,* Iverson and Turner (1974) indicated that mean weight declined by 30–40% in winter, but found that this decrease was due primarily to loss of weight of individuals and to small size of immigrants, and that heavier animals entering the winter population actually survived better than smaller ones. These and similar findings on meadow voles (Brown, 1973; Petterborg, 1978) and other microtines provide a sharp contrast to our results.

However, there is also considerable evidence that some microtines gain weight in autumn or winter. The apparent causes of these increases are diverse. In some species, winter weight was maintained or increased when the natural food supply was abundant or was supplemented artificially (Tast, 1972; Andrzejewski, 1975; Cole and Batzli, 1978; Ferns, 1979). In *Microtus montanus,* chemical cues in new plant growth appeared to trigger reproductive activity and led to an increase in body weight in late winter or early spring (Negus and Berger, 1977). However, several studies have reported that winter increases in body size occur when food supply should be most depleted (Krebs, 1964; Iverson and Turner, 1974; Petterborg, 1978) and others have noted that food appeared abundant when weight declines occurred (Tast, 1972; Andrzejewski, 1975; Merritt and Merritt, 1978; Petterborg, 1978).

Individual increases in weight in autumn and early winter have also been related to season of birth, with voles born in autumn gaining or maintaining weight in autumn and winter, and those born in spring or early summer losing weight in autumn (Zejda, 1971; Brown, 1973; Petterborg, 1978) although Bujalska and Gliwicz (1968) found all cohorts of *Clethrionomys glareolus* followed similar growth curves and maintained weight in winter.

Initiation of reproductive activity seems to be the factor associated most consistently with increases in body size. In late winter, microtines, especially males, rapidly gain weight coincident with the development of the reproductive organs (Fuller, 1969; Fuller et al., 1969; Zejda, 1971; Huminski and Krajewski, 1977; Negus and Berger, 1977; Merritt and Merritt, 1978; Wiger, 1979). In our study, males in period 5 were larger than females and larger than males or females in periods 3 and 4, but the differences were not significant. Similarly, increases of body weight in winter often have been associated with winter breeding (Frank, 1964; Tast, 1972; Iverson and Turner, 1974; Steele, 1977). Zejda (1971) has stated that sexual activity and not age is the most important determinant of growth and body size. Collared lemmings (*Dicrostonyx groenlandicus*) gain weight more rapidly and grow larger in a short photoperiod than in a long photoperiod (Hasler et al., 1976; Mallory et al., 1981), and there is some evidence that this species regularly breeds in winter (Krebs, 1964; Fuller et al., 1975) and may grow larger in winter than in summer (Fuller et al., 1975).

Nevertheless, winter breeding is unusual in northern microtines, and when it occurs both sexes should show reduced social tolerance (Webster and Brooks, 1981). This reduced tolerance will prevent group huddling, a behavior observed in nonbreeding winter microtines (Wolff and Lidicker, 1980; Webster and Brooks, 1981) that reduces energy expenditure of individuals (Gorecki, 1968; Gebczynska and Gebczynski, 1971) for thermoregulation. Presumably, this would allow each vole to maintain a smaller body size because the loss of energy from the increased surface area to body mass ratio (Riesenfeld, 1980) would be compensated by the gains from huddling. However, Wunder et al. (1977) pointed out that if the voles were in an insulated nest, and if they increased metabolic turnover rate, then energy costs would be the same for winter and summer, and there would be no advantage to smaller size. Perhaps, the advantage in this case comes from several individuals being confined to a smaller area with limited food. These problems can not be solved without more information on winter behavior, particularly the amount of time spent away from groups and from nests. If reproductive activity occurs, it is costly because it requires the animal to forego the energetic economies of huddling and small body size. Presumably, the increased size concomittant with reproduction provides advantages in reproductive competition and energy storage for lactation (Bekolay, 1979).

This overall view gains support from observations on the pine vole (*M. pinetorum*). This species breeds in winter, but breeding individuals are not heavier than nonbreeders (Cengel et al., 1978). However, pine voles are also unusual in that family units nest communally while breeding (FitzGerald and Madison, 1981), and therefore there would be reduced advantages in larger body size.

Wiger (1979) predicted that if adults lose weight in winter to reduce energy consumption, then winter weights for a species should be less at higher latitudes and nonbreeders should survive better in winter. There is some evidence to support this (Huminski and Krajewski, 1977) and this may be one explanation for the differences between our results and those of Brown (1973) and Iverson and Turner (1974) that were derived from populations of *M.*

pennsylvanicus at higher latitudes than our study area.

These considerations raise another interesting problem. Why does *Dicrostonyx,* an arctic microtine, gain weight and breed in winter, when the conventional view of winter adaptation in microtines suggests that this species should do exactly the opposite? Winter weight dynamics of microtine rodents is not a one-dimensional problem founded on efficient thermoregulation. There are many factors affecting body size (Boonstra and Krebs, 1979) and several solutions to these problems. These solutions need to be examined in terms of energy efficiency in the context of reproductive and somatic costs. For example, thermoregulation may involve several mechanisms including digestive efficiency, metabolic turnover rate, social behavior and insulation.

We hypothesize that there are two basic breeding strategies arranged as two extremes of a continuum. In one case, exemplified by the collared lemming, a large proportion of breeding effort occurs in winter. This has advantages in regions with prolonged, stable subnivean conditions. Larger body size can have both energetic and direct reproductive advantages, and group huddling can have disadvantages in terms of increased competition for food and increased, individual susceptibility to predation (D. M. Madison, personal communication). On the other hand, many microtines are adapted to summer breeding and not only do not breed in winter, but fail to survive winter if they breed in summer (Barbehenn, 1955; Tast, 1972; Petterborg, 1978; Wolff and Lidicker, 1980). These species forego reproduction and growth in winter and minimize the energetic costs of maintenance by reducing total body size and increasing metabolic turnover rate. Between these extremes, are combinations of these strategies, and probably most species have a flexible strategy that allows them to breed or not breed depending on the current winter's severity. At present, a synthesis of the vast literature on the relation between population dynamics and the mechanisms of increasing the efficiency of energy expenditure is sorely needed if we are to make progress in answering these questions.

ACKNOWLEDGMENTS

We thank D. Galbraith, J. Malcolm, L. Schwarzkopf, and G. Stephenson for their help during this study. E. G. Nancekivell provided invaluable field and technical assistance. Funding was provided by N.S.E.R.C., grant no. A5990 to R.J.B.

LITERATURE CITED

ANDRZEJEWSKI, R. 1975. Supplementary food and the winter dynamics of bank vole populations. Acta Theriol., 20:23–40.

BARBEHENN, K. R. 1955. A field study of growth in *Microtus pennsylvanicus*. J. Mamm., 36:533–543.

BEKOLAY, R. P. 1979. Effect of various levels of sodium and potassium on growth, reproduction and adrenal morphology in the meadow vole, *Microtus pennsylvanicus*. Unpublished M.Sc. dissert., Univ. Guelph, Guelph, Ontario, 117 pp.

BOONSTRA, R., and C. J. KREBS. 1979. Viability of large- and small-sized adults in fluctuating vole populations. Ecology, 60:567–573.

BROWN, E. B., III. 1973. Changes in patterns of seasonal growth of *Microtus pennsylvanicus*. Ecology, 54:1103–1110.

BUJALSKA, G., and J. GLIWICZ. 1968. Productivity investigation of an island population of *Clethrionomys glareolus* (Schreber, 1780). III. Individual growth curve. Acta Theriol., 13:427–433.

CENGEL, D. J., J. E. ESTEP, and R. L. KIRKPATRICK. 1978. Pine vole reproduction in relation to food habits and body fat. J. Wildlife Mgmt., 42:822–833.

COLE, F. R., and G. O. BATZLI. 1978. Influence of supplemental feeding on a vole population. J. Mamm., 59:809–819.

DAPSON, R. W. 1968. Growth patterns in a post-juvenile population of short-tailed shrews (*Blarina brevicauda*). Amer. Midland Nat., 79:118–129.

FERNS, P. N. 1979. Growth, reproduction and residency in a declining population of *Microtus agrestis*. J. Anim. Ecol., 48:739–758.

FITZGERALD, R. W., and D. M. MADISON. 1981. Spacing, movements and social organization of a free-ranging population of pine voles *Microtus pinetorum*. Pp. 54–59, *in* Proceedings of fifth eastern pine and meadow vole symposium (R. E. Byers, ed.), Gettysburg, Pennsylvania, 144 pp.

FRANK, F. 1964. Die Feldmaus, *Microtus arvalis* (Pallas) im Nordwestdeutschen Rekordwinter 1962/63. Zeit. Saügetierk., 29:146–152.

FULLER, W. A. 1969. Changes in numbers of three species of small rodent near Great Slave Lake, N.W.T. Canada, 1964–1967, and their significance for general population theory. Ann. Zool. Fennici, 6:113–144.

———. 1977. Demography of a subarctic population of *Clethrionomys gapperi*: size and growth. Canadian J. Zool., 55:415–425.

FULLER, W. A., A. M. MARTELL, R. F. C. SMITH, and S. W. SPELLER. 1975. High-arctic lemmings, *Dicrostonyx groenlandicus*. II. Demography. Canadian J. Zool., 53:867–878.

FULLER, W. A., L. L. STEBBINS, and G. R. DYKE. 1969. Overwintering of small mammals near Great Slave Lake Northern Canada. Arctic, 22:34–55.

GEBCZYNSKA, Z., and M. GEBCZYNSKI. 1971. Insulating properties of the nest and social temperature regulation in *Clethrionomys glareolus* (Schreber). Ann. Zool. Fennici, 8:104–108.

GORECKI, A. 1968. Metabolic rate and energy budget in the bank vole. Acta Theriol., 13:341–365.

GOURLEY, R. S., and F. J. JANNETT. 1975. Pine and montane vole age estimates from eye lens weights. J. Wildlife Mgmt., 39:550–556.

HAGEN, A., N. C. STENSETH, E. OSTBYE, and H.-J. SKAR. 1980. The eye lens as an age indicator in the root vole. Acta Theriol., 25:39–50.

HANSSON, L. 1971. Habitat, food and population dynamics of the field vole *Microtus agrestis* (L.) in south Sweden. Viltrevy, 8:267–378.

HASLER, J. F., A. E. BUHL, and E. M. BANKS. 1976. The influence of photoperiod on growth and sexual function in male and female collared lemmings (*Dicrostonyx groenlandicus*.) J. Reprod. Fert., 46:323–329.

HUMINSKI, S., and J. KRAJEWSKI. 1977. The growth process of the vole, *Microtus arvalis* (Pallas, 1779) during autumn and winter. Zool. Poloniae, 26:103–111.

HYVARINEN, H., and K. HEIKURA. 1971. Effects of age and seasonal rhythm on the growth patterns of some small mammals in Finland and in Kirkenes, Norway. J. Zool. London, 165:545–556.

IVERSON, S. L., and B. N. TURNER. 1969. Under-snow shelter for small mammal trapping. J. Wildlife Mgmt., 33:722–723.

———. 1974. Winter weight dynamics in *Microtus pennsylvanicus*. Ecology, 55:1030–1041.

KREBS, C. J. 1964. The lemming cycle at Baker Lake, Northwest Territories, during 1959–62. Arctic Inst. North Amer. Tech. Papers, 15:1–104.

KREBS, C. J., and J. H. MYERS. 1974. Population cycles in small mammals. Pp. 267–399, *in* Advances in ecological research, Academic Press, London, 8.

MALLORY, F. F., J. R. ELLIOTT, and R. J. BROOKS. 1981. Changes in body size in fluctuating populations of the collared lemming: age and photoperiod influences. Canadian J. Zool., 59:174–182.

MERRITT, J. F., and J. M. MERRITT. 1978. Population ecology and energy relationships of *Clethrionomys gapperi* in a Colorado subalpine forest. J. Mamm., 59:576–598.

MIHOK, S., and W. A. FULLER. 1981. Morphometric variation in *Clethrionomys gapperi*: are all voles created equal? Canadian J. Zool., 59:2275–2283.

MILLAR, J. S., and S. L. IVERSON. 1976. Weight of the eye lens as an indicator of age in *Peromyscus*. Canadian Field-Nat., 90:37–41.

MYERS, K., J. CARSTAIRS, and N. GILBERT. 1977. Determination of age of indigenous rats in Australia. J. Wildlife Mgmt., 41:322–326.

MYERS, K., and N. GILBERT. 1968. Determination of age of wild rabbits in Australia. J. Wildlife Mgmt., 32:841–849.

NEGUS, N. C., and P. J. BERGER. 1977. Experimental triggering of reproduction in a natural population of *Microtus montanus*. Science, 196:1230–1231.

ÖSTBYE, E., and A. SEMB-JOHANSSON. 1970. The eye lens as an age indicator in the Norwegian lemming (*Lemmus lemmus* (L.)). Nytt Mag. Zool., 18:239–243.

PETTERBORG, L. J. 1978. Effect of photoperiod on body weight in the vole, *Microtus montanus*. Canadian J. Zool., 56:431–435.

RIESENFELD, A. 1980. Body build and temperature tolerance: an experimental analysis of ecological 'rules.' Acta Anat., 107:35–45.

SCHWARZ, S. S., A. V. POKROVSKI, V. G. ISTCHENKO, V. G. OLENJEV, N. A. OVTSCHINNIKOVA, and O. A. PIASTOLOVA.

1964. Biological peculiarities of seasonal generations of rodents, with special reference to the problem of senescence in mammals. Acta Theriol., 8:11–43.

SPITZ, F. 1974. Démographie due campagnol des champs *Microtus arvalis* en vendée. Ann. Zool.-Ecol. Anim., 6:259–312.

STEELE, R. W. 1977. A demographic study of *Microtus pennsylvanicus* in relation to habitat structure. Unpublished M.Sc. dissert., Univ. Guelph, Guelph, Ontario, 124 pp.

TAST, J. 1972. Annual variations in the weights of wintering root voles, *Microtus oeconomus,* in relation to their food conditions. Ann. Zool. Fennici, 9:116–119.

THOMAS, R. E., and E. D. BELLIS. 1980. An eye-lens weight curve for determining age in *Microtus pennsylvanicus.* J. Mamm., 61:561–563.

WEBSTER, A. B., and R. J. BROOKS. 1981. Social behavior of *Microtus pennsylvanicus* in relation to seasonal changes in demography. J. Mamm., 62:738–751.

WIGER, R. 1979. Demography of a cyclic population of the bank vole *Clethrionomys glareolus.* Oikos, 33:373–385.

WOLFF, J. O., and W. Z. LIDICKER, JR. 1980. Population ecology of the taiga vole, *Microtus xanthognathus,* in interior Alaska. Canadian J. Zool., 58:1800–1812.

WUNDER, B. A., D. S. DOBKIN, and R. D. GETTINGER. 1977. Shifts of thermogenesis in the prairie vole (*Microtus ochrogaster*): strategies for survival in a seasonal environment. Oecologia, 29:11–26.

ZEJDA, J. 1971. Differential growth of three cohorts of the bank vole, *Clethrionomys glareolus* Schreb. 1780. Zool. Listy, 20: 229–245.

Address: Department of Zoology, University of Guelph, Guelph, Ontario N1G 2W1, Canada.

POPULATION ECOLOGY AND HOME RANGE UTILIZATIONS OF TWO SUBALPINE MEADOW RODENTS (*MICROTUS LONGICAUDUS* AND *PEROMYSCUS MANICULATUS*)

Jack A. Cranford

ABSTRACT

Population density, home range size and habitat quality for *Microtus longicaudus* and *Peromyscus maniculatus* were determined at two subalpine study sites in the Wasatch Mountain Range of Utah. Study sites were trapped every three weeks when snow free (May–Oct.) and every six weeks when snow covered by using trap chimneys to access the ground surface. Trap and track board data were utilized to increase the data base for home range analysis. Radioactive tantalum implanted in *M. longicaudus* permitted seasonal monitoring of activity patterns and recapture of animals dispersed from the grid. *Peromyscus maniculatus* population densities varied from 2.2 to 22.4 ha, whereas *M. longicaudus* varied from .4 to 28.5 ha over the three year study period. Population densities of both species were lowest in the November–February sample interval as were home range sizes except for male *M. longicaudus* whose home ranges were smallest in the snow free July–October period. Activity patterns of radioactive tantalum tagged *M. longicaudus* were principally diurnal during the snow free season and non-photoperiodic during the snow cover season. Lactating females spent significantly more time at their nests when compared to other non-breeding females. During early winter, poor snow cover, both sexes increased the number of foraging bouts per day (7.2–13) but decreased individual foraging bout length (67–38 min). Home range changes to areas of greater cover density (forb to grass) and changes in activity patterns during harsh environmental periods reduced individual exposure time in better environmentally buffered habitats which potentially reduced total energy needs.

INTRODUCTION

Recently Merritt and Merritt (1980) and Vaughan (1969, 1974) have published reports on subalpine ecology of various rodent species in the southern Rocky Mountains. However, year round ecological studies are limited to Minnesota (Brown, 1971, 1973) Colorado (Stinson, 1977; Merritt and Merritt, 1978a, 1978b) and Utah (Cranford, 1978) within the continental United States. Year round studies in Alaska (Whitney, 1973), Canada, and Europe are more numerous (for review see Merritt and Merritt, 1980).

Merritt and Merritt (1978a, 1978b, 1980) have reported on *Peromyscus maniculatus* and *Clethrionomys gapperi* from the southern Rocky Mountains, but little data exists on the subalpine ecology of *Microtus longicaudus*. Habitat preferences of *M. longicaudus* have been positively correlated with available water and mesic vegetation (Warren, 1942; Findley and Jones, 1962; Brown 1967, 1970; Colvin, 1973). Others have shown strong associations between *M. longicaudus* and shrub habitats specifically when sympatric with other microtine species, particularly *Microtus montanus* (Rickard, 1960; Randall, 1978; Randall and Johnson, 1979). Habitat modification in forested areas results in higher population densities of *M. longicaudus* in early successional stages and in permanent meadows which form later (Ramirez and Hornocker, 1981). Population densities have been reported as rare in alpine habitats (Vaughan, 1969), 6.5/ha in subalpine willow sedge habitats and 8/ha in grass shrub habitats (Randall and Johnson, 1979). Randall (1978) recently demonstrated that habitat preference by *M. longicaudus* is strongly influenced by the presence of a competitor *M. montanus*, with *M. longicaudus* having the broadest ecological niche and its habitat selection dependent upon the syntopic and or sympatric occurrence of *M. montanus*. Colvin (1973) concluded that these species mutually avoided each other, resulting in habitat segregation. Habitat preference could be strongly influenced by cover density since Beck and Anthony (1971) demonstrated that a covered nest reduced metabolic costs of *M. longicaudus* at 0°C by 33%. This research will report on home range size, cover densities, activity patterns and population densities of *M. longicaudus* and *P. maniculatus* in the subalpine habitat of the Wasatch Range of Utah.

METHODS

Three study sites were located within 80 km of Salt Lake City, in the Wasatch National Forest, Salt Lake Co., Utah. Sites were located at 2,500 m and 2,900 m in canyons with east to west drainage patterns. Annual precipitation ranged from 62–75 cm per year. Snow disappearance and the beginning of spring varied by 55 days between years (16 May to 10 July) with the highest variation above 2,500 m (Cranford, 1977a). This variability depends on total snow accumulation and its subsequent melt rate in spring, but leads into a phenologically predictable period of summer and fall. Study sites were established in the summer of 1974 and were sampled at a minimum interval of three weeks from May through October and at six week intervals from November through April. Sites were located in Albion Basin (2,900 m) of Little Cottonwood Canyon (AB-1 and AB-3) and at 2,500 m in Little Cottonwood Canyon (LC-2). Sites AB-1 and AB-3 were in subalpine meadow communities with scattered clumps of alpine fir (*Abies lasiocarpa*) and spruce (*Picea engelmannii*) with the predominant meadow vegetation composed of *Geranium viscosissimum*, *Potentilla gracilis*, *Veratrum californicum*, *Mertensia arizonica*, *Carex* spp., *Hackelia floribunda*, *Lupinus* spp., *Valeriana occidentalis*, *Castilleja miniata*, and *Artemisia ludoviciana*. Site LV-2 was a meadow community surrounded by a fir forest (*Abies concolor* and *Pseudotsuga menziesii*) with predominant meadow vegetation composed of *Carex occidentalis*, *Stipa columbiana*, *Mertensia arizonica*, *Thalictrum fendleri*, *Potentilla glandulosa*, *Rudbeckia occidentalis*, *Geranium viscosissimum*, and *Veratrum californicum*. The vegetation on the grids was sampled 40 to 50 days after becoming snow free with 100 randomly selected one m² sample plots to measure the percent occurrence of all species. Vegetation was subdivided into four classes: forbs, grasses, shrubs and sedges. Percent cover was determined at the same time by measuring the total area of a meter-square frame which encompassed bare ground or vegetation less than 5 cm tall (Phillips, 1959).

Trap grids AB-1 and AB-3 (10 by 10) were 1.1 ha and LC-2 (10 by 15) was 1.8 ha. Folding Sherman traps were baited with oatmeal on permanently staked grids (10 m intervals), but track boards (5 m intervals) were unbaited (Justice, 1961). During the snow free period traps and track boards were run every three weeks but during the snow cover season were sampled every six weeks. The subnivean activity of mice was monitored by using trap chimneys (200 by 35 cm) with one trap (cotton and food provided) and one track board per site (Merritt and Merritt, 1978a). Traps were checked every 8 hrs and all animals were toe clipped for specific identification. For each capture, data on location, age, sex, weight and reproductive condition were recorded.

Following the methods of Pruitt (1959), only one continuous ski or snow shoe trail was used to prevent damage to subnivean runways. Population density was assessed by the minimum number known alive (MNKA; Krebs, 1966). Home ranges were analyzed by minimum area method (Cranford, 1977b). Home range shape, defined as the ratio of the minor axis to the major axis, was divided into three classes: linear (0–.35), elliptical (.36–.69) and oval (.70–1.0), and examined to determine whether seasonal changes occurred.

Habitat quality values (proportion grasses times cover density) for each sample station were calculated, and these values were assigned to each trap station. Home ranges of *M. longicaudus* were then plotted on these grid values. For each home range a mean quality value was generated. Home range overlap between individuals was calculated on the basis that 1.00 equals total overlap and .00 equals no overlap (Cranford, 1977b).

Data on location and activity were gathered from 26 *M. longicaudus* tagged with gamma-emitting radioactive tantalum-182 wires (Cranford, 1978). In addition to trap and track board data, 12 gamma probes were placed at 60 different locations to gather additional home range and activity pattern data. Radioactively labelled animals were located both on and off the grid and were separated by at least 50 m as determined by their prior closest capture points. A single channel recorder was modified to record from each input lead for one minute every 12 min with a fixed event marker permitting each probe to be identified and transcribed. Additional activity patterns were determined by placing a probe, which recorded continuously the presence or absence of the resident, directly over a nest (N = 14). A resident more than 120 cm from the nest could not be detected and when between 50 and 80 cm the signal intensity was reduced by 50%. With snow cover on the ground, detection distance was reduced to 30 cm until the subnivean space developed, at which time voles could be detected 80 cm from the nest.

Soil and air temperatures were continuously recorded throughout the study with a Yellow Springs Instrument thermometer modified to record from eight separate input leads. Air temperatures were recorded from the subnivean space and at 1, 3, and 10 meters above the ground surface. Soil temperatures were recorded at the surface, 5, 15, and 30 cms below ground (see Cranford, 1978).

RESULTS

POPULATION DENSITIES

Population densities of *P. maniculatus* on all grids varied from a low of 2.2 to a high of 22.4 per ha (Table 1). Populations peaked in October or November and steadily declined until the onset of breeding in late April and early May. Densities calculated from both track boards and trap-revealed census methods did not differ for March–June and July–October sample periods but did differ significantly (t = 2.84, P < .05) for November–February. Trap-determined densities were lower (AB-1, 1.8; AB-3, 4) for this census interval and could not be attributed to trap occupancy by other species.

March–June densities on grid AB-1 differed significantly (t = 3.13, P < .025) from grids AB-3 and LC-2 but did not differ during other census periods. Analysis of March–April and June–July periods on grid AB-1 indicates that *P. maniculatus* densities declined during the June–July period as population densities of *Zapus princeps* increased from zero to 27 per ha while *Z. princeps* densities on AB-3 and LC-2 never exceeded 5 per ha.

Population density of *M. longicaudus* on all grids varied from a low of 0.4 per ha to a high of 28.5 per ha (Table 1). Trap and track board census methods did not differ between seasons, within or be-

Table 1.—*Population densities in hectares of* P. maniculatus *and* M. longicaudus *determined by minimum number known alive for all sites from 1974 to 1976.*

Site	Peromyscus maniculatus			Microtus longicaudus		
	March–June	July–October	November–February	March–June	July–October	November–February
AB-1	2.2–4.1	12.8–16.8	3.5–6.2	11.5–12.6	5.0–28.5	2.2–6.5
AB-3	10.4–12.6	15.3–22.4	3.4–8.4	8.4–11.6	9.0–23.5	3.7–7.1
LC-2	8.6–14.5	13.8–18.4	—	3.6–6.5	6.1–15.8	0.4–4.2

tween sites. Population densities peaked in September in each of the sample years and steadily declined until the onset of breeding in late March or April. Grids AB-1 and AB-3 did not differ in population density between seasons and years, but grid LC-2 always had lower densities and differed in habitat quality.

HOME RANGE

Mean seasonal home range size of male and female *P. maniculatus* differed only during the snow free period when males occupied significantly larger ($t = 2.23$, $P < .05$) home ranges (Table 2). Both sexes had significantly smaller home ranges (males $t = 2.67$, $P < .02$; females $t = 2.06$, $P < .05$) in early winter, but after snow cover was permanently established, home range size increased. Small sample size for juveniles precludes analysis of late winter range sizes, but snow free home ranges were larger than those of early winter and were also larger than adult home ranges. Adult home range data from grid AB-1 during April–July were excluded from general analysis due to extreme density changes. However, a comparative analysis of data for adults on grids AB-1 and AB-3 indicated that when *Z. princeps* were abundant, home range sizes of *P. maniculatus* were significantly larger ($N = 12$, $t = 3.26$, $P < .01$) on AB-1; 2 to 2.7 times those of the low density *Zapus* grid (AB-3). Adult home range overlap of *P.*

maniculatus was seasonally different with the mean overlap of males on females highest during the snow free (.39) and early winter period (.41). Mean overlap of male/male home ranges was .18, and female/female overlap was .27, while juvenile/adult overlap values (.59) were extremely high during all time periods. Home ranges were elliptical in shape, except in summer when male ranges became extremely linear (.28) and in early winter when female ranges were oval (.79).

Mean seasonal home ranges for *M. longicaudus* differed significantly ($t = 3.18$, $P < .01$) between sexes only during the Nov.–Feb. period (Table 2). Mean home range sizes during the year differed significantly ($t = 2.23$, $P < .05$) only for females between the early and late snow cover periods. Home ranges were largest during the late snow cover period and smallest for females during early snow cover but smallest for males during the snow free period. Juveniles showed no significant differences in home range size between seasons. Juvenile home ranges were always smaller than those of adults and were highly transitory. Adult home range overlap during all seasons was low (<.23), except for males during the late snow bound and snow free period when they overlapped females (>.49). During this time female overlap was extremely low (<.12). Home ranges during the snow free period were elliptical but became linear during the snow cover period. Male

Table 2.—*Mean and standard deviation of home range size (m²) for* P. maniculatus *and* M. longicaudus *based on a minimum of eight recaptures per sample interval and calculated by boundary strip method.*

Species, sex, and age	Snow bound		Snow free		Snow bound	
	N	March–June	N	July–October	N	November–February
Peromyscus maniculatus						
Male	23	276 ± 82	21	390 ± 54	14	189 ± 65
Female	18	242 ± 35	22	265 ± 47	9	137 ± 50
Juvenile	3	75 ± 64	16	446 ± 95	8	252 ± 135
Microtus longicaudus						
Male	18	527 ± 124	16	395 ± 136	8	488 ± 85
Female	20	444 ± 62	28	353 ± 25	9	290 ± 73
Juvenile	13	250 ± 35	8	285 ± 41	4	275 ± 62

Table 3.—*Mean activity patterns of* M. longicaudus *determined from continuously monitored radioactively tagged voles.*

Statistic	Snow bound March–June	Snow free July–October	Snow bound November–February
Percent diurnal			
Trap determined	47	68	53
Radioactive label	61	75	50
Percent of time at nest per day			
All voles	60	57	64
Lactating females (N = 4)	—	72	—
Number of bouts per day			
Away from nest	8.3 ± 2.5	6.1 ± 2.4	12.8 ± 4.2
Around nest	9.5 ± 4.3	18.4 ± 6.2	6.3 ± 2.2
Bout length in minutes			
Away from nest	63.4 ± .3	71.2 ± .4	38.3 ± .2
Around nest	6.7 ± 1.6	8.8 ± 2.6	5.1 ± 1.1

home ranges from March to July were highly variable in shape but always incorporated portions of at least three female home ranges.

ACTIVITY PATTERNS

Analysis of capture data by time of day indicates that *P. maniculatus* was crepuscular and nocturnal with less than 18% of captures occurring during the day light phase. Trap capture data for *M. longicaudus* showed seasonal changes varying from nocturnal to diurnal. Activity patterns for radioactively tagged voles showed clear changes during the year (Table 3). Seasonal differences were all significant ($P < .05$), and the changes were from diurnal during snow free periods to equally spaced bouts when permanent snow cover developed. During all seasons all voles were at their nest 57% of the time except for lactating females who spent 65 to 80% of their time at the nest. Activity around (<80 cm) or away (>80 cm) from the nest differed by both frequency of occurrence and time spent during all seasons.

Voles spent significantly less time (t = 2.63, $P < .05$) away from the nest from November–February, and each bout was significantly shorter (t = 3.58, $P < .01$) than for other periods; however, the total number of bouts was significantly greater (t = 4.62, $P < .01$) than during the snow free period.

HABITAT QUALITY

M. longicaudus home ranges were positively correlated (r = .84, $P < .05$) with grass sedge, negatively correlated with forbs (r = −.54, $P, < .10$) and positively correlated (r = .92, $P < .02$) with cover density. Overwintering home ranges of *M. longicaudus* were in areas which had high cover density prior to snow fall. Seven of 12 home ranges were used in summer and fall, but the remaining five had only been intermittently occupied by juveniles and subadults during the summer. These were colonized in winter by voles which previously occupied central meadow sites high in forb diversity. These adult voles (three males, two females) had home ranges in excess of 420 m^2 for males and 360 m^2 for females in summer, and all were smaller than 250 m^2 in winter. By comparison, these summer values were high relative to mean values for the population and low for males in winter (Table 2). Three of these voles (two males, one female) remained as residents within these areas over winter as did five of the voles which overwintered in ranges they had continuously occupied. Eleven other voles occupied sites which had low cover density in late fall, but were of the grass-sedge habitat type. Of these, six voles were within the study area at snow melt but had occupied at least two home ranges during the winter. Grid LC-2 always had lower densities (Table 1), but the home range correlation values were similar to those for residents on grids AB-1 and AB-3. The total suitable area on grid LC-2 was less than 20% of the total area, while the other two grids had up to 60% of the total area suitable for home ranges.

DISCUSSION

Population densities of *P. maniculatus* peaked in October on all grids and were lowest during February. Both Stinson (1977) and Merritt and Merritt (1980) have reported low densities due to reduced trappability in winter, and if only trap data were utilized, the densities reported here would be reduced by 50%. Track boards indicated that individuals seldom or never captured in traps were present and were not trapped because they did not enter available traps. Although the traps were provisioned with both food and cotton and were within 5 cm of track boards, they were not entered. Seventeen trackboards on grid AB-1 and 13 on grid AB-3, during the November through February sample interval, were tracked when adjacent traps were available (not frozen open) and remained available during that sample day. Over 35 percent of the *P. maniculatus* density and home range data during winter was from track boards when traps were avoided. The use of the Jolly estimate by Merritt and Merritt (1980) indicated that their data underestimated densities, but using track and trap data, these estimators were in closer agreement in winter. The populations densities are similar to those of Merritt and Merritt (1980) in Colorado subalpine forest and of Negus and Findley (1959) from a Wyoming spruce-fir forest. The habitats studied in Utah were a mosaic of forest and meadow and, although not identical to either cited study, were structurally very similar. Home ranges varied in both size and shape but differed principally in summer when breeding female ranges did not overlap. Male home ranges changed during the breeding season, increasing in size and becoming linear to overlap several female home ranges. Juveniles overlapped on both adult sexes, but adults of the same sex seldom had overlapping home ranges. Early winter home ranges were small until permanent snow cover and subnivean spaces developed.

The presence of *Z. princeps* in high densities on one grid (AB-1) resulted in sudden distribution and density changes in *P. maniculatus*. Two grids had low density populations of *Z. princeps,* and home range sizes of *P. maniculatus* were 20–40% larger when *Z. princeps* was present. On grid AB-1 where *Z. princeps* was seasonally very abundant, most *P. maniculatus* dispersed from the meadow area, but the few which remained had home ranges twice the size (550–725 m?) of residents on grids AB-3 and LC-2. During early spring and late fall when *Z. princeps* was not present, the home ranges of *P. maniculatus* were equivalent on all grids.

Early winter population density of *M. longicaudus* on all grids varied from year to year. In 1974 and 1976 permanent snow cover (>30 cm) occurred in November; however, in 1975 permanent snow cover occurred one month later, and winter densities were extremely low. Low temperatures at the ground surface (−10 to −20°C) probably resulted in extreme early winter mortality. Breeding occurred from April through October, but peak densities of juveniles occurred from June through August (snow free), coincident with high vegetative growth and high cover density. Juveniles which entered the trappable population in April resulted from breeding which occurred while the habitat was totally snow covered. Peak population densities occurred in September on all grids. Grids AB-1 and AB-3 did not differ in

population density, but grid LC-2 always had significantly lower densities. Although these three grids were selected to be ecologically similar, grid LC-2 differed in the types of forbs present and in the relative dominance of the forb community. Comparisons of home range sizes between grids were not different, but the uncolonized portion of grid LC-2 was dominanted by forbs instead of the preferred grass-sedge habitat type. The resulting densities on grid LC-2 reflect the small suitable area available for *M. longicaudus.*

Although others (Randall, 1978; Randall and Johnson, 1979) have demonstrated competitive displacement of *M. longicaudus* by *M. montanus* from grassy habitats, this was not observed at these sites. *M. montanus* densities were very low at these sites through the course of this research. In the relative absence of *M. montanus, M. longicaudus* preferred the more mesic grass-sedge habitat as other studies have indicated (Findley and Jones, 1962; Brown 1967, 1970; Colvin, 1973).

Habitat selection reflected by home range size and shape indicated that *M. longicaudus* preferred the grass-sedge habitat but would colonize forb areas if cover density was high. However, residents in habitat principally composed of forbs had home ranges larger than those which lived in grass-sedge habitat. Additionally, with the onset of winter, forb habitats were abandoned, and adjacent areas which had higher amounts of grasses and sedges were colonized. Forbs provided good cover density in summer, but as winter set in they died back and became prostrate on the ground surface. Conversely, many grasses and sedges remained green and, due to their bunch-like growth pattern, remained above the ground surface. These sites then provided suitable space for winter nests, a food resource probably of higher quality, and a series of spaces which would remain as snow cover accumulated. Beck and Anthony (1971) demonstrated that a nest results in 33% reduction in metabolic costs, and with the addition of a large vegetative clump and a cap of snow, metabolic costs should be further reduced due to the increased insulative capacity of these additional components. Home range switches which occurred were from forb to grass-sedge habitats, and animals which remained in areas low in cover density or low in the grass sedge component frequently switched home ranges in winter. Whether these animals switched for reasons of diet, nest, or insulative capacity of the site remains unknown, but all could potentially increase survivorship and decrease an individual's total energy budget during the most severe environmental period. During the November–December period of 1975, resident individuals in forb dominated or forb-grass mixed areas disappeared more frequently than animals which were residents in grass-shrub habitat. Although the sample sizes were small, it was consistent on all three grids and lends support to the habitat preference speculation.

Activity patterns of *M. longicaudus* changed over the course of the year as a function of both photoperiod, reproductive condition and snow cover. Lactating females spent more time at the nest than nonbreeding individuals of either sex. Other major differences occurred principally in early winter when activity away from the nest was shorter in duration but occurred more frequently. When snow cover was established, activity patterns were no longer circadian in structure but were equally spaced ultradian bouts throughout the 24 hr day. Shorter, more frequent bouts in a cold environment would be adaptive if the nest remained warm, thereby minimizing animal heat loss on return to the nest. Colonized overwintering sites were consistent with this concept.

ACKNOWLEDGMENTS

I want to thank Dr. J. Merritt of the Powdermill Nature Preserve of the Carnegie Institute for the fine symposium on "Winter Ecology of Small Mammals" from which this paper results. Additional thanks to Dr. Craig C. Black of the Carnegie Institute and other members of the board and staff who made the symposium possible.

LITERATURE CITED

BECK, L. R., and R. G. ANTHONY. 1971. Metabolic and behavioral thermoregulation in the long-tailed vole, *Microtus longicaudus*. J. Mamm., 52:404–412.

BROWN, E. B., III. 1971. Some aspects of the ecology of the small, winter-active mammals of a field and adjacent woods in Itasca State Park, Minnesota. Unpublished Ph.D. thesis, Univ. Minnesota, Minneapolis, 173 pp.

———. 1973. Changes in patterns of seasonal growth of *Microtus pennsylvanicus*. Ecology, 54:1103–1110.

BROWN, L. N. 1967. Ecological distribution of mice in the Medicine Bow mountains of Wyoming. Ecology, 48:676–680.

———. 1970. Population dynamics of the western jumping mouse (*Zapus princeps*) during a four-year study. J. Mamm., 51:651–658.

COLVIN, D. V. 1973. Agonistic behavior in males of five species of voles *Microtus*. Anim. Behav., 21:471–480.

CRANFORD, J. A. 1977a. The ecology of the western jumping mouse, *Zapus princeps*. Unpublished Ph.D. thesis, Univ. Utah, Salt Lake City, 118 pp.

———. 1977b. Home range and habitat utilization by *Neotoma fuscipes* as determined by radiotelemetry. J. Mamm., 58:165–172.

———. 1978. Hibernation in the western jumping mouse (*Zapus princeps*). J. Mamm., 59:496–509.

FINDLEY, J. S., and C. J. JONES. 1962. Distribution and variation of voles of the genus *Microtus* in New Mexico and adjacent areas. J. Mamm., 43:154–167.

JUSTICE, K. E. 1961. A new method for measuring home ranges of small mammals. J. Mamm., 42:462–470.

KREBS, C. J. 1966. Demographic changes in fluctuating populations of *Microtus californicus*. Ecol. Monogr., 36:239–273.

MERRITT, J. F., and J. M. MERRITT. 1978a. Population ecology and energy relationships of *Clethrionomys gapperi* in a Colorado subalpine forest. J. Mamm., 59:576–598.

———. 1978b. Seasonal home ranges and activity of small mammals of a Colorado subalpine forest. Acta. Theriol., 23:195–202.

———. 1980. Population ecology of the deer mouse (*Peromyscus maniculatus*) in the Front Range of Colorado. Ann. Carnegie Mus., 49:113–130.

NEGUS, N. C., and J. S. FINDLEY. 1959. Mammals of Jackson Hole, Wyoming. J. Mamm., 40:371–381.

PHILLIPS, E. A. 1959. Methods of vegetation study. Henry Holt and Co., 107 pp.

PRUITT, W. O., JR. 1959. A method of live-trapping small taiga mammals in winter. J. Mamm., 40:139–143.

RAMIREZ, J. P., and M. HORNOCKER. 1981. Small mammal populations in different-aged clearcuts in Northwestern Montana. J. Mamm., 62:400–403.

RANDALL, J. A. 1978. Behavioral mechanisms of habitat segregation between sympatric species of *Microtus*: Habitat preference and interspecific dominance. Behav. Ecol. Sociobiology., 3:187–202.

RANDALL, J. A., and R. E. JOHNSON. 1979. Population densities and habitat occupancy by *Microtus longicaudus* and *M. montanus*. J. Mamm., 60:217–219.

RICKARD, W. H. 1960. The distribution of small mammals in relation to the climax vegetation mosaic in eastern Washington and northern Idaho. Ecology, 41:99–106.

STINSON, N. S., JR. 1977. Species diversity and resource partitioning and demography of small mammals in a subalpine deciduous forest. Unpublished Ph.D. thesis, Univ. Colorado, Boulder, 238 pp.

VAUGHAN, T. A. 1969. Reproduction and population densities of a montane small mammal fauna. Pp. 51–74, *in* Contributions in Mammalogy (J. K. Jones, Jr., ed.), Misc. Publ. Mus. Nat. Hist., Univ. Kansas, 51:1–428.

————. 1974. Resource allocation in some sympatric subalpine rodents. J. Mamm., 55:764–795.

WARREN, E. R. 1942. The mammals of Colorado. Univ. Oklahoma Press, Norman, Oklahoma, xviii + 330 pp.

WHITNEY, P. H. 1973. Population biology and energetics of three species of small mammals in the taiga of interior Alaska. Unpublished Ph.D. thesis, Univ. Alaska, Fairbanks, 254 pp.

Address: Department of Biology, Virginia Polytechnic Institute, and State University, Blacksburg, Virginia 24061.

BEHAVIORAL STRATEGIES OF SMALL MAMMALS UNDER WINTER CONDITIONS: SOLITARY OR SOCIAL?

Stephen D. West and Holly T. Dublin

ABSTRACT

In cold regions non-hibernating small mammals employ a variety of morphological, physiological, and behavioral adaptations to increase winter survival. One such adaptation is winter aggregation, a phenomenon in which several individuals coinhabit a small area for periods of days to months. This review discusses the general nature of these aggregations, their occurrence across small mammal taxa, and their advantages and disadvantages. Winter aggregations have been reported for many rodent species, particularly those of the family Cricetidae, but rarely as yet for the Insectivora. This apparent dichotomy is used to provide insight into the factors promoting winter aggregations. Differences in seasonal cycles of aggression, territoriality, and foraging mode are compared among rodent and insectivore species. In winter-social species, aggression levels are linked to reproductive state such that non-breeding individuals are more tolerant of conspecifics than are breeding individuals. The resultant social climate in winter is relatively amicable. In winter-solitary species, aggression remains high during both the breeding and non-breeding seasons, and winter aggregations do not form. We suggest that these patterns of aggression and spacing reflect different foraging modes and their associated constraints.

INTRODUCTION

Individuals of some winter-active small mammal species are known to form social groups or aggregations during winter, whereas individuals of other species apparently do not. The existence of such winter groups has been known for a number of years (Cory, 1912), but their composition and the circumstances of their formation for most species are yet to be described. The most often cited explanation for their formation is that group members might reduce the amount of body surface exposed to low air temperature by huddling and thereby conserve body heat. Sealander (1952) undertook the first experimental measurements of the survival value of huddling for individuals of *Peromyscus leucopus* and *P. maniculatus*. Related data have been collected for several other species (Pearson, 1947, 1960; Howard, 1951; Sealander, 1952; Wiegert, 1961). In these studies, survival at low temperature was generally higher when caged individuals were held in groups and allowed to huddle than when they were held alone. Considering these experimental demonstrations of the survival value of huddling, one might expect that winter-active species of smaller body size would benefit more from this behavior than larger species. Apparently, however, winter aggregations are rare among temperate and arctic insectivores (Sealander 1952), which are smaller, on the whole, than the rodents thus far examined.

This paper reviews the occurrence of winter aggregations across small, winter-active mammalian species, and speculates on the reasons for the observed pattern and the mechanisms underlying it.

RESULTS

THE PATTERNS

With the notable exceptions of nest boxes used by Nicholson (1941) in his investigations of *Peromyscus* and the recent use of radiotracers and radiotransmitters, our understanding of winter dispersion in small mammals has been based upon trapping returns. Although there are difficulties in reconstructing true animal movement and space-use patterns from trap location data, such data can indicate the broad nature of winter dispersion. All species of winter-active rodents and insectivores for which winter dispersion data could be found are included in Table 1. Species were considered to be winter-solitary if winter location records of individuals showed a pattern of uniform dispersion (winter territoriality). Species were considerd winter-social if winter location records showed a highly clumped pattern, or if individuals were actually found coinhabiting a winter nest. Species for which individuals are known to share the same winter nest are designated with asterisks in Table 1. The intent in assembling this information is primarily to reveal the general taxonomic distribution of winter dispersion patterns. The very interesting questions of winter

Table 1.—*Taxonomic distribution of the tendency to aggregate in winter. A. Colonial species. B. Non-colonial species forming conspecific winter aggregations. C. Non-colonial species forming interspecific winter aggregations. Asterisks mark species for which direct evidence of winter nest sharing has been found.*

Species	References
Winter-social species	
A. *Microtus brandti**	Kucheruk and Dunayeva, 1948; Reichstein, 1962
*M. pinetorum**	Fitzgerald and Madison, 1981
B. *M. arvalis**	Fenyuk, 1941, *in* Naumov, 1972; Frank, 1957*a*
*M. xanthognathus**	Wolff and Lidicker, 1980, 1981
*M. pennsylvanicus**	Webster, 1979; Webster and Brooks, 1981*a*, 1981*b*
M. oeconomus	Tast, 1966
M. agrestis	Myllymäki, 1977
Clethrionomys rutilus	Sealander, 1966, 1967; West, 1977
C. gapperi	Criddle, 1932
Meriones meridianus	Rall', 1938, 1939, *in* Naumov, 1972
Apodemus sylvaticus	Zimmerman, 1952
Micromys minutus	Frank, 1957*b*
Glis glis	Koenig, 1960
*Cryptotis parva**	Hamilton, 1934; Davis and Joeris, 1945
C. *Peromyscus leucopus**	Nicholson, 1941; Thompsen, 1945; Howard, 1949, 1951; Wolff and Hurlbutt, 1982
*P. maniculatus**	Nicholson, 1941; Howard, 1949, 1951; Metzgar, 1979; Merritt and Merritt, 1978; Wolff and Hurlbutt, 1982
Winter-solitary species	
Sorex araneus	Crowcroft, 1957; Shillito, 1963; Michielson, 1966
S. minutus	Michielson, 1966
S. arcticus	Clough, 1963
S. cinereus	Buckner, 1966
S. palustris	Sorenson, 1962
S. obscurus	Hawes, 1975
S. vagrans	Hawes, 1975
Lemmus lemmus	Arvola et al., 1962
Onychomys leucogaster	Ruffer, 1968; Hafner and Hafner, 1979

group membership and seasonal social dynamics await investigation.

Species have been grouped to acknowledge apparent differences in seasonal social organization

(Table 1). The major division is between those species that are winter-solitary and those that may form winter groups. With the exception of *Lemmus lemmus,* discussed below, winter-solitary species tend to be small and insectivorous. Although *Onychomys* is not as small as most shrews it is primarily insectivorous. Within the winter-social groupings, there are further differences in social organization. Some species are colonial year-round (A, Table 1); others are territorial, or at least more evenly dispersed while breeding, but may be winter-social (B, Table 1); and some are territorial while breeding but may form short-term, interspecific winter groups (C, Table 1). Body sizes of all three groups of winter-social species tend to be larger than winter-solitary species, and their diets are primarily composed of plant material. *Cryptotis parva* (B, Table 1) is an obvious exception to this pattern. Little field work has been done on this shrew, but two groups of three and five shrews in New York (Hamilton, 1934) and one group of 12 shrews in Texas (Davis and Joeris, 1945) were caught in the same winter nests. All shrews were apparently adult, although not reproductively active. Such mutual tolerance is unusual for insectivores in temperate regions (Eisenberg, 1966), and invites inquiry into the circumstances allowing or promoting it. While the data in Table 1 are sparse, the patterns of winter sociality, body size, and diet are consistent enough to warrant an explanation.

An Hypothesis

The point elaborated below is simply that winter-solitary species must maintain winter separation to ensure a sufficient food supply to overwinter (Hawes, 1975). The quality, quantity, renewal rate (Waser, 1981), and dispersion of their foods dictates a uniform dispersion of individuals (Crowcroft, 1957). Winter-social species typically exploit foods that do not impose such constraints (Fig. 1).

The line of argument used here for most winter-solitary species parallels that of Jarman (1974) for the social organization of African antelopes. As with small insectivorous species, the social system of small antelopes is tightly constrained by the relatively sparse abundance and uniform distribution of their foods. Food scarcity promotes the evolution of small body size, which reduces total energy requirements, and the sparse distribution of food promotes uniform dispersion of individuals to minimize competition for food. Thus, although small insectivores might reduce heat loss considerably with a social

Fig. 1.—Adaptive behavioral complexes of subniveally winter-active small mammals.

heat conservation strategy, competition for food probably precludes this adaptation.

By exploiting abundant albeit low quality foods, winter-social small mammals reach larger body size than winter-solitary species and a more favorable body surface to volume ratio (Fig. 1). More importantly, because food competition is not a consistent winter constraint as it is for small insectivores, the option of winter gregariousness and a social heat conservation strategy becomes possible.

DISCUSSION

THE NATURE OF WINTER AGGREGATIONS

This paper views winter aggregation in the broad sense of a group of individuals which coinhabit a small area during winter for periods of days to months. Within this theme there are many variations. The variety of terms that have been applied to such groups reflects the underlying physiological and social diversity: winter aggregations, torpid aggregations (Howard, 1951), winter associations, nonbreeding aggregations (Nicholson, 1941), overwintering colonies (Kalela, 1957), survival groups (Brown, 1975), nesting aggregations (Sealander, 1967), and communal groups or midden groups (Wolff, 1980). When dealing with a broad spectrum of species, one expects details of social organization to vary across taxa and geographical location. The telling point, however, is that all winter-social species are energetically capable of coinhabitation. Once this possibility is realized, a range of social organizations of increasing interindividual dependence, including year-round coloniality, also become possible. The degree of sociality is then responsive to the balance of advantages and disadvantages associated with group living (Alexander, 1974).

It is important to consider the adaptive value of winter aggregations and the factors governing them. Species that can aggregate might obtain a selective advantage by gaining access to a good overwintering site, by conserving heat socially, or by escaping predation. One would expect potential overwintering sites to vary with respect to food supply, insulation, and protection or escape from predators. When suitable overwintering sites are in short supply (West, 1977), the survivorship of individuals sharing a good site may exceed that of single individuals residing elsewhere. In such cases individuals may gain the greater benefit from habitat characteristics, rather than from other individuals (Alexander, 1974). In contrast, heat conservation through huddling behavior is a straightforward advantage derived directly from other individuals, and huddling animals might also conserve energy simply by being less active than solitary animals (Wiegert, 1961). Additional economies may be realized by the cooperative construction of large winter nests (Frank, 1957a; Wolff and Lidicker, 1981). Enhanced predator detection and subsequent avoidance (Alexander, 1974), or once detected, predator confusion or selfish herd effects (Hamilton, 1971) are further possible advantages or winter aggregations.

The disadvantages of winter aggregation include competition for food, predator attraction, and increased exposure to diseases and parasites (Alexander, 1974). This paper hypothesizes that food competition is the reason why many winter-active small mammals do not aggregate. Among individuals of species that do aggregate, competition for food still may be a factor limiting the time spent together. Further, in the case of species such as *Microtus brandti* (Naumov, 1972) and *M. xanthognathus* (Wolff and Lidicker, 1981), which cache food communally, it is possible for individuals who

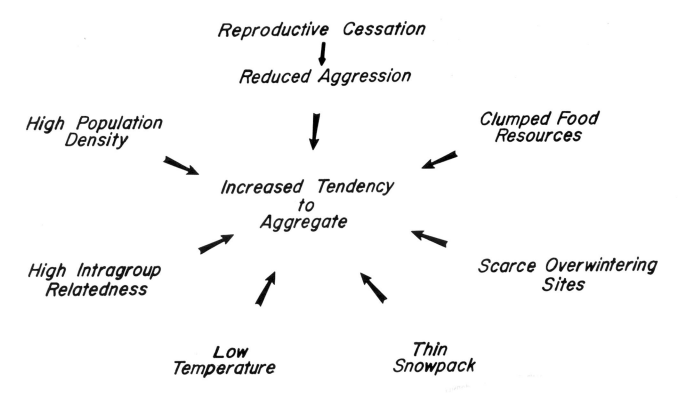

Fig. 2.—Factors promoting midwinter aggregation of small mammals.

use more food than they cache to cheat their nest-mates. The cost of this form of cheating to nestmates could be mitigated by a number of factors including high intragroup genetic relatedness, the value of having the cheater's warm body in the communal nest, the cheater's contribution to nest construction, and the cheater's value in reducing predation either by alerting nestmates or by being eaten first. Aggregations of individuals, especially where extensive signs of presence result from group activities, may enhance detection by predators. And finally, increased body contact raises the risk to individuals of contracting diseases and parasites.

The difficulty in identifying which of the foregoing advantages and disadvantages are the most critical will vary interspecifically. The simplest cases are probably the interspecific aggregations formed by *Peromyscus leucopus* and *P. maniculatus,* where short-term aggregations frequently change composition and locations (Nicholson, 1941; Howard, 1951; Wolff and Hurlbutt, 1982). For *Peromyscus,* heat conservation appears to be the key factor. Such identifications become more difficult as the complexity of social groups increases. Now that appropriate remote sensing tools are available, analyzing the relative importance of these factors should be a productive area for future research.

The formation of winter aggregations appears to be a facultative phenomenon (Metzgar 1979), with the exception of strictly colonial species. Whether or not aggregations form in a given winter depends upon the circumstances of individual animals and their local environment (Fig. 2). Most winter social species are more uniformly dispersed when breeding than when not breeding (Table 1). Intraspecific aggression or mutual intolerance is high during the breeding season, and without reproductive cessation, aggregations cannot form. This topic will be discussed further below, but this observation alone may account for the solitary nature of *Lemmus lemmus* (Table 1). Although the Norway lemming is certainly one of the larger, herviborous species, it may necessarily be winter solitary because it frequently breeds during winter (Hansson, 1984). When breeding, aggression levels remain high, and uniform dispersion is maintained. It would be interesting to know if *Lemmus lemmus* aggregates during winters of no reproduction.

Clumped food resources (large patch size) permit the formation of aggregations and may interact with other habitat features to delimit suitable overwintering sites (Fig. 2). Because heat and energy conservation are hypothesized to be major advantages for huddling individuals, the formation of aggregations should be highly correlated with periods of severe ground-level conditions. In species in which individuals cooperatively construct overwintering nests and cache food, one would expect some degree of group closure, presumably along lines of genetic relatedness, thereby protecting the investments of time and energy of group members (Fig. 2). Furthermore, one would also expect that the most stable groups compositionally and temporally would share high genetic relatedness. For *Microtus xanthognathus,* however, winter nestmates were not immediate family members (Wolff and Lidicker, 1981). To thoroughly evaluate this question for *M. xanthognathus,* it is important to know whether or not the relatedness of individuals from juxtaposed nests is significantly higher than the relatedness of a random sample of individuals. If not, the threat of cheating may be offset by other mitigating factors mentioned previously. Lastly, it has been shown for *Microtus arvalis* (Frank, 1957a) and *M. montanus* (Jannett, 1978) that high population density may lead to the formation of "extended families." These groups are composed of mothers and their offspring which are tolerated when population density is high, but not when it is low. These cohesive groups would be predisposed to aggregate in winter (Fig. 2).

Seasonal Aggression, Territoriality, and Winter Aggregation

Varying levels of intraspecific aggression are obviously important mechanisms influencing seasonal dispersion patterns. Seasons of mutual intolerance are periods of more uniform dispersion than seasons of mutual tolerance. Consequently, the establishment and maintenance of winter territories may involve a fundamental difference in the development of agonistic behavior between winter-solitary and winter-social species.

Summer-born, sexually immature individuals of *Sorex vagrans* and *S. obscurus* establish winter territories in the late summer and fall (Hawes, 1975). Unlike most winter-social species, immatures of winter-solitary species may be mutually intolerant before reaching sexual maturity. In fact, Hawes (1975) found that the most aggressive individuals of *Sorex vagrans* and *S. obscurus* were sexually immature. Early agonistic behavior and territory establishment may be necessary to prevent depletion of food reserves, which are characterized in temperate and arctic regions by an extremely low replacement rate during winter. Strict winter territoriality by sexually immature individuals appears necessary to insure winter survival and to have the subsequent opportunity to breed (Hawes, 1975). In contrast, sexually immature individuals of winter-social species typically show very little aggression. It is common in these species, as in many winter-solitary species, for late summer and fall generation individuals to forego sexual maturation until the following spring. Unlike winter-solitary species, territories are not established, and amicable relationships characterize the overwintering behavior of immature individuals. Further, it is often the case that amicable behavior also prevails among sexually mature individuals once reproduction ceases. Mutual tolerance permits formation of wintering groups until sexual recrudescense and the accompanying increased aggression causes their dissolution at the end of winter (West, 1977).

While aggression is a primary determinant of dispersion, the reasons underlying its expression may vary. In most winter-solitary species it seems likely that the maintenance of winter territories is a direct consequence of food scarcity. But is this the primary function of their breeding territories as well, or are other resources [potential mates (Madison, 1980), safe nesting sites (Webster and Brooks, 1981b), breeding sites, cover] of equal or greater importance? Far more work has been done on the social biology of breeding rodents than on insectivores, but at least for *Sorex vagrans* and *S. obscurus,* indications are that, like many rodent species, the morphology of breeding territories is intersexually variable, and probably responsive to factors other than just food availability. In these species the fivefold increase of male summer territory size over that of winter is difficult to reconcile with the much greater availability of food during summer, but not perhaps with competition among males for mates (Hawes, 1975). Although the smaller, two-fold increase of female summer territory size (Hawes, 1975) superficially appears more in line with the increased food requirements of gestation and lactation, the importance of safe, alternative nesting sites as a determinant of territory size needs to be systematically investigated, particulary considering the cannabalistic tendencies of soricids (Crowcroft, 1957). Thus, insectivorous small mammals may maintain seasonally different territories, one type to maximize winter survivorship, and the other to maximize reproduction.

ACKNOWLEDGMENTS

We thank the Carnegie Museum of Natural History and the Powdermill Nature Reserve for sponsoring the colloquium, and K. Aubry, B. Bowen, and R. Koford for critically reading the manuscript.

LITERATURE CITED

ALEXANDER, R. D. 1974. The evolution of social behavior. Ann. Rev. Ecol. Syst., 5:325–383.

ARVOLA, A., M. ILMEN, and K. TERTTU. 1962. On the aggressive behaviour of the Norwegian lemming (*Lemmus lemmus*) with special reference to sounds produced. Arch. Soc. Zool. Bot. Fennicae, 17:80–101.

BROWN, J. L. 1975. The evolution of behavior. W. W. Norton and Co., New York, xix + 760 pp.

BUCKNER, C. H. 1966. Populations and ecological relationships of shrews in tamarack bogs of southeastern Manitoba. J. Mamm., 47:181–194.

CLOUGH, G. C. 1963. Biology of the arctic shrew, *Sorex arcticus.* Amer. Midland Nat., 69:69–81.

CORY, C. B. 1912. The mammals of Illinois and Wisconsin. Field Mus. Nat. Hist., Zool. Ser., 11:1–505.

CRIDDLE, S. 1932. The red-backed voles (*Clethrionomys gapperi loringi* Bailey) in southern Manitoba. Canadian Field-Nat., 46:178–181.

CROWCROFT, P. 1957. The life of the shrew. M. Reinhardt, London, viii + 166 pp.

DAVIS, W. B., and L. JOERIS. 1945. Notes on the life-history of the little short-tailed shrew. J. Mamm., 26:136–138.

EISENBERG, J. F. 1966. The social organization of mammals. Handbuch der Zoologie, 10:1–92.

FITZGERALD, R. W., and D. M. MADISON. 1981. Spacing, movements, and social organization of a free-ranging population of pine voles *Microtus pinetorum.* Pp. 54–59, *in* Proceedings of fifth eastern pine and meadow vole symposium (R. E. Byers, ed.), Gettysburg, Pennsylvania, 144 pp.

FRANK, F. 1957a. The causality of microtine cycles in Germany. J. Wildlife Mgmt., 21:113–121.

———. 1957b. Zucht und Gefangenschafts–Biologie der Zwergmaus. Z. Säugetierk., 22:1–43.

HAFNER, M. S., and D. J. HAFNER. 1979. Vocalizations of grasshopper mice (genus *Onychomys*). J. Mamm., 60:85–94.

HAMILTON, W. D. 1971. Geometry for the selfish herd. J. Theoret. Biol., 31:295–311.

HAMILTON, W. J., JR. 1934. Habits of *Cryptotis parva* in New York. J. Mamm., 15:154–155.

HANSSON, L. 1984. Winter reproduction of small mammals in relation to food conditions and population dynamics. This volume.

HAWES, M. L. 1975. Ecological adaptations in two species of shrews. Unpublished Ph.D. dissert., Univ. British Columbia, Vancouver, 211 pp.

HOWARD, W. E. 1949. Dispersal, amount of inbreeding, and longevity in a local population of prairie deermice on the George Reserve, Southern Michigan. Contrib. Lab. Vert. Biol., Univ. Michigan, 43:1–50.

———. 1951. Relation between low temperature and available food to survival of small rodents. J. Mamm., 32:300–312.

JANNETT, F. J., JR. 1978. The density-dependent formation of the extended maternal families of the montane vole, *Microtus montanus nanus.* Behav. Ecol. Sociobiol., 3:245–263.

JARMAN, P. J. 1974. The social organization of antelope in relation to their ecology. Behaviour, 48:215–267.

KALELA, O. 1957. Regulation of reproduction rate in subarctic populations of the vole *Clethrionomys rufocanus* (Sund.). Ann. Acad. Sci. Fennicae, Ser. A4, 34:1–60.

KOENIG, L. 1960. Das Aktionssystem des Siebenschläfers (*Glis glis* L.). Z. Tierpsychol., 17:427–505.

KUCHERUK, V. V., and T. N. DUNAYEVA. 1948. Materials on the abundance dynamics of Brandt's vole. Mater. po Gryzunam, 3.

MADISON, D. M. 1980. Space use and social structure in meadow voles. Behav. Ecol. Sociobiol., 1:65–71.

MERRITT, J. F., and J. M. MERRITT. 1978. Seasonal home ranges and activity of small mammals of a Colorado subalpine forest. Acta Theriol., 23:195–202.

METZGAR, L. H. 1979. Dispersion patterns in a *Peromyscus* population. J. Mamm., 60:129–145.

MICHIELSON, N. C. 1966. Intraspecific aggression and social organization in the shrews *Sorex araneus* L. and *S. minutus* L. Arch. Néerlandaises Zool., 17:73–174.

MYLLYMÄKI, A. 1977. Intraspecific competition and home range dynamics in the field vole *Microtus agrestis.* Oikos, 29:553–569.

NAUMOV, N. P. 1972. The ecology of animals [English translation (N. D. Levine, ed.)], Univ. Illinois Press, Urbana, x + 650 pp.

NICHOLSON, A. J. 1941. The homes and social habits of the woodmouse (*Peromyscus leucopus novaboracensis*) in southern Michigan. Amer. Midland Nat., 25:196–223.

PEARSON, O. P. 1947. The rate of metabolism of some small mammals. Ecology, 28:127–145.

———. 1960. The oxygen consumption and bioenergetics of harvest mice. Physiol. Zool., 33:152–160.

REICHSTEIN, H. 1962. Beiträge zur Biologie eines Steppennagers, *Microtus (Phaeomys) brandti* (Radde, 1861). Z. Säugetierk., 27:146–163.

RUFFER, D. G. 1968. Agonistic behavior of the northern grasshopper mouse (*Onychomys leucogaster breviauritus*). J. Mamm., 49:481–487.

SEALANDER, J. A. 1952. The relationship of nest protection and huddling to survival of *Peromyscus* at low temperature. Ecology, 33:63–71.

———. 1966. Seasonal variation in hemoglobin and hematocrit values in the northern red-backed mouse, *Clethrionomys rutilus dawsoni* (Merriam) in interior Alaska. Canadian J. Zool., 44:213–224.

———. 1967. Reproductive status and adrenal size in the northern red-backed vole in relation to season. Int. J. Biometeorol., 2:213–220.

SHILLITO, J. F. 1963. Observations on the range and movements of a woodland population of the common shrew *Sorex araneus.* Proc. Zool. Soc. London, 140:533–546.

SORENSON, M. W. 1962. Some aspects of water shrew behavior. Amer. Midland Nat., 68:445–462.

Tast, J. 1966. The root vole, *Microtus oeconomus* (Pallas), as an inhabitant of seasonally flooded land. Ann. Zool. Fennici, 3:127–171.

Thompsen, H. P. 1945. The winter habits of the northern white-footed mouse. J. Mamm., 26:138–142.

Waser, P. M. 1981. Sociality or territorial defense? The influence of resource renewal. Behav. Ecol. Sociobiol., 8:231–237.

Webster, A. B. 1979. A radiotelemetry study of social behavior and activity of free-ranging meadow voles, *Microtus pennsylvanicus*. Unpublished M.S. thesis, Univ. Guelph, Guelph, 110 pp.

Webster, A. B., and R. J. Brooks. 1981a. Daily movements and short activity periods of free-ranging meadow voles *Microtus pennsylvanicus*. Oikos, 37:80–87.

———. 1981b. Social behavior of *Microtus pennsylvanicus* in relation to seasonal changes in demography. J. Mamm., 62:738–751.

West, S. D. 1977. Midwinter aggregation in the northern red-backed vole, *Clethrionomys rutilus*. Canadian J. Zool., 55:1404–1409.

Wiegert, R. G. 1961. Respiratory energy loss and activity pattern in the meadow vole, *Microtus pennsylvanicus*. Ecology, 42:245–253.

Wolff, J. O. 1980. Social organization of the taiga vole (*Microtus xanthognathus*). The Biologist, 62:34–45.

Wolff, J. O., and B. Hurlbutt. 1982. Day refuges of *Peromyscus leucopus* and *Peromyscus maniculatus*. J. Mamm., 63:666–668.

Wolff, J. O., and W. Z. Lidicker, Jr. 1980. Population ecology of the taiga vole, *Microtus xanthognathus*, in interior Alaska. Canadian J. Zool., 58:1800–1812.

———. 1981. Communal winter nesting and food sharing in taiga voles. Behav. Ecol. Sociobiol., 9:237–240.

Zimmerman, K. 1952. Guttungstypische Verhaltensformen von Gelbhals-, Wald- und Brandmaus. Zool. Garten, 22:162–171.

Address: Wildlife Science Group, College of Forest Resources AR-10, University of Washington, Seattle, Washington 98195.

Present address (Dublin): Department of Zoology, University of British Columbia, Vancouver, British Columbia V6T 1W5, Canada.

OVERWINTERING ACTIVITY OF *PEROMYSCUS MANICULATUS,* *CLETHRIONOMYS GAPPERI, C. RUTILUS,* *EUTAMIAS AMOENUS,* AND *MICROTUS PENNSYLVANICUS*

LUCIUS L. STEBBINS

ABSTRACT

Seasonal changes in circadian rhythms and short term rhythms of activity were studied at 3 latitudes in Western Canada for *Clethrionomys gapperi, C. rutilus, Peromyscus maniculatus, Eutamias amoenus,* and *Microtus pennsylvanicus.* All species tested were housed individually in cages placed in natural settings and provided with insulated nest boxes and food and water *ad libitum.* Cages were left undisturbed except for feeding and cleaning through summer, fall, and winter seasons for all species and through spring for all except *M. pennsylvanicus.* Snow covered and filled cages in winter.

Three types of seasonal changes in circadian rhythms of activity

were observed: decreases in levels of activity in winter, changes in their nocturnal or diurnal nature, and changes in the shape of the daily peak of activity. Three types of changes in short term rhythms of activity were observed in *C. gapperi*—decreases in duration of short active periods in winter, a decrease in number of short active periods causing the daily peak of activity in winter, and changes in the duration of certain short active periods as a function of photoperiod. Changes in activity that are related to strategies of survival in winter involved foraging, hoarding, nesting, thermogenesis, and reproduction.

INTRODUCTION

Small mammals are adapted to seasonal environmental changes at northern latitudes in a variety of behavioural and physiological ways, collectively known as acclimatization. They are primarily oriented toward two goals: survival in colder months when energy is more scarce and in greater demand for thermogenesis; growth and reproduction in warmer months when energy is more easily available and in greater demand for these functions. Acclimatization must also be oriented toward other requirements, including predation, competition, nesting, nutrition, and territoriality. Among the mechanisms of acclimatization are seasonal adjustments in activity. As all other behaviour occurs within the context of daily patterns of activity, it is a basic aspect of acclimatization, yet it has had but little study, and this primarily oriented toward descriptions of circadian rhythms of activity rather than toward their relationships to behaviour and ecology. Such studies are difficult to do on unrestrained mammals in subnivean environments, but once sufficient descriptive information has been

gained on activity, ecological and behavioural inferences can be suggested.

Two separate and endogenous rhythms of activity exist, one with a period of about 24 hr and one with a period of 2 to 3 hr, called circadian and short term rhythms, respectively. Both are controlled by a continuously active physiological clock mechanism whose basic action is environmentally independent, but whose various adjustments are responsive to environmental cues. The same physiological clock is thought to be involved with both rhythms. Its nature is highly speculative.

The purpose of this paper is to describe seasonal changes in circadian and short term rhythms of activity of 5 species of small northern mammals, to discuss a causal relationship between the two separate rhythms, and to correlate changes to aspects of feeding behavior, thermogenesis, growth, reproduction, and "Population rhythms" (Daan and Slopsema, 1978). The species studied were *Clethrionomys gapperi, C. rutilus, Peromyscus maniculatus, Microtus pennsylvanicus,* and *Eutamias amoenus.*

METHODS

Conditions and locations of each study are presented in Table 1. The general kinds of seasonal changes observed in circadian rhythms found are listed in Table 2. The general description of a short term rhythm is given in Fig. 1. The general types of seasonal changes observed in short term rhythms are given in

Table 3. The general ways in which circadian rhythms and their seasonal changes are thought adaptive are given in Table 4. The general ways in which short term rhythms and their seasonal changes are thought adaptive are given in Table 5.

Table 1.—*Reference to experimental conditions reported in this paper.*

Species	Dates	Location	Recording	Food	Conditions of experiment	Published reference
Clethrionomys gapperi	December 1964–June 1965	Edmonton, Alta. 53°N, 113°W	At nest	Seeds and prepared pellets	Caged in natural environment	Stebbins, 1972, 1974
Clethrionomys gapperi	December 1965 May–June 1966	Heart Lake, N.W.T. 69°N, 116°W	At nest	Seeds and prepared pellets	Caged in natural environment	Stebbins, 1972
Clethrionomys rutilus	December 1965 May–June 1966	Heart Lake, N.W.T. 60°N, 116°W	At nest	Seeds and prepared pellets	Caged in natural environment	Stebbins, 1972
Peromyscus maniculatus	November 1965–June 1966	Heart Lake, N.W.T. 60°N, 116°W	At nest	Seeds and prepared pellets	Caged in natural environment	Stebbins, 1971
Peromyscus maniculatus	August 1968– March 1969	Lethbridge, Alta. 49°N, 112°W	At nest At food	Seeds and prepared pellets	Caged in seminatural environment	Stebbins, 1978
Eutamias amoenus	August 1968– May 1969	Lethbridge, Alta. 49°N, 112°W	At nest At food At wheel	Seeds and prepared pellets	Caged in seminatural environment	Stebbins, 1977
Microtus pennsylvanicus	August–December 1977	Lethbridge, Alta. 49°N, 112°W	Anywhere in cage	Greenery and prepared pellets	Caged in natural environment	Stebbins, in preparation

RESULTS

The seasonal changes observed in circadian rhythm of *P. maniculatus* at Heart Lake are given in Fig. 2. Points to note are the nocturnal timing of the daily peak of activity in late fall, winter, and early spring, the extension of the daily peak of activity into the first 6 hr of morning daylight in late spring, the increase in amplitude of the daily peak in April, and the bimodal nature of the peak in spring. The animals were regularly observed in torpor in winter when snow covered and filled their cages.

The seasonal changes observed in circadian rhythms of *P. maniculatus* tested at Lethbridge, Alberta are given in Fig. 3. Points to note are nocturnal occurrence of the daily peak of activity, and the higher amplitude of the daily peak of activity in fall and spring. Torpor was also observed in these mice in winter.

The seasonal changes in circadian rhythms of *E. amoenus* tested at Lethbridge are given in Fig. 4. Points to note are the diurnal timing of the daily periods of high activity, the absence of any night-time activity at all, the absence of a single obvious peak of activity, and the decrease in amount of activity in winter months. When checked at infrequent intervals, chipmunks were almost always alert and active in their nest boxes in winter months, and nest boxes were filled with husked seeds. Hibernation was observed only once, and torpor was not observed.

Seasonal changes in circadian rhythms of *C. gapperi* tested at Heart Lake are given in Fig. 5. Points to note are the bimodal, nocturnal daily peak of

Table 2.—*Seasonal changes in circadian rhythms.*

1. Nocturnal or diurnal nature.
2. Amplitude of daily peak of activity.
3. Duration of daily peak of activity.
4. Modality of daily peak of activity.
5. General level of activity.

Table 3.—*Seasonal changes in short term rhythms.*

1. Length of short rest periods.
2. Length of short active periods.
3. Time of day when they are highly synchronized.
4. Number per day which are highly synchronized.

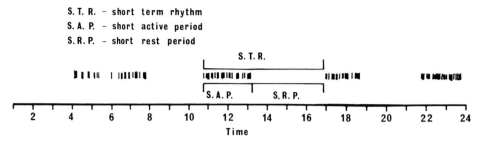

S. T. R. - short term rhythm
S. A. P. - short active period
S. R. P. - short rest period

Fig. 1.—Components of a short term rhythm.

activity in November, December, and May, the extension of the daily peak of activity into the first 6 hr of daylight in June, and the tendency for the daily patterns of activity in sequential test periods to closely follow each other, not just as expected during the daily peak of activity but throughout the remaining hours of the day. It should also be noted that extension of the daily peak of activity into the

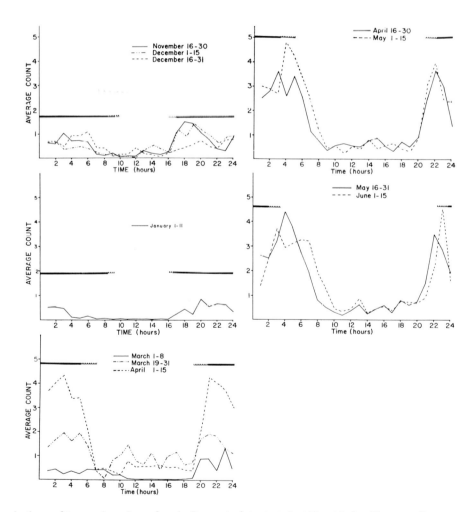

Fig. 2.—Circadian rhythms of two male and one female *P. maniculatus* tested at Heart Lake. The term "average count" refers to the average number of times a gate at the entrance of the best box was crossed by the animals. Time is given with hour 0100 through hour 2400 abbreviated to hour 1 through hour 24. Periods of darkness are indicated by vertical striped lines across the top of the graph. The short line with loops indicates the range of variation in photoperiod during the test period (from Stebbins, 1971).

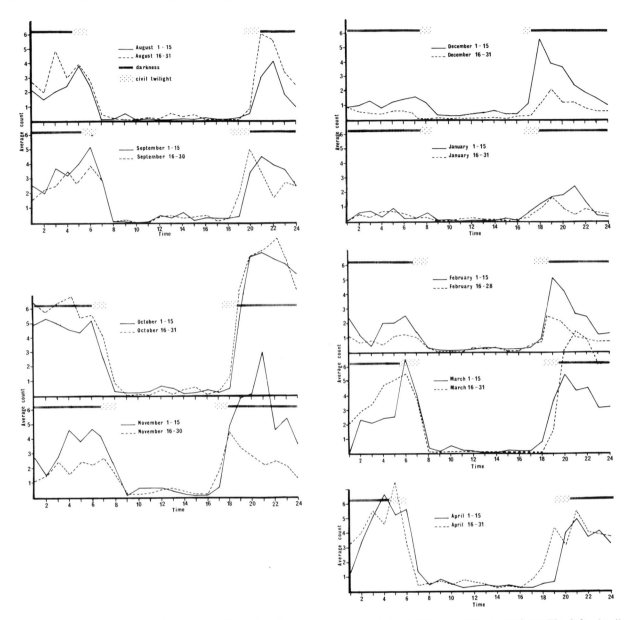

Fig. 3.—Circadian rhythms of two male and two female *P. maniculatus* tested at Lethbridge, Alberta. Refer to Fig. 2 for details.

daylight hours of morning involved the heightening of amplitude of a minor peak of activity which began to be seen in May. This species thus followed a pattern similar to that of *P. maniculatus* at Heart Lake, being basically nocturnal in winter and spring but, in late spring, when darkness lasted but a few hours and activity was at its highest, extending the peak of activity into the bright daylight of morning for several hours.

Seasonal changes in circadian rhythms of activity of *C. gapperi* tested at Edmonton are presented in Figs. 6 and 7. Points to note are the diurnal timing of the daily peak of activity in all seasons, the extension of the daily peak from a unimodal to a bimodal one in June, the generally higher level of activity in late spring than in late fall or winter, and the tendency for patterns of activity in sequential test periods to closely resemble each other.

Seasonal changes in circadian rhythms of *C. rutilus* tested at Heart Lake are given in Fig. 8. Points to note are the nocturnal timing of the daily peak of activity in May and June, the absence of an obvious peak in November and December, the development of amplitude of the daily peak in May,

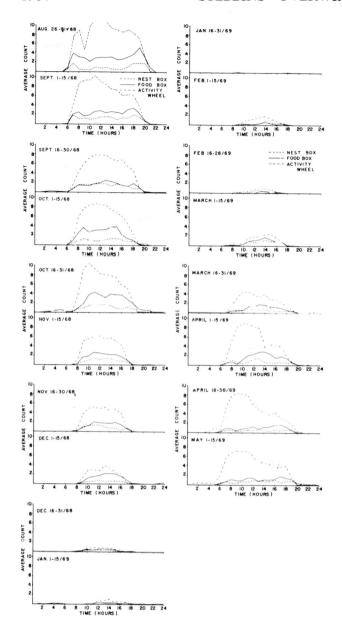

Fig. 4.—Patterns of activity of two male and two female *E. amoenus*. The term 'average count' refers to the average number of 5-min periods in which activity occurred in any hour. Time is given with hour 0100 through hour 2400 shortened to hour 1 through hour 24. Lightly stippled areas indicate periods of darkness (from Stebbins, 1977).

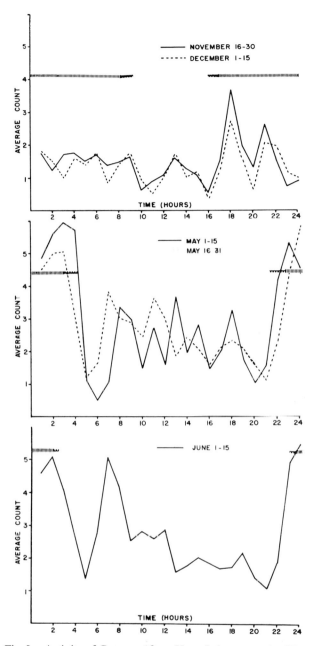

Fig. 5.—Activity of *C. gapperi* from Heart Lake as tested at Heart Lake. Two males and two females housed individually were tested. Refer to Fig. 2 for details (from Stebbins, 1972).

and again, the tendency for patterns of activity to closely follow each other throughout the entire day.

Seasonal changes in circadian rhythms of *M. pennsylvanicus* tested at Lethbridge are given in Fig. 9. Points to note are the absence of an obvious daily peak of activity in any season, the generally lower level of activity in colder months, and, once more,

the tendency for patterns of activity to closely resemble each other throughout the day in sequential test periods.

Seasonal changes in short term rhythms of activity were only studied in two species of voles. Characteristics of these rhythms for *C. gapperi* studied at Heart Lake at given in Fig. 10. Points to note are the two sequential pulses of short active periods

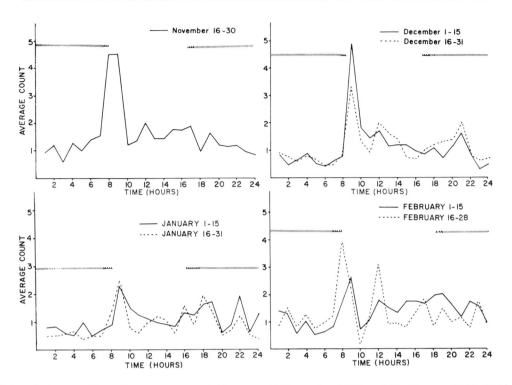

Fig. 6.—Activity of *C. gapperi* from Edmontson as tested between 16 November 1964, and 28 February 1965, at Edmonton. Three males and three females housed individually were tested. Refer to the legend of Fig. 2 for details (from Stebbins, 1972).

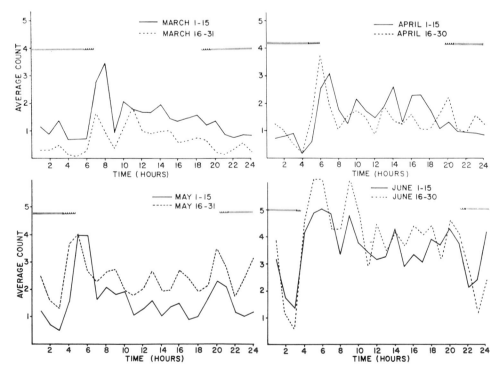

Fig. 7.—Activity of *C. gapperi* from Edmonton as tested between 1 March and 30 June 1965, at Edmonton. Three males and three females housed individually were tested. Refer to the legend of Fig. 2 for details (from Stebbins, 1972).

Table 4.—*Proposed adaptive values of circadian rhythms.*

1. Conserve energy by decreasing amount used in activity (Stebbins, 1977; Weigert, 1961).
2. Conserve energy by increasing insulative affect of nests (Glaser and Rustick, 1975; Menaker, 1969).
3. A component of periodic relations of community (Miller, 1955; Allee et al., 1949:545).
4. Help decrease susceptibility to predation (Cloudsley-Thompson, 1960; Madison, 1978a; Daan and Slopsema, 1978).
5. Affect competition (Kikkawa, 1964; Hamilton, 1937).
6. Affect reproduction (Cloudsley-Thompson, 1960; Herman, 1977; Madison, 1978b).
7. Avoidance of unfavorable environmental conditions (Cloudsley-Thompson, 1960).
8. Affect longevity (Saint Paul and Aschoff, 1978).
9. Affect feeding (Cloudsley-Thompson, 1960; Park, 1935; Lehmann, 1976).
10. Help capture prey (Cloudsley-Thompson, 1960; Eckert and Kracht, 1978).
11. Adjustment of behavior to changing density of population (Ambrose, 1973).
12. Maintenance of synchrony with rhythmic external events (Menaker, 1969).
13. Navigation (Menaker, 1969).
14. Use as biological clocks (Menaker, 1969; Pittendrigh and Minis, 1964).
15. Maintenance of synchrony with various internal physiological events (Menaker, 1969).

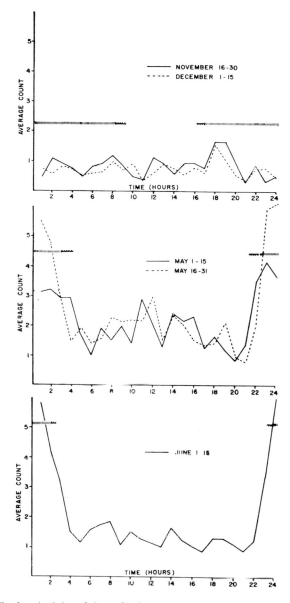

Fig. 8.—Activity of *C. rutilus* from Heart Lake as tested at Heart Lake. Two males and two females housed individually were tested. Refer to the legend of Fig. 2 for details (from Stebbins, 1972).

beginning after dark in November and December, the synchrony of these pulses on sequential days, the occurrence of nearly continuous activity in darkness in May and June, which is followed by a short rest period and short active period after dawn, and the minor tendency for short active periods and short rest periods to occur at the same time on sequential days in daylight hours.

Seasonal changes in short active periods of *C. gapperi* studied at Edmonton are given in Fig. 11. Points to note are the synchronous occurrence of a

Table 5.—*Proposed adaptive values of short term rhythms of activity.*

1. Meeting basic metabolic needs of feeding, drinking, excretion (Lehmann, 1976).
2. Fundamental units of circadian organizations (Honma, 1977).
3. Adaptive mechanism allowing seasonal and latitudinal adjustments in circadian rhythms (Stebbins, 1968, 1974; Erkinaro, 1969).
4. Synchronization of activity among members of a population (Stebbins, 1974; Daan and Slopsema, 1978).
5. Decrease susceptibility to predation (Daan and Slopsema, 1978).

short active period at dawn in winter, the relatively short duration of short active periods in winter, relative to spring, and the continuously increasing duration of the short active periods after dawn throughout April, May and June. The duration of short rest periods varies inversely with that of short active periods, while total length of short term rhythms tends to remain unchanged (Fig. 12). It should also be noted that a short rest period occurs prior to dawn in all seasons.

Characteristics of the short term rhythms of *M.*

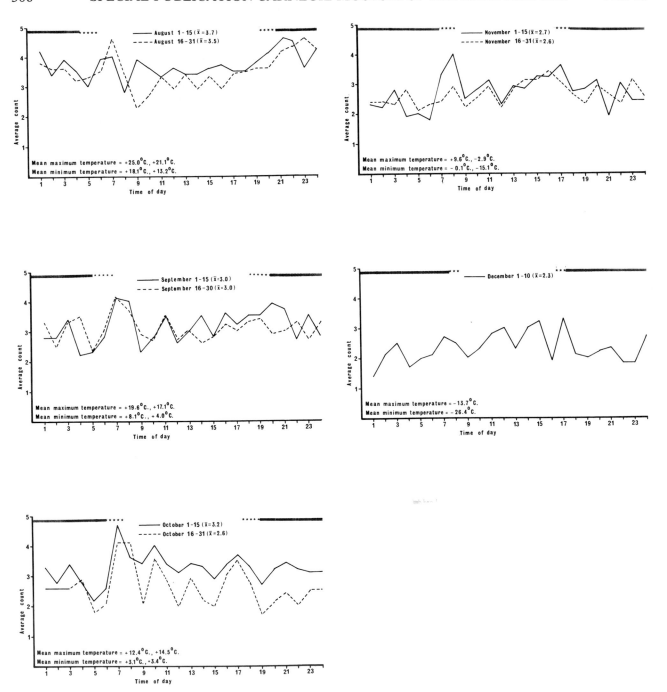

Fig. 9.—Circadian rhythms of activity of *M. pennsylvanicus* expressed as the mean number of 10 minute periods in which any activity occurred for each hour of the day. Mean levels of activity for each test interval are expressed as the mean number of 10 minute periods per hour in which any activity occurred during the interval. Darkness and civil twilight are indicated by the horizontal lines and stars, respectively.

pennsylvanicus are given in Fig. 13. Points to note are the lack of change evident as fall progresses, the tendency toward synchrony of two sequential short term rhythms in the first few hours of daylight, 0600 to 1100 hrs each day, and the tendency for this synchrony to lessen as the day proceeds.

Table 6.—*Detailed changes observed in patterns of activity.*

Species and latitude	Characteristics of pattern of activity	
	Winter	Summer
Clethrionomys gapperi—60°N	bimodal peak lower amplitude peak nocturnal about 6 hours of peak activity 2 synchronized short term rhythms	trimodal peak higher amplitude peak nocturnal and diurnal about 10 hours of peak 4 synchronized short term rhythms
Clethrionomys gapperi—53°N	unimodal peak higher amplitude peak without snow lower amplitude peak with snow diurnal peak about 2 hours of peak activity generally lower level of activity 1 synchronized short term rhythm	bimodal peak diurnal peak about 6 hours of peak activity generally higher level of activity 2 synchronized short term rhythms higher amplitude peak
Clethrionomys rutilus—60°N	unimodal peak lower amplitude peak nocturnal about 2 hours of higher activity	unimodal peak higher amplitude peak nocturnal about 4 hours of higher activity
Peromyscus maniculatus—60°N	unimodal peak lower amplitude peak nocturnal about 5 hours of peak activity	bimodal peak higher amplitude peak about 10 hours of peak activity generally higher levels of activity nocturnal and diurnal
Peromyscus maniculatus—49°N	unimodal peak lower amplitude peak about 10 hours of peak activity nocturnal	bimodal peak higher amplitude peak about 4 hours of peak activity nocturnal
Eutamias amoenus—49°N	no single peak of activity diurnal nearly inactive all day	no single peak of activity diurnal nearly inactive all day
Microtus pennsylvanicus—49°N	no obvious daily peak nocturnal and diurnal 2 synchronized short term rhythms	no obvious daily peak nocturnal and diurnal 2 synchronized short term rhythms

DISCUSSION

A summary of the seasonal changes in patterns of activity noted in each species is given in Table 6. The most complete and interesting observations in this table are of *C. gapperi*. I believe a single mechanism is associated with both the synchrony and seasonal and latitudinal flexibility in this species; specifically that all of the changes noted in amplitude, duration, modality, timing and synchrony of circadian patterns of activity result from a biological clock mechanism controlling three characteristics of its short term rhythms. If a single short term rhythm occurs synchronously among members of a population at about the same time each day, it produces a unimodal daily peak of activity. If two sequential short term rhythms occur synchronously in about

the same time interval each day, they may produce a bimodal daily peak of activity, with the trough between the peaks representing the intervening short rest period. Extension of a shorter unimodal peak to produce a longer, bimodal peak, is accomplished by addition of a second short term rhythm. Changes in amplitude of daily peaks are accomplished by an increase in duration of the short active period and an equal decrease in duration of the short rest period of the same short term rhythms. Differences in nocturnal or diurnal nature of circadian rhythms depend on whether synchronization of one or more short term rhythms occurs at dawn or at dusk. Evidence that these are the actual mechanisms involved is limited to *C. gapperi,* and, to a lesser degree, to

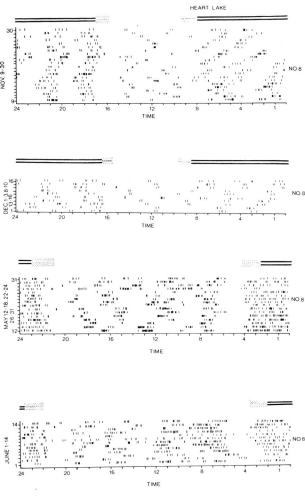

Fig. 10.—A transcription of the raw data for test animal No. 8 at Heart Lake for November, December 1965 and May, June 1966. Time is given with hour 0100 through 2400 abbreviated to hour 1 through 24. Each division on the ordinate is for positioning of data for a separate day. Each small vertical line indicates one crossing in either direction of a gate at the entrance to the nest box. The dark double line across the top of the graphs indicate time of darkness. The stipled areas terminating each double line indicate the twilight period between sunrise or sunset and first or last light of an intensity of about 21×. No data was available for several days in December and 19 and 25 May (from Stebbins, 1975).

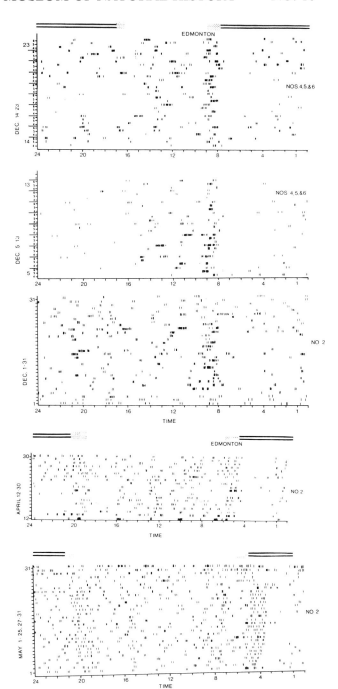

Fig. 11.—A transcription of raw data for test animals for December 1964 and April 1965 at Edmonton. Data for No. 4–6 are grouped to demonstrate synchrony of occurrence among individuals of the short activity period between 0800 and 0900. Refer to legends of Fig. 2 for details.

M. pennsylvanicus. All seasonal and latitudinal changes in circadian rhythms of these species are explained by the appropriate changes in their short term rhythms noted above. Similar observations and conclusions have been made by Erkinaro (1969) for several other species.

Thus it seems that at least some species have become adapted to the environmental extremes, imposed by seasonal and latitudinal differences in

northern environments, through development of this behavioural mechanism involving interactions of two separate rhythms of activity, one with a 24 hr period and one with a 2 to 3 hr period. Using this

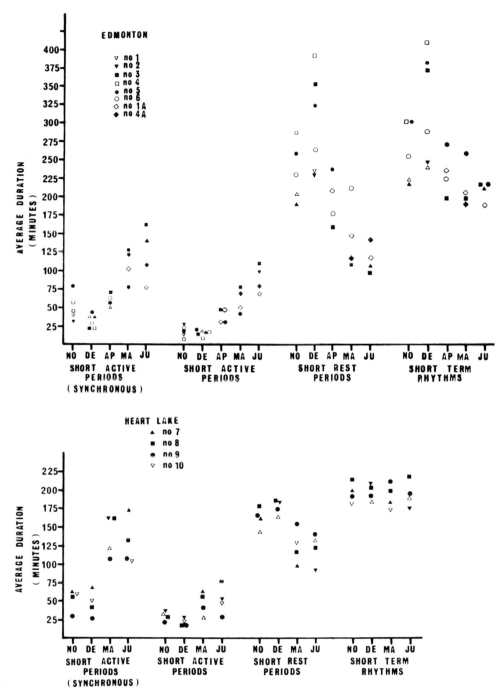

Fig. 12.—A summary of mean duration of short term rhythms, short active periods, and short rest periods for each month for each animal tested at Heart Lake and Edmonton. Values for ∝ short active periods at Edmonton and ∝ and β short active periods at Heart Lake are labelled as "synchronous" and presented apart from the others.

system at least one species, *C. gapperi,* can be either nocturnal or diurnal and can have a daily peak of activity of varying modality, duration, and amplitude at different latitudes and seasons. Studies of the other species are less complete, but are suggestive in that they could involve the same sort of changes.

In addition to the flexibility in patterns of activity observed, the nature of the patterns themselves are presumably adaptive and relate to natural history

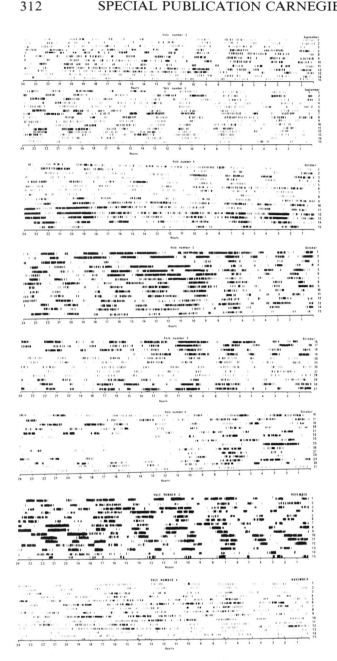

Fig. 13.—A direct transcription of raw activity charts of *Microtus Pennsylvanicus* at Edmonton to demonstrate existence and degree of synchronization of short term rhythms of activity.

of the species. Decreases in total amount of activity in winter conserves energy needed for thermogenesis, as do the corresponding increases in amount of time spent in insulative nests. This change in activity must be related to foraging, torpor, and hibernation. The species in these studies used highly divergent strategies in this regard. They form a continuum from active, non-hoarding species through non-torpid hoarders and torpid hoarders to intermittently hibernating hoarders.

M. pennsylvanicus stored no food in the study and thus depended on regular feeding during short active periods day and night for maintenance. Its general level of activity was highest of the species studied in winter. It had no obvious daily peak of activity, but instead had its activity distributed more evenly throughout the light and dark hours. *Clethrionomys* spp. stored nearly all food available during the study in all seasons. Such food stores are presumably used in winter and allow the species to minimize foraging trips in winter, when food is hardest to obtain and microclimate is relatively cold. They had an obvious daily peak of activity, as did all other hoarding species except *E. amoenus*. Their various indicators of levels of activity, including amplitude, duration and modality of daily peaks of activity were lower in winter as were general levels of activity throughout the nonpeak hours of the day. *P. maniculatus* stored nearly all food available to it in all seasons. It was frequently seen in torpor at Heart Lake and occasionally at Lethbridge. Its various indicators of levels of activity were all considerably lower than those of *Clethrionomys* spp. in winter at both latitudes. *E. amoenus* filled its nests with husked seeds and seldom left the nests in colder months. Though it does not store fat like typical hibernators, it can become torpid in winter and enter typical hibernation though it was seen to do so only once in these studies. Its levels of activity in winter were lowest of all species studied. Unlike the other species which stored food, it did not have a single daily peak of activity, but was active nearly continuously during daylight hours in warmer months and was never active outside the nests at night.

This suggests a relationship between feeding habits, nesting, thermogenesis, and activity. The species, which has no torpor, does no hoarding and must, therefore, forage at regular intervals, has no obvious daily peak of activity, is equally active in daylight and darkness, and is more nearly equally active in summer and winter. The species which can hibernate and does hoard could be considered at the other extreme of the spectrum of behaviour reported here; it is highly active during all hours of daylight, but not at all in darkness and is highly active in summer, but nearly completely inactive outside its nest in winter. The hoarding, non-torpid species is intermediate, but is closest to the non-hoarding species, and the hoarding torpid species is closest to the hoarding occasional hibernator. Thus the continuum of feeding behaviour and of thermogenic behaviour complement the differences in patterns of activity of each species.

The adaptive value of these continua presumably reflect other aspects of the physiology and energetics. Each of these species is known to stop growing and reproducing in winter months and to start rapid growth again prior to or coinciding with onset of reproduction in spring (Fuller et al., 1969; Iverson and Turner, 1974; Stebbins, 1977, 1978). *E. amoenus* and *P. maniculatus* are known to require less food in winter months and to require more food during rapid growth and reproduction in spring (Stebbins, 1977, 1978). Thus the seasonal changes observed in activity complement these seasonal changes in growth, reproduction, and food requirements. It is generally known that small mammals conserve energy in winter by being less active and staying longer in insulative nests. In spring, activity increases coincidently with physiological and environmental changes.

Both circadian and short term rhythms of activity presumably serve another general adaptive function, that of synchronizing periods of activity among members of a population. The phenomenon of synchronization is commonly documented in studies of circadian rhythms, but less frequently so among studies of short term rhythms. Such synchronization during the daily peaks of activity common to circadian rhythms are presumably adaptive for various aspects of territoriality, reproduction, predation, and competition for food, space, and nesting sites. Any adaptive advantage accruing from synchronous activity during the single daily peak of activity is likely to be similarly adaptive during synchronous short active periods. Several authors have noted a tendency for such synchronization, always with the following pattern (Daan and Slopsema, 1978); each day synchrony is greatest for one brief period of time, coinciding with the daily peak of activity; it usually occurs at dawn or dusk and may involve more than one short term rhythm. Synchrony tends to fall off during the remainder of the day but is renewed at the same time the next day. Though the tendency for synchronous occurrence of short term rhythms is lessened during non-peak hours, upon averaging data from several animals over several days it is sufficient to produce the remarkable similarity of pattern reported here in *C. gapperi* and *M. pennsylvanicus*. Presumably the same mechanism was operating to produce the synchrony observed in *C. rutilus* and *P. maniculatus*. Experimental evidence demonstrates one adaptive advantage gained by such a synchrony is a decreased susceptibility to predators (Daan and Slopsema, 1978). Other intraspecific interactions are at least potentially affected by such synchrony, but experimental studies on this subject have yet to be done.

Thus, in summary circadian and short term rhythms of activity are correlated with patterns of feeding, hoarding, thermogenesis, growth, reproduction and nesting. They also serve to produce daily and hourly pulses of synchronous activity among members of a population, the adaptive value of which requires further study. Short term rhythms also provide a mechanism whereby internal structure of circadian rhythms can be adjusted to season and latitude.

LITERATURE CITED

ALLEE, W., A. EMERSON. O. PARK, T. PARK, and K. SCHMIDT. 1949. Principles of Animal Ecology. Saunders, Philadelphia, 837 pp.

AMBROSE, H. 1973. An experimental study of some factors affecting the spatial and temporal activity of *Mirotus pennsylvanicus*. J. Mamm., 54:79–110.

CLOUDSLEY-THOMPSON, J. 1960. Adaptive functions of circadian rhythms. Proceedings of the Cold Springs Harbor Symposium in quantitative biology, 25:345–357.

DAAN S., and S. SLOPSEMA. 1978. Short-term rhythms in foraging behaviour of the common vole, *Microtus arvalis*. Jour. Comp. Physiol., 127:215–227.

ECKERT, H., and S. KRACHT. 1978. Evidence for ecological adoption of circadian systems. Ecologia, 32:71–78.

ERKINARO, E. 1969. Der Phasenwechsel der lokomotorischen aktivität bei *Microtus agrestis* (L.), *M. arvaslis* (Pall.), und *M. oeconomus* (Pall.). Aquilo, Ser. Zool., 8:1–31.

FULLER, W. A., L. L. STEBBINS, and G. R. DUKE. 1969. Overwintering of small mammals near Great Slave Lake, Northern Canada. Arctic, 22:34–55.

GLASER, H., and S. RUSTICK. 1975. Energetics and nesting behaviour of the northern white-footed mouse, *Peromyscus leucopus nova boracensis*. Physiol. Zool., 48:105–113.

HAMILTON, W. 1937. Activity and home range of the field mouse, *Microtus pennsylvanicus pennsylvanicus*. Ecology, 18:255–263.

HERMAN, T. 1977. Activity patterns and movements of subarctic voles. Oikos, 29:434–444.

HONMA, K. 1977. The mechanism of synchronization of endogenous biological rhythms. Hokkaido J. Med. Sci., 53:213–236.

IVERSON, S. L., and B. N. TURNER. 1974. Winter weight dynamics in *Microtus pennsylvanicus*. Ecology, 55:1030–1041.

KIKKAWA, J. 1964. Movement, activity, and distribution of the small rodents *Clethrionomys glareolus* and *Apodemus sylvaticus* in woodland. J. Anim. Ecol., 33:259–295.

LEHMANN, V. 1976. Short-term and circadian rhythms in the behaviour of the vole, *Microtus agrestis* (L.). Oecologia, 23: 185–199.

MADISON, D. 1978a. Behavioural and sociochemical susceptibility of meadow voles (*Microtus pennsylvanicus*) to snake predators. Amer. Midland Nat., 100:23–28.

———. 1978b. Movement indications of reproductive events among female meadow voles as revealed by radiotelemetry. J. Mamm., 59:835–843.

MENAKER, M. 1969. Biological Clocks. Bioscience, 19:681–689.

MILLER, R. S. 1955. Activity rhythms in the wood mouse, *Apodemus sylvaticus,* and the bank vole, *Clethrionomys glareolus.* Proc. Zool. Soc. London, 125:505–519.

PARK, O. 1935. Studies in nocturnal ecology. III. Recording apparatus and further analysis of activity rhythms. Ecology, 16:152–163.

PITTENDRIGH, C., and D. MINIS. 1964. The entrainment of circadian oscillations by light and their role as photoperiod clocks. Amer. Nat., 98:261–294.

SAINT PAUL, U., and J. ASCHOFF. 1978. Longevity among blow flies, *Phormia terraenovae* R.D. kept in non-24-hour light-dark cycles. J. Comp. Physiol., 127:191–195.

STEBBINS, L. L. 1968. Seasonal and latitudinal variations in circadian rhythms of three species of small rodents in northern Canada. Unpublished Ph.D. thesis, Univ. Alberta, Edmonton, 68 pp.

———. 1971. Seasonal variations of circadian rhythms of deermice in northwestern Canada. Arctic, 24:124–131.

———. 1972. Seasonal and latitudinal variations in circadian rhythms of red-backed voles. Arctic, 25:216–224.

———. 1974. Response of circadian rhythms in *Clethrionomys* mice to a transfer from 60°N to 53°N. Oikos, 25:108–113.

———. 1975. Short activity periods in relation to circadian rhythms in *Clethrionomys gapperi.* Oikos, 26:32–38.

———. 1977. Some aspects of overwintering in the chipmunk, *Eutamias amoenus.* Canadian J. Zool., 55:1139–1146.

———. 1978. Some aspects of overwintering in *Peromyscus maniculatus.* Canadian J. Zool., 56:386–390.

WEIGERT, R. 1961. Respiratory energy loss and activity patterns in the meadow vole, *Microtus pennsylvanicus pennsylvanicus.* Ecology, 42:245–253.

Address: Department of Biological Sciences, University of Lethbridge, Lethbridge, Alberta T1K 3M4, Canada.

OVERWINTERING BEHAVIORAL STRATEGIES IN TAIGA VOLES (*MICROTUS XANTHOGNATHUS*)

JERRY O. WOLFF

ABSTRACT

During winter, taiga voles (*Microtus xanthognathus*) form groups of 5 to 10 individuals and construct an underground midden, which consists of a communal nest and food store. The communal nest contains an average of 7.1 individuals and includes both sexes and all age groups. Midden groups may contain sisters, but most midden inhabitants do not appear to be closely related. Taiga voles remain active all winter, but over 90% of the diet consists of *Equisetum* and *Epilobium* rhizomes stored in the food cache. Animals lose about 25% of their body weight from September to December. The physiological and behavioral characteristics associated with communal nesting and food sharing are discussed with regard to adaptations for overwintering in severe boreal climates.

INTRODUCTION

The taiga vole (*Microtus xanthognathus*) is an uncommon microtine rodent which has a patchy distribution throughout the taiga of Northwest Canada and interior Alaska (Hall and Kelson, 1959; Youngman, 1975). Recent investigations on the population ecology and reproductive biology (Wolff and Lidicker, 1980), social organization (Wolff, 1980), and overwintering behavior (Wolff and Lidicker, 1981) of taiga voles have provided insight into the behavioral ecology and social biology of the species. The social biology of taiga voles is characterized by male territoriality and a polygynous mating system during spring and summer (Wolff, 1980), followed by a period of cooperative food storing and communal winter nesting (Wolff and Lidicker, 1981). This social organization, which had not previously been reported for a small mammal, is an apparent adaptation for survival in the severe boreal climate. In this paper I review the social biology of taiga voles, making special reference to behavioral adaptations associated with overwinter survival. The results were obtained from a series of studies conducted on a population of taiga voles in interior Alaska from 1975 to 1978 (Wolff and Lidicker, 1980; Wolff, 1980; Wolff and Lidicker, 1981).

The studies were conducted in an eleven-year old, post-burn, upland black spruce (*Picea mariana*) community 150 km northwest of Fairbanks, Alaska. The yearly climate is characterized by a short, but highly productive growing season from mid-May until mid-August and a long, cold winter with temperatures frequently reaching −40°C. A 70 to 80 cm snowcover typically persists from late September until early May. A variety of field and laboratory methods were used in the four-year study. These methods included live- and snap-trapping, excavating burrows and nest sites, monitoring nest temperatures with thermistors, identifying family groups by radio-isotope labeling, observing behavior in the laboratory, and conducting nesting studies in environmental chambers. For details on methods, see references previously cited.

RESULTS

SUMMER: SPACING AND SOCIAL STRUCTURE

In early spring following snowmelt, males are intrasexually aggressive and actively defend exclusive territories in riparian habitat (Wolff, 1980). Anywhere from one to four females establish overlapping home ranges within the territory of a given male. The average home range size for adult males and females during the breeding season is about 625 m² and does not differ significantly between sexes. The breeding season extends from early May until mid-July, with females having a maximum of two litters. All juveniles are weaned by late July. Dispersal of juveniles occurs during mid and late summer, with males having a tendency to move farther than females (Wolff and Lidicker, 1980). In late July and August, territorial boundaries break down, and both adults and juveniles work cooperatively to prepare for winter.

WINTER: COMMUNAL NESTING AND FOOD SHARING

During fall, five to 10 individuals form a group and dig a communal burrow system which includes a winter nest and an underground food cache (Wolff, 1980). The underground nest and food cache are referred to as a midden. The midden is located about 30 cm under the surface and often under a tangled mass of branches, roots, or a fallen log. Burrowing activities produce a mound 30 to 50 cm high and up to 2 m across. Midden density varies, but approximates 15–20/ha in good habitat. Nests have up to five entrance tunnels. The food cache, which is adjacent to the nest, has a single access from the nest. The nest is made of dried grass (*Calamagrostis*) and is about 25 to 30 cm in diameter. The food cache may be 80 cm² by 30 cm high, but is frequently smaller. The largest cache excavated contained 3.6 kg (dry weight) of rhizomes (about one bushel). The primary food items stored are rhizomes of horsetail (*Equisetum*) and fireweed (*Epilobium angustifolium*). These two foods are taken in proportion to their frequency in the habitat. *Equisetum* and *Epilobium* have a higher nutrient content than other plant foods in the study area (Wolff, unpublished data). Food is collected and stored from mid-August to mid-September. Taiga voles remain active all winter, but over 90% of their food consists of rhizomes stored in the food cache, and the remainder is obtained from foraging under the snow (Wolff and Lidicker, 1980).

During construction of middens in fall, each midden group contains an average of 7.1 individuals (Wolff and Lidicker, 1981). The composition of each group is random with respect to age and sex. The adult/juvenile ratio averages 1:6.1, but 40% of the midden groups contain no adults. The overall sex ratio of midden groups is 1:1, but Wolff and Lidicker (1981) reported one group with only juvenile females. Using radioisotopes to determine genetic relationships in taiga voles (Wolff and Holleman, 1978), Wolff and Lidicker (1981) concluded that most midden groups do not consist of family units. They followed the distribution of 8 radioactively-marked juveniles into winter middens. One male and three females overwintered in separate nests with nonsiblings. In two cases, pairs of full sibling sisters occupied the same midden. One adult female was caught at three different adjacent middens, where she had one daughter in each of the three middens. Other than this instance most middens appear to be exclusive, with minimal if any movement between groups. The degree of relatedness of most midden inhabitants has not been possible to obtain, so the possibility of half-sibs or more distant relatedness can not be ruled out.

Prior to entering middens in early winter, adult voles lose body weight from a mean of 120 g in July to less than 100 g in October (Wolff and Lidicker, 1980). No adults have been caught after October, so it is not known how much weight they lose or even if they survive until mid-winter. Juvenile and subadult animals also lose body weight in winter from a mean of 48 g in August to 30 g by December. A loss in body weight is accompanied by a reduction in size of reproductive organs and scent glands. Winter reproduction has not been reported for taiga voles.

Wolff and Lidicker (1981) monitored nest, ground, and air temperatures at two field locations from early October until December 1977. They found that mean daily air temperatures ranged from −5° to −23°C, whereas ground temperatures ranged from −3° to −5°C. The mean daily nest temperatures throughout the study period ranged from +4° to +7°C and were never as low as ground temperatures. These data suggest that nests were never totally vacated and that one or several voles were in the nests at all times. This conclusion was supported by results from a nesting study in the laboratory. In an environmental chamber set at 0°C, 5 voles were placed in one nest and one vole in a separate nest. Food was provided *ad libitum* about 1 m from the nest. Nest temperature in the communal nest was higher than in the single nest and never reached ambient levels. When the single vole left its nest to feed (mean of five times a day), the nest temperature quickly approached ground or ambient temperature. Voles from the communal nest foraged one at a time, while the remainder stayed in the nest. The communal nest temperature ranged from 6° to 8°C above ambient temperature and never cooled to ambient levels. Thus, animals returned to a warm nest following a foraging bout.

DISCUSSION

Communal winter nesting and food sharing are cooperative behaviors which are apparently adaptive to surviving long, cold, winters in the boreal forest. See West and Dublin (1984) for an excellent review of behavioral strategies of small mammals under winter conditions. Large, well-insulated, grass nests with several entrances are constructed under logs, branches, or roots which provide protection from predation. The rhizomes of *Equisetum* and *Epilobium*, which are collected in late summer and early fall and stored close to the nest, are high in nutrients and provide a stable food source through the winter. The availability of these and other essential foods may be, in part, responsible for the patchy distribution and microhabitat preferences of taiga voles (Douglass and Douglass, 1977; Wolff and Lidicker, 1980). Readily accessible food stores reduce the time spent actively foraging and allow voles to remain in the nest most of the time.

The adaptive significance of communal winter nesting is twofold. Monitoring of nest temperatures in the field and in the laboratory showed that the nest was never totally vacated. Voles forage one at a time and return to a warm nest. This reduces heat loss which would normally be dissipated in warming up the nest area around the vole's body. Secondly, group huddling reduces the surface area of each animal which is exposed to the ambient temperature. This in turn, reduces heat loss to the environment by thermal conductance. Wolff (1980) concluded that the reduction of body weight was an adaptation to reduce the absolute food requirement per individual during winter. The increased surface area to volume ratio of a smaller animal, which would normally increase heat loss by thermal conductance, is apparently compensated for by group huddling. This and other studies have concluded that communal winter nesting is a mechanism for behavioral thermoregulation (Wolff and Lidicker, 1981; Howard, 1951; Webster and Brooks, 1981; West and Dublin, 1984).

Wolff and Lidicker (1981) have shown that most midden group members are not close relatives (siblings or parent-offspring units). Although they can not rule out the possibility of more distant relatives nesting together, their data on differential dispersal of male and female juveniles suggest that middens can potentially be established by a random sample of the population. Wolff (1980) has provided two explanations for the evolution of communal nesting and food sharing among nonrelatives. Predators,

such as martens (*Martes americana*) and short-tail weasels (*Mustela erminea*), are apparently the major winter predators of taiga voles. If either of these or other subnivean predators located a midden, they could potentially kill all the occupants. If the midden consisted of a family group, the whole genetic lineage could be eliminated. By differential dispersal of littermates and overwintering in separate nests, a higher proportion of genes may be passed on than if the family overwintered together as a unit.

An alternative, but not mutually exclusive, explanation for communal nesting with nonrelatives is inbreeding avoidance. During spring, dominant males claim territories within the range of their midden and will mate with females which overwintered in that home range area. Thus, differential pre-winter dispersal and subsequent overwintering with nonrelatives becomes a mechanism which reduces the probability of inbreeding. The evolutionary mechanisms for the origin of communal winter nesting and food sharing have been discussed in more detail by Wolff and Lidicker (1981).

Although winter social biology similar to that of taiga voles has not been described in any other small mammal, it may not be unique. Communal winter nesting has recently been described for meadow voles (*Microtus pennsylvanicus*) (Webster and Brooks, 1981), pine voles (*Microtus pinetorum*) and white-footed mice (*Peromyscus leucopus*) (Madison, 1984), and probably occurs in red-backed voles (*Clethrionomys rutilus*) (West, 1977). Food storing has been reported for meadow voles (Gates and Gates, 1980) and the singing vole (*Microtus miurus*) (Youngman, 1975). and I have found food caches of tundra voles (*Microtus oeconomus*) which contain rhizomes of *Equisetum*. The physiological and behavioral characteristics which have been described here and elsewhere in this volume are apparent adaptations for overwintering in an environment characterized by long and cold winters. I have no doubt that further investigations on the winter ecology of small mammals will reveal similar adaptations for living in boreal or arctic climates.

ACKNOWLEDGMENTS

I thank Joe Merritt and the Director of the Carnegie Museum for inviting me to participate in this symposium and to share these ideas with my fellow biologists. This work was supported by Cooperative Aid Agreements 24-PNW-76, 32-PNW-77, and 41-PNW-78 from the Pacific Northwest Forest and Range Experiment Station, USDA Forest Service, in Fairbanks, Alaska. I thank two anonymous reviewers for helpful comments on earlier drafts of this manuscript.

LITERATURE CITED

DOUGLASS, R. J., and K. S. DOUGLASS. 1977. Microhabitat selection of chestnut-cheeked voles (*Microtus xanthognathus*). Canadian Field-Nat., 91:72–74.

GATES, J. E., and D. M. GATES. 1980. A winter food cache of *Microtus pennsylvanicus*. Amer Midland Nat., 103:407–408.

HALL, E. R., and K. R., KELSON. 1959. The mammals of North America. Ronald Press, New York, 1,162 pp.

HOWARD, W. E. 1951. Relation between low temperatures and available food to survival of small rodents. J. Mamm., 32: 300–312.

MADISON, D. M. 1984. Group nesting and its ecological and evolutionary significance in overwintering microtine rodents. This volume.

WEBSTER, A. B., and R. J. BROOKS. 1981. Social behavior of *Microtus pennsylvanicus* in relation to seasonal changes in demography. J. Mamm., 62:738–751.

WEST, S. D. 1977. Midwinter aggregation in the northern red-backed vole (*Clethrionomys rutilus*). Canadian J. Zool., 55: 1404–1409.

WEST, S. D., and H. T. DUBLIN. 1984. Behavioral strategies of small mammals under winter conditions. This volume.

WOLFF, J. O. 1980. Social organization of the taiga vole (*Microtus xanthognathus*). The Biologist, 62:34–45.

WOLFF, J. O., and D. F. HOLLEMAN. 1978. Use of radioisotope labels to establish genetic relationships in free-ranging small mammals. J. Mamm., 59:859–860.

WOLFF, J. O., and W. Z. LIDICKER, JR. 1980. Population ecology of the taiga vole, *Microtus xanthognathus*, in interior Alaska. Canadian J. Zool., 48:1800–1812.

———. 1981. Communal winter nesting and food sharing in taiga voles. Behav. Ecology and Sociobiol., 9:237–240.

YOUNGMAN, P. M. 1975. Mammals of the Yukon Territory. Nat. Mus. Nat. Sci., Ottawa, 192 pp.

Address: Department of Biology, University of Virginia, Charlottesville, Virginia 22901.

MICROTUS RICHARDSONI MICROHABITAT AND LIFE HISTORY

Daniel R. Ludwig

ABSTRACT

Microtus richardsoni, the water vole, is unique among the microtines because of its large body size, small populations, limited geographic range and a high degree of habitat specificity. Animals restrict their activity to areas of *Salix*, grasses, sedges and mesic forbs in close proximity to alpine and subalpine streams and exist in a subnivean environment 7–8 months each year.

Water voles occurred repeatedly at certain streamside sites, but extensive areas of seemingly suitable habitat adjacent to occupied areas were not utilized. Data collection was designed to provide a quantitative profile of the type of stream occupied by water voles and identify the characteristics important in the selection of a home area.

Three major questions regarding water vole distribution were examined: What limits the species distribution to streamside montane meadows? What types of sites are selected within these macrohabitats? Why are such microhabitats selected? Snow, temperature and microhabitat data are discussed in relation to the species' life history traits.

INTRODUCTION

Several characteristics make the water vole, *Microtus richardsoni* (DeKay, 1842), an interesting species for the study of the evolution of population characteristics and life history traits, and for comparison with more widespread and thoroughly studied microtine rodents.

The water vole is among the largest of the North American microtines. Total body lengths ranging from 198 to 274 mm and body weights ranging from 66 to 132 g have been reported for adult animals (Banfield, 1958; Hall, 1981; Hooven, 1973).

The species' geographic range is restricted to the mountains of northwestern North America (Hall, 1981). Within this limited range, water voles demonstrate a strong preference for small habitat patches of dwarf willow, birch, grasses, mesic forbs and mosses within close proximity to alpine and subalpine streams. *M. richardsoni* generally occurs between 1,524 to 2,378 m in Canada (Banfield, 1974) and 914 to 3,201 m in the United States (Hall, 1981).

The vole's geographic distribution is highly discontinuous, and large expanses of coniferous forest, mountain slopes and valleys present seemingly insurmountable barriers to overland movement between occupied habitats. Small groups (8–40) of these animals are distributed linearly along streams (Anderson et al., 1976; Hollister, 1912; Hooven, 1973; Pattie, 1967; Racey and Cowan, 1935). The crude densities reported in the literature are quite low by microtine standards.

Prior to this study, the species' population biology, habitat specificity and evolutionary strategy had not been approached in a holistic or multifactorial manner. The information available concerning social behavior and structure (Anderson et al., 1976; Jannett and Jannett, 1974; Pattie, 1967; Skirrow, 1969), habitat specificity and resource utilization (Koplin and Hoffmann, 1968), population characteristics and life history traits (Anderson et al, 1976; Brown, 1977; Pattie, 1967; Jannett et al., 1979; Taber et al., unpublished manuscript) was either rather speculative or tentative.

The main objective of this study (Ludwig, 1981) was to examine and describe the population biology and life history of a number of populations of *Microtus richardsoni richardsoni* in southwestern Alberta, Canada. Special emphasis was placed on: 1) examining the species' social organization, dispersion pattern, and resource and habitat utilization, 2) examining habitat and microhabitat characteristics to explain the local and scattered occurrence of the water vole along mountain streams, and 3) providing an explanation for the low population numbers characteristic of the species. This paper is a brief outline of the species' life history traits, habitat preferences and social organization, and the interaction of factors assumed to have occurred during the species' evolution.

MATERIALS AND METHODS

The four populations studied were located along the edges of gently winding and gradually sloping spring-fed streams at or below timberline 1,875 to 2,219 m in the eastern front ranges of the Canadian Rocky Mountains, 74 to 99 km to the southwest of Calgary, Alberta. Animals were live-trapped between June and September or October 1977 to 1979. The number of the sites trapped varied (4, 3, and 2 during 1977, 1978 and 1979, respectively).

Sherman and Elliott folding traps (26 cm and 33 cm in length, respectively), supplied with cotton batting nest material and baited with rolled oats, were set for 72 hr intervals and checked approximately every eight hours starting at 0600, 1300 and 2000 hr. Since the objective of the trapping program was a total enumeration of all animals within a study area, traps were set wherever evidence of water vole activity was found. The distance between individual traps and groups of traps varied. Early in the trapping periods, traps were placed to determine the distribution of animals over the length (1,030–1,950 m) of the study areas. With seasonal recruitment, or if an area showed new signs of water vole activity, the number of traps set was increased.

Each time a water vole was captured, its body weight (g), body length (mm) and reproductive condition were recorded. Males were classified as active reproductively (testes scrotal), possibly active reproductively (testes palpable in the abdomen) or inactive reproductively (testes abdominal but not palpable). Females were classified as inactive reproductively (nipples small = 2.0–2.5 mm diameter, pubic symphysis closed, vagina imperforate or perforate), possibly active reproductively (nipples moderate = 3.0–4.0 mm diameter and vagina perforate or imperforate; or if the nipples were small, the pubic symphysis open and the vagina perforate), and active reproductively (nipples large = 4.5–5.0 mm diameter, or if moderate, showing a hair-free area around them, pubic symphysis open or closed, vagina perforate), or pregnant or parturient.

Vole distribution and movement patterns were determined by fitting water voles with 5.0 g radio collar transmitters (SMI, Collar Type R) manufactured by AVM Instrument Company, Champaign, Illinois. Animals were collared only if the transmitter package represented no more than 10% of an animal's body weight (Brander and Cochran, 1971). Only overwintered adults or ani-

mals from the first litter of the season which attained a weight ≥50.0 g were collared.

Snow and thermal microclimate data were collected at one study site. Snow depths were recorded on three stakes (15 m apart) set in a line running diagonally across a 45 m wide portion of habitat. Snow stakes were located at the stream bank and in the middle and at the outer edge of a streamside meadow.

Air temperatures were measured 1 m above the soil surface by a Bacharach seven-day hygrothermograph (No. 22-4504) and maximum-minimum thermometer (C) housed in a Stevenson screen. Data was gathered for at least one seven day period in each of nineteen months between March 1978 and November 1979. Subnivean temperatures were measured with a system of Yellow Springs Instrument Company, Inc. thermistor probes (No. 418) connected to a Tele-Thermometer (No. 42SC). Probes were set along the soil surface prior to the establishment of a snow cover.

Ground level temperatures were measured by a Taylor seven-day thermograph (No. 2350) and a maximum-minimum thermometer housed in a Stevenson screen. Screens were placed on the ground at four points along a 100 m line crossing and oriented perpendicular to the study stream. The thermograph locations were: 1) 30 m away from the stream in a *Salix* forb meadow, 2) on the stream bank, 3) in a moss bed 30 m from the stream and 4) 70 m from the stream on a larch (*Larix lyallii*) covered ridge.

Soil, stream, vegetation and topographic characteristics were examined at sampling locations spaced at 50 m intervals over at least 1,000 m at three study areas. The right and left sides of the stream were considered separate sampling areas. Data were collected within a 1 m by 10 m strip on each bank and along a transect line laid perpendicular to the bank and extending to the outer boundary of streamside meadow habitat. Each 50 m mark was the center of the bank sampling area and the origin of the transect. Data were collected after vegetative growth was completed during a two and a half week period in August 1979.

RESULTS

Individual records of each vole were maintained, and captures were recorded in a calendar of captures (Petrusewicz and Andrzejewski, 1962). The following analyses are based on data obtained during 1,099 captures of 323 water voles.

Reproductive Biology

Previous studies indicate the water vole's reproductive biology is similar to other microtines (review by Hasler, 1975). The species displays induced estrus and ovulation, a 22-day gestation period, postpartum estrus and lactational pregnancy (Brown, 1977; Jannett et al., 1979). A large annual reproductive output would be expected if overwintered adult females produced several large litters per breeding season and a large proportion of the young of the year entered the breeding population in their first summer. However, this was not the case.

Reproductive condition was classified as inactive

(I), possibly active (P) and active (A) reproductively based on the series of morphological characteristics described in the methods. The data from all sites and years combined indicated that of all the overwintered adults captured from June through September, 84.6–98.2% of overwintered adult females and 74.9–100.0% of adult males were either P or A.

Only a small number of young entered the breeding population (P or A) in their first year. Monthly values of active individuals ranged from 6.2–11.3% of the young female population and 5.9–20.0% of the young male population (Fig. 1). Of the 219 young voles known alive during this study, the number of young considered P or A was only 25.7% (27, N = 105) of the female young, and 26.3% (30, N = 114) of the male young.

Some overwintered adult females produced two litters per breeding season. During this study, the reproductive condition of 32 overwintered adult fe-

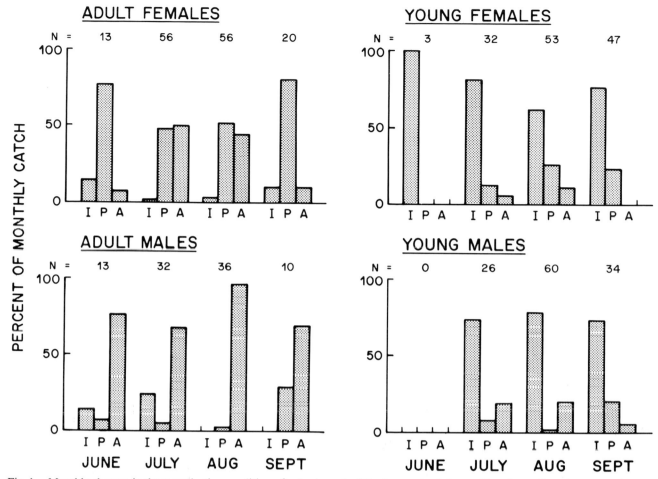

Fig. 1.—Monthly changes in the reproductive condition of voles in each of the four sex-age classes. Data from all study areas and years are lumped. Voles were classified as reproductively inactive (I), possibly active (P), or active (A).

males was assessed from June through early September. Twenty adult females (62.5%) were found to be possibly pregnant or pregnant in one breeding season. Only four of these 20 adult (20.0%) may have produced two litters in one breeding season.

Litter size estimates were obtained from voles

Table 1.— *Water vole litter size estimates.*

Statistics	N	Mean ± SE	Range	Mode
Study data				
Embryos	8	5.00 ± 0.80	2–9	4
Placental scars	9	5.33 ± 0.58	3–8	5
All Alberta data				
Embryos	19	5.17 ± 0.44	2–9	4
Placental scars	12	5.42 ± 0.58	2–8	5
All data				
Embryos	86	5.52 ± 0.18	2–9	5
Placental scars	26	6.11 ± 0.43	2–10	8

collected during my study and from museum specimens captured throughout the species' range. Litter size estimates based on embryo and placental scar counts show a mean litter size of five was the most common (Table 1).

Consequently, the annual reproductive output in a water vole population is low due to the production of only two moderately sized litters per overwintered adult female each breeding season and the limited participation in breeding activity by young in their first year. In addition, a prolonged snow cover limits the availability of high quality food and the length of the breeding or snow-free season to several months.

THERMAL MICROCLIMATE AND MICROHABITAT

Water voles live a subnivean existence 7 or 8 months each year (Fig. 2). Several centimeters of snow usually fell in September and October, but

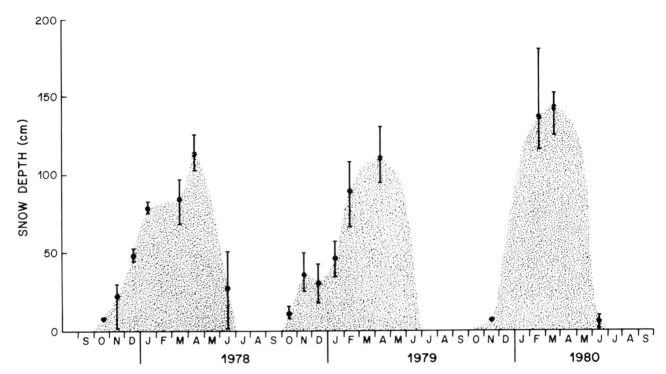

Fig. 2.—Annual period of snow cover and snow depths (cm) recorded during the winters of 1977 to 1979 on three snow stakes. Mean snow depths are denoted by closed circles, and the vertical line represents the range of snow depths recorded. The stipled areas indicate the period of snow cover.

snow usually began to accumulate during November. Peak accumulations of 125, 130, and 180 cm were recorded in March and April. In addition to the snow falling directly on an area, snow was blown off surrounding slopes and ridges and re-deposited in the lower meadows and stream beds.

The thermal microclimate experienced annually by water voles is illustrated in Fig. 3. This figure is a composite of mean monthly air temperatures recorded 1 m above ground, at ground level along the stream bank and in a subnivean runway. During the snow-free period, the ground level temperatures fluctuated with the air temperature. However, beneath a snow cover water voles experienced temperatures between −1.0 and 1.0°C. A mean depth

of 14 cm or more of new snow effectively shielded the underlying soil from the colder supranivean air and the wind. The annual range of mean ambient temperatures experienced by water voles was 9.30°C (−0.40 to 8.90°C), while air temperature 1 m above the soil ranged over 25.96°C (−16.35 to 9.61°C).

The stream and banks were among the first areas to melt out. By the last week of June, most streamside areas were free of snow. Large snow drifts persisted until late July. Water voles experienced the greatest fluctuations in ambient temperature during the snow-free period. The voles' preference for streamside microhabitats and absence from seemingly suitable areas adjacent to the streams might be explained by large differences in ground surface

Table 2.—*Minimum (Min) and maximum (Max) temperatures (C) recorded in four microhabitats during the 1978 snow-free season.*

Month	*Salix* meadow		Stream bank		Moss		Larch	
	Min	Max	Min	Max	Min	Max	Min	Max
June	−8.0	22.0	−6.0	22.0	−7.0	19.0	−7.0	23.0
July	−4.0	24.0	−3.0	25.0	−1.0	24.0	−3.0	23.0
August	−4.0	24.0	−4.0	27.0	−3.0	26.0	−2.0	24.0
September	−10.0	21.0	−8.0	23.0	−9.0	17.0	−6.0	16.0
October	−17.0	17.0	−13.0	16.0	−12.0	14.0	−9.0	19.0
November	−16.0	11.0	−16.0	11.0	−18.0	4.0	−28.0	7.0

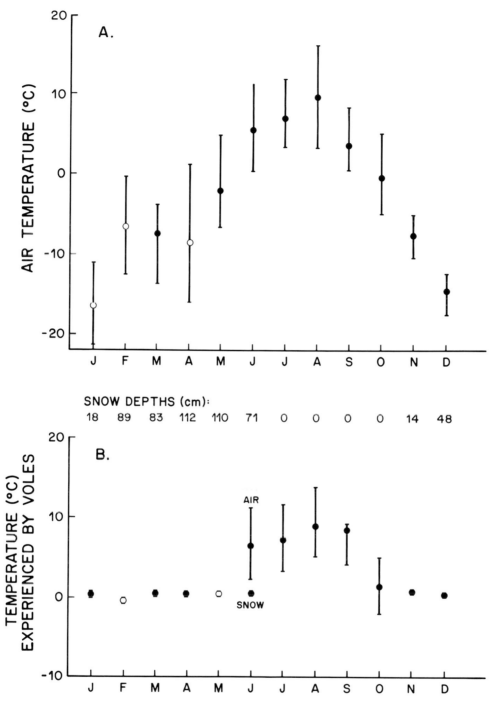

Fig. 3.—A comparison of air temperatures 1 m above the ground (A) with the range of ambient temperatures experienced by water voles during a year (B). Circles indicate mean temperatures, and the ends of the vertical lines indicate the mean minimum and maximum recorded in 1978 (A) one meter above the ground, (B) at ground level along the stream bank and in a subnivean runway. Some 1979 data are included (open circles). Snow depths (cm) are indicated.

air temperatures between microhabitats. To examine this possibility, air temperatures were recorded at ground level by thermographs. Unfortunately, due to instrument failure comparative 24 hour thermograph records from all stations were obtained for only a 2 to 6 day period each month (Fig. 4). The figure indicates a small range (3–6°C) of mean hourly temperatures among microhabitats.

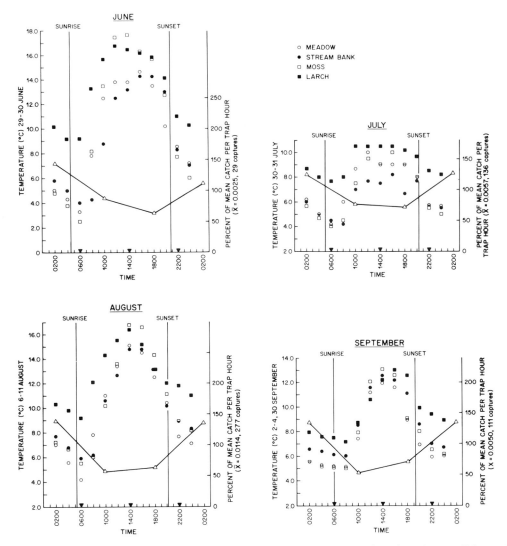

Fig. 4.—Mean ground-level air temperatures recorded every 2 hr in four different microhabitats from June until September 1978. Vole activity expressed as percent of mean catch per trap hour is plotted (connected open triangles) as the midpoint of 8-hr intervals beginning at 0600 for 32 hr.

A distinct pattern of temperature differences among microhabitats was not evident at any time over a 24-hr period.

Maximum-minimum temperatures recorded at each microhabitat showed only very slight temperatures differences (Table 2). The surface air temperature data from both sources indicated a very small difference in the ambient temperatures voles might experience in the four microhabitats. On the basis of temperature data alone, I feel the observed differences in ground air temperatures between microsites were not large enough and did not persist for a long enough period to explain the species' preference for streamside areas.

Streamside habitat patches were narrow. The

mean ± SD habitat width measured on one side of a stream at four study sites was 14.7 ± 8.0 (N = 146). Water vole captures were usually restricted to within 5 to 10 m of the stream banks (mean = 2.3 m, N = 1,078). Eighty-seven percent of all captures occurred within 5 m of the stream bank. My field observations indicated water voles occurred repeatedly at some microsites along streams and seldom, if ever, at other sites. Collection of habitat data was designed to: 1) provide a quantitative profile of the type of alpine or subalpine stream occupied by water voles, and 2) identify the characteristics important in the selection of a home area. Twenty habitat characteristics were recorded over 1,000 m along transects spaced at 50 m intervals and set perpendicular to

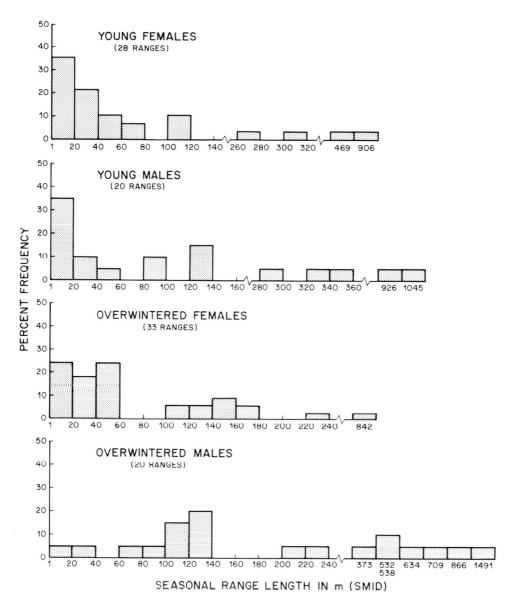

Fig. 5.—Frequency distribution of the seasonal maximum intercapture distances (SMID) for the four sex-age classes of water voles. Data from four study sites and 1977 to 1979 are lumped.

the three streams. Habitat characteristics were chosen because of their assumed biological importance and relationship to the type and availability of food and cover, the accessibility to subterranean spaces, ease of burrow or nest construction and micro-topographic influences on vegetation, soil moisture and drainage (Table 3). A microhabitat sampling location was considered occupied if an overwintered adult was captured two or more times within 5 m, that is, either up or downstream of a habitat sampling station during a breeding season.

Discriminant analysis showed occupied sites could

be distinguished from unoccupied sites on the basis of stream gradient, the number of openings in the stream bank and soil depth (Table 4). Habitat occupied by water voles tended to be more level sites with a deeper, more continuous soil layer and more openings in the stream banks than unoccupied areas.

VOLE SPACING AND MOVEMENTS

Home range, as used here, was defined as the area over which an animal moves in its normal daily activity (Burt, 1943). The distance between the two most widely separated capture points was used to

Table 3.—*Habitat variables used in the microhabitat analysis and their suggested biological significance.*

Availability of cover:

Percent of the stream bank undermined

Number of openings in stream bank (that is, not produced by rodent activity)

Visual (horizontal) obstruction afforded by vegetation

Availability and type of food and cover:

Percent cover of ground, herbaceous vegetation, shrub and tree cover

Height (cm) of herbaceous vegetation, shrub and tree layers

Accessibility to the subterranean environment and ease of burrow and nest chamber construction:

Number of rocks in the stream bank

Ease of surface soil penetration (cm)

Soil depth (cm)

Bank height (cm)

Topographic influences on vegetation, soil moisture and drainage and the effect of these factors on the location of nest chambers and burrows:

Percent slope of the banks down to the stream

Habitat width (m) or the amount of potential streamside living space available

Percent stream gradient (downstream slope of stream)

Ease of movement along and across streams:

Stream width (cm)

Stream depth (cm)

Stream velocity (cm/sec)

estimate a vole's home range length. Trapping session maximum intercapture distance (MID) was calculated for each animal captured two or more times within a trapping session and a seasonal maximum intercapture distance (SMID) was calculated for animals captured in more than one session.

The mean SMID of the sex-age classes were ordered from low to high as follows: overwintered adult females, young females, young males and overwintered adult males (Table 5). A large proportion of the SMID of young females (67.9%, N = 28), young males (50.0%, N = 20) and overwintered

Table 4.—*Standardized discriminant function coefficients and group centroids (direct method).*

Variable	Standardized discriminant function coefficients
Stream gradient	0.662
Number of openings	−0.611
Soil depth	−0.506
Group centroids or mean ± SE	
Unoccupied areas	0.174 ± 0.098
Occupied areas	−0.512 ± 0.161

Table 5.—*Seasonal maximum intercapture distances (m).*

Sex-age class	N	Mean ± SE
Young females	28	102.2 ± 35.9
Young males	20	200.8 ± 79.6
Adult females	33	93.8 ± 25.6
Adult males	20	332.0 ± 82.7

adult female (66.3%, N = 33) were ≤60 m. These seasonal values contrasted sharply with those of adult males for whom SMID values ≤60 m represented only 10.0% of 20 values (Fig. 5).

Kruskal-Wallis single factor analysis of variance by ranks ($a = 0.05$) indicated a significant difference among SMIDs of the four sex-age classes ($\chi^2_3 = 15.962$, $0.005 < P < 0.001$). A series of Mann-Whitney tests ($a = 0.01$) was run for each of the six possible pair-wise comparisons of sex-age classes to locate the intergroup differences (Table 6).

For overwintered adult males, SMIDs were larger than for either group of females, but were the same as those of young males. However, an inconsistency exists, since the SMID value of young males did not differ from those of young and overwintered adult females. The differences between U′ and $U_{0.01(2)}$ values of the young male × adult male comparison was small and very close to being significantly different, whereas these same values for comparisons of young males with young females and adult females were much larger and not significantly different.

Fig. 6 displays the SMIDs of overwintered adults captured at the Marmot Creek study site over three years. Replacement of one individual by another is shown by stacking the range length (line or box) of one vole upon another of the same sex. Spacing and movement patterns similar to those described below were observed by trapping in three other study populations (Ludwig, 1981).

Some streamside sites were occupied annually by overwintered adult females (325–375 m, north slope 20–50 m) and were apparently preferred. Other streamside locations supported animals for one or two years.

When population numbers were low (1977), overwintered females were scattered widely along streams. During years of high vole densities (1978, 1979), adult females clustered in some areas, but home range overlap was minimal or non-existent. Less preferred sites were also occupied.

Overwintered females remained within smaller areas than other sex-age classes and maintained exclusive home areas. Overwintered males moved over

MARMOT

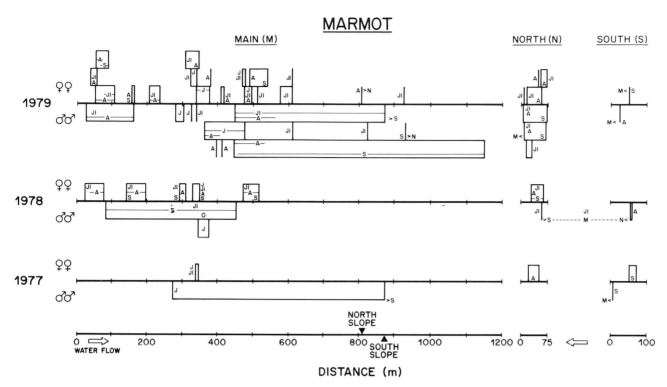

Fig. 6.—Seasonal maximum intercapture distances (boxes) and monthly trapping session maximum intercapture distances (vertical or horizontal lines) of overwintered adults at Marmot Creek by year. The month is indicated by a capital letter with or without horizontal lines, or outside boxes, or vertical lines (single capture). Use of > indicates a movement to another stream.

longer portions of stream and overlapped the seasonal ranges of both adult females and adult males (1978, 1979).

Depending upon the number and distribution of females along a stream, males demonstrated two types of movement patterns ("mobile" and "stationary"). When voles were widely spaced along a stream and crude densities were low, males were less site loyal and moved over wide areas between females (1977). During years of high density (1979), a male was usually found repeatedly in one area near a group of adult females.

Radio tracking was used to observe the movements and spatial distribution of free-ranging overwintered adult water voles along the Marmot stream in 1979. Over 730 hours between 27 June through 4 October, 577 positional fixes were obtained for 16 individuals (four males, 12 females). The spatial distribution of animals during each radio tracking session was reconstructed by plotting the telemetry-revealed home range of radio-collared voles and the location of other adults on a diagram of the Marmot study area.

A representative set of telemetry-revealed home

Table 6.—*Results of Mann-Whitney U Tests ($\alpha = 0.01$) comparing the maximum intercapture distances within one breeding season (SMID) of pairs of water vole sex-age classes.*

Comparisons and sample sizes A with B	Mean rank		U'	U or Z	$U_{0.01(2)}$ or P	Conclusion*
	A	B				
Adult males (20) with adult females (33)	36.13	21.47	147.5	512.5	470	Reject H_0
Adult males (20) with young females (28)	33.72	17.91	95.5	464.5	403	Reject H_0
Adult males (20) with young males (20)	24.65	16.35	117.0	283.0	295	Accept H_0
Young females (28) with young males (20)	23.48	25.92	251.5	308.5	403	Accept H_0
Young males (20) with adult females (33)	26.70	27.18	324.0	336.0	470	Accept H_0
Young females (28) with adult females (33)	28.04	33.52	379.0	−1.2014	0.230	Accept H_0

* H_0: The SMIDs of each member of a pair (A and B) are the same length.

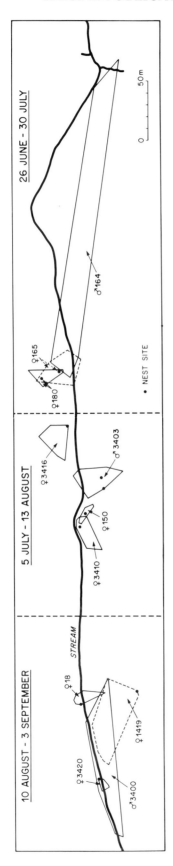

Fig. 7.—Home range areas of water voles during two to three 72 hr radio tracking sessions between 26 June to 3 September 1979 at Marmot Creek. The positions of 11 voles (three males, eight females) along the stream, dates of tracking and animal identification numbers are shown.

Table 7.—*Home range area and cumulative distance (mean ± SE) moved (m) during 72 hr radio telemetry sessions. Mann-Whitney U tests comparing male and female values.*

Sex	N	Home range area (m²)	Distance moved (m)
Females	16	221.6 ± 75.9 18.9–1,018.0	100.7 ± 15.8 24.5–204.0
Males	9	770.3 ± 358.8 128.6–3,418.8	241.0 ± 76.2 83.1–772.3
Mann Whitney U $U_{0.05(2)16,9} = 107$		U = 27, U′ = 117 Reject H_0	U = 34, U′ = 110 Reject H_0

H_0: The home range (cumulative distance moved within a home range) of males and females are equal.

ranges consisting of two or three 72-hr periods of observations per vole is shown in Fig. 7. The spacing and movement patterns indicated agree with the trap-revealed patterns.

Individual water voles typically remained solitary in a single nest site located inside small rises in the microtopography and beneath clumps of willow or subalpine fir.

Overwintered adult females maintained exclusive non-overlapping home ranges. The three female "overlaps" shown in Fig. 7 actually represent movements of females during different telemetry sessions (18 and 1,419), or a range expansion (165 and 3,410) following the death (180) or disappearance (150) of a neighboring female. Adult females usually remained within the same area throughout the breeding season and overlapped minimally, if at all, the home area of other adult females.

Solitary overwintered adult males remained in close proximity to groups of one to four overwintered adult females and extensively overlapped the females' home areas. Trapping indicated "non-resident" males entered female areas, but appeared to remain only a short time (72 hr). Males which were site loyal also made occassional exploratory movements. For example, male 164 usually concentrated his activity near females 165 and 180 (470–550 m), but also made two round trips (220 m one way) downstream to a spring-fed slope within one 24-hr period (Fig. 7). This type of movement is assumed to allow males to determine the position of other males and reproductively active females.

Twenty-five 72-hour periods (eight males and 17 female records, 399 fixes) were used to calculate the mean range and mean cumulative distance moved within home ranges by each sex (Table 7). Overwintered adult males maintained home area 3.5 times larger than overwintered adult females. With-

in a home area, males traveled 2.5 times as far as females.

DISCUSSION

HABITAT PREFERENCE AND THERMAL MICROCLIMATE

It is still unclear which abiotic and/or biotic factors determined the strong habitat preference of *M. richardsoni*. In current ecological time, the water vole exists within a limited geographic range and displays a strong habitat preference.

During the Pleistocene, the ice fields and expanded glacier systems composing the Cordilleran glacier complex extended from the Brooks Range in Alaska to the southern Rocky Mountains and connected with the larger Laurentide ice sheet to the east. A narrow belt of treeless tundra bordered these ice sheets and gradually moved further north and into higher elevations as the glaciers receded and the continental climate warmed (Flint, 1971; Matsch, 1976).

The species' habitat requirements suggest it may have been specializing as an occupant of the tundra region bordering the Cordilleran glacial front. The water vole's geographic range was presumably limited during the post-Pleistocene period as the species followed the retreating tundra-like vegetation into high montane areas (Banfield, 1958; Dalquest, 1948; Findley and Anderson, 1956; Hoffmann and Taber, 1967). In ecological time, the species' local distribution is restricted to small patches within the narrow strips of mesic habitat bordering alpine and subalpine streams. However, it is not known if the habitat specificity for streamside areas evolved prior to or after its move to high elevations. Over evolutionary time, interspecific competitors, such as the muskrat or beaver, may have limited the water vole's range of habitat preferences to areas bordering montane streams or prevented the species' movement down from higher elevations subsequent to its post-Pleistocene range contraction. Gene flow was undoubtedly limited due to the surrounding inhospitable habitat. Limited gene flow may have resulted in decreased genetic variability and a limited range of possible genetic and phenotypic responses to environmental challenges.

Some streamside areas are occupied by water voles over many generations, and certain sites are apparently preferred. Occupied sites are located in level areas where openings in the stream banks provide access to subterranean spaces and the soil is easily

penetrated. However, my study did not find evidence to suggest the thermal microclimate near the stream banks was sufficiently different to explain the water vole's absence from adjacent, seemingly habitable meadows. I suspect flooding of these adjacent meadows and a saturated soil combined with freezing temperatures during the period of snow melt prevent the occupancy of these areas and the initiation of breeding early in the snow-free season.

The period prior to and after the establishment of the heimal threshold in the fall and during the spring melt, respectively, are the periods of greatest thermal stress for microtine rodents (Fuller et al., 1969; Merritt and Merritt, 1978; Pruitt, 1957; Whitney, 1976). Other than during these two periods, the deep snow accumulations in the low lying streamside areas occupied by *M. richardsoni* provide a stable thermal microclimate for the species during its subnivean existence. The shallow and compact snow layers and the resulting large fluctuations in subnivean temperatures common to wind-swept slopes reduce the suitability of the slopes as overwintering sites.

LIFE HISTORY TRAITS

The life history traits characterizing a species can be viewed as a coadapted set designed by natural selection to solve particular ecological problems (Stearns, 1976, 1977). Since a habitat imposes a specific selective regime on the organisms occupying it (Partridge, 1978), a species' life history traits are the result of evolutionary trade-offs of costs versus benefits evolved in the process of adaptation to habitats (Southwood, 1977).

I feel the short alpine breeding season, a brief and rapid seasonal burgeoning of high quality plant food, and the limited availability of habitat patches have been the major factors influencing the evolution of *M. richardsoni* subsequent to its restriction to high elevations in western North America. The range of possible adaptive responses were undoubtedly limited by the high metabolic rate, short life span, and limited mobility of these small herbivorous mammals (review by Bourliere, 1975; Kleiber, 1961). *M. richardsoni* appears to possess the reproductive traits (induced ovulation, post-partum estrus and lactational pregnancies) normally associated with the high reproductive potential of microtine rodents (Brown, 1977; Jannett et al., 1979; Hasler, 1975), but it does not often reach high population densities.

Production of only two litters by most overwintered adult females per breeding season and reproduction by only a small proportion of young during their first summer are the most noteworthy features distinguishing *M. richardsoni's* life history traits from the "typical microtinc" traits (review by French et al., 1975) characteristic of more widespread species such as *M. pennsylvanicus* and *Arvicola terrestris.* Overwintered adult female water voles produce a surprisingly smaller number of moderately sized litters (French et al., 1975; Innes, 1978). My analysis of museum specimen data (Ludwig, 1981) suggests a litter size of five to six is consistent throughout the species latitudinal and altitudinal range.

In *M. richardsoni* natural selection has apparently favored life history traits which fill limited local vacancies along streams and provide emigrants to re-colonize openings in nearby areas. Iteroparity has been de-emphasized in favor of the production of a few high quality competitive young per adult female. If habitat patches and the resources they contain are limited, the production of an "excessive number" of young might debilitate habitat quality and the chances of survival of all young. Therefore, reduced reproductive output may have been selected for due to the decrease in fitness experienced by voles investing heavily in iteroparous reproduction.

Local events such as the removal of reproductives and young by mustelids (MacLean et al., 1974; Maher, 1967) can inflict substantial losses on the small water vole populations, thus opening habitat spaces and favoring dispersal. The presumed scarcity of habitat spaces and the potential for catastrophic events to open habitat patches should favor the production of high quality emigrants. that is, young with high survival potential.

Maintenance of non-overlapping home areas by adult females and female site loyalty suggest a home area containing a nest site and an adequate food supply are the key resources in a female's reproductive success. Locally, the limited availability of female home areas and the small reproductive output per breeding female set the upper bounds of population size. Home area size is assumed to be a reflection of habitat quality, measured in terms of food quality and quantity, and the availability of cover and nest sites.

Male movement patterns are dependent on the density and distribution of reproductively receptive females. Reproductively active males move between females and apparently compete for copulations during periods of low populations numbers, but restrict their activity to a group of females when population numbers are high.

LITERATURE CITED

ANDERSON, P. K., P. H. WHITNEY, and J. P. HUANG. 1976. *Arvicola richardsoni:* ecology and biochemical polymorphism in the front ranges of southern Alberta. Acta Theriol., 21:425–468.

BANFIELD, A. W. F. 1958. The mammals of Banff National Park, Alberta. Nat. Mus. Canada Bull. Biol. Ser., 57:1–53.

———. 1974. The mammals of Canada. Univ. Toronto Press, Toronto, 438 pp.

BOURLIERE, F. 1975. Mammals small and large: the ecological implications of size. Pp. 1–8, *in* Small mammals: their productivity and population dynamics (F. B. Golley, K. Petrusewicz and L. Ryszkowski, eds.), Cambridge Univ. Press, Cambridge, 451 pp.

BRANDER, R. B., and W. W. COCHRAN. 1971. Radio-location telemetry. Pp. 95–104, *in* Wildlife management techniques (R. H. Giles, Jr., ed.), Edwards Brothers Inc., Ann Arbor, 633 pp.

BROWN, L. N. 1977. Litter size and notes on reproduction in the giant water vole (*Arvicola richardsoni*). Southwestern Nat., 22:281–282.

BURT, W. H. 1943. Territoriality and home range concepts as applied to mammals. J. Mamm., 24:346–352.

DALQUEST, W. W. 1948. Mammals of Washington. Univ. Kansas Publ., Mus. Nat. Hist., 2:1–444.

DEKAY, J. E. 1842. Natural History of New York. Part I; Zoology. Thurlow Weed, Albany, 146 pp.

FINDLEY, J. S., and S. ANDERSON. 1956. Zoogeography of the montane mammals of Colorado. J. Mamm., 32:118–120.

FLINT, R. F. 1971. Glacial and Quarternary Geology. John Wiley and Sons, Inc., New York, 892 pp.

FRENCH, N. R., D. M. S. STODDART, and B. BOBEK. 1975. Patterns of demography in small mammal populations. Pp. 73–102, *in* Small mammals: their productivity and population dynamics (F. B. Golley, K. Petrusewicz, and L. Ryszkowski, eds.), Cambridge Univ. Press, Cambridge, 451 pp.

FULLER, W. A., L. L. STEBBINS, and G. R. DYKE. 1969. Overwintering of small mammals near Great Slave Lake, northern Canada. Arctic, 22:34–55.

HALL, E. R. 1981. The Mammals of North America. John Wiley and Sons, New York, 1,270 pp.

HASLER, J. F. 1975. A review of reproduction and sexual maturation in the microtine rodents. Biologist, 57:52–86.

HOFFMANN, R. S., and R. D. TABER. 1967. Origin and history of Holartic tundra ecosystems with special reference to their vertebrate faunas. Pp. 143–170, *in* Arctic and Alpine Environments (H. E. Wright. Jr. and W. H. Osborn, eds.), Indiana Univ. Press, Bloomington, 308 pp.

HOLLISTER, N. 1912. Mammals of the alpine club expedition to the Mount Robson Region. Canadian Alp. J., Sp. No., 1912:1–75.

HOOVEN, E. F. 1973. Notes on the water vole in Oregon. J. Mamm., 54:751–753.

INNES, D. G. L. 1978. A reexamination of litter size in some North American microtines. Canadian J. Zool., 56:1488–1496.

JANNETT, F. J., JR., and J. A. JANNETT. 1974. Drum-marking by *Arvicola richardsoni* and its taxonomic significance. Amer. Midland Nat., 92:230–234.

JANNETT, F. J., JR., J. A. JANNETT, and M. E. RICHMOND. 1979. Notes on reproduction in captive *Arvicola richardsoni*. J. Mamm., 60:837–838.

KLEIBER, M. 1961. The fire of life: an introduction to animal energetics. Wiley, New York, 454 pp.

KOPLIN, J. R., and R. S. HOFFMANN. 1968. Habitat overlap and competitive exclusion in voles (*Microtus*). Amer. Midland Nat., 80:494–507.

LUDWIG, D. R. 1981. The population biology and life history of the water vole, *Microtus richardsoni*. Unpublished Ph.D. thesis, Univ. Calgary, Alberta, 266 pp.

MACLEAN, S. F., JR., B. M. FITZGERALD, and F. A. PITELKA. Population cycles in Arctic lemmings: winter reproduction and predation by weasels. Arctic Alp. Res., 6:1–12.

MAHER, W. J. 1967. Predation by weasels on a winter population of lemmings, Banks Island, Northwest Territories. Canadian Field Nat., 81:248–250.

MATSCH, C. L. 1976. North America and the Great Ice Age. MacGraw-Hill, New York, 131 pp.

MERRITT, J. F., and J. M. MERRITT. 1978. Population ecology and energy relationships of *Clethrionomys gapperi* in a Colorado subalpine forest. J. Mamm., 59:576–598.

PARTRIDGE, L. 1978. Habitat selection. Pp. 351–376, *in* Behavioural Ecology. An evolutionary approach (J. R. Krebs and N. B. Davies, eds.), Sinauer Associates, Sunderland, 494 pp.

PATTIE, D. J. 1967. Dynamics of alpine small mammal populations. Unpublished Ph.D. thesis. Univ. Montana, Missoula, 103 pp.

PETRUSEWICZ, K., and R. ANDREZEJEWSKI. 1962. Natural history of a free-living population of house mice (*Mus musculus* Linnaeus) with particular reference to groupings within the population. Ekologia Polska Ser. A, 10:85–122.

PRUITT, W. D., JR. 1957. Observations on the bioclimate of some taiga mammals. Arctic, 10:130–138.

RACEY, K., and I. M. COWAN. 1935. Mammals of the Alta Lake Region of Southwestern British Columbia. Rept. Prov. Mus. British Columbia for the year 1935, pp. 15–29.

SKIRROW, M. 1969. Behavioural studies of five microtine rodents. Unpublished Ph.D. thesis, Univ. of Calgary, Alberta, 211 pp.

SOUTHWOOD, T. R. E. 1977. Habitat, the templet for ecological strategies? J. Anim. Ecol., 46:337–365.

STEARNS, S. C. 1976. Life-history tactics: A review of the ideas. Quart. Rev. Biol., 51:3–47.

———. 1977. The evolution of life history traits. A critique of the theory and a review of the data. Ann. Rev. Ecol. Syst., 81:145–171.

WHITNEY, P. 1976. Population ecology of two sympatric species of subarctic microtine rodents. Ecol. Monogr., 46:85–104.

Address: Forest Preserve District of Du Page County, P.O. Box 2339, Glen Ellyn, Illinois 60138.

DISPERSION OF INSULAR *PEROMYSCUS MANICULATUS* IN COASTAL CONIFEROUS FOREST, BRITISH COLUMBIA

Thomas B. Herman

ABSTRACT

Populations of the deer mouse *Peromyscus maniculatus* were studied by monthly capture-mark-recapture trapping between January 1976 and October 1977 on the west coast of Vancouver Island and on three small (13–174 ha) forested islands in adjacent Barkley Sound. The area is characterized by mild, but excessively rainy winters and relatively dry summers. Although mice remained active all winter, reproduction was limited to summer, except in 1976 when breeding in island populations extended into winter.

Dispersion of resident adult males and of resident adult females tended slightly toward a regular distribution in all locations. Dispersion of males vs. females appeared contagious on the small islands, and random on Vancouver Island. Although monthly variations occurred, there were no clear changes in dispersion in any group in relation to season or reproductive activity. Indices of dispersion of resident adults do not reflect the seasonal and geographic differences in demography and behaviour observed among these populations.

INTRODUCTION

Dispersion is a fundamental parameter in the social organization and population dynamics of any species. In mice of the genus *Peromyscus,* various parameters of social organization and population dynamics are well documented.

Theoretical considerations suggest that these parameters (for example, dispersion, density, density oscillations, reproduction, recruitment, dispersal, agonistic behaviors, mortality) may differ between island and mainland populations of a given species (Grant, 1965; Soulé, 1966; MacArthur and Wilson, 1967; Krebs et al., 1969; Van Valen, 1971; MacArthur, 1972; Crowell, 1973; Lidicker, 1975; Sullivan, 1977; Case, 1978; Halpin and Sullivan, 1978). Experimental evidence for such differences exists in *Peromyscus* populations (Redfield, 1976; Sullivan, 1977; Halpin and Sullivan, 1978; Herman, 1979; Nadeau et al., 1981) and in numerous populations of other small mammal species (for a review see Gliwicz, 1980).

In mainland populations of *Peromyscus* aggregated, random and even dispersions have all been reported (McCabe and Blanchard, 1950; Davenport, 1964; Stickel, 1968; Myton, 1974; Metzgar, 1979; Mihok, 1979; Metzgar, 1980). This variability may result from differences in the physical environment (climatic severity and seasonality, habitat structure, etc.) and/or the biotic environment (predation, interspecific competition, intraspecific competition, social factors, and others), and perhaps should be expected in a widespread genus known for its phenotypic plasticity.

In contrast, few data are available on dispersion in island populations of *Peromyscus.* The data presented here were collected as part of a study comparing the population ecologies of *Peromyscus maniculatus* (Wagner) from islands in Barkley Sound and from adjacent coastal Vancouver Island, British Columbia (Herman, 1979). Herein I examine dispersion of resident adult mice for differences among seasons and among locations in relation to other demographic and behavioral components in these populations.

MATERIALS AND METHODS

Study Area

Barkley Sound lies on the west coast of Vancouver Island (48°50′N, 125°15′W) and contains numerous islands which range in size from less than 1 to more than 1,700 ha. These islands are relatively homogeneous, comparable in habitat structure, and undisturbed by human activity. This area lies within the "Coastal Western Hemlock Zone" of the "Pacific Coastal Mesothermal Forest Region" (Krajina, 1965) and is dominated by uneven-aged stands of western hemlock (*Tsuga heterophylla*), western red cedar (*Thuja plicata*), and sitka spruce (*Picea sitchensis*). Large tracts of similar undisturbed lowland forest are available on adjacent Vancouver Island for comparative purposes. Topography is generally rugged, soils are shallow, poorly developed and acidic.

The area is characterized by cool, relatively dry summers and mild, excessively rainy winters (Fig. 1). Precipitation may vary from year to year and, due to the rugged topography, from place to place within a year. Generally, the islands in Barkley Sound

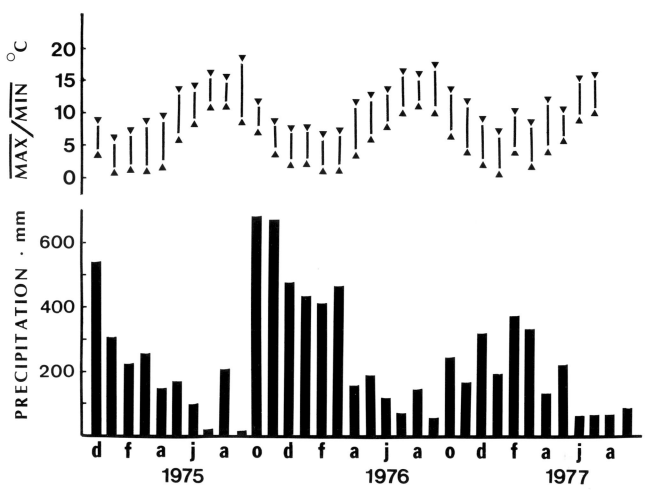

Fig. 1.—Precipitation (mm) and mean monthly maximum and minimum temperatures (°C) at Bamfield, British Columbia, from December 1974 to September 1977 (from P. Janitis, unpublished).

receive slightly less precipitation than adjacent Vancouver Island (Herman, 1979).

Islands in the sound have been isolated in their present configurations for at least 5,500 years, but were joined previously to the mainland (Mathews et al., 1970; Carter, 1971). Distributional evidence from Barkley Sound (Herman, 1979) and genetic and distributional evidence from other west coast populations (Redfield, 1976) suggest that populations of deer mice on near shore islands are probably not in dynamic equilibrium, but are relicts from times of lowered sea level, and have therefore been isolated for a relatively long time.

Sampling Regime

Pairs of live-trapping transects were established on three islands in the sound (Haines, 13 ha; Helby, 71 ha; Diana, 174 ha) and at two areas on adjacent Vancouver Island. Due to its large size (3.21 × 10⁶ ha), Vancouver Is. is treated as a mainland in this study and will be referred to as such. Each transect contained 16 trap stations evenly spaced at 10m intervals. This trapping configuration has proven highly efficient and appropriate for work with *P. maniculatus* in coastal British Columbia (Petticrew and Sadleir, 1970). Members of each transect pair were well separated,

and all transects were located a minimum of 75 m from prominent edges such as shorelines and roads. These lines were monitored for two successive nights monthly from February 1976 through October 1977. Four Longworth® traps containing sunflower seeds and terylene bedding were placed within a 3m radius of each trap station.

All mice captured were individually marked by toe-clipping. At each capture the date, trap station, body weight to the nearest 0.5 g, pelage, and reproductive condition were recorded. Visible wounds and/or scars and infestations of ectoparasites were also noted. Males with large, fully or partially descended testes and females with perforate vagina, pregnancy plug, palpable embryos, or prominent mammae were considered to be reproductively active.

In addition to maintenance of live trapping transects, the study included systematic monthly dead-trapping on three islands and the mainland (Vancouver Island) from February 1976 through January 1977, a "blitz" dead-trapping during early summer 1977 on 22 islands in the sound and at 10 locations on the adjacent mainland, and arena experiments involving resident adult reproductive males from the island and mainland live-trapping transects (Herman, 1979).

DISPERSION ANALYSES

The statistics used to measure dispersion in the live-trapped populations are derived and described in detail by Metzgar (1979, 1980) and Metzgar and Hill (1971). They are based on the frequency distribution obtained by arranging capture data according to the number of stations in which 0, 1, 2, ... , n individuals are recorded over a given period. This frequency distribution varies with dispersion; as a population becomes increasingly aggregated, the number of stations containing few or many mice increases, and the variance of the frequency distribution increases. Dispersion is measured by comparing the variance (S^2), based on the observed capture patterns of n individuals at t stations, with the variance (σ^2) of the theoretical frequency distribution that would be obtained if those capture patterns were randomly distributed over the study area.

In order to measure within-group dispersion of capture patterns and the degree to which that dispersion diverges from a totally random association, independent of random changes in density, I employed Metzgar's (1979) index I_2 where

$$I_2 = \frac{(S^2/\sigma^2 - 1)}{(\text{Density} - 1)} \times 100$$

I_2 values <0, =0, and >0 indicate even, random, and aggregated dispersion, respectively.

In order to measure the degree of association between sexes, I employed Metzgar's (1980) index I_3 where

$$I_3 = \frac{(S^2_{mf}/\sigma^2_{mf} - 1)}{(N_m + N_f - 1)} \times 100$$

I_3 is a between-group dispersion index, based on the ratio of the variance (S_{mf}^2) of the frequency distribution from capture patterns of all individuals (males and females) to the theoretical variance (σ_{mf}^2) of the frequency distribution obtained by randomly superimposing the separate observed frequency patterns of males and females ($\sigma_{mf}^2 = S_m^2 + S_f^2$). I_3 values <0, =0, and >0 indicate negative association, lack of association, or positive association respectively.

RESULTS

DEMOGRAPHIC AND BEHAVIORAL PARAMETERS

Detailed analyses of demography, reproduction, and behavior from these populations will be published elsewhere. A summary of comparisons among locations from which dispersion data are also available is provided in Table 1.

Data were combined for each pair of live-trapping transects due to their similarity (Herman, 1979). Average monthly captures of deer mice were similar in all locations except Helby Island, where numbers of both males and females were relatively low (Fig. 2). Two things are apparent: numbers did not vary

greatly seasonally (assuming low capture success at beginning of study); and adults constituted a large proportion of most populations throughout the study.

In 1976 breeding continued on islands through winter with no apparent interruption from the previous summer, but on the mainland breeding did not commence until May (Fig. 3). In 1977 no winter breeding occurred, and the season was more or less synchronous at all locations.

Survival of adults, calculated as the percentage of adults captured in month n known to be alive in

Table 1.—*A summary of comparisons of demographic and behavioral parameters from locations in Barkley Sound for which dispersion data are available (from Herman, 1979).*

Statistics	Mainland	Islands		
		Diana	Helby	Haines
Population density	⸺⸺ no consistent differences* ⸺⸺			
Reproductive activity	seasonal (spring/summer)	⸺⸺ extended into winter 1975/76 ⸺⸺		
Litter size	larger	smaller	smaller	smaller
Juvenile recruitment	concentrated at end of breeding season	⸺⸺ dispersed throughout breeding season ⸺⸺		
Aggression in adult males	less	less	more	more
Mortality				
Prenatal	low	low	low	low
Preweaning	lower	⸺⸺ higher (1976) ⸺⸺		
Average age of adults				
Males	younger	younger	older	older
Females	⸺⸺ no differences ⸺⸺			
Sex ratio	⸺⸺ no consistent differences (slightly favour males) ⸺⸺			
Body size	small	mod.	large	large

* With exception of Helby Is. live-trap plots.

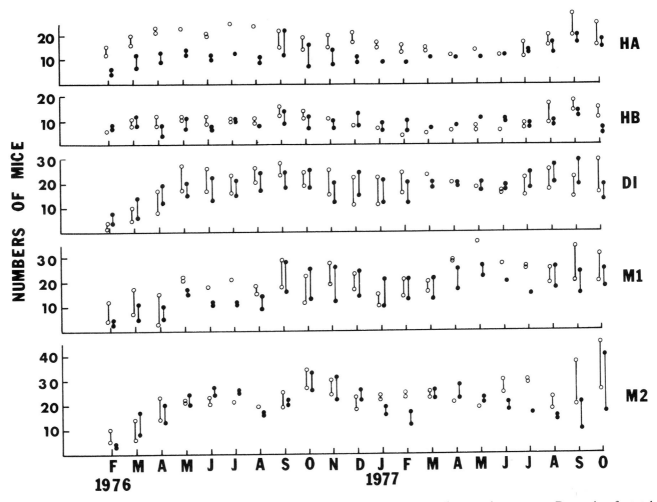

Fig. 2.—Monthly captures of male (open circles) and female (closed circles) deer mice on live-trapping transects. Data points for total captures (above) and adult captures (below) joined by vertical lines for both sexes in each month. HA = Haines Island, HB = Helby Island, DI = Diana Island, M1 and M2 = mainland.

month n+1, contains no apparent seasonal component at any location (Fig. 3). Within sexes, overall averages of monthly survival are high and vary little among locations (females 0.83–0.87; males 0.77–0.84); however, in all locations values for females exceed those for males.

DISPERSION

I considered all adults present on a transect for a minimum of 2 mo to be residents. I chose to look at dispersion in this group for several reasons: 1) I assume that these are the key individuals in terms of social organization and reproduction; 2) this is generally the largest single sub-group of the populations in this study; and 3) other studies (Metzgar, 1979, 1980) have shown that established resident adults may differ in their dispersion from newly arrived adults.

Overall, I predicted that individuals of the same sex (particularly males) would be evenly dispersed during the breeding season (summer) and aggregated during the non-breeding season (winter), and that males and females would be more strongly associated during the breeding season than during the non-breeding season. Specifically, I expected the strongest even dispersion in males on Haines and Helby Islands, where I found that resident breeding adults were highly aggressive (Herman, 1979).

An examination of data by month for within-sex dispersion of resident adult males and females reveals no startling pattern (Fig. 4). The majority of I_2 values are negative (indicating even dispersion) in both sexes at all locations; but when the accompanying I_1 values (S^2/σ^2) are analyzed according to Lefkovitch (1966), few are significant. If a seasonal trend is present, it should become obvious by com-

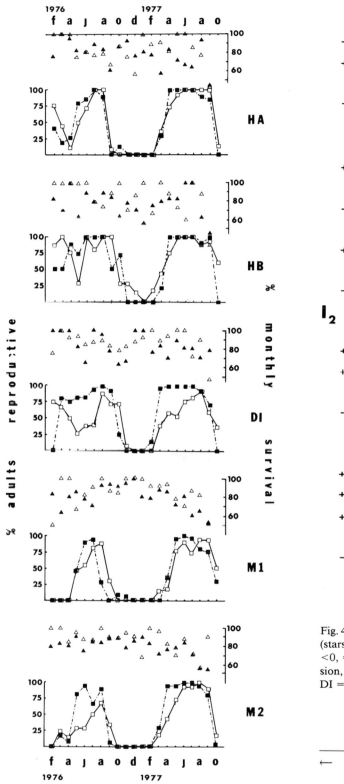

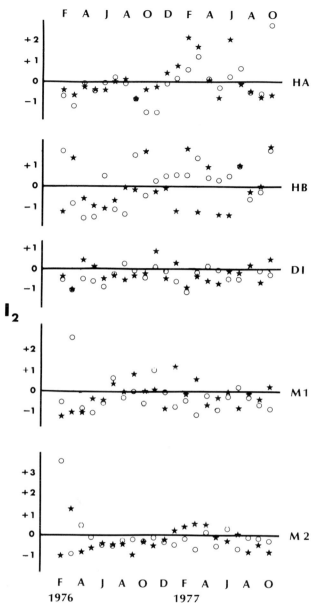

Fig. 4.—I_2 values for within-sex dispersion of resident adult males (stars) and females (circles) on live-trapping transects. I_2 values <0, =0, and >0 indicate even, random, and aggregated dispersion, respectively. HA = Haines Island, HB = Helby Island, DI = Diana Island, M1 and M2 = mainland.

Fig. 3.—Reproductive activity (squares), expressed as percentage of adults reproductively active, and adult survival (triangles), expressed as percentage of adults captured in month n known to

←

be alive in month n + 1, in males (closed symbols) and females (open symbols) on live-trapping transects. HA = Haines Island, HB = Helby Island, DI = Diana Island, M1 and M2 = mainland.

Table 2.—*Within-group dispersion values (I_2) for resident adult males, from samples combined according to reproductive activity. Mainland locations combined.*

Location	Season	N[1]	I_2
Mainland[2]	Non-breeding (February–April 1976)	35	−0.74
	Breeding (May–August 1976)	150	−0.41*
	Non-breeding (September–March 1977)	248	−0.03
	Breeding (April–September 1977)	244	+0.08
	Non-breeding (October 1977)	24	−0.26
Diana Is.	Non-breeding (February 1976)	2	−0.18
	Breeding (March–October 1976)	121	+0.35
	Non-breeding (November–February 1977)	52	−0.21
	Breeding (March–September 1977)	117	−1.91*
	Non-breeding (October 1977)	12	+0.40
Helby Is.	Breeding (February–October 1976)	76	−0.38
	Non-breeding (November–March 1977)	33	−0.47
	Breeding (April–September 1977)	45	−0.34
	Non-breeding (October 1977)	5	+1.49
Haines Is.	Breeding (February–October 1976)	131	−0.24
	Non-breeding (November–March 1977)	83	+0.08
	Breeding (April–September 1977)	90	−0.14
	Non-breeding (October 1977)	10	−0.61

* $P < 0.05$ (for I_1 analyzed according to Lefkovitch, 1966).
[1] N = number of capture patterns (see text).
[2] Mainland locations were combined.

paring data combined from all months during the breeding season with those combined from all months during the non-breeding season. This analysis considers capture patterns at one location as independent from month to month and thus increases sample sizes substantially.

In such an analysis of males, dispersion diverged significantly from randomness only during breeding season 1976 on the mainland and breeding season 1977 on Diana Is. (Table 2). In both cases dispersion was even. Surprisingly, there was no indication of a strong regular seasonal component, and dispersion

Table 3.—*Within-group dispersion values (I_2) for resident adult females, from samples combined according to reproductive activity. Mainland locations combined.*

Location	Season	N[1]	I_2
Mainland[2]	Non-breeding (February–April 1976)	34	+0.52
	Breeding (May–September 1976)	158	−0.32*
	Non-breeding (October–March 1977)	196	−0.35*
	Breeding (April–October 1977)	222	−0.39
Diana Is.	Breeding (February–October 1976)	112	−0.47
	Non-breeding (November–February 1977)	49	−0.51
	Breeding (March–October 1977)	137	−0.22
Helby Is.	Breeding (February–November 1976)	72	−0.16
	Non-breeding (December–February 1977)	20	+0.20
	Breeding (March–October 1977)	61	+0.03
Haines Is.	Breeding (February–March 1976)	11	−1.01
	Non-breeding (April 1976)	9	+0.58
	Breeding (May–August 1976)	42	−0.26
	Non-breeding (September–February 1977)	51	−0.59
	Breeding (March–September 1977)	86	−0.11
	Non-breeding (October 1977)	8	+2.42

* $P < 0.05$ (for I_1 analyzed according to Lefkovitch, 1966).
[1] N = number of capture patterns (see text).
[2] Mainland locations were combined.

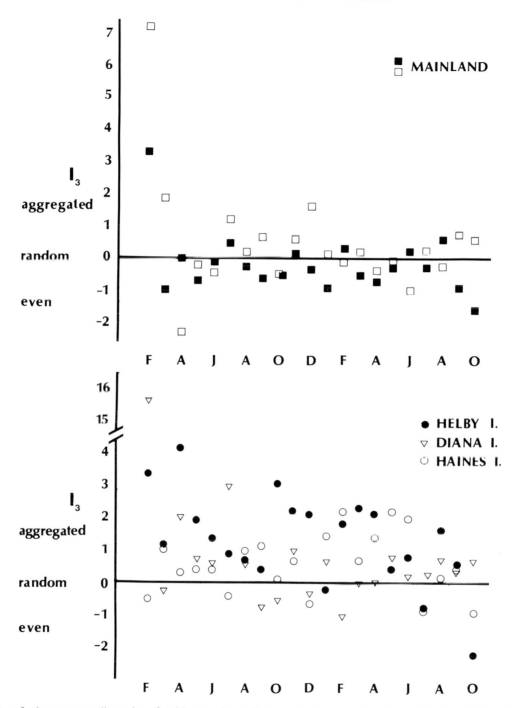

Fig. 5.—I_3 values for between-sex dispersion of resident adults on live-trapping transects. I_3 values <0, $=0$, and >0 indicate negative, random and positive association, respectively.

in the two island populations with relatively aggressive mice (Haines and Helby) never diverged significantly from randomness.

In females, dispersion diverged significantly from randomness only on the mainland, during breeding season 1976 and the following non-breeding season (Table 3). In both cases females were evenly dispersed.

Association between sexes also showed no seasonal pattern (Fig. 5). Overall, association in mainland populations does not diverge greatly from randomness. In island populations, especially Haines and Helby, males and females appear more positively associated, but no seasonal component is apparent.

DISCUSSION

Is the apparent randomness in dispersion due simply to small sample sizes? In some monthly samples one might argue that the number of capture patterns is too low to detect significant deviations from randomness. However, many combined within-group dispersion samples are relatively large (see Tables 2 and 3) in relation to other published data that show greater evenness in the breeding season and far greater aggregation in the non-breeding season (Metzgar, 1979, 1980). Grids are probably more appropriate than transects for dispersion analyses. Transects were used in this study because of their efficiency (Petticrew and Sadleir, 1970) and because of the often impenetrable habitat.

There is little indication of seasonality in dispersion patterns of either within-group or between-group samples, despite seasonal changes in reproductive activity. Increased aggressiveness among adult males during the breeding season is well documented in *P. maniculatus* from coastal British Columbia, and is believed to limit recruitment of juveniles (Sadleir, 1965; Healey, 1967; Petticrew and Sadleir, 1974) and of adults (Fairbairn, 1976, 1978) to the population. In light of this, the apparent lack of avoidance among reproductive males in aggressive populations in Barkley Sound (Haines and Helby islands) is especially surprising.

There is also no clear seasonal pattern in adult survival. Estimates from these populations include both mortality and emigration, and are relatively high for *P. maniculatus*. The apparent differences between sexes are likely due to greater vagility in males rather than differences in mortality (Herman, 1979).

Although aggregations among all sex and age classes during autumn are probably normal in *Peromyscus* populations elsewhere (Nicholson, 1941; Howard, 1949; Merritt and Merritt, 1978; Metzgar, 1979, 1980), they are not apparent in this study. The increases in population size of *Peromyscus* that normally co-occur at this time are also lacking in Barkley Sound populations, whose numbers remained relatively constant throughout the study.

This relative constancy may reflect the climatic regime. I am aware of no climatically stressful period for these populations, as there may be for montane or high latitude continental populations of *Peromyscus* (Fuller, 1969; Merritt and Merritt, 1980). Temperatures are mild year-round, and sufficient refuges from flooding are available in most areas during periods of heavy rainfall. In such a climate, large communal nests would be of little advantage. Studies from other areas with moderate climates have also reported random dispersion in *Peromyscus* populations (Davenport, 1964; Myton, 1974).

Food may affect dispersion, but unfortunately there are no data on absolute availability or patchiness of food resources in these populations. Winter breeding in island populations during 1976 was coincident with an abundant tree seed crop (Herman, unpublished) and presumably indicated that food was more abundant than in the following winter, since reproduction in *P. maniculatus* is sensitive to food supply (Sadleir et al., 1973; Gashwiler, 1979). However, there were no apparent differences in dispersion between the two winters.

Although moderate climate may account for the lack of aggregation during winter, the lack of any strong avoidance within sexes during summer is still difficult to explain. Two factors may be involved — habitat complexity and habitat specificity. A complex three-dimensional environment such as the coastal rain forest may require too much energy to defend. As well, the microhabitat requirements of mice from these populations may be less specific than those of their truly mainland counterparts, since there are virtually no potentially competitive rodent species and few or no quadrupedal predators in any of the study areas.

Random dispersion in a population should not be disturbing; it does not necessarily indicate absence of social organization. This type of analysis is only one of several available to determine the social interactions and relationships within a species. It is by necessity crude, since it ignores spatial and temporal detail. Additional techniques that are more sensitive to these details, such as radio telemetry, would complement live trapping studies and make the interpretation of dispersion less ambiguous.

ACKNOWLEDGMENTS

I extend appreciation to Fu-Shiang Chia for support and advice during the study. The Bamfield Marine Station kindly allowed me use of their facilities. Drew Harvell willingly assisted in the field. Thanks also go to Peter Smith for comments on the manuscript. Financial support was provided by an NSERC postgraduate scholarship to the author and by an NSERC operating grant to F.-S. Chia.

LITERATURE CITED

CARTER, L. 1971. Surficial sediments of Barkley Sound and adjacent continental shelf, Vancouver Is., B.C. Unpublished Ph.D. dissert., Univ. British Columbia, xii + 196 pp.

CASE, T. J. 1978. A general explanation for insular body size trends in terrestrial vertebrates. Ecology, 59:1–18.

CROWELL, K. L. 1973. Experimental zoogeography: Introduction of mice to small islands. Amer. Nat., 107:535–558.

DAVENPORT, L. B., JR. 1964. Structure of two *Peromyscus polionotus* populations in oldfield ecosystems at the AEC Savannah River Plant. J. Mamm., 45:95–113.

FAIRBAIRN, D. J. 1976. The spring decline in deer mice: death or dispersal? Canadian J. Zool., 55:84–92.

———. 1978. Behaviour of dispersing deer mice (*P. maniculatus*). Behav. Ecol. Sociobiol., 3:265–282.

FULLER, W. A. 1969. Changes in numbers of three species of small rodent near Great Slave Lake, N.W.T., Canada, 1964–1967 and their significance for general population theory. Ann. Zool. Fennici, 6:113–144.

GASHWILER, J. S. 1979. Deer mouse reproduction and its relationship to the tree seed crop. Amer. Midland Nat., 102:95–104.

GLIWICZ, J. 1980. Island populations of rodents: their organization and functioning. Biol. Rev., 55:109–138.

GRANT, P. R. 1965. The adaptive significance of some size trends in island birds. Evolution, 19:255–267.

HALPIN, Z. T., and T. P. SULLIVAN. 1978. Social interactions in island and mainland populations of the deer mouse, *Peromyscus maniculatus*. J. Mamm., 59:395–401.

HEALEY, M. C. 1967. Aggression and self-regulation of population size in deer mice. Ecology, 48:377–392.

HERMAN, T. B. 1979. Population ecology of insular *Peromyscus maniculatus*. Unpublished Ph.D. dissert., Univ. Alberta, Edmonton, 155 pp.

HOWARD, W. E. 1949. Dispersal, amount of inbreeding and longevity in a local population of prairie deermouse on the George reserve Southern Michigan. Contr. Lab. Vert. Biol., Univ. Michigan, 43:1–50.

KRAJINA, V. J. 1965. Biogeoclimatic zones and classification of British Columbia. Ecol. West. N. Amer., 1:1–17.

KREBS, C. J., B. L. KELLER, and R. H. TAMARIN. 1969. *Microtus* population biology: Demographic changes in fluctuating populations of *M. ochrogaster* and *M. pennsylvanicus* in southern Indiana. Ecology, 50:587–607.

LEFKOVITCH, L. P. 1966. An index of spatial distribution. Res. Popul. Ecol. (1966)8:89–92.

LIDICKER, W. Z., JR. 1975. The role of dispersal in the demography of small mammals. Pp. 105–128, *in* Small mammals: their productivity and population dynamics (F. B. Golley, K. Petrusewicz and L. Ryszkowski, eds.), Cambridge Univ. Press, New York and London, xxv + 451 pp.

MACARTHUR, R. H. 1972. Geographical ecology. Harper and Row, New York, 269 pp.

MACARTHUR, R. H., and E. O. WILSON. 1967. The theory of island biogeography. Princeton Univ. Press, Monogr. Popul. Biol., 1:1–203.

MATHEWS, W. H., J. G. FYLES, and H. W. NASMITH. 1970. Postglacial crustal movements in southwestern British Columbia and adjacent Washington State. Canadian J. Earth Science, 7:690–702.

MCCABE, T. T., and B. D. BLANCHARD. 1950. Three species of *Peromyscus*. Rood Associates, Santa Barbara, California, 136 pp.

MERRITT, J. F., and J. M. MERRITT. 1978. Seasonal home ranges and activity of small mammals of a Colorado subalpine forest. Acta Theriol., 23:195–202.

———. 1980. Population ecology of the deer mouse (*Peromyscus maniculatus*) in the front range of Colorado. Ann. Carnegie Mus., 49:113–130.

METZGAR, L. H. 1979. Dispersion patterns in a *Peromyscus* population. J. Mamm., 60:129–145.

———. 1980. Dispersion and numbers in *Peromyscus* populations. Amer. Midland Nat., 103:26–31.

METZGAR, L. H., and R. HILL. 1971. The measurement of dispersion in small mammal populations. J. Mamm., 52:12–20.

MIHOK, S. 1979. Behavioral structure and demography of subarctic *Clethrionomys gapperi* and *Peromyscus maniculatus*. Canadian J. Zool., 57:1520–1535.

MYTON, B. 1974. Utilization of space by *Peromyscus leucopus* and other small mammals. Ecology, 55:277–290.

NADEAU, J. H., R. T. LOMBARDI, and R. H. TAMARIN. 1981. Population structure and dispersal of *Peromyscus leucopus* on Muskeget Island. Canadian J. Zool., 59:793–799.

NICHOLSON, A. J. 1941. The homes and social habits of the woodmouse (*Peromyscus leucopus noveboracensis*) in southern Michigan. Amer. Midland Nat., 25:196–223.

PETTICREW, B. G., and R. M. F. S. SADLEIR. 1970. The use of index trap lines to estimate population numbers of deermice (*Peromyscus maniculatus*) in a forest environment. Canadian J. Zool., 48:385–389.

———. 1974. The ecology of the deermouse *Peromyscus maniculatus* in a coastal coniferous forest. I. Population dynamics. Canadian J. Zool., 52:107–118.

REDFIELD, J. A. 1976. Distribution, abundance, size and genetic variation of *Peromyscus maniculatus* on the Gulf Islands of British Columbia. Canadian J. Zool., 54:463–474.

SADLEIR, R. M. F. S. 1965. The relationship between agonistic behaviour and population changes in the deer mouse *Peromyscus maniculatus* (Wagner). J. Anim. Ecol., 34:331–352.

SADLEIR, R. M. F. S., K. D. CASPERSON, and J. HARLING. 1973. Intake and requirements of energy and protein for the breeding of wild deermice, *Peromyscus maniculatus*. J. Reprod. Fertil., Suppl., 19:237–252.

SOULÉ, M. 1966. Trends in the insular radiation of a lizard. Amer. Nat., 100:47–64.

STICKEL, L. F. 1968. Home range and travels. Pp. 373–411, *in* Biology of *Peromyscus* (J. A. King, ed.), Spec. Publ. Amer. Soc. Mamm., 2:XII + 593 pp.

SULLIVAN, T. P. 1977. Demography and dispersal in island and mainland populations of the deer mouse, *Peromyscus maniculatus*. Ecology, 58:964–978.

VAN VALEN, L. 1971. Group selection and the evolution of dispersal. Evolution, 25:591–598.

Address: Department of Biology, Acadia University, Wolfville, Nova Scotia B0P 1X0, Canada.

THE POPULATION DYNAMICS AND THE INFLUENCE OF WINTER ON THE COMMON SHREW (*SOREX ARANEUS* L.)

Kalevi Heikura

ABSTRACT

Shrews were snap-trapped on double-line method (50 + 50 traps) 1966–1972 once a month in spruce forest and hayfield at three localities in Northern Finland and on 12 by 12 trap-grids 1972–1978 at two localities four times a year on five-day periods, only. The traps were located on the ground with 10 m intervals and baited with Edam cheese. The snow-covered period at the westernmost locality is on the average 6.5 months and at the easternmost locality 7.5 to 8 months, the maximum depths of snow being about 40 and 90 cm, respectively.

The common shrew shows a four year cycle in the population fluctuation on the whole study area in both habitats, the average density being in spruce forest about 2/5 of that in hayfields. The density during the peak phase is about twice as big as during the low phase, and the peak is formed by the increased amount of females having two litters during the reproduction period (June–September) though the amount of embryos then (about 6.5 e/f) is smaller than during low phases (about 8.5 e/f). Both sexes reach their sexual maturity after overwintering, males at the end of March, females 2 to 3 weeks later, and that is not regulated by the ambient temperature. The individuals lose their weight and length during the winter, in field less than in forest. The leanness-index (body/weight) begins to grow when the ground temperature sinks to about +1.5°C and goes higher in forest than in field, and also the males seem to be more sensitive to cold during the winter than females. The individuals seem to form colonies on family-basis, but changes take place in late autumn and before the reproduction period.

INTRODUCTION

The common shrew (*Sorex araneus* L.) is the most numerous and widely spread insectivore in Europe (Miller, 1912; Ognev, 1950; Siivonen, 1977), which gives a possibility to presume that it as a species must be well adapted to various circumstances, and on that basis should have been widely and thoroughly studied. There has been various studies performed on this species outside Scandinavia, but inside this area the restricted knowledge about the common shrew and its relations to prevailing circumstances made the team, which was put up in the Department of Zoology and the Zoological Museum of the University of Oulu, Finland, to take this species, among others, under special interest.

The team was founded in 1966 to study the changes in small mammal populations with special interest in the phenomena connected to overwintering.

This study is a part of these investigations.

MATERIAL AND METHODS

Common shrew specimens were captured with snaptraps in (Fig. 1) three localities within the administrative district of Oulu, at Tupos (64°55′N, 25°30′E, about 10 m above sea level), Muhos (64°50′N, 26°05′E, about 30–50 m asl) and Paljakka (64°40′N, 28°05′E, about 380 m asl) in spruce forests and hayfields. The biotopes were selected to be as similar as possible in all the three localities.

The trapping was carried out by the double-line method in which two 50-trap lines were located at 10 m distance from each other. The distance between traps was also 10 m. The trapping was carried out regularly once a month in 1966–1972, and the lines were moved 20 m aside from the former months' sites, and the length of a trapping period was one night at a time. To control the trapping method, 5 to 7 days of line-trapping was carried out four times a year at least. The traps were baited with Edam cheese.

The 1972–1978 trapping method was 12 by 12 trap grid (distances between the traps still 10 m), and trapping took palce at Tupos and Paojakka four times a year, five days at a time, only. To compare the results of the recent and previous methods, line method was used simultaneously in 1972.

The material of this study is composed by 2,400 common shrews from 1966–1972 (Fig. 2) and 781 from that onwards.

Each specimen was stored in a polythene bag in a deep-freeze, and the laboratory treatment was carried out at the Zoological Museum of the University of Oulu. The normal measurements as well as skull, skin, stomach, and parasite samples were taken (Erkinaro and Heikura, 1977), and the sexual status as well as the age of the individuals was determined at the same time. In the age determination, all overwintered individuals caught after March were considered as old ones.

Temperature recording was carried out at the height of 2 m, at the earth surface (Fig. 3) and 10 cm below it at all the trapping sites. Snow analyses and observations about the movements of small mammals above the snow cover were made during each trapping.

343

RESULTS

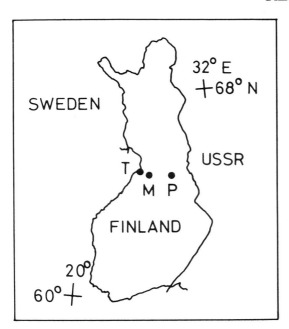

Fig. 1.—The trapping localities (T = Tupos, M = Muhos and P = Paljakka).

TRAPPABILITY OF THE SPECIES

Because the trapping period in double-line method is only one night, the catch of this period has been compared to the 5-day period trapping results to relate this "first day catch" with the whole catchable amount of shrews living and moving around at the trapping area, both with the double-line and grid methods.

The 5-day period can be considered to give very uniform figures by both methods (Fig. 4), and thus the first day catch represents about 50% of the whole catchable amount of shrews, on average. This general view gets its variation by the seasons, in the way that during the snow covered period the caught number of shrews stays underestimated when compared to the real population, and by the population density in the way that during high densities the first day catch is relatively higher (58.4%) than during low densities (44.6%). During the winter the influence of the snow depth on the trapping procedure can be seen already at the depth of 20 cm but is at its clearest at the depth of 40 cm (Figs. 5 and 6).

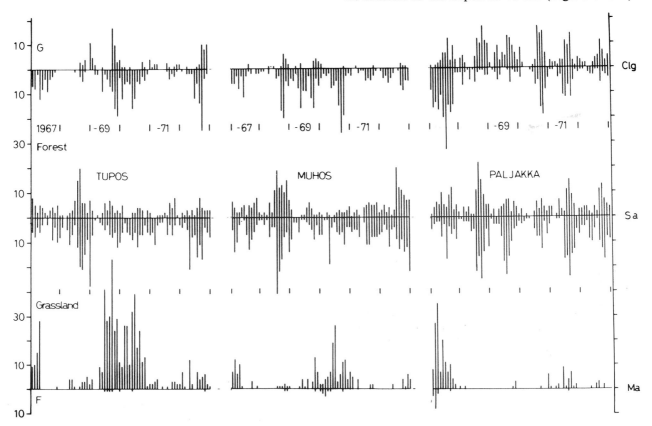

Fig. 2.—The monthly trapping results of the three most common species in 1967–1972. Individuals per 100 trap nights. Clg = *Clethrionomys glareolus*, Sa = *Sorex araneus*, and Ma = *Microtus agrestis*.

field   forest

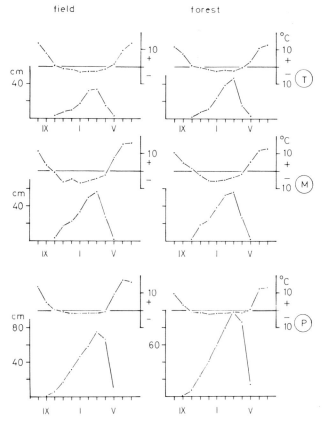

Fig. 3.—The snow depth and temperature at earth surface in each trapping site as monthly means of 1966–1972.

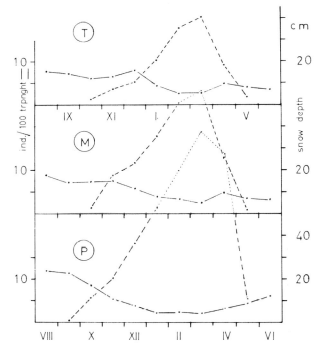

Fig. 5.—The monthly variation in the catch of *Sorex araneus* and the depth of snow.

The trappability of both sexes follows a very uniform formula by seasons and 90% of the individuals of both sexes are caught during the three first days of the trapping period (Fig. 7). During the winter the first day percentage decreases from about 56%

at September to about 48% of the next summer, and this is most clearly visible among the males.

THE OCCURRENCE AND THE VARIATION IN THE NUMBER OF THE SHREWS BY THE BIOTOPES

The variation in the number of the shrew month by month in 1967–1972 is shown in Fig. 2. Generally speaking there are two peaks in the number of the shrews, in 1968 and 1972, when the amount

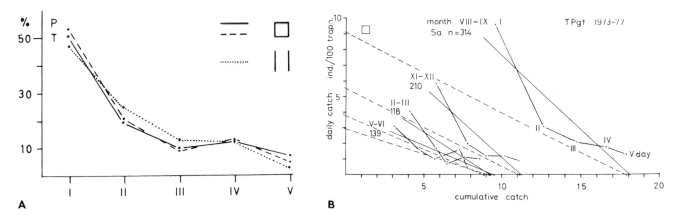

Fig. 4.—Comparison of daily trapping results of *Sorex araneus* as the percentage of the total catch in 5-day trapping period between the trapping methods (A) (see the text), and the trapping successs on the 5 consecutive days of grid trapping during different seasons (B). The solid lines show the linear regression between the daily and cumulative catches, and the broken ones show "Hayne's regression", which is evaluated to show the size of the whole population on the point where it intersects the x-axis.

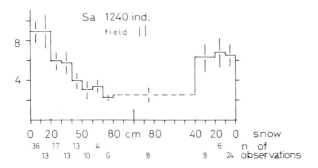

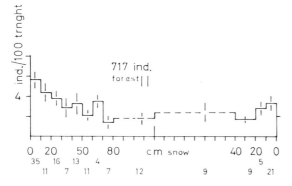

Fig. 6.—The change of the catch of the common shrew in respect to the change of snow depth in forest and field (mean ± SE ind./100 trap nights) TMP combined in 1967–1972.

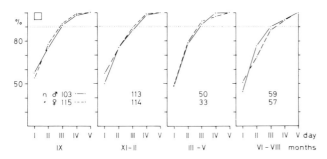

Fig. 7.—The comparison of the trapping success of the males and the females of the common shrew in grid trapping by seasons.

of individuals is higher than average in all the localities and on both biotopes. In addition high densities are met at Paljakka in 1971 both in forest and field and in 1969 in the field. The high densities among the shrew do not occur simultaneously with voles, in general, but between them, with exceptions in 1972 and at Paljakka in 1971.

It seems to be that the high densities are met between 4 years intervals at the latitude of Oulu, and they are formed by an increase in the number of young individuals (Fig. 8) (in Poland between 6 years; Pucek, 1959). High densities were met in the grid trappings also in 1975–1976 at Paljakka and in 1976 at Tupos.

During the monthly trappings in 1966–1972, 63.3% of the total catch was obtained from field biotopes. There is a difference of about 8% between the localities such that percentages at Tupos, Muhos and Paljakka are 68.2, 61.8, and 60.4%, respectively. Hayfields, especially if being reserved fields, are more favoured by the shrew than spruce forests.

There are no differently directed changes in numbers of the shrew along the seasons in separate biotopes, but the percentage of the forest stays at about the same level around the year (Fig. 9). Further, there are no differences in the biotopes visible according to age groups or sexes, which means that no migration between the biotopes takes place.

The change in the number of the shrew in respect to age, or the survival curve, is seen in Fig. 10. The amount of young individuals is at its highest soon after the reproduction period both in the forest and field and decreases from that during the fall. The death rate during the winter seems to be greater in forest than in field, where the amount of individuals seems to stay at the same level until the beginning of the reproduction period in April. The deep pit in the trapping curve in the mid-winter is

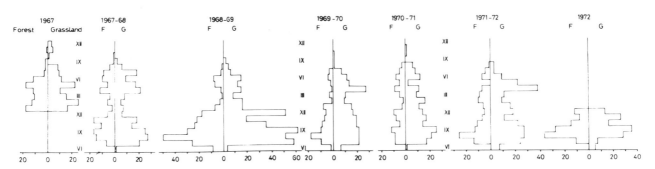

Fig. 8.—The population pyramids of the common shrew in 1967–1972. In the bases of the pyramids are the newly born juveniles and at the top are old adults. The pyramids are constructed on the basis of monthly trapping results (males and females combined).

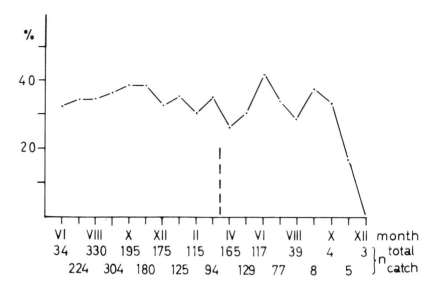

Fig. 9.—The portion of the catches in forest as the percentage of the total monthly catches in both habitats.

caused by the deep snow, which makes the trapping difficult at the time. In the forest in the beginning of the reproduction period, the density of shrews is about 40% of that of the end in the previous reproduction period, the percentage in the field being about 60%.

Sex Ratio

The sex ratio among young unoverwintered shrews is rather evenly 1:1 until the snowcovered period, when some variation from month to month takes place (Fig. 11). This variation most believably reflects the trapping success rather than the real sex ratio in the population. After March the portion of males increases sharply until July (during the reproduction period), after which the overwintered part of the population turns to female dominance. On average, the portion of males in unwintered animals is 49.4% and among overwintered, 57.8%.

Among the oldest individuals, after September

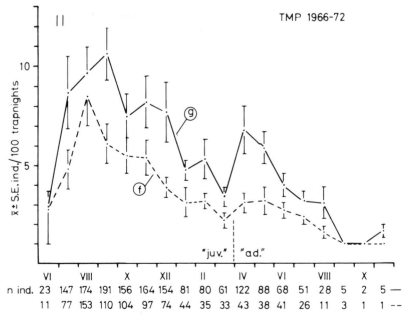

Fig. 10.—The trapping results in forest and field with respect to the lifetime of the shrew.

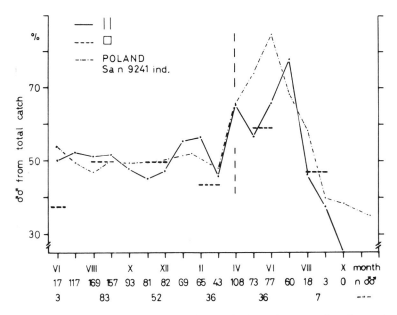

Fig. 11.—Sex ratio of the common shrew according to both trapping methods used compared to the results in Poland (Pucek, 1959).

almost no males are met, but females are found even until December, which means that females live longer than males.

When taking the overall view in the sex ratio of the shrew, the portion of males is 53.0% (in Poland 51.9%; Pucek, 1959). According to Pucek (1959), the sex-ratio in peak years among young individuals is female dominant. The same observation can be seen also in this material, though the differences between years are not statistically significant (Fig. 12).

REPRODUCTION

Males

Common shrew males do not reach their sexual maturity before overwintering, during and before which the length of the testes stays at about the same level until February in both biotopes and all localities (Fig. 13). After February very rapid growth of testes takes place, and the length of reproductive capability is reached at the turn of March and April, when the length of reproductive limit is exceeded (testes length over 5 mm, Saure et al., 1971). This happens simultaneously in both biotopes and all localities, and there has been no differences between the years either.

The length of testes reaches its maximum in June (about 7.2 mm) after which a decrease in the length takes place. It must, however, be pointed out that there exist reproductively capable old males still in August–September.

During the maturation phase in March–April, there is no correlation between the testes length and the ambient (on earth surface) temperature of the same or previous month (Fig. 14).

Females

The females of the common shrew differ from the males in the respect that they can reach their sexual maturity already before overwintering (Fig. 15), and a part of them is in active status also during the winter (see Skaren, 1979). The portion of these is, however, very small (4.6% of 451 females).

Usually, the young females do not reproduce before their overwintering (in this material there have

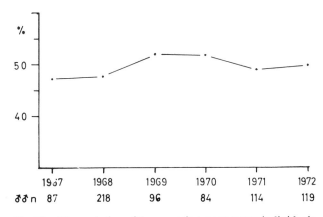

Fig. 12.—The variation of the sex ratio among young individuals in 1967–1972.

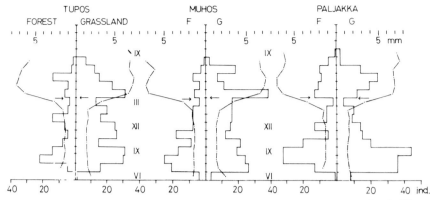

Fig. 13.—The change of the testes length of the common shrew by months (years 1966–1972 combined). The arrows show the time of the reaching of reproductive maturity.

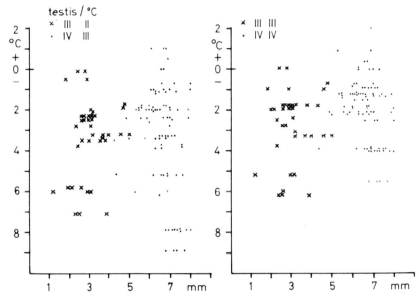

Fig. 14.—The mean monthly values of the testes length compared to the earth temperature (TMPfg 1967–1972) during the late winter (February–April).

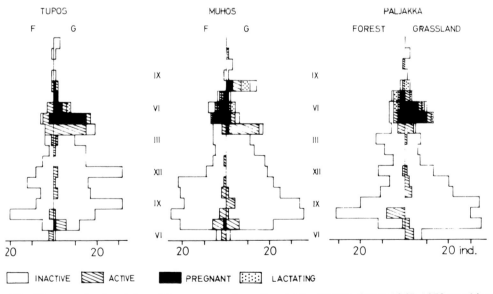

Fig. 15.—The reproductive status of the common shrew females by months. (years 1967–1972 combined).

349

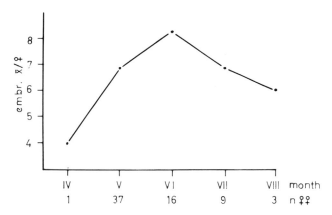

Fig. 16.—The amount of embryos per female during the reproduction period.

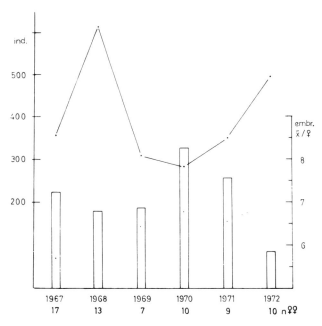

Fig. 17.—The average amount of embryos per female in respect to the population cycle.

been met two individuals in July), and the reproduction period begins in April (see also Skaren, 1973) (as an exception there has been met one lactating female in March in this material). In May 60% of females have been pregnant and only 11.1% inactive. In June–August 18.2% of overwintered females have been inactive and from then to the end of their lifetime, November–December, reproduction is very infrequent.

The length of the reproduction period at Tupos and Muhos is longer than at Paljakka; for instance, there were pregnant females encountered in the middle of April at Muhos but not before May at Paljakka. Pregnant females were not present at Paljakka at the end of the summer, as was the case at Tupos and Muhos. According to Skaren and Kaikusalo (1966), the reproduction period lasts from April to September in southern Finland as well as in Poland (Pucek, 1960) and in England (Crowcroft, 1957).

The average amount of embryos per a female is, in this material, 7.1 (range 4–11, n = 66) and there is no difference between the biotopes (7.11 in field and 7.10 in forest). The amount of embryos per female is at its highest (8.3 ± 0.4 e) in the middle of the reproduction period, in June, after which a decrease takes place (Fig. 16). The similar fact has been noticed both in Poland (Pucek, 1960) and England (Crowcroft, 1957).

There are differences in the amount of embryos between separate years (Fig. 17) such that during years of low densities, there are more embryos per female (for instance 1970, 8.5 ± 0.45 e) than in the years of high densities (1968, 6.8 ± 0.38 and 1972, 5.7 ± 0.62 e). The differences are statistically significant (t 1970/1968 = 2.98** and 1970/1972 = 3.68**).

A female gives birth to one to three litters during a reproduction period (Siivonen, 1972) and if it is possible to be pregnant while simultaneously lactating (Skaren, 1979) the previous litter, the theoretical maximum in these circumstances is four litters. However, according to the observations made during the laboratory treatment of this material, it seems to be that during the peak years a bigger part of the females gets two to three litters than during the low years, and an average number of litters per female is smaller than two.

Theorical calculations give a litter amount for a peak year as 2.6 (1968 and 1972) and for a low year as 1.5, the average being 1.8 (1967–1972).

VARIATION IN THE BODY LENGTH AND WEIGHT

The monthly variation in the body length and weight is represented in Fig. 18. The growth of the young, unwintered shrews to the wintering size takes place until September, after which the length and weight decrease until January–February (Hyvärinen and Heikura, 1971). After February a very strong and rapid growth takes place, and the growth goes on almost to the end of the reproduction period. It can be noticed that though there is visible decrease in the winterly length and weight among both sexes in both habitats and in all localities, the individuals living in field are on average bigger and heavier than

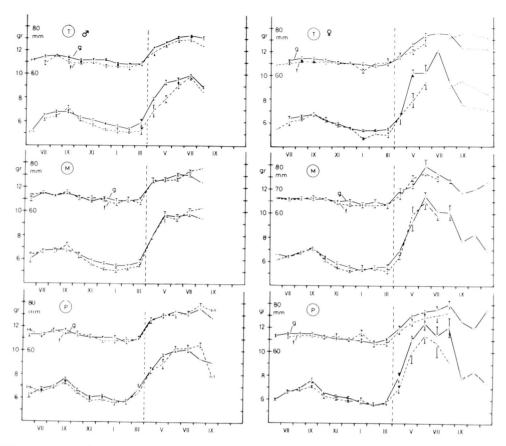

Fig. 18.—The variation in the body length and the weight along the lifetime of the shrew (years 1966–1972 combined).

in forest, and the variation is bigger in forest than in field. The decrease in weight of females is on average 24.1% in field and 26.6% in forest and in length, 4.9% and 5.5%, respectively. Corresponding values for males are 21.6%, 26.9%, 5.3%, and 6.9%, respectively. The decrease in the length is of same size class in all localities, but in the weight the decrease at Tupos (max. 30.9% in forest) is bigger than at Paljakka (max. 21.1% in forest). The differences in field are smaller than in forest.

The heavy peak in the curves of weight in September is caused by moulting, which is taking place at this time.

Generally, the body length and weight are highly significant and positively correlated (males in forest $r = +0.963***$, in field $+0.965***$ and females in forest $+0.036***$ and in field $+0.955***$) in the population level.

Variation in the "Leanness-Index" (B/W)

By dividing the body length (mm) of an animal by its weight (g) a "leanness-index" (B/W) (Heikura,

1977) is formed. This is done in order to simplify the comparisons of animals by localities, biotopes and sexes by using one figure per animal, and this figure is the bigger, the "leaner" the animal is. Further, the changes taking place in the animal itself and in its surrounding circumstances are more easily handled than using weight and length separately.

According to the combined material (TMPfg), there has been formed a mean B/W-curve for males

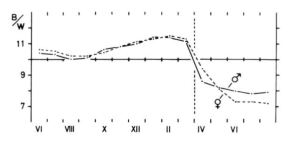

Fig. 19.—The mean curves for the leanness-index (B/W) of the common shrew as mean values of the combined material from TMPfg 1966–1972.

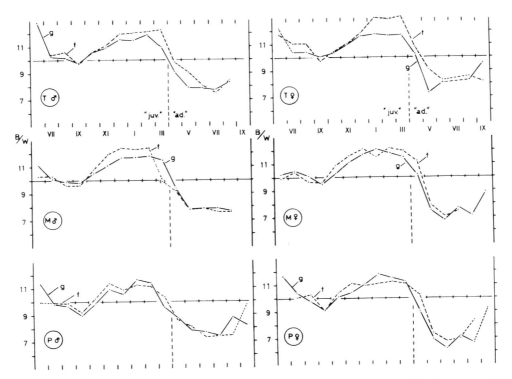

Fig. 20.—The monthly variation in the leanness-index (years 1966–1972 combined, f = forest, g = field).

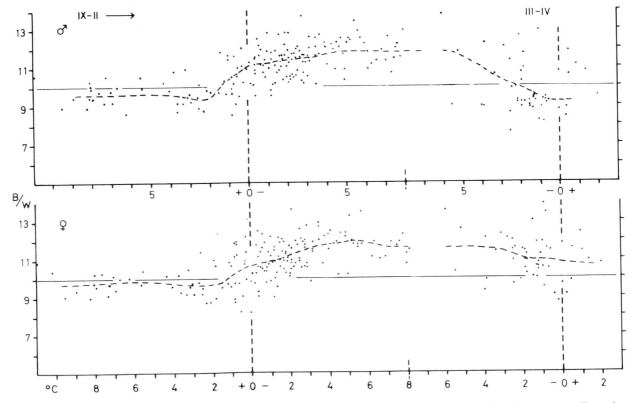

Fig. 21.—The monthly leanness-indexes of males and females from the winters 1966–1972 plotted against the corresponding values of the temperature at earth surface.

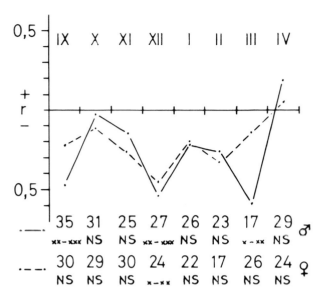

Fig. 22.—The monthly variation in the correlation coefficient between the leanness-index and the earth temperature.

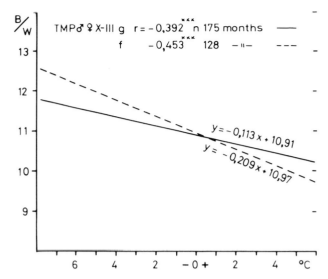

Fig. 24.—The regression of the monthly values for the leanness-index and earth temperature by habitats.

and females (Fig. 19). There is a slight difference in the leanness-index among young individuals during the autumn, so that males are a little heavier in respect to their length than females. This difference disappears in September. Another difference is again visible in late winter (early spring) between March and April, when males gain weight more rapidly than females. This takes place at the time when the sexual maturation of males is at its strongest. The

smaller B/W-index of females in June–August is caused by pregnant females which have not been separated but counted in the whole material as a fact (as well as the growing testes of males).

In comparing localities (Fig. 20), there are differences at the level of the curves, on average, so that at Tupos the index is higher than at Paljakka. Muhos stays between them though, being very similar to the level of Tupos. That means that the animals at Paljakka are heavier in respect to their length than at Tupos.

The common feature for all the localities and both sexes is that among the individuals living in forest habitat the B/W-index is higher, on average, than among individuals in field habitats, and this difference is clearest among females at Tupos. An exception is formed by Paljakka during the latter half of the winter.

CHANGES IN THE B/W-INDEX IN RESPECT TO TEMPERATURE

By plotting the B/W-indexes and temperatures at earth surface against each other month by month, it has been found out that the change in the B/W-index to the winter-level takes place at the time when the temperature sinks near +2°C both among males and females (Fig. 21). This happens faster among males than among females. According to Hyvärinen (unpublished), this expresses a change in the level of metabolism of the common shrew, and the average winterly level is, in this material, reached

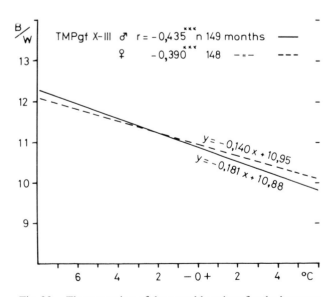

Fig. 23.—The regression of the monthly values for the leanness-index and earth temperature by sexes.

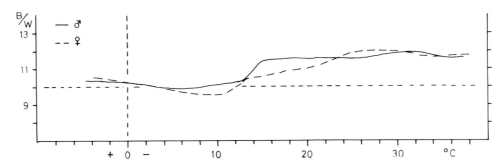

Fig. 25.—The leanness-indexes of males and females in respect to the coldest night temperature in air. See also the legend of Fig. 21.

in October–November though the values of B/W-index go on rising until December in some cases, and the increase to this level begins soon after the moult in September.

The comparison of monthly values of earth temperature and B/W-index (Fig. 22) shows that among males there is at least significant negative correlation between them in September, December and March. This occurs among females in December only. Combining the values of the whole winter period (September–March) to a regression analysis (Fig. 23),

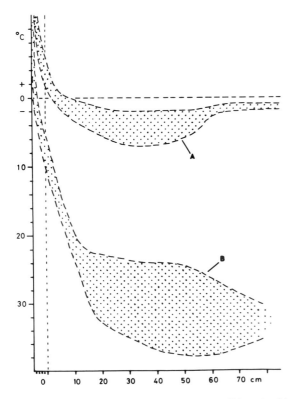

Fig. 26.—The changes of the earth temperature (A) and coldest night temperature (B) in respect to the snow depth in 1966–1972 (TMPfg combined).

the B/W-index in both sexes is in highly significant negative correlation with earth temperature (males $r = -0.435***$, females $r = -0.390***$), and the regression line in males is declining slightly steeper than in females.

The comparison of habitats (sexes and localities combined) (Fig. 24) during the same period shows correspondingly negative correlation and that the regression line in forest is on a clearly steeper decline than in field.

Comparing the B/W-index to the coldest night temperature in air (Fig. 25), it is visible that among females the rise of B/W-values to the winter level begins about 2°C earlier (at about -11°C) than among males (at about -13°C), but in males this rise happens faster.

When the changes taking place in B/W-index in respect to the temperatures are moved to the circumstances in nature (Fig. 26), the change to the winter level happens in autumn at the time when the earth get its snowcover, whereas the temperature-circumstances normally are as described above and which in time scale fits with October most often.

THE LOCATION OF THE COMMON SHREW INDIVIDUALS IN THE TRAPPING-FIELD AND RELATIONS TO THE INDIVIDUALS OF ITS OWN AND OF THE MOST COMMON VOLE SPECIES

The five-day grid trapping method gives an opportunity, by locating each trapped individual to grid-map, to evaluate group formation of a species, distances between individuals and overlapping of possible groups of different species. As seen in the examples (Fig. 27), both the common shrew and the most common vole species (the bank vole, *Clethrionomys glareolus* and the field vole, *Microtus agrestis*) appear in colonies of varying size and form. The overlapping of the area of colonies of different species is evaluated by calculating the percentage of the individuals of separate species obtained from

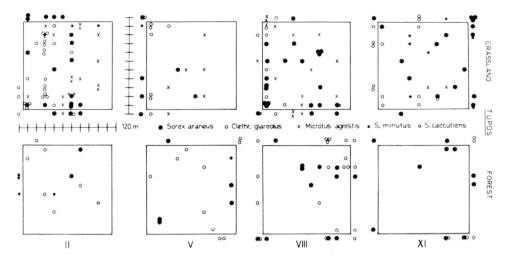

Fig. 27.—An example of the location of the individuals of different species in the grid trapping.

the same trap, which most often happens on the edges of the colonies.

The overlapping of the common shrew individuals with the vole species (Fig. 28) is about 19–21%

on average, and it is at its smallest (11.2%) in the beginning of the reproduction period.

Within the common shrew, most of the trapped individuals have been trapped each from its "own"

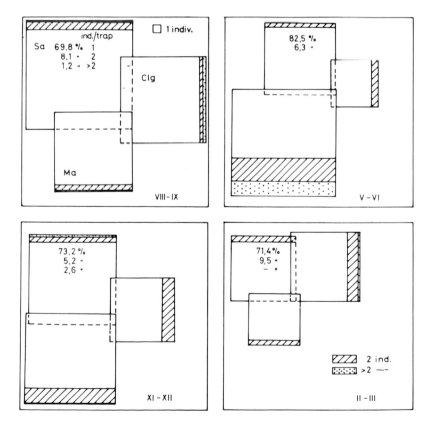

Fig. 28.—The overlapping of the individuals (as percentages of the trapped individuals of the species) according to the grids in 1972–1978. See text.

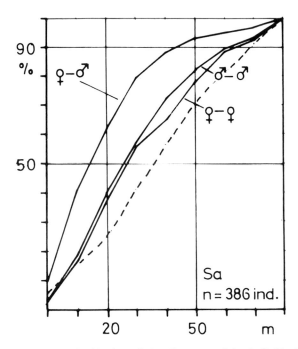

Fig. 29.—The distribution of the distances of the individuals according to the sex (in grids). The broken line represents the theoretical distribution in the case of evenly distributed individuals but accepting the known percentage to be caught from "the same trap."

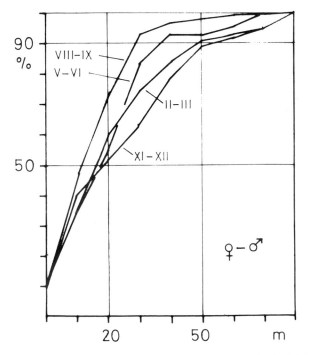

Fig. 30.—The distribution of the distances between the individuals of opposite sexes during the seasons.

trap, but a part of them have come also from the same trap. Two or more shrews from the same trap have been trapped in August–September (9.3% of all individuals) and in November–December (7.8%) and two in February–March (9.5%) as well as in the first half of the reproduction period May–June (6.3%).

On the basis of grid trapping, it has also been possible to calculate the distribution of the distances between individuals. The distance between opposite sexes (Fig. 29) is on average smaller than the distance between the individuals of the same sex, which is very similar in its distribution among males and females and is always smaller than if the individuals were evenly distributed in the trapping field. Fifty percent of the distances between opposite sexes are less than 20 m during all the seasons, and of the remaining 50%, the distances between opposite sexes are at their longest in November–December (Fig. 30) and at their shortest in August–September, when all of the family groups have not yet dispersed. In the beginning of the reproduction period, the distances are smaller than in winter as well.

GROUP SIZE OF THE SPECIES

For the determination of the group size and the average distances between groups, it would have been necessary to use the results of the line trappings because of the small size of the trapping grids. In that case, the group has been imagined in the form of circle to be able to make comparisons between different cases, though it is in fact known that the groups are of varying shape. It has to be noted that a double line may just by-pass a group, and by that way there may be uncontrolled factors in the results.

The variation in the distances between the groups (Fig. 31) is very similar in field and forest during the snowless period (June–September), and there is no significant difference in the distances of the groups between the habitats. During the winter, however, there is bigger variation in the distances of the groups in forest than in field. Accordingly, there is difference in the size of the group in the forest and field so that in forest habitats, groups are smaller than in field both in the size and in the number of individuals. When at their smallest, the groups are of about the same size both in forest and field, however, by the amount of individuals (in field about 3.5 ind./group in January and in forest about 2.9 ind./group in December). There is the clearest difference in group size between the habitats during the reproduction period, when there are more individuals in

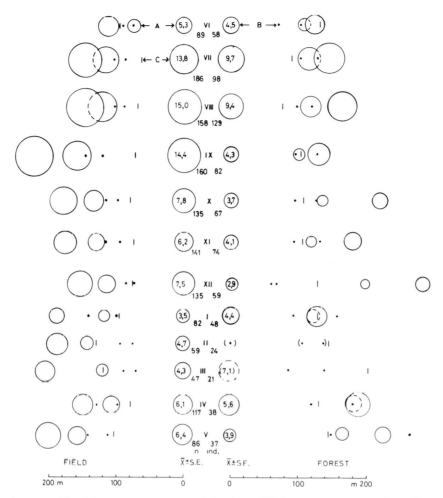

Fig. 31.—The distances (mean ± SE m) between two groups of the shrew (A), between a group and a solitary animal (B) and the theoretical distance between individuals if they were evenly distributed (C) with the diameter (mean ± SE m) of a group (in the form of circle) and the mean amount of members of a group. Results from the double-lines in 1966–1972 TMP combined.

a group in field than in forest. The difference may be caused by the smaller "starting groups" in the forest, which leaves the secondary groups smaller, too. The decrease in the size of the group takes place earlier in forest, in August–September, than in field where it happens about one month later.

The diameter of a group and the amount of its individuals follow each other very closely (Fig. 32), and both the forest and the field behave very similarly in that respect. The distances between the individuals of a group also behave uniformly in both habitats, being at the smallest when the diameter of the group is smaller than 30 m and the amount of individuals is smaller than 6 ind./group.

When imagining the area used by one individual in the form of a circle, it can be calculated that the size of it varies in the limits of 64 to 167 m², being on average about 102 m² in both habitats.

Both in grid and double-line trapping, there can be observed solitary individuals in addition to groups. Most of them occur in February–March (72.3 and 72.1%, respectively), in June (46.9%) and in December (44.3% of all individuals). The number of them is smallest in July (23.2%) and August (24.4%). It must be noticed, however, that a part of them belongs to a group which is just by-passed by the double-lines, which can be seen in Fig. 31 where the distance between a group and a solitary animal is the same as the distance between two groups, on average (for instance, in June in field and in August both in forest and in field). When the sex-ratio of these solitary individuals is compared to the sex-ratio of the whole population (Fig. 33) in the corresponding time, it can be seen that among the solitary animals there are more males than in the whole population. This is at its clearest in August–Septem-

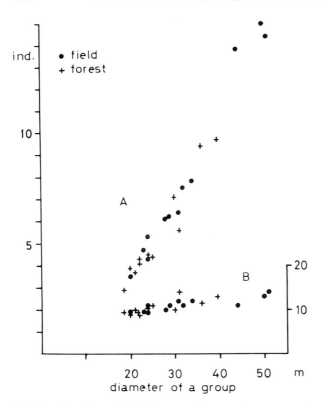

Fig. 32.—The change in the diameter of a group in respect to the amount of the members in it (A) and the mean distance between the members of the group in respect to the diameter of it (B).

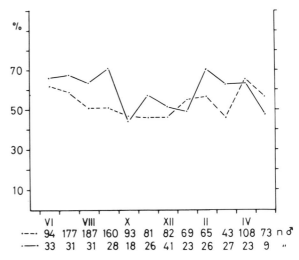

Fig. 33.—The monthly percentages of the male shrews among the solitary animals (solid line) and the corresponding percentages in the whole population (broken line).

distances of the groups (Fig. 31), in the distances between individuals (Fig. 30) and in the correlation coefficient of earth temperature and B/W-values (Fig. 22). In March and thereafter, the males tend to ag-

ber (71.8% versus 63.3%) and in February–March (70.3%/62.8%). In December there is no clear difference in the sex-ratio among these two fractions, however, which indicates that both sexes are moving around equally at that time.

GROUP FORMATION OF THE SPECIES

In the beginning of the reproduction period, the groups seem to be formed by a few pairs of opposite sexes (Fig. 34) which are situated in places optimal in respect to shrews' requirements (Fig. 35). The individuals of the first litter seem to stay rather near their mother, thus forming a bigger group. The dispersion from this group takes place slowly by young females with young males following them, but the group does not seem to split totally. Most obviously the rearrangements in the group compositions take place in September, when there are more free-running males around and the young females have taken their home-ranges already. These groups seem to wander around according to the food resources and during that time some straying happens. Perhaps in December the individuals are moving around most eagerly, which is seen in the changes in the size and

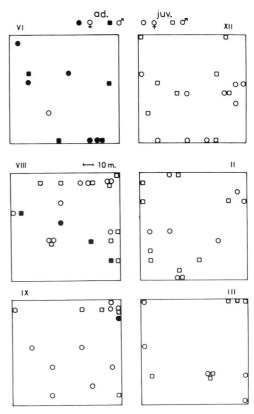

Fig. 34.—An example of the location of the common shrew individuals in grids at Tg in 1973–1974.

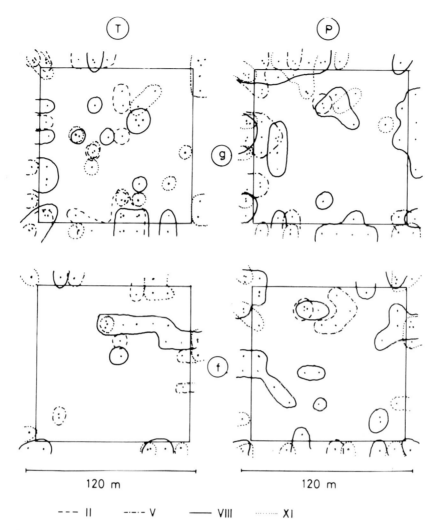

Fig. 35.—An example of the location of the individuals in one grid site trapped four times a year (control trappings Tfg and Pfg) in different seasons.

gregate around females as their sexual ripening begins to grow, and in the beginning of the reproduc-

tion period there again are small groups of shrew pairs on suitable microhabitats.

DISCUSSION

The common shrew is a species which has an extensive occurrence and which thus is well adapted to prevailing circumstances on very large geographical areas and varying habitats as well.

This gives us a possibility to suppose that it is endowed with very flexible mechanisms in its morpho-physiology, among other things, especially because it seems to be rather conservative in its use of food resources (eating mainly Lumbricidae, Mollusca, Arachnida, and Coleoptera species on a large

geographical area and during all the seasons; Sulkava, 1980).

It seems to be sensitive to changes in ambient temperature to a certain extent, especially outside the summer-period, but it is difficult to find any circumstances under which it could not survive, perhaps except during its autumn molt if there happens to be heavy daily rains combined with very cold nights and in spring during very rapid thaw. This is due to its shrinking-mechanism (Hyvärinen, 1969;

Hyvärinen and Heikura, 1971), which buffers the influence of increased energy requirement during the cold period, and the lack of fat deposits as energy resources (brown fat exists, only; Pasanen, 1971) by decreasing the relative need of energy. This can even be seen as a difference in the morphological characteristics between habitats, so that the individuals living in "worse" habitat, here in spruce forest, are smaller and lighter in general than in "better" habitat, in field, and they also lose their weight and length more in forest than in field during the winter. The change into the winter level in the body length/ weight-ratio starts soon after the autumn molt when the earth temperature sinks below about $+2°C$ and the coldest night minimum falls below about $-10°C$, which happens on the study area in October, on average. This takes place more rapidly among males than among females, and the males also are a little more sensitive to the changes in temperature during the winter than females.

The males begin to grow earlier than females, already during the winter in March, when their sexual maturation begins. That means that their maturation is not controlled by ambient temperature in the first place but by some other factors, most likely by photoperiodism.

The common shrew has a cycle of about 4 years at the latitude of Oulu in its variation in the number of individuals, and the peak years are formed by increased numbers of young individuals, which is not caused by increased numbers of embryos of reproducing overwintered females but by increased mean numbers of litters in population. The peaks in this 4 year cycle most often are situated between the peaks of the voles living in same habitat, but they also can be simultaneous because the population cycle of these voles (*Microtus agrestis* and *Clethrionomys glareolus*) is 3 to 4 years by frequency. The basic level (density during a low phase) among the common shrew is always higher than among voles, and that is why the change in number of shrews in consecutive years is not as easily observed as in voles. The density during a high phase is on average about twice as high as during a low phase only. The situation of the peak years of the shrew in between the peaks of the voles cannot be caused by trapping results, and thus be artificial, because the simultaneous low and high numbers of all the species were unexplainable then.

The voles and the shrews seem to live in colonies in both habitats and at all the localities. These colonies are neither situated completely separately nor on each other, but have an overlapping of about 20% of their size. The colonies of the shrews seem to stay at about the same place in the terrain, just drifting around a little along the seasons, which means that the individuals tend to stay in places where there are the best possibilities to meet the changing environmental circumstances as well as availability of suitable food resources.

Most obviously, the sites for the formation of colonies are originally "chosen" by young females, and that takes place already during the summer, and at that time the individuals of a litter form a basic group. The individuals of successive litters of same female seem to stay rather near their mother and even form a united colony. The basic groups seem to scatter little by little during the winter because of the disappearance of its members for different reasons, but groups are rearranged again so that their size very seldom is smaller than three individuals, but most often is more than four. The movements of individuals leading to this are in mid-winter performed by both sexes, but during late winter/early spring females are less movable than males. The role of males seem to be mostly a role of "follower," and females are the directing fraction in the population.

The beginning of the reproduction period is regulated by females because males become reproductively active earlier, during the thickest snowcover at the end of March, and are thus ready, waiting for females to reach the same status. The overwintered females continue reproducing until September and have 1 to 3 litters during a reproduction period. The young females also form a reproduction potential, though without any remarkable significance, in the summer of their birth, which young males do not.

This mechanism of colonization and reproduction assures the reproductively important fraction of the population to have the most suitable environmental circumstances, and at the same time makes sure that there is always enough reproduction potential of both sexes present at the time needed. It also saves the new reproduction potential to be used after overwintering, which is important considering both the short lifetime of a shrew generation and the limited possibilities of an individual to move around in respect to the energy and time available. It can also be thought that this formula, added to changes in the number of individuals (especially to peak years) and the morpho-physiological variability, gives to the common shrew, as a species, good adaptability to a large scale of circumstances on different geographical areas and even to rather rapidly changing environmental factors.

The fact that common shrew individuals are not evenly distributed in the terrain, trying to survive each on its own more or less optimal home range formed by possible competition, but that they appear as rather tight groups and that they tend to keep the groups existing, leads to the conclusion that there must be some advantages, in addition those mentioned above, in that way of living. A possible one is that a group uses the available food resources from one place at a time most effectively, but at the same time leaves another place untouched for use to come. This is how a group, keeping the contact between its members (for instance by voices and scent marking, which in fact are noticeable even by human senses) and existing as a unit, assures continuity in very varying circumstances, not allowing a pop-

ulation to drive into almost total suicide year after year. This could happen if the individuals lived separately and, for instance, ate their food resources in the run of the winter below the limit of being able to find it in short enough time. The group behavior may be fatal for a part of the individuals (slowest in learning?) of the group, but that is not a disadvantage in the point of view of the entirety.

The group behaviour within and between different bird species as a form of cooperation in the utilization of food resources and use of space and time (not only as a result or form of competition) has been recently studied by E. Merilä (unpublished), and the results concerning birds seem to have many connections to phenomena met among small mammals and, according to my judgement, to point in the direction of taking the theories of competition into new consideration.

LITERATURE CITED

CROWCROFT, P. 1957. The life of the shrew. The Stellar Press Ltd., London, 166 pp.

ERKINARO, E., and K. HEIKURA. 1977. Dependence of Porrocaecum sp. (Nematoda) occurrence on the sex and age of the host (Soricidae) in northern Finland. Aquilo Ser. Zool., 17:37–41.

HEIKURA, K. 1977. Effects of climatic factors on the field vole, *Microtus agrestis*. Oikos, 29:607–615.

HYVÄRINEN, H. 1969. On the seasonal changes in the skeleton of the common shrew (*Sorex araneus* L.) and their physiological background. Aquilo Ser. Zool., 7:1–32.

HYVÄRINEN, H. and K. HEIKURA. 1971. Effects of age and seasonal rhythm on the growth patterns of some small mammals in Finland and in Kirkenes, Norway. J. Zool. London, 165:545–556.

MILLER, G. S. 1912. Catalogue of mammals of Western Europe. British Mus. (Nat. Hist.), London, 1019 pp.

OGNEY, S. I. 1950. Mammals of USSR and the Adjacent countries. (Translated from Russian in Jerusalem, 1964).

PASANEN, S. 1971. Seasonal variations in interscapular brown fat in three species of small mammals wintering in an active state. Aquilo Ser. Zool., 11:1–32.

PUCEK, Z. 1959. Some biological aspects of the sex ratio in the common shrew (*Sorex a. araneus* L.). Acta Theriol., 3:43–73.

———. 1960. Sexual maturation and variability of the reproductive systems in young shrews (*Sorex* L.) in first calendar year of life. Acta Theriol., 3:269–296.

SAURE, L., K. HEIKURA, A. PELTTARI, and T. TALMAN. 1971. The functioning of the testes in the common shrew (*Sorex araneus* L.) in northern Finland. Aquilo Ser. Zool., 13:81–86.

SIIVONEN, L. 1972. Suomen nisäkkäät 1. Otava, Helsinki, 474 pp.

———. 1977. Pohjolan nisäkkäät (Mammals of northern Europe). Keuruu, 196 pp.

———. 1979. Variation, breeding and moulting in *Sorex isodon* Turov in Finland. Acta Zool. Fenn. 159:1–30.

SKAREN, U., and A. KAIKUSALO. 1966. Suomen pikkunisäkkäät. Otava, Helsinki, 227 pp.

SKAREN, U. 1973. Spring moult and onset of the breeding season of the common shrew (*Sorex araneus* L.) in central Finland. Acta Theriol., 18:443–458.

SULKAVA, S. 1980. Food of the common shrew (*Sorex araneus* L.) in the vicinity of Oulu, Northern Finland. Manuscript.

Address: Zoological Museum, University of Oulu, 90100 Oulu, Finland.

A POSSIBLE SUBNIVEAN FOOD CHAIN

C. W. AITCHISON

ABSTRACT

The invertebrate portion of a possible subnivean food chain would consist of winter-active invertebrates such as collembolans, cicadellid larvae, mites, spiders, and to a lesser extent in Manitoba, carabid and staphylinid beetles. The fungivorous collembolans and possibly the herbivorous cicadellid larvae would comprise the primary consumers, while the predatory mites, spiders and possibly beetles would form the secondary consumers in such a chain.

In the vertebrate portion of such a food chain the insectivorous shrew is the most probable tertiary consumer active in subnivean runways. Nevertheless it in turn may also be eaten by a quaternary consumer. The high metabolic rates of shrews necessitate their almost constant foraging for prey, even during winter.

The soricid eats what it encounters, especially active prey, and is dependent on tactile, auditory, kinaesthetic and olfactory sensory cues. Being an opportunist, the hungry shrew probably consumes all invertebrate prey which it can subdue. Diets of shrews are closely correlated with invertebrate catch in pitfall traps and are dependent on the invertebrate and sometimes the vertebrate prey available on the soil surface.

INTRODUCTION

This paper is mainly an expository one concerning subnivean shrews preying upon active invertebrates in northern, holarctic areas. The winter-active invertebrate fauna of North America and of northern Europe is discussed, mentioning their general activity at low temperatures and their positions in a food chain. A major source is a continuing study in southern Manitoba of thermal restrictions to activity of invertebrates (Aitchison 1978a, 1978b, 1979a, 1979b, 1979c, 1979d, 1979e, 1979f), together with analyses of stomachs of shrews in various habitats and an investigation of foraging behaviour and natural history. A comprehensive literature review of diets of *Sorex* spp. throughout the year provides insight to winter diets, which are dependent upon the availability of local epigean, prey fauna.

During winter months small mammals are commonly known to be active under snow cover, while less is known about the activity of invertebrate prey in the same period. The omnivorous shrew has a high metabolic rate and must feed almost constantly to stay alive (Seton, 1909; Crowcroft, 1957), while other winter-active small mammals are generally herbivorous (Bergeron, 1972). In the past the general consensus of opinion has been that shrews fed upon hibernating insects and other invertebrates which they happened upon during winter months (Hamilton, 1930, 1940; Kisielewska, 1963). However more recent literature indicates that shrews do not feed on hibernating and quiescent invertebrates, but rather on winter-active species (Ackefors, 1964; Pernetta, 1976).

Soricids attack anything that they encounter, eating most invertebrate prey (insects, their larvae and pupae, other invertebrates) and seasonally other food such as coniferous seeds (Yudin, 1962; Platt and Blakley, 1973). All of these animals are found on the soil surface, usually with some litter or humus present (Crowcroft, 1957; Yudin, 1962; Kisielewska, 1963; Ackefors, 1964; Pernetta, 1976). As an example, *Sorex cinereus* Kerr is possibly a key predator in litter-detritus food webs of the grassland ecosystem in Iowa, because of its culling the dominant, large-sized invertebrate prey species (Platt and Blakley, 1973).

WINTER-ACTIVE INVERTEBRATES

In areas with 20 or more cm of snow cover lasting for two or more months, invertebrate fauna can be active in the subnivean space, where temperatures range from 0° to −10°C. Subnivean pitfall traps (Näsmark, 1964) capture a winter-active fauna consisting of basically Acari, Araneae, Collembola and Coleoptera, all of which may be extracted from winter litter (Holmquist, 1926; Agrell, 1941; Chapman, 1954; Heydemann, 1956; Wolska, 1957; Näsmark, 1964; Mason, 1972; Aitchison, 1978a, 1979a, 1979b, 1979c). Generally species diversity and the number of species decrease in autumn, are low in winter and increase gradually in late spring to reach a vernal maximum.

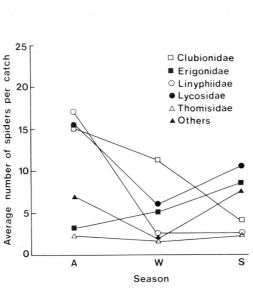

Fig. 1.—The average number of active spiders per catch at different seasons from 1974 to 1975 (A—autumn, W—winter, and S—spring).

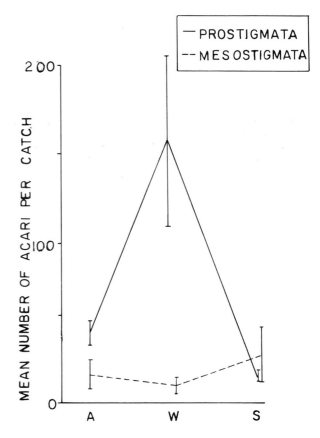

Fig. 2.—The mean number of active Acari per catch versus seasons from 1973 to 1974, showing the standard error for each point (Aitchison, 1979c).

Araneae

Low temperature activity by winter-active spiders has been documented in northern Europe (Wolska, 1957; Huhta, 1971; Hågvar, 1973) and in North America (Holmquist, 1926; Bertram, 1935; Aitchison, 1978a). *Linyphia phrygiana* Westring was active on snow between −3° and 9°C in Poland (Wolska, 1957), while in Norway *Bolyphantes index* (Thorell) built webs in snow crevices to capture collembolans at 0° to −2°C (Hågvar, 1973). In Greenland five species of spiders were recorded active between −3° and 8°C (Bertram, 1935). Even though not under snow, these spiders exhibited activity and feeding at low temperatures.

The cursorial families Clubionidae, Erigonidae, Linyphiidae, Lycosidae and Thomisidae formed between 80 to 95% of the bulk of 19 winter-active species in the subnivean conditions of southern Manitoba, with Agelenidae, Hahniidae and Tetrag-

nathidae occasionally collected. This mobile order Araneae has activity decreasing as subnivean temperatures approach −8°C (Aitchison, 1978a). In Sweden spiders comprised about 10% of the winter-active catch in pitfall traps (Näsmark, 1964).

Most families in the Manitoba study were represented by a mixture of juveniles and adults in late autumn and early winter, except for the linyphiid *Centromerus sylvaticus* (Blw.) with only adults present. Linyphiids generally predominated in autumn (September–October), followed by lycosids *Pardosa* spp. and clubionids *Agroeca* spp. (Fig. 1). During winter clubionids were the dominant family, with a number of active lycosids and the erigonid, *Ceraticelus laetus* (O. P.-Cambridge). Lycosids were most active in spring (Aitchison, 1978a).

Each spider family in Manitoba had its activity limited by temperature in winter conditions. Because temperature measurements and emptying of traps were weekly, catch reflected the thermal conditions of the previous week. However, an indication of minimal temperatures down to which specimens were active, referred to hereafter as the mean lower thermal threshold (LTT), is as follows: in linyphiids it was about $-5.3°C$; other families $-5.5°C$; lycosids $-6.2°C$; thomisids $-6.4°C$; clubionids $-6.8°C$; and erigonids $-8.0°C$ (Aitchison, 1978a). These LTTs seem to be correlated with winter activity of the different families.

Weather conditions of late autumn and early winter may determine the numbers of spiders found in winter (Huhta, 1971). During the first winter of study in Manitoba, snow arrived in late October after much frost, while in the second winter a warm, snowfree November was associated with a higher number of active spiders. Spider catches during the first winter were also positively correlated with temperature, while in the second winter they were not (Aitchison, 1978a).

Hence it may be seen that certain families of cursorial spiders are winter-active at subzero temperatures in the subnivean space in southern Canada. They possibly eat other winter-active subnivean invertebrates such as Collembola and are potential prey for shrews.

ACARI

There is little documentation of winter activity in this order, although articles on its cold-hardiness can be found. Many litter-dwelling mites have preferences for low temperatures and move downwards in humus in early winter (Wallwork, 1970). On sub-Antarctic islands with a mean air temperature of about 0°C, 21 species of oribatids were mobile at subzero temperatures (Dalenius and Wilson, 1958), and 20 species of prostigmatids also occurred there (Wallwork, 1973).

Irrespective of temperature, the acarine families Eupodidae, Rhagidiidae and Parasitidae were some of the most abundant, winter-active subnivean groups in southern Canada, with catch not correlated with subnivean temperatures (Aitchison, 1979c). Acari occurred infrequently in pitfall traps during Swedish winters (Näsmark, 1964).

The predatory and parasitic Acari were the most numerous winter-active mites in Manitoba, with prostigmatids dominant and more numerous, while the mesostigmatids were secondarily dominant and less abundant. Numbers were low in autumn when many families were present; prostigmatid numbers increased sharply in February–March

and declined in April (Fig. 2) (Aitchison, 1979c). The prostigmatid genus *Evadorhagidia* was primarily responsible for the winter maximum (98% of the acarine catch), which fell to a spring minimum. Fifteen winter-active species were mainly represented by eupodids *Eupodes* spp. and *Linopodes* spp., rhagidiid *Evadorhagidia quinqueseta* Zacharda and parasitid *Pergamasus crassipes* L. (Aitchison, 1979c).

With the establishment of lasting snow cover in the second winter, the number of mites increased. Nonetheless no correlations were found between numbers and subnivean temperature. The mean LTT for activity of prostigmatids was approximately $-7.8°C$ and of mesostigmatids $-7.6°C$, although some individuals were active down to $-10°C$ (Aitchison, 1979c).

Other researchers have noted the acarine increase in numbers in February and March (Mason, 1972; Willard, 1973). From February to March the incident solar radiation increases from 40 to 113 langleys/day at the latitude of southern Manitoba (50°N) (Budyko, 1969), and simultaneously light penetration into snow cover increases (Geiger, 1965; Evernden and Fuller, 1972; O'Neill and Gray, 1972). Possibly the mites and some other insects, such as collembolans, are sensitive to these changes in subnivean light intensity and quality (Aitchison, 1979a, 1979c).

Consequently it can be seen that subnivean winter-active mites are numerous during winter and tolerant of cold temperatures, down to $-10°C$. Most of these species are predatory or parasitic and in turn may be eaten by insectivores.

COLLEMBOLA

The most abundant subnivean invertebrates trapped in Sweden were the Collembola, with isotomids most numerous, followed by the entomobryids (Näsmark, 1964). In Canada the number of mites trapped exceeded that of the springtails but the same dominance patterns of collembolan families was seen (Aitchison, 1979a). Both these families moved about under snow in the Chicago region (Holmquist, 1926), while *Isotoma viridis* Bourlet and other species were active on snow at temperatures near 0°C (Chapman, 1954). Some people even call them "Schnee- und Gletschflöhe" (Strübing, 1958).

Sixteen species were winter-trapped in Manitoba and mainly consisted of the isotomids *Isotoma manitobae* Fjellberg, *I. blufusata* Fjellberg, *I. viridis*, *I. violacea* Tullberg, and the entomobryids *Orche-*

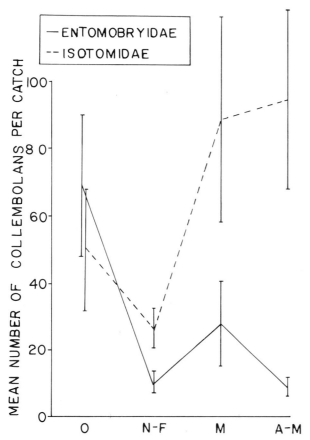

Fig. 3.—The mean number of active collembolans per catch versus months from 1973 to 1974 (Aitchison, 1979a).

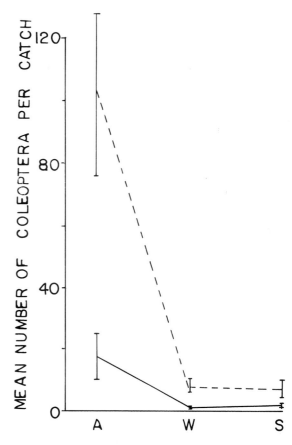

Fig. 4.—The mean number of active Coleoptera per catch versus seasons from 1974 to 1975 (Aitchison, 1979b) (dotted line, staphylinids, and black line, carabids).

sella ainslei Folsom and *Tomocerus flavescens* Tullberg. These animals were trapped at all times of the winter but were most numerous in October and March, and few in numbers in January and February. The increase in March appears to be the same phenomenon seen in the mites and is illustrated in Fig. 3 (Aitchison, 1979a). Similar peaks in collembolan numbers on the soil surface in February and March have been noted elsewhere (Wallwork, 1970; Mason, 1972; Willard, 1973).

The mean LTT was estimated for each family as −7.8°C for isotomids and as −7.9°C for entomobryids (Aitchison, 1979a). European springtails examined in thermal gradients were active down to between −4° and −7°C (Agrell, 1941; Wolska, 1957; Simon, 1961).

Predators of collembolans include carabid and staphylinid beetles, spiders and mites (Simon, 1964; Ernsting and Joosse, 1974), all present in the winteractive subnivean fauna of Manitoba. Thus the mostly fungivorous Collembola might form a base for a

subnivean food chain, if predators do feed at subzero temperatures. Sufficient numbers of collembolans occur in winter to be prey for either invertebrate or vertebrate predators.

COLEOPTERA

Beetle activity at near-freezing temperatures is well documented, and "Schneewürme," or carnivorous cantharid larvae, winter-active lathridiid larvae and cold-hardy staphylinids may run about under snow (Holmquist, 1926; Chapman, 1954; Heydemann, 1956; Renken, 1956; Wolska, 1957; Strübing, 1958; Näsmark, 1964; Wallwork, 1970; Kaufmann, 1971; Mason, 1972; Aitchison, 1979b). The cold-tolerant carabid, *Pterostichus brevicornis* Kirby, even feeds in winters on rotten wood of stumps when temperatures are as low as −4°C (Kaufmann, 1971). Winter-active genera of Europe are also winter-active in Canada (Heydemann, 1956; Aitchison, 1979b).

In Manitoba 62 species of beetles were wintertrapped, represented by carnivorous staphylinids

(*Atheta* spp., *Falagria dissecta* Er., *Heterothops fusculus* Leconte, *Hyponygrus* spp., *Philonthus* spp., and *Tachyporus nitidulus* (Fab.)), carabids (*Bembidion canadianum* Casey, *B. graphicum* Casey, *Pterostichus femoralis* Kirby and *Trichocellus cognatus* Gyllenhal), lathridiids and cantharids in that order of abundance (Aitchison, 1979*b*). Most activity occurred in October and November, primarily by staphylinids and secondarily by carabids, with no midwinter catch at all, and with a few beetles trapped in March as subnivean temperatures rose. Fig. 4 demonstrates the high autumn catch of staphylinids and lower one of carabids, with few beetles taken during winter and spring (Aitchison, 1979*b*).

Heydemann (1956) found staphylinid activity inhibited at and below −4°C, and the estimates for mean LTT for activity in Manitoba were −4.4°C for staphylinids and −3.25°C for carabids (Aitchison, 1979*b*).

Thus coleopterans are active in early and late winter in the subnivean space at temperatures near freezing, possibly feeding on small invertebrates in the litter, and available as prey for other predators, such as beetle-loving shrews.

DIPTERA AND HYMENOPTERA

Some fly species are associated with snow (the tipulid genus *Chionea* and other winter-active families) and include many fungivores and detritivores (Holmquist, 1926; Renken, 1956; Hågvar and Østbye, 1973; Jonsson and Sandlund, 1975; Aitchison, 1979*d*). In Montana and Norway *Chionea* spp. were active down to −5.6°C and −6°C respectively (Chapman, 1954; Hågvar, 1971). *Diamesa permacer* (Walk.) was recorded active at −4.5°C (Hågvar and Østbye, 1973), and *D. thienemanni* Kieffer recorded flying at an air temperature of about 0°C (Jonsson and Sandlund, 1975), while mycetophilids and other families were active in the subnivean space (Holmquist, 1926; Näsmark, 1964; Aitchison, 1979*d*). The mean LTT of activity by 21 species of dipterans in Manitoba was −3.5°C (Aitchison, 1979*d*).

Some hymenopteran families are active at temperatures near 0°C (Renken, 1956), and ants, ichneumonids, diapriids and braconids may be mobile on the soil surface under snow (Holmquist, 1926; Aitchison, 1979*d*). In Sweden few wasps were active in the subnivean space (Näsmark, 1964).

Activity by 27 winter-active species of hymenopterans in Manitoba was closely correlated with that of dipterans and was represented by ceraphronids, diapriids and scelionids; possibly the correlation may be explained by hymenopteran predation on the dipterans, and to a lesser extent on other insects and spiders. Most dipteran and hymenopteran activity occurred in October and November, followed by a sharp decline during winter, and a resumption again in April. The mean LTT of the hymenopterans was −5.5°C, below that of the dipterans (Aitchison, 1979*d*).

Although present in the subnivean space, these groups were barely active in Manitoba at subzero temperatures and were few in numbers. Thus they are probably not commonly taken as prey during Canadian winters.

OTHER WINTER-ACTIVE INVERTEBRATES

Subnivean molluscs and oligochaetes were rarely collected in Sweden (Näsmark, 1964) or in Canada (Aitchison, 1979*e*). Carnivorous arthropods, not already mentioned and sometimes active at subzero temperatures but infrequently under snow, include centipedes and pseudoscorpions (Näsmark, 1964; Aitchison, 1979*e*, 1979*f*). Other subnivean herbivorous arthropods were aphids and cicadellid nymphs (Holmquist, 1926; Näsmark, 1964; Aitchison, 1978*b*), active down to −5.5°C and −7.0°C, respectively, in Manitoba (Aitchison, 1978*b*). Possibly the herbivores may help to form the basis of a subnivean food chain, and all of these invertebrates may ultimately fall prey to the voracious shrews.

SHREWS

The thermally stable, relatively warm and humid subnivean habitat is home for the energetic shrews, the dominant insectivores. There many prey items are checked physiologically by the low or subzero temperatures, becoming easy prey for shrews if they can be detected, or are winter-active and must be pursued.

The species of shrews present at the Manitoba study site are the shorttail shrew, *Blarina brevicauda* (Say), the masked shrew, *Sorex cinereus* Kerr, the arctic shrew, *Sorex arcticus* Kerr, and the pygmy shrew, *Microsorex hoyi* (Baird). To provide a comprehensive overview, the foods of other species of *Sorex* are discussed.

The high metabolic rate of soricids necessitates their constant search for food the year round (Platt and Blakley, 1973). In Tennessee Randolph (1973) stated, "Winter shrews at winter temperatures have a higher metabolic rate than do summer shrews at summer temperatures." With reference to *Blarina* Seton (1909) said that shrews must be "eternally vigilant under the snow," because they do not hibernate or store food and because of their rapid digestion.

Foraging Behavior

Much of the food of shrews is taken from the epigean habitat under litter (Platt and Blakley, 1973). Except for large *Blarina* which makes shallow surface runs, these animals generally cannot burrow well or at all due to their small size (Crowcroft, 1957), but do prefer to utilise vole runways of suitable size.

Shrews are constantly hunting for prey items, pushing their long proboscs into litter, taking prey which they touch and not pursuing escaped prey (Yudin, 1962). The dependence of shrews upon the use of vibrissae and upon sound, with mobile prey attracting the attention of shrews, emphasises the importance of tactile, kinaesthetic (movement-detecting) and auditory senses (Crowcroft, 1957; Rudge, 1968; Pernetta, 1977). Nonetheless the olfactory sense may be important, with *Sorex araneus* L. detecting prey up to 5 cm away, though probably not during winter (Tupikova, 1949; Mezhzherin, 1958; Rudge, 1968). However, sound under snow may be more significant, and a spider or beetle brushing against delicate hoarfrost crystals may alert shrews to its presence by the tinkling of falling crystals.

Thus it appears that the shrew attacks active prey by using tactile, auditory, olfactory and kinaesthetic senses and with little dependence on sight.

Food Items

Shrews catch and overpower small animals, such as insects, spiders, crustaceans and worms (Crowcroft, 1957), and prey size is determined by the size of the predator (Platt and Blakley, 1973). Prey species vary seasonally on the soil surface, with seasonal changes in the availability of prey items (Lavrov, 1943; Mezhzherin, 1958; Yudin, 1962; Platt and Blakley, 1973). Ackefors (1964) in Sweden found that there seemed to be an obvious relationship between the winter-active subnivean invertebrates taken in pitfall traps (Näsmark, 1964) and the stomach contents of shrews. He found that *S. araneus*

caught larger prey such as coleopterous larvae, worms, homopterans and especially molluscs, while *Sorex minutus* L. took smaller victims (Ackefors, 1964). The same correlation was documented in England (Pernetta, 1976). Those invertebrates inactive in winter are most likely to be in soil crevices and therefore unavailable to shrews (Wallwork, 1970).

Winter diets of shrews are stated to be carnivorous (Hamilton, 1940) or more omnivorous, with plants to supplement the fewer invertebrates found (Hamilton, 1940; Lavrov, 1943; Yudin, 1962). This is corroborated by seasonal changes in salivary gland secretions (Rudge, 1968). Winter food items of *S. araneus* in Poland differ very little from those of the spring period, taking carabid prey from easily available hibernaculaae in stumps and deep litter layers (Renken, 1956; Kisielewska, 1963). Spiders may be important summer food (Lavrov, 1943), important winter food (Mezhzherin, 1958), or important foods for both seasons (Rudge, 1968).

Hoarding of food items has been observed in some instances for several species of shrews, despite the literature not being unanimous on this point (Crowcroft, 1957). *B. brevicauda* may hoard snails (Shull, 1907; Hamilton, 1930), *Sorex vagrans* Baird and *S. araneus* may cache oligochaetes in balls, probably for short periods (Broadbooks, 1939; Dehnel, 1961), and *S. araneus* and *B. brevicauda,* coniferous seeds and beechnuts (Hamilton, 1930; Dehnel, 1961).

Hence the active shrew constantly searches under snow for prey and feeds on the epigean fauna encountered, especially the mobile species. It may even hoard abundant seeds, snails or worms. The inactive invertebrates are probably inaccessible to the shrew.

Specific Examples of Winter Food Habits of Shrews

1. *Blarina brevicauda* consumes beechnuts, fungi, insects, worms, slugs, myriapods, isopods, spiders, and vole and mouse flesh in summer and in winter (Shull, 1907; Hamilton, 1940; Whitaker and Mumford, 1972), but needs extra amounts of food during winter in order to compensate for its elevated metabolic rate (Randolph, 1973, 1980). Two winter analyses of diet were made in New York State, one from scats and another from stomachs; insects, annelids, myriapods and vertebrates were important foods in the scat analyses (Eadie, 1944), while insects and vegetation were found in the stomach analyses (Hamilton, 1930).

2. The foods eaten by European *S. araneus* in-

Table 1.—*Analysis of 66 guts of* Sorex araneus *from Scotland showing the percentage frequency of occurrence of each food item (after Rudge, 1968).*

Number of guts	Lumbricids	Slugs	Acari	Spiders	Opilionids	Hymen.	Hemi.	Diptera	Coleop.	Lepid.	Plant	Vert.
Spring 27	48%	40	0	14	4	0	28	24	72	24	96	36
Summer 30	17	10	7	17	13	3	20	30	50	30	60	10
Autumn 3	33	0	0	66	0	33	0	66	66	0	100	100
Winter 6	17	50	17	17	33	17	33	50	84	0	100	33
Average	29	24	5	14	12	5	23	30	61	23	77	26

clude annelids, molluscs, arachnids, orthopterans, coleopterans, lepidopteran larvae, dipterans, ant larvae and vertebrates, including *Phoxinus, Rana, Lacerta, Sorex, Apodemus, Microtus,* and pine seeds (Yudin, 1962; Kisielewska, 1963; Pernetta, 1976). Beetles comprise most of the food consumed in a year-round analysis of stomach contents, with other invertebrates important in other seasons (Hamilton, 1940; Mezhzherin, 1958; Yudin, 1962; Rudge, 1968; Grainger and Fairley, 1978).

A comparison of 265 stomachs collected in August of two different years revealed that the shrew's diet consisted of 25.8 and 9.4% spiders and 8.6 and 23.5% carabids respectively (Mezhzherin, 1958). Different proportions of prey may be consumed in various seasons and habitats and represent a "crude scale of importance" to the animal (Rudge, 1968). "The general consensus suggests that dietary differences within shrew species reflect differences in availability which change with season, weather and habitat . . . ." (Pernetta, 1976).

Table 1 collates the gut contents of wild-trapped *S. araneus* from three sites in Great Britain and shows the seasonal variations of food components (Rudge, 1968), demonstrating the importance of slugs, opilionids, hemipterans, dipterans, coleopterans, plant parts and vertebrate remains in the winter diet of *S. araneus.* In the subnivean space in the U.S.S.R., Mezhzherin (1958) reported spiders to be important winter food items for shrews, including mites and opilionids as well. The spiders are a dominant, winter-active order in the subnivean environment of Manitoba (Aitchison, 1978a).

3. *Sorex cinereus* preys on all invertebrates longer than 3 mm long, with the exception of ants, collembolans and mites (Platt and Blakley, 1973). This species has a diet including annelids, molluscs, is-

opods, arachnids, coleopterans, large lepidopteran larvae, dipteran larvae, orthopterans, fungi, plant material, and vertebrates (Hamilton, 1930; Whitaker and Mumford, 1972; Platt and Blakley, 1973).

Of seven autumn gut contents of *S. cinereus* from the aspen parkland of Manitoba, four contained coleopteran adults, two coleopteran larvae, one each had mites, trichopterans, orthopterans and other insects, and three plant material (possibly incidental). Winter gut contents of two shrews included shrew flesh, adult beetles, dipteran larvae, thrips and spiders. Analyses of February and March gut contents from southern Manitoba revealed a diet with a mean of 87.7% insects and of 8.6% plants; there were no differences between the volume of the gut contents in summer or in winter. This species was labelled a "fortituitous feeder" (Bergeron, 1972).

4. *Sorex fumeus* of the northeastern United States and southeastern Canada eats earthworms, snails, centipedes, isopods, spiders, insects, salamanders, mammals, birds and plant material, and feeds in leaf mould and loose litter. Table 2 presents its food habits determined from stomachs taken throughout the year (Hamilton, 1940). Its winter foods include many invertebrates (beetles, centipedes, spiders) and vegetation. In a February catch, stomachs contained 20% each of spiders, snails and beechnuts, and 80% insects (mostly coleopterans). In March with the ground still frozen its diet consisted of centipedes and insects, especially dipterans, and in April it took more spiders again (14.4%) (Hamilton, 1940).

5. *Sorex minutus* of Europe consumes a diet of lumbricids, molluscs, arachnids, isopods, collembolans and other small insects, *Phoxinus, Sorex, Lacerta, Apodemus* and microtines (Mezhzherin, 1958; Yudin, 1962; Pernetta, 1976; Grainger and Fairley, 1978). Small beetles comprised the majority

Table 2.—*Food habits of* Sorex fumeus, *determined from examination of 168 stomach analyses throughout the year. The figure below the month indicates the number examined. Other figures denote the percentage frequency of occurrence of different food items (Hamilton, 1940).*

Food	Jan. 4	Feb. 5	Mar. 14	Apr. 27	May 20	June 16	July 19	Aug. 8	Sept. 17	Oct. 21	Nov. 11	Dec. 6	Total 168
Insects	75	80	85.7	77.7	70	93.7	94.7	100	88.2	52.4	81.8	66.6	80.8
Worms	—	—	7.1	14.4	15	6.2	10.5	12.5	—	9.5	23.3	—	10.1
Plants	50	20	14.1	14.4	25	18.7	5.3	50	—	—	18.2	50	14.9
Centipedes	25	—	21.4	18.5	5	18.7	15.8	—	5.9	9.5	18.2	16.6	13.1
Snails	—	20	—	11.1	10	—	10.5	—	17.6	23.8	9.1	—	10.1
Salamanders	—	—	—	3.7	5	—	5.3	—	—	—	—	—	1.8
Mammals	—	—	—	3.7	10	—	10.5	—	—	—	—	—	3.0
Isopods	—	—	—	11.1	—	6.2	21.1	—	—	—	9.1	—	5.3
Spiders	—	20	—	14.4	10	6.2	—	12.5	—	—	9.1	—	5.9
Birds	—	—	—	—	5	—	—	—	—	—	—	—	0.5

of its food in the U.S.S.R. and in Ireland (Yudin, 1962; Grainger and Fairley, 1978). This diet very closely corresponds to the mobile invertebrate fauna taken in pitfall traps, and it is the smaller members of the epigean litter fauna which are important to this species, that is mites, small linyphiid spiders, collembolans, aphids, the opilionid *Nemastoma* sp., and small carabids and staphylinids (Yudin, 1962; Pernetta, 1976; Grainger and Fairley, 1978).

Two winter diets have been determined. In Ireland isopods, mites, adult beetles, flies, and plant matter are the major food components during winter (Grainger and Fairley, 1978), while in Siberia they are spiders, carabids, and other adult beetles (Yudin, 1962). The Siberian prey corresponds closely to the winter-active, subnivean litter fauna in Sweden and in Canada (Näsmark, 1964; Aitchison, 1978a, 1979b).

CONCLUSION

Winter-active invertebrates such as collembolans, cicadellid nymphs, mites, spiders and to a lesser extent in Manitoba, carabid and staphylinid beetles probably comprise the invertebrate portion of a possible subnivean food chain. The fungivorous collembolans and possibly the herbivorous cicadellid nymphs would comprise the primary consumers, while the predatory mites, spiders and possibly beetles would form the secondary consumers in such a chain.

In the vertebrate portion of this food chain the insectivorous shrew is the most probable tertiary consumer active in the subnivean runways. The high metabolic rates of shrews necessitate their almost constant foraging for prey, even during winter. In areas with snow cover, the subnivean environment offers them protection from the vagaries of the elements. Probably most rodents eat only a small proportion of insects and other arthropods during the winter months.

The soricid eats what it encounters and probably does not hoard much food. This animal uses tactile, auditory, kinaesthetic and olfactory sensory cues to search for prey, with possibly less dependence upon scent during winter. Being an opportunist, the hun-

gry shrew probably consumes all easily accessible invertebrate and vertebrate prey which it can subdue, with beetles as important food. The soricid diet at any season closely corresponds to the catch in pitfall traps.

Winter diets of shrews are dependent on the invertebrate and sometimes vertebrate prey that is available on the soil surface, with prey varying regionally and with the amount of frost and/or snow cover. In Sweden and England, winter diets of *S. araneus* and *S. minutus* were correlated with the active, epigean invertebrate fauna (Ackefors, 1964; Pernetta, 1976), and in Manitoba winter diets of *S. cinereus* contained a high percentage of insects (Bergeron, 1972).

Although the soricids may be considered as secondary predators in epigean food webs, they themselves are subject to predation by small and medium-sized tertiary predators, such as raptorial birds, mustelids, and canids. During winter, plunge holes of owls into snow and the use of vole (and shrew) runways by small weasels facilitate the predation of shrews, further incorporating them into food chains and webs.

ACKNOWLEDGMENTS

I wish to thank L. B. Smith, E. Huebner and P. W. Aitchison for suggestions to improve the manuscript.

LITERATURE CITED

ACKEFORS, H. 1964. Vinteraktiva näbbmöss under snö. Zool. Revy, 26:16–22.

AGRELL, I. 1941. Zur Ökologie der Collembolen. Untersuchungen in Schwedisch Lappland. Opuscula Ent. (Lund), Suppl., 3:236 pp.

AITCHISON, C. W. 1978a. Spiders active under snow in southern Canada. Symp. Zool. Soc. London, 42:139–148.

———. 1978b. Notes on low temperature and winter activity of Homoptera in Manitoba. Manitoba Entomol., 12:58–60.

———. 1979a. Winter-active subnivean invertebrates in southern Canada. I. Collembola. Pedobiologia, 19:113–120.

———. 1979b. Winter-active subnivean invertebrates in southern Canada. II. Coleoptera. Pedobiologia, 19:121–128.

———. 1979c. Winter-active subnivean invertebrates in southern Canada. III. Acari. Pedobiologia, 19:153–160.

———. 1979d. Winter-active subnivean invertebrates in southern Canada. IV. Diptera and Hymenoptera. Pedobiologia, 19:176–182.

———. 1979e. Notes on low temperature activity of oligochaetes, gastropods and centipedes in southern Canada. Amer. Midland Nat., 102:399–400.

———. 1979f. Low temperature activity of pseudoscorpions and phalangids in southern Manitoba. J. Arachnol., 7:85–86.

BERGERON, J.-M. 1972. The Role of Small Mammals in the Population Dynamics of the Semiothisa Complex, Lepidoptera: Geometridae: Ennominae. Unpublished Ph.D. thesis, Univ. Manitoba, Winnipeg, 268 pp.

BERTRAM, G. C. L. 1935. The low temperature limit of activity of arctic insects. J. Anim. Ecol., 4:35–42.

BROADBOOKS, H. E. 1939. Food habits of the vagrant shrew. The Murellet, 20:62–66.

BUDYKO, M. I. 1963. The heat balance of the surface of the Earth, Table 6. In Weather and Climate (M. Flohn), McGraw-Hill, New York and Toronto, 253 pp.

CHAPMAN, J. 1954. Observations of snow insects in western Montana. Canadian Entomol., 86:357–363.

CROWCROFT, P. 1957. The Life of the Shrew. Max Reinhardt, London, 166 pp.

DALENIUS, P., and O. WILSON. 1958. On the soil fauna of the Antarctic and of the Sub-Antarctic Islands. The Oribatidae (Acari). Arkiv för Zool., 11:393–425.

DEHNEL, A. 1961. Aufspeichung von Nahrungsvorräten durch Sorex araneus Linnaeus 1758. Acta Theriol., 4:265–268.

EADIE, W. R. 1944. The short-tailed shrew and field mouse predation. J. Mamm., 25:359–364.

ERNSTING, G., and E. N. G. JOOSSE. 1974. Predation on two species of surface dwelling Collembola. A study with radio-isotope labelled prey. Pedobiologia, 14:222–231.

EVERNDEN, L. N., and W. A. FULLER. 1972. Light penetration caused by snow and its importance to subnivean rodents. Canadian J. Zool., 50:1023–1032.

GEIGER, R. 1965. The Climate near the Ground. Harvard Univ.

Press, Cambridge, Massachusetts, (translated from German by Scripta Technica, Inc.), 611 pp.

GRAINGER, J. P., and J. S. FAIRLEY. 1978. Studies on the biology of the pygmy shrew Sorex minutus in the west of Ireland. J. Zool. London, 186:109–141.

HÅGVAR, S. 1971. Field observations on the ecology of a snow insect, Chionea araneoides Dalm. (Dipt., Tipulidae). Norsk Ent. Tidsskr., 18:33–37.

———. 1973. Ecological studies on a winter-active spider Bolyphantes index (Thorell) (Araneida, Linyphiidae). Norsk Ent. Tidsskr., 20:309–314.

HÅGVAR, S., and E. ØSTBYE. 1973. Notes on some winter-active Chironomidae. Norsk Ent. Tidsskr., 20:253–257.

HAMILTON, W. J., JR. 1930. The food of the Soricidae. J. Mamm., 11:26–39.

HAMILTON, W. J., JR. 1940. The biology of the smoky shrew (Sorex fumeus fumeus Miller). Zoologica, 25:473–492.

HEYDEMANN, V. B. 1956. Untersuchungen über die Winteraktivität von Staphyliniden auf Feldern. Entomol. Blätter, 52:138–150.

HOLMQUIST, A. M. 1926. Studies in arthropod hiberation. I. Ecological survey of hibernating species from forest environments of the Chicago region. Ann. Ent. Soc. Amer. 19:395–426.

HUHTA, V. 1971. Succession in the spider communities of the forest floor after clear-cutting and prescribed burning. Ann. Zool. Fennicae, 8:483–542.

JONSSON, B., and O. T. SANDLUND. 1975. Notes on winter activity of two Diamesa spp. (Dipt., Chironomidae) from Voss, Norway. Norwegian J. Ent., 22:1–6.

KAUFMANN, T. 1971. Hibernation in the arctic beetle Pterostichus brevicornis in Alaska. J. Kansas Entomol. Soc., 44:81–92.

KISIELEWSKA, K. 1963. Food consumption and reproduction of Sorex araneus Linnaeus, 1758, in light of parasitological research. Acta Theriol., 7:127–153.

LAVROV, N. F. 1943. On the biology of the common shrew (Sorex araneus L.). Zool. Zh., 22:361–365.

MASON, J. L. 1972. Ecological Studies on Coleoptera of a Kentish Woodland. Unpublished Ph.D. thesis, Univ. London.

MEZHZHERIN, V. A. 1958. On feeding habits of Sorex araneus L. and Sorex minutus L. Zool. Zh., 37:948–953.

NÄSMARK, O. 1964. Vinteraktivitet under snö hos landlevande evertebrater. Zool. Revy, 26:5–15.

O'NEILL, A. D. J., and D. M. GRAY. 1972. Solar radiation penetration through snow. Pp. 227–241, in The role of snow and ice in hydrology. UNESCO-WMO-IAHS, 1.

PERNETTA, J. C. 1976. Diets of the shrews Sorex araneus L. and Sorex minutus L. in Wytham grassland. J. Anim. Ecol., 45:899–912.

———. 1977. Anatomical and behavioural specialisations of shrews in relation to their diet. Canadian J. Zool., 55:1442–1453.

PLATT, W. J., and N. R. BLAKLEY. 1973. Short-term effects of shrew predation upon invertebrate prey set in prairie ecosystems. Proc. Iowa Acad. Sci., 80:60–66.

RANDOLPH, J. C. 1973. Ecological energetics of a homeothermic predator, the short-tailed shrew. Ecology, 54:1166–1187.

———. 1980. Daily metabolic patterns of short-tailed shrews (*Blarina*) in three natural seasonal temperature regimes. J. Mamm., 61:628–638.

RENKEN, W. 1956. Untersuchungen über Winterlager der Insekten. Z. Morph. Ökol. Tiere, 45:34–106.

RUDGE, M. R. 1968. The food of the common shrew *Sorex araneus* L. (Insectivora: Soricidae) in Britain. J. Anim. Ecol., 37:565–581.

SETON, E. T. 1909. Life histories of northern Animals. Charles Scribner's Sons, New York, vol. II, 590 pp.

SHULL, A. F. 1907. Habits of the short-tailed shrew *Blarina brevicauda* (Say). Amer. Midland Nat., 41:495–522.

SIMON, H. R. 1961. Beobachtungen zum Verhalten arthropleoner Collembolen (Apterygota). Dtsch. Entomol. Z., 8:215–221.

———. 1964. Zur Ernährungsbiologie collembolenfangender Arthropoden. Biol. Zentralblatt, 83:273–296.

STRÜBING, H. 1958. Schneeinsekten. Die Neue Brehm-Bücherei. Z. Ziemsen Verlag, Wittenberg, 47 pp.

TUPIKOVA, N. V. 1949. [The diet and nature of the daily cycle of activity of shrews from the central region of the U.S.S.R.]. Zool. Zh., 28:561–672.

WALLWORK, J. A. 1970. Ecology of soil animals. McGraw-Hill, London, 283 pp.

———. 1973. Zoogeography of some terrestrial micro-arthropods in Antarctica. Biol. Rev., 48:233–259.

WHITAKER, J. O., JR., and R. E. MUMFORD. 1972. Food and ectoparasites of Indiana shrews. J. Mamm., 53:329–335.

WILLARD, J. R. 1973. Soil invertebrates. III. Collembola and minor insects: populations and biomass. I.B.P., Technical Report, 19:1–75.

WOLSKA, H. 1957. Preliminary investigations on the thermic preferendum of some insects and spiders encountered on snow. Folia Biol. Krakow, 5:195–208. (In Polish, English summary.)

YUDIN, B. S. 1962. [Ecology of shrews (genus *Sorex*) in western Siberia]. Akad. Nauk SSSR, Siberian Section, Trudy, Inst. Biol., 8:33–134.

Address: Department of Entomology, University of Manitoba, Winnipeg, Manitoba R3T 2N2, Canada.

SUBNIVEAN ACCUMULATION OF CO_2 AND ITS EFFECTS ON WINTER DISTRIBUTION OF SMALL MAMMALS

Cheryl E. Penny and William O. Pruitt, Jr.

ABSTRACT

Subnivean CO_2 and small mammal populations were monitored in six habitats in the taiga of southeastern Manitoba from May 1974 to May 1976. Subnivean CO_2 accumulated consistently in some habitats, but not in others; accumulation was similar in the same habitats in both winters. Accumulation of subnivean CO_2 was significantly associated with a change in distribution of small mammals, especially in the aspen upland. Concentration of subnivean CO_2 is therefore an additional dimension to the ecological niche of *Clethrionomys gapperi* and must be taken into account when considering winter distribution of this species.

INTRODUCTION

A disagreement was found in the literature regarding subnivean accumulations of CO_2 and its effects on small mammal distributions. Bashenina (1956), Pichler (in Geiger, 1965), Kelley et al. (1968), and Havas and Mäenpää (1972) found subnivean CO_2 accumulated to levels greater than ambient. Reiners (1968) found no soil CO_2 production in winter (January). Kelley et al. (1968) and Coyne and Kelley (1971) found subnivean CO_2 levels decreased to ambient once a snowcover of 20 cm was established.

Bashenina (1956) found that, in areas with greater concentrations of CO_2, subnivean mouse nests were constructed higher above the ground than in areas with lesser concentrations of CO_2. In contrast, Fuller and Holmes (1972) stated that subnivean CO_2 did not increase sufficiently to affect small mammal distributions in the taiga.

Small mammals such as *Clethrionomys* and *Microtus* have been reported to exhibit a reduction in heart rate of 12 to 25% when exposed to CO_2 levels above ambient (Galantsev and Tumanov, 1969). In laboratory testing of *Mus musculus,* Schlenker and Herreid (1981) found oxygen consumption to be significantly reduced at levels of 0.14 to 0.50% CO_2 in the air. They also found the respiratory quotient increased from 0.7 to 1.0 and higher when CO_2 levels were increased. Soholt et al. (1973) and Withers (1975) have also recorded physiological reactions of small mammals to increased levels of CO_2.

After the Hiemal Threshold (Pruitt, 1957) is established, small mammals can change their distributions within the habitat (West, 1977). Small mammal distributions are not necessarily correlated with changes in temperature (Fuller, 1977) or food availability (Gorecki and Gebczynska, 1962; Grodzinski, 1963; Chitty et al., 1968; Flowerdew, 1972; Andrzejewski and Mazurkiewicz, 1976; Fairbairn, 1977). The changes in distribution of small mammals can be related to their response to increase in CO_2 concentrations.

STUDY AREA

This study was made at Taiga Biological Station (TBS) (51°02'40"N Lat., 95°20'40"W Long.), about 280 km northeast of Winnipeg, Manitoba, near Bissett. All habitats were on the Precambrian Shield and included: black spruce bog, aspen upland, alder-tamarack bog, alder-ridge ecotone, jackpine ridge and jackpine sandplain. The climate of the area is continental with January mean temperatures ranging from −22.8 to −19.8°C and July mean from +18 to +19.5°C and 410 to 435 mm annual precipitation (Woo et al., 1977).

METHODS

Square grids of 0.4 ha in each of the six habitats were sampled for both CO_2 and small mammals from May 1974 to May 1976. At each CO_2 sampling site the upper end of a 2 m piece of rubber tubing (0.6 cm inside diameter) was taped to one of the wooden stakes of the grid, while the lower end lay on the ground-air interface, away from the regular grid pathways. CO_2 testing sites were established to sample as much as possible of the microtopographic and habitat variation. Thus, there were five stations

373

on the jackpine ridge, ten on the aspen upland, ten on the alder ridge ecotone, five on the black spruce bog, ten on the alder tamarack bog and two on the jackpine sandplain. All stations were tested at least three times a month from the onset of the Fall Critical Period to the end of the Spring Critical Period (Pruitt, 1957).

Percentage of CO_2 by volume in ambient air was measured using a Dräger multigas analyzer and analyzing tubes with a range of 0.01 to 0.3 ± 0.005 (Penny, 1978).

Small mammals were live-trapped throughout the study and toe-clipped for individual recognition. Wooden trap chimneys and masonite traps were used for all winter trapping (Pruitt, 1959). During the two winters of the study, the alder-tamarack bog, alder-ridge ecotone, black spruce bog and aspen upland were trapped for a three-night session each month. Twenty-five sites per habitat were trapped between December 1974 and April 1975; 35 sites were trapped in the alder-tamarack bog and aspen upland and 30 sites in the black spruce bog and alder-ridge ecotone from October 1975 to April 1976. All trap chimneys were pre-baited with raw oatflakes at least two different times before winter trapping commenced, thereby inducing small mammals to include the chimneys in their food search pattern and thus open subnivean tunnels to them in winter. In an effort to minimize trapping mortality, traps were checked every 6 to 8 hr throughout each three-day session, and large quantities of peanut butter, bacon grease and rolled oat bait were placed in each trap.

RESULTS

Carbon dioxide accumulations were considered to be important to small mammals when subnivean CO_2 concentrations were greater than ambient levels for two or more consecutive weekly samplings. This restriction meant that CO_2 had been greater than ambient for at least one week and not just momentarily. It is important to note that we consistently measured ambient CO_2 at TBS to be 0.02%, whereas other studies have reported 0.028–0.035% CO_2. We used 0.02%/volume to define ambient conditions.

CO_2 accumulated to values greater than ambient at 22 of 39 CO_2 sampling stations on the six habitats in winter 1974–1975, and 23 of 44 in the six habitats in winter 1975–1976. The proportion of CO_2 measurements within the concentration ranges differed between habitats, but was similar from year to year within a given habitat. This was especially true in the aspen upland, alder-tamarack bog, alder-ridge ecotone and jackpine ridge (Table 1). The black spruce bog and jackpine sandplain exhibited a greater variation of CO_2 measurements between years. We shall discuss the results from the aspen upland and alder-ridge ecotone, with the alder-tamarack bog for comparison.

The aspen upland exhibited the most marked increase in CO_2 concentrations in both years. The alder-tamarack bog did not accumulate CO_2. The other four plots accumulated CO_2 in one year, while not necessarily in the other.

In 1974–1975 a total of 1,700 trap/nights resulted in 160 small mammals captured, none of which was unmarked. In 1975–1976 a total of 2,260 trap/nights resulted in 300 small mammals captured, 79 of which were new arrivals. More red-backed voles (*Clethrionomys gapperi*) were captured in each trapping period than all other species combined, so our discussion concerns this species.

The microtopography of the aspen upland was varied with areas of rock, thick soil (>25 cm) and thin soil (<25 cm). The greatest CO_2 accumulation was associated with the thick soils. Before the CO_2 concentrations started to increase in January, there was a significant number of captures on the area of thick soil. After January, captures decreased in the areas of thick soil and increased in the areas of rock and thin soil which had lesser CO_2 concentrations. Once the snow melted and subnivean air mixed with environmental air, a marked animal that had frequented the area of thick soil before CO_2 concentrations increased was recaptured there. In the aspen plot the distribution of four frequently-captured voles was monitored throughout the winter. They were caught quite frequently from October to January or later. *Clethrionomys* #530 was a young female caught first in August 1975. During August and November this animal was found in the area of the plot which later would have the greatest CO_2 concentration. In October it was captured in areas with thin soil. In February and March it was captured only in one area with thin soil at the edge of a bare rock. The area with thick soil and greater CO_2 concentrations was part of its home range in the late autumn prior to the Fall Critical Period, but not once CO_2 had started to accumulate.

Clethrionomys #9 was a young female first captured in October 1975. She was eventually captured 12 times from October to December in the area where an unusually great CO_2 concentration would eventually be measured. In January when CO_2 started to accumulate in the areas of thick soil, this vole was captured on the rock ridge where no CO_2 accumulation occurred. Unfortunately, she died in the trap.

Clethrionomys #41 was a female captured first in

Table 1.—*Measured CO$_2$ concentrations (%/volume) and accumulations from six habitats during winters 1973–1974, 1974–1975, and 1976–1976.*

| Plot and period | CO$_2$ concentrations (%/volume) | | | | | | | | | | Total no. of readings | No. of stations with accumu-lation | No. of stations without accumulation |
| | <0.015 | | 0.015–0.025 | | 0.025–0.045 | | 0.045–0.065 | | >0.065 | | | | |
	No.	%	No.	%	No.	%	No.	%	No.	%			
Black spruce bog													
1973–1974	0	0	2	29	3	43	1	14	1	14	7	1	0
1974–1975	0	0	52	68	24	32	0	0	0	0	76	3	2
1975–1976	8	13	26	43	19	31	8	13	0	0	61	5	0
Aspen													
1973–1974	0	0	3	25	4	33	3	25	2	17	12	2	0
1974–1975	5	4	57	45	55	43	9	7	1	1	127	8	1
1975–1976*	4	3	59	47	37	30	15	12	10	8	125	11	1
1975–1975**	4	4	52	50	33	32	13	13	1	1	103	9	1
1975–1976***	4	4	45	49	29	32	12	13	1	1	91	8	1
Alder-tamarack bog													
1973–1974	3	20	8	53	4	27	0	0	0	0	15	1	1
1974–1975	7	5	110	81	19	14	0	0	0	0	136	3	6
1975–1976	5	5	85	85	10	10	0	0	0	0	100	0	10
Alder-ridge ecotone													
1974–1975	7	6	90	71	27	21	2	2	0	0	126	7	2
1975–1976	8	8	73	70	20	19	3	3	0	0	104	4	6
Jackpine ridge													
1973–1974	0	0	9	82	2	18	0	0	0	0	11	1	1
1974–1975	1	2	54	92	4	7	0	0	0	0	59	0	5
1975–1976	2	4	43	81	8	15	0	0	0	0	53	1	4
Jackpine sandplain													
1974–1975	1	4	23	82	4	14	0	0	0	0	28	1	1
1975–1976	1	5	13	59	6	27	0	0	2	9	22	2	0

* With soil.
** Without soil.
*** Without A6 and soil.

Table 2.—*Fienberg's contingency table test for aspen upland 1975–1976: χ^2 values.*

| Cell | Observed no. of captures | Model 1* | | Model 2** | | Model 3*** | | Model 4**** | |
		Expected no. of captures	χ^2	Expected no. of captures	χ^2	Expected no. of captures	χ^2	Expected no. of captures	χ^2
111	9	8.4	0.0429	9.4	0.0170	8.9	0.0011	12.5	0.9800
112	8	5.4	1.2519	6.0	0.6667	3.6	5.3778	5.1	1.6490
113	6	3.4	1.9882	1.9	8.8474	3.4	1.9882	5.3	0.0924
121	13	14.9	0.2423	14.0	0.0714	13.1	0.0008	9.5	1.2895
122	1	9.6	7.7042	9.0	7.1111	5.4	3.5852	3.9	2.1564
123	11	6.1	3.9361	7.7	1.4143	13.6	0.4971	11.7	0.0418
211	94	83.7	1.2675	93.6	0.0017	94.1	0.0001	90.5	0.1354
212	58	54.0	0.2963	60.0	0.0667	62.4	0.3103	60.9	0.1381
213	15	34.3	10.8598	19.1	0.8801	17.6	0.3841	15.7	0.0312
221	140	148.0	0.4324	139.0	0.0072	139.9	0.0001	143.5	0.0854
222	98	95.4	0.0709	90.0	0.7111	93.6	0.2068	95.1	0.0884
223	73	60.7	2.4924	76.3	0.1427	70.4	0.0960	72.3	0.0068
			30.5849 (df = 8)		19.9374 (df = 5)		12.4476 (df = 3)		6.6944 (df = 2)
			$\chi^2 = 10.6474$ (df = 3, P = 0.05)			$\chi^2 = 7.4898$ (df = 2, P = 0.05)			$\chi^2 = 5.7810$ (df = 1, P = 0.05)

* Model 1: 1 + A + B + C.
** Model 2: 1 + A + B + C + BC.
*** Model 3: 1 + A + B + C + BC + AC.
**** Model 4: 1 + A + B + C + BC + AC + AB.

Table 3.—*Fienberg's contingency table test for aspen upland 1975–1976: observed values.*

	C_1	C_2	C_3	C^+
A_1B_1	9	8	6	23
	111	112	113	
A_1B_2	13	1	11	25
	121	122	123	
A_2B_1	94	58	15	167
	211	212	213	
A_2B_2	140	98	73	311
	221	222	223	
A_1B^+	22	9	17	48
A_2B^+	234	156	88	478
A^+B_1	103	66	21	190
A^+B_2	153	99	84	336
A^+B^+	256	165	105	526

A = trap captures, A_2 = trap failures, B = CO_2 (B_1 high, B_2 low), C = habitat (C_1 deep, C_2 shallow, C_3 rock), A+, B+, C+ = total numbers in each of those letters.

November. She was caught nine times in November and December in the area which later showed the greatest CO_2 accumulation, but in January and March she was captured a total of three times, only on rock ridges. In April she was captured once on a ridge and once back in the area where she had been captured originally nine times before CO_2 had started to accumulate.

Clethrionomys #1 was a female first captured in October 1975. In October and November she was captured four times in the area where CO_2 would accumulate to the greatest extent. In December she was captured once in the area with thin soil and in January on a rock ridge.

These four voles avoided the area where CO_2 was to accumulate to its maximum extent as soon as it started to do so. One vole was found dead, two others disappeared entirely from the plot, and the one that survived through the Spring Critical Period returned to the area that had once held the greatest amount of CO_2. The area of greatest CO_2 concentration was part of their home range, but was utilized only when CO_2 was low.

CO_2 accumulation in the aspen upland occurred in both winters (Table 1), and microtopographic variation was obvious. This variation resulted in three distinct types of areas; rock, thin soils and thick soils. There were not enough captures in 1974–1975 to allow statistical analysis. For the 1975–1976 data, Fienberg's (1970) test was used (Tables 2 and 3). Model 1 (1 + A + B + C) of complete independence was rejected. Model 2 (1 + A + B + C + BC) shows the relationship of CO_2 with habitat, which accounted for a significant amount of the variation ($\chi^2 = 10.6$, df = 3). Model 3 (1 + A + B + C + BC + AC + AB), showing the effect of CO_2 on captures again, accounted for a significant portion of the variation ($\chi^2 = 7.5$, df = 2). When the effect of habitat on captures was also included (Model 4: 1 + A + B + C + BC + AC + AB), it accounted for a significant part of the variation ($\chi^2 = 5.8$, df = 1). CO_2 levels and habitat were related with both having significant effects on the numbers of captures, but some other factor causing the significant remainder of unexplained variation ($\chi^2 = 6.7$, df = 2) was also present. This could have been due to "trap addict"

Table 4.—*Fienberg's contingency table test for alder-ridge ecotone 1975–1976: χ^2 values.*

Cell	Observed no. of captures	Model 1* Expected no. of captures	χ^2	Model 2** Expected no. of captures	χ^2	Model 3*** Expected no. of captures	χ^2	Model 4**** Expected no. of captures	χ^2
111	2	1.6	0.1000	2.6	0.1385	4.8	1.6333	1.6	0.1000
112	0	3.5	3.5000	2.4	2.4000	1.5	1.5000	0.4	0.4000
121	6	2.8	3.6571	1.8	9.8000	3.2	2.4500	6.4	0.0250
122	6	6.2	0.0065	7.2	0.2000	4.5	0.5000	5.6	0.0286
211	70	41.6	19.3885	69.4	0.0052	67.2	0.1167	70.4	0.0023
212	66	91.4	7.0586	63.6	0.0906	64.5	0.0349	65.6	0.0024
221	42	74.1	13.9057	46.3	0.3994	44.8	0.1750	41.6	0.0038
222	192	163.0	5.1595	190.8	0.0170	193.5	0.0116	192.4	0.0008
			52.7759 (df = 4)		13.0502 (df = 3)		6.4215 (df = 2)		0.5629 (df = 1)
				$\chi^2 = 39.7252$ (df = 1, $P = 0.01$)		$\chi^2 = 6.6292$ (df = 1, $P = 0.05$)		$\chi^2 = 5.8585$ (df = 1, $P = 0.05$)	

* Model 1: 1 + A + B + C.
** Model 2: 1 + A + B + C + BC.
*** Model 3: 1 + A + B + C + BC + AC.
**** Model 4: 1 + A + B + C + BC + AC + AB.

Table 5.—*Fienberg's contingency table test for alder-ridge ecotone 1975–1976: observed values.*

	C_1		C_2	C^+
A_1B_1	2	Number in that cell	0	2
	111	Cell number	112	
A_1B_2	6		6	12
	121		122	
A_2B_1	70		66	136
	211		212	
A_2B_2	42		192	234
	221		222	
A_1B^+	8		6	14
A_2B^+	112		258	370
A^+B_1	72		66	138
A^+B_2	48		198	246
A^+B^+	120		264	384

A = trap captures, A_2 = trap failure, B_1 = CO_2 > .02%/volume, B = CO_2 ≤ 0.2%/volume, C = habitats (C_1 bog, C_2 rock), A+, B+, C+ = combined values of letter involved.

animals, predators, nonfunctioning traps or any number of unknowns.

The alder-ridge ecotone was microtopographically divided into two areas, ridge and bog. CO_2 did accumulate to some extent in 1975–1976 in the ridge area. In January there were two single captures in the part of the ecotone area that later had more CO_2, but these captures occurred prior to the increased CO_2 measurements. In February the same area still had high concentrations of CO_2, but no small mammals. The two captures recorded were on the ridge areas with lower amounts of CO_2. Most of the greater concentrations of CO_2 occurred in March, and no small mammals were captured then in either area. The small mammals may have moved into the adjacent alder-tamarack bog area, which had relatively less CO_2. There was a corresponding increase in numbers from 5 in February to 12 in March in the alder-tamarack bog. In April the snowcover disappeared, and small mammals were captured again in the alder-ridge ecotone plot. The alder-ridge was not a good overwintering habitat because only one of the marked animals captured was there during the winter.

The alder-ridge ecotone had an accumulation of CO_2 at certain times (Table 1) and a marked microtopographic division into two areas. Because of these factors, Fienberg's (1970) test was used to determine the effect of CO_2 on distributions of small mammals for 1975–1976. Greater CO_2 concentrations are defined as readings greater than ambient levels more than 33.3% of the time in the entire area of the habitat being studied. Table 4 gives the observed values, and Table 5 shows the test. Model 2 showing the effect of CO_2 on habitat accounted for a significant amount ($P < 0.01$) of the variation ($\chi^2 = 39.7$, df = 1). That is, CO_2 was closely related to habitat. Model 3 (1 + A + B + C + BC + AC), showing the effect of CO_2 on captures, again accounted for a significant portion ($P < 0.01$) of the variation ($\chi^2 = 6.6$, df = 1). When the effect of habitat on captures was included (Model 4: 1 + A + B + C + BC + AB), it accounted for a significant part ($P < 0.05$) of the variation ($\chi^2 = 5.9$, df = 1); an insignificant remainder of variation was left. CO_2 and habitat were related, and both had significant effects (<0.01) on numbers of captures.

As a contrast to these areas of subnivean accumulation of CO_2, we can examine the alder-tamarack bog, which had no marked microtopographic or vegetation zones nor any significant CO_2 accumulation (Table 1). Small mammals exhibited no distinct change of areas occupied in the habitat. Therefore, to test for a change in distribution of small mammals, each trapping line (A–J) was designated as a different area, and the captures along each trapping line were tested using Chi Square. No significant change was observed in small mammal distributions either from December 1974 to April 1975 or November 1975 to April 1976.

DISCUSSION

CO_2 ACCUMULATION

In this study we found a consistent variation between habitats in the accumulation of subnivean CO_2. Such variation could explain why some authors found CO_2 accumulated under the snow (Bashenina, 1956; Kelley et al., 1968; and others), while others did not (Fuller and Holmes, 1972; Reiners, 1968; Coyne and Kelley, 1971). The CO_2 concentrations measured in this study agreed with those found by other authors (Coyne and Kelley, 1971; Kelley et al., 1968; Bashenina, 1956). Havas and Mäenpää (1972) recorded accumulations of three to four times the atmospheric levels (0.023%) in their area, while we measured maximum concentrations up to four or five times the atmospheric levels (0.02%) recorded at the Taiga Biological Station and at the Experimental Lakes Area near Kenora, Ontario. The low values for atmospheric CO_2 in our study area may be due to the large mass of continually green conifers available for photosynthesis above the snow and to the lack of air pollution. The atmospheric levels recorded by Havas and Mäenpää (1972) are less than those of other authors from areas without evergreens (0.035% CO_2, Kelley et al., 1968).

We found that CO_2 concentrations in some areas, such as the aspen upland and the alder-ridge ecotone, showed sustained accumulations until the Spring Critical Period and ambient mixing. In contrast, CO_2 concentrations in the alder tamarack bog and jackpine ridge sometimes increased above ambient and then usually returned quickly to ambient concentrations. Reiners (1968), Coyne and Kelley (1971), and Kelley et al. (1968) found that CO_2 concentrations decreased to ambient levels or less once the Hiemal Threshold has been reached, but they only studied a few habitats intensively and thus may not have encountered habitats with sustained accumulations.

Photosynthesis by mosses (Hagerup in Kalela, 1962) or other plants (Yashina, 1961) may have produced the much lower CO_2 concentrations that we found. Photosynthesis is known to occur at temperatures as low as 0°C (Havas and Mäenpää, 1972) with the light available under a snowcover of 15 cm (Evernden, 1966). This could use up subnivean CO_2 and thus lower its concentration. The transitory occurrence of the lowered CO_2 levels, always in association with a thick snowcover, supports this suggestion.

The aspen upland and the alder-ridge ecotone exhibited the greatest diversity of microtopography, significant CO_2 accumulations and significant effects of CO_2 ($P < 0.01$) on small mammal distributions. The specific movements of marked small mammals away from greater CO_2 concentrations, and their return once the concentrations had lessened in the aspen upland, indicate the effect of CO_2 accumulations.

In spite of the few captures in relation to trapping effort (1974–1975; 0 new animals/1,700 trap nights), a statistically significant relationship emerged between small mammal distribution and accumulations of subnivean CO_2.

Small mammals respond physiologically to concentrations of CO_2 as low as 0.14 to 0.50% (Schlenker and Herreid, 1981) at temperatures about 20°C. When stressed by lower temperatures, Dawson (1955) and Baudinette (1974) found a greater response to

low levels (less than 1 to 2%) of CO_2. The values we measured in the subnivean space (up to 0.12%/volume), while not as great as these, were much greater than the ambient (0.02%/volume). In artificial *Dipodomys merriami* burrow nests, CO_2 has been recorded at 0.2 to 1.5% (Soholt, 1974). In an enclosed subnivean spot, such as a nest, CO_2 accumulations could also be more marked than those measured in the subnivean space.

Mammals respond to increased CO_2 by increasing respiratory function, decreasing oxygen consumption and decreasing heart rate and rectal temperatures in an apparent effort to lower stress caused by increased CO_2 concentration. Altered body functions may lead in turn to lessened activity, decreased winter metabolic rates and related lower calorie requirements. However, these adaptive responses may themselves cause problems for the animals, whose ability to perceive temperature differences, food and predators may be decreased with a resulting decrease in their survival rate. Temporarily leaving an area of greater CO_2 concentration would constitute a solution to these problems. This is supported by the aforementioned changes in distribution. It would be advantageous for the small mammal to frequent areas which elicit less physiological stress (that is, those with less CO_2).

If food shortages caused the changes in distribution shown in the aspen upland and alder-ridge ecotone, one would also expect marked changes in the distribution of small mammals in the alder-tamarack bog. This did not happen. The standard error of mean subnivean temperatures was small in all four plots (Penny, 1978). Small mammal numbers in an area are not necessarily related to available food (Gorecki and Gebczynska, 1962; Chitty et al., 1968) or supplementary food (Watts, 1968; Flowerdew, 1972; Fairbairn, 1977) or warmer subnivean temperatures (Fuller, 1977). Thus, these factors do not explain all the observed variations in distribution. Given the facts that subnivean CO_2 does accumulate, that small mammals do react physiologically to CO_2, that areas of homogeneous habitat with increased concentrations of CO_2 have few small mammals, it is, therefore, reasonable to conclude that small mammals avoid areas of greater CO_2 concentrations. CO_2 thus affects small mammal distribution. In the case of *Clethrionomys gapperi*, concentration of subnivean CO_2 must be considered as an additional dimension to its ecological niche and taken into account when considering winter distribution of this species.

SUMMARY

1. Subnivean CO_2 accumulated consistently in some habitats, but not in others. The aspen upland, black spruce bog, jackpine sandplain and part of the alder-ridge ecotone exhibited CO_2 accumulations. However, the extent and amount of accumulation varied between these habitats, being greatest in the aspen upland. CO_2 did not accumulate in alder-tamarack bog or the jackpine ridge. CO_2 accumulation was found to be similar in the same habitats in both winters.

2. A reduction in small mammal numbers was associated with the Fall Critical Period. Lack of accumulation of subnivean CO_2 was significantly associated with no change in small mammal distribution, especially in alder-tamarack bog. Accumulation of subnivean CO_2 was significantly associated with a change in distribution of small mammals especially in the aspen upland.

Given the facts that subnivean CO_2 does accumulate, mammals do react physiologically to CO_2 and areas of homogeneous habitat that have increased concentrations of CO_2 have few small mammals, it is reasonable to conclude that small mammals avoid areas of above ambient CO_2 concentrations, and thus subnivean CO_2 affects small mammal distributions.

ACKNOWLEDGMENTS

This study was supported by an operating grant to Pruitt from the National Research Council, to which organization we express gratitude. We are grateful to Dr. N. Arneson for advice on statistical treatment of the data.

LITERATURE CITED

ANDRZEJEWSKI, R., and M. MAZURKIEWICZ. 1976. Abundance of food supply and size of the bank vole's (Clethrionomys glareolus) home range. Acta Theriol., 21:237–253.

BASHENINA, N. V. 1956. Influence of the quality of subnivean air on the arrangement of winter nests of voles. Zool. Zh., 35:940–942.

BAUDINETTE, R. V. 1974. Physiological correlations of burrow gas conditions in the California ground squirrel. Comp. Biochem. Physiol., 48A:733–743.

CHITTY, D., D. PIMENTEL, and C. J. KREBS. 1968. Food supply ot overwintered voles. J. Anim. Ecol., 37:113–120.

COYNE, P. I., and J. J. KELLEY. 1971. Release of CO_2 from frozen soil to the arctic atmosphere. Nature, 234:407–408.

DAWSON, W. R. 1955. The relation of oxygen consumption to temperature in desert rodents. J. Mamm., 36:543–553.

EVERNDEN, L. L. N. 1966. Light alteration caused by snow and its importance to subnivean rodents. Unpublished MSc. thesis, Univ. Alberta, Edmonton, 111 pp.

FAIRBAIRN, D. J. 1977. The spring decline in deer mice: death or dispersal? Canadian J. Zool., 55:84–92.

FIENBERG, S. E. 1970. An analysis of multidimensional contingency tables. Ecology, 51:419–433.

FLOWERDEW, J. R. 1972. The effect of supplementary food on a population of wood mice (Apodemus sylvaticus). J. Anim. Ecol., 41:533–566.

FULLER, W. A. 1977. Demography of a subarctic population of Clethrionomys gapperi: numbers and survival. Canadian J. Zool., 55:42–51.

FULLER, W. A., and J. G. HOLMES. 1972. The life of the far north. McGraw-Hill, New York, 232 pp.

GALANTZEV, V. P., and I. L. TUMANOV. 1969. Ecological-physical characteristics of the cardiac activity of rodents. Zool. Zh., 48:1066–1073.

GEIGER, R. 1965. The climate near the ground. Harvard Univ. Press, Cambridge, Massachusetts, 494 pp.

GORECKI, A., and Z. GEBCZYNSKA. 1962. Food conditions for small rodents in a deciduous forest. Acta Theriol., 10:275–295.

GRODZINSKI, W. 1963. Can food control the number of small rodents in the deciduous forest? Proc. XVI Internat. Congress Zool., 1:257.

HAVAS, P., and E. MÄENPÄÄ. 1972. Evolution of carbon dioxide at the floor of a Hylocomium-Myrtillus type spruce forest. Aquilo. Ser. Bot., 11:4–22.

KALELA, O. 1962. On the fluctuations in the numbers of arctic and boreal small rodents as a problem of production biology. Ann. Acad. Sci. Fennici, Ser. A, IV Biologica, 66:1–38.

KELLEY, J. J., D. F. WEAVER, and B. P. SMITH. 1968. The variation of CO_2 under the snow in the Arctic. Ecology, 49:358–361.

PENNY, C. E. 1978. Subnivean accumulation of CO_2, its effects on the distribution of small mammals. Unpublished MSc. thesis, Univ. Manitoba, Winnipeg, 106 pp.

PRUITT, W. O., JR. 1957. Observations on the bioclimate of some taiga mammals. Arctic, 10:131–139.

———. 1959. A method of live trapping small taiga mammals in winter. J. Mamm., 40:139–143.

REINERS, W. A. 1968. CO_2 evolution from the floor of three Minnesota forests. Ecology, 49:471–483.

SCHLENKER, E. H., and C. F. HERREID, II. 1981. The effect of low levels of carbon dioxide on metabolism of Mus musculus Comp. Biochem. Physiol., 68:673–676.

SOHOLT, L. F. 1974. Environmental conditions in an artificial burrow occupied by Merriam's kangaroo rat, Dipodomys merriami. J. Mamm., 55:859–866.

SOHOLT, L. F., M. K. YOUSEF, and D. B. DILL. 1973. Responses of Merriam's kangaroo rat (*Dipodomys merriami*) to various levels of CO_2 concentrations. Comp. Biochem. Physiol., 45A: 455–462.

WATTS, C. H. S. 1968. The foods eaten by wood mice (*Apodemus sylvaticus*) and bank voles (*Clethrionomys glareolus*) in Wytham Woods, Berkshire. J. Anim. Ecol., 37:25–41.

WEST, S. D. 1977. Midwinter aggregations in the northern redbacked vole, (*Clethrionomys rutilus*). Canadian J. Zool., 55: 1404–1409.

WITHERS, P. C. 1975. A comparison of respiratory adaptations of a semi-fossorial and surface dwelling Australian rodent. J. Comp. Physiol., 98:193–203.

WOOD, V., G. F. MILLS, H. VELDUIS, and D. B. FORRESTER. 1977. A guide to biophysical land classification Hecla-Carroll Lake, 62P-52M Manitoba. Northern Resource Information Program, Canada-Manitoba Soil Tech. Report, 77-3:1–32.

YASHINA, A. V. 1961. Subnivean growth of plants. Pp. 137–165, *in* Role of Snowcover in natural processes (M. I. Iveronova, ed.), Acad. Sci. Press, Moscow, 272 pp.

Address: Department of Zoology, University of Manitoba, Winnipeg, Manitoba R3T 2N2, Canada.
Present address (Penny): Riding Mountain National Park, Wasagaming, Manitoba R0J 1T0, Canada.